INTRODUCTION TO GAUGE FIELD THEORY

Graduate Student Series in Physics

Other books in the series

Weak Interactions
D BAILIN

Collective Effects in Solids and Liquids
N H MARCH and M PARRINELLO

Hadron Interactions
P D B COLLINS and A D MARTIN

Supersymmetry, Superfields and Supergravity: an Introduction
P P SRIVASTAVA

Symmetry Principles and Magnetic Symmetry in Solid State Physics
S J JOSHUA

Mechanics of Deformable Media
A B BHATIA (deceased) and R N SINGH

Physics of Structurally Disordered Matter: an Introduction
N E CUSACK

Gauge Theories in Particle Physics (2nd edition)
I J R AITCHISON and A J G HEY

Geometry, Topology and Physics
M NAKAHARA

Superfluidity and Superconductivity (3rd edition)
D R TILLEY and J TILLEY

GRADUATE STUDENT SERIES IN PHYSICS

Series Editor: Professor Douglas F Brewer, M.A., D.Phil.
Professor of Experimental Physics, University of Sussex

INTRODUCTION TO GAUGE FIELD THEORY

DAVID BAILIN
School of Mathematical and Physical Sciences
University of Sussex

ALEXANDER LOVE
Department of Physics
Royal Holloway and Bedford New College, University of London

Revised edition

INSTITUTE OF PHYSICS PUBLISHING
BRISTOL AND PHILADELPHIA
Published in association with the University of Sussex Press

British Library Cataloguing in Publication Data. A catalogue record for this book is available from the British Library

Library of Congress Cataloging-in-Publication Data are available

ISBN 0-85274-817-5
ISBN 0-85274-818-3 Pbk
ISBN 0-7503-0281-X Revised edn

First published 1986
Revised edition 1993 (pbk)

Published by IOP Publishing Ltd, a company wholly owned by the Institute of Physics, London

IOP Publishing Ltd
Techno House, Redcliffe Way, Bristol BS1 6NX, England

US Editorial Office: IOP Publishing Inc., The Public Ledger Building, Suite 1035, Independence Square, Philadelphia, PA 19106

Originally typeset by Mid-County Press, London
Revised edition typeset by P & R Typesetters, Salisbury

Printed in Great Britain by J W Arrowsmith Ltd, Bristol

To Anjali
and
To Christine

PREFACE TO FIRST EDITION

In the course of the 1970s important developments in the substance and form of particle physics have gradually rendered the excellent field theory texts of the 1950s and 1960s inadequate to the needs of postgraduate students. The main development in the substance of particle physics has been the emergence of gauge field theory as the basic framework for theories of the weak, electromagnetic and strong interactions. The main development on the formal side has been the increasing use of path (or functional) integral methods in the manipulation of quantum field theory, and the emphasis on the generating functionals for Green functions as basic objects in the theory. This latter development has gone hand-in-hand with the former because the comparative complexity and subtlety of non-Abelian gauge field theory has put efficient methods of proof and calculation at a premium.

It has been our objective in this book to introduce gauge field theory to the postgraduate student of theoretical particle physics entirely from a path integral standpoint without any reliance on the more traditional method of canonical quantisation. We have assumed that the reader already has a knowledge of relativistic quantum mechanics, but we have not assumed any prior knowledge of quantum field theory. We believe that it is possible for the postgraduate student to make his first encounter with scalar field theory in the path integral formalism, and to proceed from there to gauge field theory. No attempt at mathematical rigour has been made, though we have found it appropriate to indicate how well-defined path integrals may be obtained by an analytic continuation to Euclidean space.

We have chosen for the contents of this book those topics which we believe form a foundation for a knowledge of modern relativistic quantum field theory. Some topics inevitably had to be included, such as the path integral approach to scalar field theory, path integrals over Grassmann variables necessary for fermion field theories, the Faddeev–Popov quantisation procedure for non-Abelian gauge field theory, spontaneous breaking of symmetry in gauge theories, and the renormalisation group equation and asymptotic freedom. At a more concrete level this enables us to discuss quantum chromodynamics (QCD) and electroweak theory. Some topics have been included as foundation material which might not have appeared if the book had been written at a slightly earlier date. For example, we have inserted a chapter on field theory at non-zero temperature, in view of the large body of literature that now exists on the application of gauge field theory to cosmology. We have also included a chapter on grand unified theory. Some topics we have omitted from this introductory text, such as an extensive discussion of the results of perturbative QCD (though some applications have been discussed in the text), non-perturbative QCD, and supersymmetry.

We owe much to Professor R G Moorhouse who suggested that we should write this book, and to many colleagues, including P Frampton, A Sirlin, J Cole, T Muta, H F Jones, D R T Jones, D Lancaster, J Fleischer, Z Hioki and G Barton, for the physics they have taught us. We are very grateful to Mrs S Pearson and Ms A Clark for their very careful and speedy typing of the manuscript. Finally, we are greatly indebted to our wives, to whom this book is dedicated, for their invaluable encouragement throughout the writing of this book.

David Bailin
Alexander Love

PREFACE TO REVISED EDITION

We are grateful to IOP Publishing Ltd for giving us the opportunity to correct the typographical and other errors which occurred in the original edition of this book. This task was greatly assisted by a careful reading of the book by G Barton. We are also grateful to I Lawrie and D Waxman who pressed us to clarify several points. Some new material on instantons and axions has been included in chapter 13, and some of the calculations in chapters 12 and 16 have been updated to make use of the more precise values of gauge coupling constants now available.

David Bailin
Alexander Love

CONTENTS

1

PATH INTEGRALS

There are two widely used approaches to quantum field theory. The first is based on field operators and the canonical quantisation of these operator fields, and will not be discussed in this book. The second approach, as we shall see in Chapter 4, involves path integrals[1] over classical fields, and it is upon this latter approach that this book relies for its derivations. In this chapter, the idea of path integrals (or functional integrals) will be developed in a very intuitive way without any attempt at mathematical precision or rigour. Instead, the analogy between vectors and functions, and between matrices and differential operators on functions will be exploited extensively. Since very few path integrals can be performed exactly, we shall concentrate on Gaussian path integrals[2]. These are important in their own right, but much more so because they can be used in approximation schemes when the exact path integral is intractable, as we shall see in later chapters.

Our starting point is the ordinary Gaussian integral

$$\int_{-\infty}^{\infty} \mathrm{d}y \exp(-\tfrac{1}{2}ay^2) = (2\pi)^{1/2}a^{-1/2} \qquad a>0. \tag{1.1}$$

This may be generalised to the integral over n real variables

$$\int_{-\infty}^{\infty} \mathrm{d}y_1 \ldots \mathrm{d}y_n \exp(-\tfrac{1}{2}\boldsymbol{Y}^{\mathrm{T}}\mathbf{A}\boldsymbol{Y}) = (2\pi)^{n/2}(\det \mathbf{A})^{-1/2} \tag{1.2}$$

where $\mathbf{A}$ is a real symmetric positive definite matrix, $\boldsymbol{Y}$ is the column vector with components $(y_1 \ldots, y_n)$, the transpose of $\boldsymbol{Y}$ is denoted by $\boldsymbol{Y}^{\mathrm{T}}$, and each integral is understood to be over the range $(-\infty, \infty)$. Equation (1.2) is easily derived by diagonalising $\mathbf{A}$, when the n-dimensional integration becomes a product of n integrals of the form (1.1). It will prove convenient to write

$$\det \mathbf{A} = \exp \ln \det \mathbf{A} = \exp \operatorname{Tr} \ln \mathbf{A} \tag{1.3}$$

where Tr denotes the trace of the matrix. The identity

$$\ln \det \mathbf{A} = \operatorname{Tr} \ln \mathbf{A} \tag{1.4}$$

is most easily proved by diagonalising $\mathbf{A}$. Equation (1.2) can now be written as

$$(2\pi)^{-n/2} \int_{-\infty}^{\infty} \mathrm{d}y_1 \ldots \mathrm{d}y_n \exp(-\tfrac{1}{2}\boldsymbol{Y}^{\mathrm{T}}\mathbf{A}\boldsymbol{Y}) = \exp(-\tfrac{1}{2}\operatorname{Tr} \ln \mathbf{A}). \tag{1.5}$$

We wish to generalise (1.5) to the case where the integration is over the continuous infinity of components of a function $\varphi(x)$ rather than over the finite number of components of the column vector $\boldsymbol{Y}$. Such an integral is called a path (or functional) integral. Proceeding intuitively[2] we write

$$\int \mathscr{D}\varphi \exp\left(-\tfrac{1}{2}\int \mathrm{d}x' \int \mathrm{d}x\varphi(x')A(x',x)\varphi(x)\right)=\exp(-\tfrac{1}{2}\operatorname{Tr}\ln \mathbf{A}) \tag{1.6}$$

where we use the symbol $\mathscr{D}$ for path integration, and we assume that the integral has been defined in such a way as to remove any normalisation factor (corresponding to the factor $(2\pi)^{-n/2}$ in (1.5)). The integrals over x' and x are assumed to be one-dimensional integrals over the range $(-\infty, \infty)$. However, the treatment generalises trivially to the case where $\mathrm{d}x$ is replaced by d^4x, and the integration is over the whole four-dimensional space. The trace in (1.6) may be evaluated by Fourier transforming. For example, consider the case

$$A(x',x)=\left(\frac{\partial}{\partial x'}\frac{\partial}{\partial x}+r\right)\delta(x'-x) \tag{1.7}$$

where r is a constant. (This is closely related to situations we shall encounter in later chapters.) The one-dimensional Dirac delta function has the integral representation

$$\delta(x'-x)=\int_{-\infty}^{\infty}\frac{\mathrm{d}p}{2\pi}\,\mathrm{e}^{\mathrm{i}p(x'-x)}. \tag{1.8}$$

Thus

$$A(x',x)=\int_{-\infty}^{\infty}\frac{\mathrm{d}p}{2\pi}\,\mathrm{e}^{\mathrm{i}p(x'-x)}\,(p^2+r) \tag{1.9}$$

and

$$\operatorname{Tr}\ln \mathbf{A}=\int \mathrm{d}x\int\frac{\mathrm{d}p}{2\pi}\ln(p^2+r) \tag{1.10}$$

where to take the trace we have set $x'=x$ and integrated over all values of x, since we have a continuous infinity of degrees of freedom.

A slight generalisation can be made by introducing a linear term in (1.5).

$$(2\pi)^{-n/2}\int_{-\infty}^{\infty}\mathrm{d}y_1\ldots\mathrm{d}y_n\exp(-\tfrac{1}{2}\boldsymbol{Y}^{\mathrm{T}}\mathbf{A}\boldsymbol{Y}+\boldsymbol{\rho}^{\mathrm{T}}\boldsymbol{Y})$$

$$=\exp(-\tfrac{1}{2}\operatorname{Tr}\ln\mathbf{A})\exp(\tfrac{1}{2}\boldsymbol{\rho}^{\mathrm{T}}\mathbf{A}^{-1}\boldsymbol{\rho}) \tag{1.11}$$

where $\boldsymbol{\rho}$ is a given column vector, and $\mathbf{A}^{-1}$ exists because $\mathbf{A}$ is positive definite. Equation (1.11) is derived from (1.5) by completing the square,

$$\boldsymbol{Y}^{\mathrm{T}}\mathbf{A}\boldsymbol{Y}-2\boldsymbol{\rho}^{\mathrm{T}}\boldsymbol{Y}=(\boldsymbol{Y}-\mathbf{A}^{-1}\boldsymbol{\rho})^{\mathrm{T}}\mathbf{A}(\boldsymbol{Y}-\mathbf{A}^{-1}\boldsymbol{\rho})-\boldsymbol{\rho}^{\mathrm{T}}\mathbf{A}^{-1}\boldsymbol{\rho} \tag{1.12}$$

and making the change of variable

$$\boldsymbol{Y}'=\boldsymbol{Y}-\mathbf{A}^{-1}\boldsymbol{\rho}. \tag{1.13}$$

The corresponding path integral is

$$\int \mathscr{D}\varphi \exp\left(-\tfrac{1}{2}\int dx' \int dx\varphi(x')A(x',x)\varphi(x) + \int dx\rho(x)\varphi(x)\right)$$
$$= \exp(-\tfrac{1}{2}\operatorname{Tr}\ln \mathbf{A}) \exp\left(\tfrac{1}{2}\int dx' \int dx\rho(x')A^{-1}(x',x)\rho(x)\right). \quad (1.14)$$

where $\rho(x)$ is a given function. In (1.14), $A^{-1}(x', x)$ is easily evaluated from the Fourier transform of $A(x', x)$. Thus, with $A(x', x)$ as in (1.7), we have from (1.9),

$$A^{-1}(x',x) = \int_{-\infty}^{\infty} \frac{dp}{2\pi}\, e^{ip(x'-x)}(p^2+r)^{-1}. \quad (1.15)$$

Equations (1.11) and (1.14) enable us to carry out somewhat more general integrals than Gaussian integrals. If we differentiate with respect to $\rho_{m_1}, \rho_{m_2}, \ldots, \rho_{m_p}$ at $\rho=0$ in (1.11) we obtain

$$(2\pi)^{-n/2}\int_{-\infty}^{\infty} dy_1 \ldots dy_n\, y_{m_1} \ldots y_{m_p} \exp(-\tfrac{1}{2}\boldsymbol{Y}^{\mathrm{T}}\mathbf{A}\boldsymbol{Y})$$
$$= \exp(-\tfrac{1}{2}\operatorname{Tr}\ln \mathbf{A})(A^{-1}_{m_1m_2} \ldots A^{-1}_{m_{p-1}m_p} + \text{permutations}) \quad (1.16)$$

when p is even, and zero when p is odd.

Generalised to the path integral case

$$\int \mathscr{D}\varphi\, \varphi(x_1) \ldots \varphi(x_p) \exp\left(-\tfrac{1}{2}\int dx' \int dx\, \varphi(x')A(x',x)\varphi(x)\right)$$
$$= \exp(-\tfrac{1}{2}\operatorname{Tr}\ln \mathbf{A})(A^{-1}(x_1,x_2) \ldots A^{-1}(x_{p-1},x_p) + \text{permutations}). \quad (1.17)$$

To carry through the differentiations in the path integral case we understand the derivatives to be functional derivatives $\delta/\delta\rho(x_1), \ldots, \delta/\delta\rho(x_p)$, where by definition

$$\frac{\delta}{\delta\rho(x_i)}\left(\int dx\, \rho(x)\varphi(x)\right) = \varphi(x_i) \qquad i = 1, \ldots, p. \quad (1.18)$$

We have been discussing a real column vector or a real function $\varphi(x)$. The discussion is easily extended to the case of complex column vectors or functions. Thus, for example,

$$(2\pi)^{-n}\int dz_1\, dz_1^* \ldots dz_n\, dz_n^* \exp(-\boldsymbol{Z}^{\dagger}\mathbf{A}\boldsymbol{Z}) = (\det \mathbf{A})^{-1} = \exp(-\operatorname{Tr}\ln \mathbf{A}) \quad (1.19)$$

where $\mathbf{A}$ is a Hermitian matrix, $\boldsymbol{Z}$ is the complex column vector with components $(z_1, \ldots, z_n)$, $\boldsymbol{Z}^{\dagger} = (\boldsymbol{Z}^*)^{\mathrm{T}}$, and

$$\int dz\, dz^* \equiv 2\int d(\operatorname{Re} z)\, d(\operatorname{Im} z). \quad (1.20)$$

The corresponding path integral is

$$\int \mathscr{D}\varphi \mathscr{D}\varphi^* \exp\left(-\int \mathrm{d}x' \int \mathrm{d}x\, \varphi^*(x')A(x',x)\varphi(x)\right) = \exp(-\mathrm{Tr}\ln \mathbf{A}). \tag{1.21}$$

So far we have been assuming that the reader is making the intuitive leap from a column vector with a finite number of components to a function with a continuous infinity of components. We can put path integrals on a (slightly) more formal basis, as follows[1]. Suppose that the x and x' integrations in (1.6) are over the finite range from X to $\bar{X}$. We can take the limit of an infinite range of integration at the end of our discussion. Divide the range up into $N+1$ equal segments of length ε

$$(N+1)\varepsilon = \bar{X} - X. \tag{1.22}$$

Let the steps begin at $x_0 = X$, x_1, x_2, ..., x_N, and adopt the notations

$$\varphi_i = \varphi(x_i) \qquad A_{jk} = A(x_j, x_k). \tag{1.23}$$

Then we may define the Gaussian path integral as follows:

$$\begin{aligned} &\int \mathscr{D}\varphi \exp\left(-\tfrac{1}{2}\int \mathrm{d}x' \int \mathrm{d}x\, \varphi(x')A(x',x)\,\varphi(x)\right) \\ &= \lim_{N\to\infty} (2\pi)^{-N/2} \prod_{i=1}^{N} \int \mathrm{d}\varphi_i \exp\left(-\tfrac{1}{2}\sum_{j,k} \varphi_j A_{jk}\varphi_k\right) \\ &= \lim_{N\to\infty} (2\pi)^{-N/2} \prod_{i=1}^{N} \int \mathrm{d}\varphi_i \exp(-\tfrac{1}{2}\boldsymbol{\varphi}^{\mathrm{T}}\mathbf{A}\boldsymbol{\varphi}) \end{aligned} \tag{1.24}$$

where $\boldsymbol{\varphi}$ is the column vector with components $(\varphi_1, \ldots, \varphi_N)$, and $\mathbf{A}$ is the matrix with entries A_{jk}. In the case where we allow the range of integration $(X, \bar{X})$ to become the interval $(-\infty, \infty)$, we may perform the Gaussian integral to obtain the result of equation (1.6). We must, of course, interpret $\lim_{N\to\infty} \exp(-\frac{1}{2}\mathrm{Tr}\ln \mathbf{A})$, where $\mathbf{A}$ is the matrix, as $\exp(-\frac{1}{2}\mathrm{Tr}\ln A)$, where A is $A(x', x)$.

Problem

1.1 Derive (1.19) from (1.5).

References

1 Feynman R P and Hibbs A R 1965 *Quantum Mechanics and Path Integrals* (New York: McGraw-Hill)

2 We follow most closely the approach of
Coleman S 1973 *Lectures given at the 1973 International Summer School of Physics, Ettore Majorana.*

2

PATH INTEGRALS IN NON-RELATIVISTIC QUANTUM MECHANICS

2.1 Transition amplitudes as path integrals

It was shown by Feynman (following a lead by Dirac) that quantum mechanics could be formulated in terms of path integrals[1,2]. We shall discuss this approach to quantum mechanics in some detail since it provides the key to the path integral formulation of quantum field theory. For simplicity, we shall consider in the first instance a system described by a single generalised coordinate Q, with a conjugate momentum P. When corresponding quantum mechanical operators for Q are required we shall use the notation $\hat{Q}_H$ in the Heisenberg picture, and $\hat{Q}_S$ in the Schödinger picture. We denote the eigenstates of $\hat{Q}_S$ by $|q\rangle_S$:

$$\hat{Q}_S|q\rangle_S = q|q\rangle_S. \tag{2.1}$$

Since $\hat{Q}_H$ is time-dependent, so are its eigenstates, which we denote by $|q, t\rangle$:

$$\hat{Q}_H(t)|q, t\rangle = q|q, t\rangle. \tag{2.2}$$

(It should, of course, be remembered that the physical state vectors, as opposed to the eigenstates of $\hat{Q}_H(t)$, are time-independent in the Heisenberg picture.) The relevant connections between the two pictures are

$$\hat{Q}_H(t) = \mathrm{e}^{\mathrm{i}\hat{H}t/\hbar}\,\hat{Q}_S\,\mathrm{e}^{-\mathrm{i}\hat{H}t/\hbar} \tag{2.3}$$

and

$$|q, t\rangle = \mathrm{e}^{\mathrm{i}\hat{H}t/\hbar}\,|q\rangle_S \tag{2.4}$$

where $\hat{H}$ is the (time-independent) Hamiltonian operator.

The probability amplitude that a system which was in the eigenstate $|q', t'\rangle$ at time t' will be found to have the value q'' of Q at time t'' is

$$\langle q'', t''|q', t'\rangle = {}_S\langle q''|\mathrm{e}^{-\mathrm{i}\hat{H}(t''-t')/\hbar}|q'\rangle_S. \tag{2.5}$$

This transition amplitude may be expressed as a path integral by dividing the time interval from t' to t'' into $N+1$ small steps of equal length ε, with

$$(N+1)\varepsilon = t'' - t'. \tag{2.6}$$

Let the steps begin at $t', t_1, t_2, \ldots, t_N$. The eigenstates of $\hat{Q}_H(t)$ form a complete set for any given value of t. Thus

$$\langle q'',t''|q',t'\rangle=\prod_{j=1}^{N}\int \mathrm{d}q_j\langle q'',t''|q_N,t_N\rangle\langle q_N,t_N|q_{N-1},t_{N-1}\rangle$$
$$\ldots\langle q_1,t_1|q',t'\rangle. \tag{2.7}$$

We need to study $\langle q_{j+1},t_{j+1}|q_j,t_j\rangle$ as N becomes very large and the step length ε becomes very small. The discussion is much simplified if the Hamiltonian is of the form

$$H(Q,P)=\frac{P^2}{2m}+V(Q). \tag{2.8}$$

(Simplification results because products of P with Q are not involved, and problems of order associated with the lack of commutativity of the corresponding operators are alleviated.) Applying equation (2.5) to first non-trivial order in ε,

$$\langle q_{j+1},t_{j+1}|q_j,t_j\rangle\approx{}_{\mathrm{S}}\left\langle q_{j+1}\left|\mathbf{1}-\frac{\mathrm{i}\hat{H}\varepsilon}{\hbar}\right|q_j\right\rangle_{\mathrm{S}}. \tag{2.9}$$

But

$${}_{\mathrm{S}}\langle q_{j+1}|\hat{H}|q_j\rangle_{\mathrm{S}}=\left(-\frac{\hbar^2}{2m}\frac{\partial^2}{\partial q_j^2}\right){}_{\mathrm{S}}\langle q_{j+1}|q_j\rangle_{\mathrm{S}} \tag{2.10}$$

and using the usual integral representation of the Dirac δ function

$${}_{\mathrm{S}}\langle q_{j+1}|q_j\rangle_{\mathrm{S}}=\delta(q_{j+1}-q_j)=\hbar^{-1}\int_{-\infty}^{\infty}\frac{\mathrm{d}p_j}{2\pi}\exp[\mathrm{i}p_j(q_{j+1}-q_j)\hbar^{-1}] \tag{2.11}$$

where we have chosen to write the integration variable as $p_j/\hbar$. Using (2.11) in (2.10) gives

$${}_{\mathrm{S}}\langle q_{j+1}|\hat{H}|q_j\rangle_{\mathrm{S}}=\hbar^{-1}\int\frac{\mathrm{d}p_j}{2\pi}\left(\frac{\hbar^2p_j^2}{2m}+V(q_j)\right)\exp[\mathrm{i}p_j(q_{j+1}-q_j)\hbar^{-1}] \tag{2.12}$$

$$=\hbar^{-1}\int\frac{\mathrm{d}p_j}{2\pi}\exp[\mathrm{i}p_j(q_{j+1}-q_j)\hbar^{-1}]H(q_j,p_j). \tag{2.13}$$

We may now rewrite (2.9) as

$$\langle q_{j+1},t_{j+1}|q_j,t_j\rangle$$
$$\approx\hbar^{-1}\int\frac{\mathrm{d}p_j}{2\pi}\exp[\mathrm{i}p_j(q_{j+1}-q_j)\hbar^{-1}](1-\mathrm{i}\varepsilon\hbar^{-1}H(q_j,p_j)). \tag{2.14}$$

Still working to first non-trivial order in ε, we write the integrand in (2.14) as an exponential

$$\langle q_{j+1}, t_{j+1}|q_j, t_j\rangle \approx \hbar^{-1}\int\frac{\mathrm{d}p_j}{2\pi}\exp\{\mathrm{i}\hbar^{-1}\varepsilon[p_j(q_{j+1}-q_j)\varepsilon^{-1}-H(q_j,p_j)]\}. \quad (2.15)$$

Returning to (2.7), the transition amplitude may now be expressed in the form

$$\langle q'', t''|q', t'\rangle$$
$$\approx \prod_{j=1}^{N}\int \mathrm{d}q_j \prod_{j=0}^{N}\frac{\mathrm{d}p_j}{2\pi\hbar}\exp\left(\mathrm{i}\hbar^{-1}\varepsilon\sum_{j=0}^{N}[p_j(q_{j+1}-q_j)\varepsilon^{-1}-H(q_j,p_j)]\right) \quad (2.16)$$

where we have written

$$q_0 = q' \qquad q_{N+1} = q''. \quad (2.17)$$

Taking the limit $N\to\infty$ with $(N+1)\varepsilon$ fixed as in (2.6), we obtain the transition amplitude as a path integral

$$\langle q'', t''|q', t'\rangle \propto \int \mathscr{D}q\int\mathscr{D}p\exp \mathrm{i}\hbar^{-1}\int_{t'}^{t''}\mathrm{d}t(p\dot{q}-H(p,q)) \quad (2.18)$$

where the integration is over all functions $p(t)$, and over all functions $q(t)$ which obey the boundary conditions

$$q(t') = q' \qquad q(t'') = q''. \quad (2.19)$$

The result is more general than the case to which we have restricted ourselves in (2.8).

When the Hamiltonian is given by (2.8), the p_j integrations in (2.15) and (2.16) may be carried out (formally). We complete the square by making the change of variables

$$\tilde{p}_j = p_j - m\varepsilon^{-1}(q_{j+1}-q_j) \quad (2.20)$$

and perform the integrations formally by pretending that $\mathrm{i}\varepsilon$ is real (continuation to imaginary time). We then have Gaussian integrals and obtain

$$\langle q_{j+1}, t_{j+1}|q_j, t_j\rangle$$
$$\approx (2\pi\mathrm{i}\varepsilon\hbar/m)^{-1/2}\exp\{\mathrm{i}\hbar^{-1}\varepsilon[\tfrac{1}{2}m\varepsilon^{-2}(q_{j+1}-q_j)^2-V(q_j)]\}. \quad (2.21)$$

Using (2.21) in (2.7) gives

$$\langle q'', t''|q', t'\rangle$$
$$\approx (2\pi\mathrm{i}\varepsilon\hbar/m)^{-(N+1)/2}\prod_{j=1}^{N}\int\mathrm{d}q_j\exp\left(\mathrm{i}\hbar^{-1}\varepsilon\sum_{j=0}^{N}[\tfrac{1}{2}m\varepsilon^{-2}(q_{j+1}-q_j)^2-V(q_j)]\right). \quad (2.22)$$

Taking the limit $N\to\infty$ with $(N+1)\varepsilon$ fixed as in (2.6) yields the path integral representation

$$\langle q'', t''|q', t'\rangle \propto \int \mathscr{D}q \exp\left(i\hbar^{-1} \int_{t'}^{t''} dt\, L(q, \dot{q})\right) \tag{2.23}$$

where

$$L(q, \dot{q}) = \tfrac{1}{2}m\dot{q}^2 - V(q) \tag{2.24}$$

is the Lagrangian, and the path integral is over all functions $q(t)$ which obey the boundary conditions of (2.19). The constant of proportionality is formally infinite, but is inessential for our purposes.

2.2 The ground-state-to-ground-state amplitude, $W[J]$

If an external source term (or driving force) $-J(t)Q$ is added to the Hamiltonian in (2.8), then the transition amplitude in the presence of this source is

$$\langle q'', t''|q, t\rangle^J \propto \int \mathscr{D}q \int \mathscr{D}p \exp i\hbar^{-1} \int_{t'}^{t''} dt(p\dot{q} - H(p, q) + Jq) \tag{2.25}$$

where $H(p, q)$ denotes the Hamiltonian for $J=0$. The integration is over all functions $p(t)$ and over functions $q(t)$ obeying the boundary conditions of (2.19). We shall see in Chapter 4 that the ground-state-to-ground-state amplitude in the presence of a source plays a central role in quantum field theory. With that application in mind, we now derive the corresponding amplitude in non-relativistic quantum mechanics. To start with we shall take $J(t)$ to be zero for times less than t_- and also for times greater than t_+. Using the completeness of the eigenstates $|q_+, t_+\rangle$ and $|q_-, t_-\rangle$ of $\hat{Q}_H$ at times t_+ and t_-,

$$\langle q'', t''|q', t'\rangle^J = \int dq_+ \int dq_- \langle q'', t''|q_+, t_+\rangle\langle q_+, t_+|q_-, t_-\rangle^J\langle q_-, t_-|q', t'\rangle \tag{2.26}$$

provided $t'' > t_+ > t_- > t'$. Let $|n\rangle$ be the energy eigenstates

$$\hat{H}|n\rangle = E_n|n\rangle \tag{2.27}$$

and introduce corresponding time-dependent wave functions

$$\psi_n(q, t) = \langle q, t|n\rangle = e^{-iE_n t/\hbar}\, {}_S\langle q|n\rangle.$$

Then

$$\langle q'', t''|q_+, t_+\rangle = \sum_n \langle q'', t''|n\rangle\langle n|q_+, t_+\rangle$$

$$= \sum_n \psi_n(q'', t'')\psi_n^*(q_+, t_+) \tag{2.29}$$

and

$$\langle q_-, t_- | q', t' \rangle = \sum_n \psi_n(q_-, t_-)\psi_n^*(q', t'). \tag{2.30}$$

The connection with the ground-state-to-ground-state amplitude is obtained by continuing to the imaginary time axis and taking the limit $t'' \to -i\infty$, $t' \to i\infty$. Then the decaying exponentials ensure that only the contribution from the ground-state wave function ψ_0 survives in (2.29) and (2.30), and we have

$$\lim_{t'' \to -i\infty} \langle q'', t'' | q_+, t_+ \rangle = \lim_{t'' \to -i\infty} \psi_0(q'', t'')\psi_0^*(q_+, t_+) \tag{2.31}$$

and

$$\lim_{t' \to i\infty} \langle q_-, t_- | q', t' \rangle = \lim_{t' \to i\infty} \psi_0(q_-, t_-)\psi_0^*(q', t'). \tag{2.32}$$

Substituting in (2.26) and returning to real time we obtain

$$\lim_{t'' \to \infty, t' \to -\infty} \langle q'', t'' | q', t' \rangle^J / \psi_0(q'', t'')\psi_0^*(q', t')$$

$$= \int dq_+ \int dq_- \psi_0^*(q_+, t_+) \langle q_+, t_+ | q_-, t_- \rangle^J \psi_0(q_-, t_-). \tag{2.33}$$

(The alert, or even half-alert, reader will, with reason, feel uneasy about the way in which we have returned to real time, where the exponentials are oscillatory and cannot unambiguously damp all but the ground-state wave function. The result only remains unambiguous if we replace real t by $e^{-i\varepsilon}t$, with ε a small positive quantity. We shall make amends after (2.38).)

The expression on the right-hand side of (2.33) is the probability amplitude to find the system in the ground state at time t_+ given that it was in the ground state at the time t_-. We are really interested in the case where $J(t)$ is non-zero not just between finite times t_- and t_+, but for all times. We can reach this case by taking t_- large and negative, and t_+ large and positive. Then the right-hand side of (2.33) is the probability amplitude to find the system in the ground state at time ∞, given that it was in the ground state at time $-\infty$. We shall denote this ground-state-to-ground-state amplitude by $W[J]$. Thus, subject to the same qualifications as (2.33),

$$W[J] \propto \lim_{t'' \to \infty, t' \to -\infty} \langle q'', t'' | q', t' \rangle^J \tag{2.34}$$

which is the required result. Notice that it does not matter what values we choose for q' and q''. Returning to (2.25) we see that

$$W[J] \propto \int \mathscr{D}q \int \mathscr{D}p \exp i\hbar^{-1} \int_{-\infty}^{\infty} dt(p\dot{q} - H(p, q) + Jq). \tag{2.35}$$

The path integration is over all functions $p(t)$, and, using (2.19), over all functions $q(t)$ obeying the boundary conditions

$$\lim_{t'\to-\infty} q(t')=q' \qquad \lim_{t''\to\infty} q(t'')=q'' \tag{2.36}$$

where q' and q'' are any chosen constants, but are often taken to be zero.

In the special case where the Hamiltonian is given by (2.8), the generalisation of (2.23) to include a source term is

$$\langle q'', t''|q', t'\rangle^J \propto \int \mathscr{D}q \exp i\hbar^{-1} \int_{t'}^{t''} \mathrm{d}t(L(q,\dot{q})+Jq) \tag{2.37}$$

and the corresponding expression for $W[J]$ is

$$W[J] \propto \int \mathscr{D}q \exp i\hbar^{-1} \int_{-\infty}^{\infty} \mathrm{d}t(L(q,\dot{q})+Jq). \tag{2.38}$$

The oscillatory path integral of (2.38) is not well defined, and it is necessary to make some more precise statement before it can be evaluated unambiguously. A convenient procedure suggested by the above derivation is to continue the integrand to imaginary time, which makes the path integral well defined, perform the path integral, and then continue back to real time (see figure 2.1). We introduce the variable

$$\bar{t}=\mathrm{i}t \tag{2.39}$$

and denote the continuation of $W[J]$ to imaginary time by $W_{\mathrm{E}}[J]$ (where the subscript E is used because the continuation will be from Minkowski to Euclidean space in the relativistic case considered in Chapter 4).

$$W_{\mathrm{E}}[J] \propto \int \mathscr{D}q \exp \hbar^{-1} \int_{-\infty}^{\infty} \mathrm{d}\bar{t}\left[L\left(q,\mathrm{i}\frac{\mathrm{d}q}{\mathrm{d}\bar{t}}\right)+Jq\right] \tag{2.40}$$

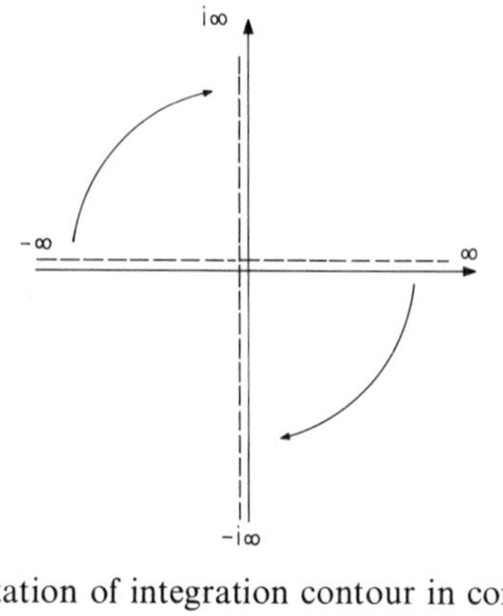

Figure 2.1 Rotation of integration contour in complex t plane.

where q and J are now to be regarded as functions of $\bar{t}$. In accordance with (2.36), the path integral is over all functions $q(\bar{t})$ which obey the boundary conditions

$$\lim_{\bar{t}\to-\infty} q(\bar{t})=q' \qquad \lim_{\bar{t}\to\infty} q(\bar{t})=q'' \tag{2.41}$$

where q' and q'' are any chosen constants. The integrand is now a damped exponential and the path integral is well defined. As has been emphasised by Coleman[3], if we proceed directly from the Hamiltonian form of (2.35), continuing to imaginary time does not immediately yield a damped exponential because the derivative term remains oscillatory. If we are to proceed unambiguously from (2.35), we must first perform the path integral over p (formally), and then continue the integrand to imaginary time as in (2.39). In the case where the Hamiltonian is of the form of (2.9) this leads to (2.40), but, more generally, an effective Lagrangian (differing from the true Lagrangian) has to be introduced to render $W_{\mathrm{E}}[J]$ into the form of (2.40).

With these words of caution, we shall continue the integrand in (2.35) to imaginary time and write for the general case

$$W_{\mathrm{E}}[J] \propto \int \mathscr{D}q \int \mathscr{D}p \exp -\hbar^{-1} \int_{-\infty}^{\infty} \mathrm{d}\bar{t}\left(-\mathrm{i}p\frac{\mathrm{d}q}{\mathrm{d}\bar{t}}+H(p,q)-Jq\right). \tag{2.42}$$

If any ambiguity arises, it is to be resolved as described above.

2.3 Ground-state expectation values from $W[J]$

So far we have been concentrating upon the ground-state amplitude $W[J]$. In the discussion of quantum field theory in Chapter 4, we shall see that Green functions, which are ground-state expectation values of products of field operators, play an important part. The analogous objects in non-relativistic quantum mechanics are ground-state expectation values of products of operators $\hat{Q}_{\mathrm{H}}$. We shall now see how these may be derived from $W[J]$. For notational simplicity, we shall consider first the case where the product involves only two operators. Thus, we begin by studying the object

$$\langle q'', t''|\hat{Q}_{\mathrm{H}}(t_{\mathrm{b}})\hat{Q}_{\mathrm{H}}(t_{\mathrm{a}})|q', t'\rangle$$

with $t_{\mathrm{b}} > t_{\mathrm{a}}$. Just as when discussing the transition amplitude $\langle q'', t''|q', t'\rangle$, we divide the time interval t' to t'' into small steps beginning at $t', t_1, t_2, \ldots, t_N$. Choosing t_{b} and t_{a} to coincide with the beginning of two of these steps, we may use completeness as in (2.7) to obtain

$$\langle q'', t''|\hat{Q}_{\mathrm{H}}(t_{\mathrm{b}})\hat{Q}_{\mathrm{H}}(t_{\mathrm{a}})|q', t'\rangle = \prod_{j=1}^{N} \int \mathrm{d}q_j \langle q'', t''|q_N, t_N\rangle\langle q_N, t_N|q_{N-1}, t_{N-1}\rangle \ldots$$

$$\langle q_{\mathrm{b}+1}, t_{\mathrm{b}+1}|\hat{Q}_{\mathrm{H}}(t_{\mathrm{b}})|q_{\mathrm{b}}, t_{\mathrm{b}}\rangle \ldots \langle q_{\mathrm{a}+1}, t_{\mathrm{a}+1}|\hat{Q}_{\mathrm{H}}(t_{\mathrm{a}})|q_{\mathrm{a}}, t_{\mathrm{a}}\rangle \ldots \langle q_1, t_1|q', t'\rangle \tag{2.43}$$

$$= \prod_{j=1}^{N} \int \mathrm{d}q_j q_\mathrm{b} q_\mathrm{a} \langle q'', t'' | q_N, t_N \rangle \langle q_N, t_N | q_{N-1}, t_{N-1} \rangle \ldots \langle q_1, t_1 | q', t' \rangle. \tag{2.44}$$

For a Hamiltonian of the form of (2.8), $\langle q_{j+1}, t_{j+1} | q_j, t_j \rangle$ is given by (2.15). Thus,

$$\langle q'', t'' | \hat{Q}_\mathrm{H}(t_\mathrm{b}) \hat{Q}_\mathrm{H}(t_\mathrm{a}) | q', t' \rangle$$

$$\approx \prod_{j=1}^{N} \int \mathrm{d}q_j \prod_{j=0}^{N} \frac{\mathrm{d}p_j}{2\pi\hbar} q_\mathrm{b} q_\mathrm{a} \exp \mathrm{i}\hbar^{-1}\varepsilon \sum_{j=0}^{N} [p_j(q_{j+1} - q_j)\varepsilon^{-1} - H(p_j, q_j)]. \tag{2.45}$$

Taking the limit $N \to \infty$ with $(N+1)\varepsilon$ fixed as in (2.6) we obtain the path integral

$$\langle q'', t'' | \hat{Q}_\mathrm{H}(t_\mathrm{b}) \hat{Q}_\mathrm{H}(t_\mathrm{a}) | q', t' \rangle$$

$$\propto \int \mathscr{D}q \int \mathscr{D}p \, q(t_\mathrm{b}) q(t_\mathrm{a}) \exp \mathrm{i}\hbar^{-1} \int_{t'}^{t''} \mathrm{d}t (p\dot{q} - H(p, q)) \tag{2.46}$$

for $t_\mathrm{b} > t_\mathrm{a}$, where the integration is over all functions $p(t)$ and over all functions $q(t)$ which obey the boundary conditions of (2.19), and the constant of proportionality is the same as in (2.18). In deriving (2.46), we have assumed that $t_\mathrm{b} > t_\mathrm{a}$. If instead we had assumed that $t_\mathrm{a} > t_\mathrm{b}$, then we would have found that

$$\langle q'', t'' | \hat{Q}_\mathrm{H}(t_\mathrm{a}) \hat{Q}_\mathrm{H}(t_\mathrm{b}) | q', t' \rangle \propto \int \mathscr{D}q \int \mathscr{D}p \, q(t_\mathrm{b}) q(t_\mathrm{a}) \exp \mathrm{i}\hbar^{-1} \int_{t'}^{t''} \mathrm{d}t (p\dot{q} - H(p, q)) \tag{2.47}$$

for $t_\mathrm{a} > t_\mathrm{b}$. We may summarise (2.46) and (2.47) as

$$\langle q'', t'' | T(\hat{Q}_\mathrm{H}(t_\mathrm{b}) \hat{Q}_\mathrm{H}(t_\mathrm{a})) | q', t' \rangle$$

$$\propto \int \mathscr{D}q \int \mathscr{D}p q(t_\mathrm{b}) q(t_\mathrm{a}) \exp \mathrm{i}\hbar^{-1} \int_{t'}^{t''} \mathrm{d}t (p\dot{q} - H(p, q)) \tag{2.48}$$

where the time ordering operation T is defined by

$$\begin{aligned} T(\hat{Q}_\mathrm{H}(t_\mathrm{b}) \hat{Q}_\mathrm{H}(t_\mathrm{a})) &= \hat{Q}_\mathrm{H}(t_\mathrm{b}) \hat{Q}_\mathrm{H}(t_\mathrm{a}) \qquad \text{for } t_\mathrm{b} > t_\mathrm{a} \\ &= \hat{Q}_\mathrm{H}(t_\mathrm{a}) \hat{Q}_\mathrm{H}(t_\mathrm{b}) \qquad \text{for } t_\mathrm{a} > t_\mathrm{b}. \end{aligned} \tag{2.49}$$

The constant of proportionality is the same as in (2.18).

The result is easily generalised to a time-ordered product of any number n of operators.

$$\langle q'', t'' | T(\hat{Q}_\mathrm{H}(t'_n) \ldots \hat{Q}_\mathrm{H}(t'_2) \hat{Q}_\mathrm{H}(t'_1)) | q', t' \rangle$$

$$\propto \int \mathscr{D}q \int \mathscr{D}p q(t'_n) \ldots q(t'_2) q(t'_1) \exp \mathrm{i}\hbar^{-1} \int_{t'}^{t''} \mathrm{d}t (p\dot{q} - H(p, q)) \tag{2.50}$$

where the time ordering operation T puts the operators in chronological order, with operators corresponding to later times to the left of operators corresponding to earlier times.

When the Hamiltonian is given by (2.8), the integrations in (2.45) may be performed (formally), as in equations (2.20) to (2.23), to yield the alternative path integral representation

$$\langle q'', t''|T(\hat{Q}_{\mathrm{H}}(t'_n) \ldots \hat{Q}_{\mathrm{H}}(t'_2)\hat{Q}_{\mathrm{H}}(t'_1))|q', t'\rangle \propto \int \mathscr{D}q\, q(t'_n) \ldots q(t'_2)q(t'_1) \exp i\hbar^{-1} \int_{t'}^{t''} \mathrm{d}t L(q, \dot{q}) \quad (2.51)$$

where the constant of proportionality is the same as in (2.23), and the path integral is over all functions $q(t)$ that obey the boundary conditions of (2.19).

To make the connection with the expectation value in the ground state of the time-ordered product of operators, we now introduce times t'_+ and t'_- with $t'' > t'_+ > t'_- > t'$. Using the completeness of the eigenstates $|q'_+, t'_+\rangle$ and $|q'_-, t'_-\rangle$ of $\hat{Q}_{\mathrm{H}}$ at times t'_+ and t'_-,

$$\langle q'', t''|T(\hat{Q}_{\mathrm{H}}(t'_n) \ldots \hat{Q}_{\mathrm{H}}(t'_1))|q', t'\rangle = \int \mathrm{d}q'_+\, \mathrm{d}q'_- \langle q'', t''|q'_+, t'_+\rangle \times \langle q'_+, t'_+|T(\hat{Q}_{\mathrm{H}}(t'_n) \ldots \hat{Q}_{\mathrm{H}}(t'_1))|q'_-, t'_-\rangle\langle q'_-, t'_-|q', t'\rangle. \quad (2.52)$$

Introducing energy eigenstates and proceeding in the fashion of equations (2.27) to (2.33), we see that

$$\begin{aligned}
&\lim_{t' \to -\infty, t'' \to \infty} \langle q'', t''|T(\hat{Q}_{\mathrm{H}}(t'_n) \ldots \hat{Q}_{\mathrm{H}}(t'_1))|q', t'\rangle / \psi_0(q'', t'')\psi_0^*(q', t') \\
&= \int \mathrm{d}q'_+ \int \mathrm{d}q'_- \psi_0^*(q'_+, t'_+)\langle q'_+, t'_+|T(\hat{Q}_{\mathrm{H}}(t'_n) \ldots \hat{Q}_{\mathrm{H}}(t'_1))|q'_-, t'_-\rangle\psi_0(q'_-, t'_-) \\
&= \langle 0|T(\hat{Q}_{\mathrm{H}}(t'_n) \ldots \hat{Q}_{\mathrm{H}}(t'_1))|0\rangle \qquad (2.53)
\end{aligned}$$

where $\langle 0|T(\hat{Q}_{\mathrm{H}}(t'_n) \ldots \hat{Q}_{\mathrm{H}}(t'_1))|0\rangle$ denotes the expectation value in the ground state of the time-ordered product of operators. (The same qualifications as after (2.33) apply.) Combining (2.53) with (2.50),

$$\langle 0|T(\hat{Q}_{\mathrm{H}}(t'_n) \ldots \hat{Q}_{\mathrm{H}}(t'_1))|0\rangle \propto \int \mathscr{D}q \int \mathscr{D}p\, q(t'_n) \ldots q(t'_1) \exp i\hbar^{-1} \int_{-\infty}^{\infty} \mathrm{d}t(p\dot{q} - H(p, q)). \quad (2.54)$$

Armed with (2.54) we may now make the connection of ground-state-to-ground-state amplitude $W[J]$ with ground-state expectation values of time-ordered products of operators $\hat{Q}_{\mathrm{H}}$. Consider the functional derivative of $W[J]$ with respect to $J(t'_1)$, $J(t'_2)$, …, $J(t'_n)$, where the differentiation is in the sense of

(1.18). Using (2.35),

$$\left.\frac{\delta^n W[J]}{\delta J(t'_n)\ldots\delta J(t'_1)}\right|_{J(t)=0}$$

$$\propto (\mathrm{i}\hbar^{-1})^n \int \mathscr{D}q \int \mathscr{D}p\, q(t'_n)\ldots q(t'_1) \exp \mathrm{i}\hbar^{-1} \int_{-\infty}^{\infty} \mathrm{d}t(p\dot{q} - H(p,q)). \quad (2.55)$$

The constants of proportionality are the same in (2.54) and (2.55), thus

$$(\mathrm{i}\hbar^{-1})^n \langle 0|T(\hat{Q}_{\mathrm{H}}(t'_n)\ldots\hat{Q}_{\mathrm{H}}(t'_1))|0\rangle = \left.\frac{\delta^n W[J]}{\delta J(t'_n)\ldots\delta J(t'_1)}\right|_{J(t)=0}. \quad (2.56)$$

We now see that once the ground-state-to-ground-state amplitude $W[J]$ is known, all the ground-state expectation values of time-ordered products of operators $\hat{Q}_{\mathrm{H}}$ may be generated from it. For this reason, $W[J]$ is referred to as the generating functional. The corresponding amplitude in quantum field theory is used to generate the Green functions, as we shall see in Chapter 4.

Problems

2.1 Derive expression (2.21).

2.2 Obtain the effective Lagrangian when $H = \frac{1}{2}p^2 f(q)$.

References

1 Feynman R P and Hibbs A R 1965 *Quantum Mechanics and Path Integrals* (New York: McGraw-Hill)
2 Our approach is closest to
Abers E S and Lee B W 1973 *Phys. Rep.* **9C** 1 and
Taylor J C 1976 *Gauge Theories of Weak Interactions* (Cambridge: Cambridge University Press)
3 Coleman S 1973 *Lectures given at the 1973 International Summer School of Physics, Ettore Majorana.*

3

CLASSICAL FIELD THEORY

3.1 Euler–Lagrange equations

So far we have been concerned with the non-relativistic quantum mechanics of a single particle with a generalised coordinate $q(t)$. Particle physics is formulated using the language of relativistic quantum field theory. We therefore address ourselves first to the derivation of the principal results of classical (i.e. non-quantum) field theory. This will facilitate the generalisation to the quantum theory which will be developed in the following chapters.

A field is a generalisation of the notion of a generalised coordinate, not just to several particles, but to a continuum of particles, with the possibility of one or more at each point in space. Thus we write a field as $\varphi(\boldsymbol{x}, t)$ in the simplest case where there is one generalised coordinate for each point $\boldsymbol{x}$ of space. The simplest example arises when we consider waves on a vibrating string. Then the generalised coordinate is the displacement $y(x, t)$ of the point on the string at x from its equilibrium position. (In this case x takes values only in the range $0 \leqslant x \leqslant l$ corresponding to the string, but we could as easily consider sound waves in space, in which case $\boldsymbol{x}$ could assume any value.) A more complicated example is provided by the electromagnetic field at $\boldsymbol{x}$ at time t. Evidently this involves more than one coordinate at each $\boldsymbol{x}$, because of the vector nature of $\boldsymbol{E}$ and $\boldsymbol{B}$.

The Lagrangian L of a field theory will evidently involve an integral over the continuum labels $\boldsymbol{x}$, instead of merely a sum over the discrete labels i, if we had had a finite number of coordinates. Thus

$$L = \int d^3x \mathscr{L} \tag{3.1}$$

where $\mathscr{L}$ is called the Lagrangian density. In the discrete case the Lagrangian is a function of the coordinates q_i and the generalised velocities $\dot{q}_i \equiv dq_i/dt$. Thus in the continuum case we expect $\mathscr{L}$ to be a function of φ and $\partial\varphi/\partial t$. We shall also expect $\mathscr{L}$ to depend on coordinate differences i.e. upon $\nabla\varphi$. In any case, in a relativistically covariant theory we must not arbitrarily select a preferred direction in space–time. Thus we shall have

$$\mathscr{L} = \mathscr{L}(\varphi, \partial_\mu \varphi) \tag{3.2a}$$

where

$$\partial_\mu \varphi \equiv \frac{\partial \varphi}{\partial x^\mu} \qquad (\mu = 0, 1, 2, 3) \tag{3.2b}$$

with $x^0 = ct$. Thus the action

$$S[\varphi] \equiv \int_{t_1}^{t_2} \mathrm{d}t L = \int_{t_1}^{t_2} \mathrm{d}^4 x \mathscr{L}(\varphi, \partial_\mu \varphi) \tag{3.3}$$

is a functional of the field φ. Note that, since L has the dimensions of energy, S has the dimensions of angular momentum, i.e. it is dimensionless if we use natural units $\hbar = c = 1$.

Lagrange's equations follow by demanding that S is stationary under variations of the generalised coordinate q. Thus we must consider an arbitrary change in the field φ:

$$\varphi \to \varphi + \delta\varphi. \tag{3.4}$$

Then

$$S \to S + \delta S$$

where

$$\begin{aligned}\delta S &= \int_{t_1}^{t_2} \mathrm{d}^4 x \left(\frac{\partial \mathscr{L}}{\partial \varphi} \delta\varphi + \frac{\partial \mathscr{L}}{\partial(\partial_\mu \varphi)} \delta(\partial_\mu \varphi) \right) \\ &= \int_{t_1}^{t_2} \mathrm{d}^4 x \left[\frac{\partial \mathscr{L}}{\partial \varphi} - \partial_\mu \left(\frac{\partial \mathscr{L}}{\partial(\partial_\mu \varphi)} \right) \right] \delta\varphi + \int_{t_1}^{t_2} \mathrm{d}^4 x \, \partial_\mu \left(\frac{\partial \mathscr{L}}{\partial(\partial_\mu \varphi)} \delta\varphi \right). \end{aligned} \tag{3.5}$$

The last term may be integrated and it involves the value of $[\partial\mathscr{L}/\partial(\partial_\mu\varphi)]\delta\varphi$ only on the surface of the space–time volume being integrated over. If we restrict the permissible variations $\delta\varphi$ to those which vanish on this surface, just as in deriving Lagrange's equations we require δq to vanish at t_1 and t_2, then this last term vanishes and the functional derivative

$$\frac{\delta S}{\delta\varphi(x)} = \frac{\partial \mathscr{L}}{\partial \varphi} - \partial_\mu \left(\frac{\partial \mathscr{L}}{\partial(\partial_\mu \varphi)} \right). \tag{3.6}$$

If we demand that S is stationary under such variations, then

$$\frac{\delta S}{\delta\varphi(x)} = 0 \tag{3.7}$$

and the Euler–Lagrange equations follow:

$$\partial_\mu \left(\frac{\partial \mathscr{L}}{\partial(\partial_\mu \varphi)} \right) = \frac{\partial \mathscr{L}}{\partial \varphi}. \tag{3.8}$$

If there are several fields φ^a associated with each point $\boldsymbol{x}$, then there is an Euler–Lagrange equation for each index a.

It is easy to construct Lagrangian densities to yield any desired field equation. For example,

$$\mathscr{L} = \tfrac{1}{2}\rho(\partial^\mu y)(\partial_\mu y) \tag{3.9}$$

with ρ a mass density, gives

$$\frac{\partial \mathscr{L}}{\partial y}=0 \qquad \text{and} \qquad \frac{\partial \mathscr{L}}{\partial(\partial_\mu y)}=\rho \partial^\mu y \tag{3.10}$$

so the Euler–Lagrange equation implies

$$\Box y \equiv \partial_\mu \partial^\mu y = \frac{1}{c^2}\frac{\partial^2 y}{\partial t^2} - \nabla^2 y = 0 \tag{3.11}$$

which is just the wave equation.

The electromagnetic field equations are also derivable from a Lagrangian, which may be expressed in terms of the Lorentz covariant vector potential A^μ:

$$\mathscr{L} = -\tfrac{1}{4}(\partial_\mu A_\nu - \partial_\nu A_\mu)(\partial^\mu A^\nu - \partial^\nu A^\mu) - j_\mu A^\mu. \tag{3.12}$$

Then the Euler–Lagrange equations (3.8) for each component A_ν give

$$-\partial_\mu(\partial^\mu A^\nu - \partial^\nu A^\mu) = -j^\nu. \tag{3.13}$$

This is just the covariant form of Maxwell's equations, since the $\nu=0$ component gives

$$\nabla \cdot \boldsymbol{E} = \rho \tag{3.14a}$$

where

$$\boldsymbol{E} = -\frac{1}{c}\frac{\partial \boldsymbol{A}}{\partial t} - \nabla A_0 \tag{3.14b}$$

and

$$\rho = j^0. \tag{3.14c}$$

The spatial components give

$$\nabla \wedge \boldsymbol{B} = \frac{1}{c}\frac{\partial \boldsymbol{E}}{\partial t} + \boldsymbol{j} \tag{3.15a}$$

where

$$\boldsymbol{B} = \nabla \wedge \boldsymbol{A}. \tag{3.15b}$$

The two remaining Maxwell's equations

$$\nabla \wedge \boldsymbol{E} = -\frac{1}{c}\frac{\partial \boldsymbol{B}}{\partial t} \tag{3.16a}$$

$$\nabla \cdot \boldsymbol{B} = 0 \tag{3.16b}$$

follow directly from (3.14b) and (3.15b).

3.2 Noether's theorem

One of the advantages of the Lagrangian formulation is that it permits the ready identification of conserved quantities by studying the invariances of the action S. The fundamental result behind this statement is Noether's theorem[1], which identifies the conserved current associated with the invariance of S under a very general infinitesimal transformation.

We consider the variation δS of

$$S=\int_{t_1}^{t_2} \mathrm{d}^4x \mathscr{L}(\varphi(x), \partial_\mu \varphi(x)) \tag{3.17}$$

under a general transformation of the coordinates x *and* the field φ. Suppose

$$x^\mu \rightarrow x'^\mu = x^\mu + \delta x^\mu \tag{3.18}$$

where the infinitesimal δx^μ is specified by a set of infinitesimal parameters $\delta\omega^i$, so

$$\delta x^\mu = X_i^\mu(x)\, \delta\omega^i. \tag{3.19}$$

As examples we shall later consider the cases when the parameters $\delta\omega^i$ specify an infinitesimal translation, or an infinitesimal Lorentz transformation. Under such a transformation the field $\varphi(x)$ will, in general, also transform. Thus

$$\varphi(x) \rightarrow \varphi'(x') = \varphi(x) + \delta\varphi(x) \tag{3.20}$$

where $\delta\varphi$ is also specified by the parameters $\delta\omega^i$, so

$$\delta\varphi(x) = \Phi_i(x)\, \delta\omega^i. \tag{3.21}$$

Evidently the total variation $\delta\varphi(x)$ derives both from the variation of the field function and from the variation of its argument:

$$\begin{aligned} \varphi'(x') &= \varphi'(x+\delta x) \\ &= \varphi'(x) + \delta x^\nu(\partial_\nu \varphi) \\ &= \varphi(x) + \delta_0\varphi(x) + \delta x^\nu(\partial_\nu\varphi) \end{aligned} \tag{3.22}$$

where $\delta_0\varphi$ is the variation of the field function alone. Thus from (3.21) and (3.22), using (3.19),

$$\delta_0\varphi(x) \equiv \delta\varphi(x) - \delta x^\nu(\partial_\nu\varphi) = [\Phi_i(x) - (\partial_\nu\varphi) X_i^\nu(x)]\, \delta\omega^i. \tag{3.23}$$

Then the variation of the action S is

$$\delta S = \int_{t_1}^{t_2} \delta(\mathrm{d}^4x)\mathscr{L} + \int_{t_1}^{t_2} \mathrm{d}^4x\, \delta\mathscr{L} \tag{3.24}$$

where $\delta\mathscr{L}$ is the variation in the Lagrangian density caused by the above variations of x and φ, and $\delta(\mathrm{d}^4x)$ is the variation of the integration measure caused by the variation (3.18). In fact

$$\begin{aligned}\mathrm{d}^4x' &= \left|\det\left[\frac{\partial x'^\mu}{\partial x^\nu}\right]\right| \mathrm{d}^4x \\ &= \left|\det[\delta^\mu_\nu + \partial_\nu(X^\mu_i\, \delta\omega^i)]\right| \mathrm{d}^4x \\ &= [1 + \partial_\mu(X^\mu_i\, \delta\omega^i)]\, \mathrm{d}^4x\end{aligned}$$

so

$$\delta(\mathrm{d}^4x) = [\partial_\mu(X^\mu_i\, \delta\omega^i)]\, \mathrm{d}^4x. \tag{3.25}$$

The variation $\delta\mathscr{L}$ derives from the variations δx and $\delta\varphi$, where $\delta\varphi$ in turn derives from $\delta_0\varphi$ and δx, as specified in (3.23). We have already evaluated the variation of the action caused by an arbitrary variation of the field function alone in (3.5). (What we are now calling $\delta_0\varphi$ was called $\delta\varphi$ in (3.5).) If φ satisfies the Euler–Lagrange equation (3.8), only the second integral on the right-hand side of (3.5) survives. Thus

$$\delta S = \int \mathrm{d}^4x\, \partial_\mu(j^\mu_i\, \delta\omega^i) \tag{3.26}$$

where

$$j^\mu_i \equiv \left(\frac{\partial\mathscr{L}}{\partial(\partial_\mu\varphi)}(\partial_\nu\varphi) - \mathscr{L}\,\delta^\mu_\nu\right)X^\nu_i - \frac{\partial\mathscr{L}}{\partial(\partial_\mu\varphi)}\Phi_i. \tag{3.27}$$

Now, suppose that S is invariant when the variations are parametrised by *constant* $\delta\omega^i$ (i.e. 'global' transformations). Then, under 'local' transformations (in which $\delta\omega^i$ depend upon x), δS must have the form

$$\delta S = \int \mathrm{d}^4x j^\mu_i\, \partial_\mu(\delta\omega^i) \tag{3.28}$$

for some j^μ_i. Thus in these circumstances we deduce from (3.27) that

$$0 = \int \mathrm{d}^4x(\partial_\mu j^\mu_i)\, \delta\omega^i$$

for arbitrary $\delta\omega^i(x)$, so the current j^μ_i is conserved:

$$\partial_\mu j^\mu_i = 0. \tag{3.29}$$

This is Noether's theorem. It follows that the 'charge'

$$Q_i(t) \equiv \int \mathrm{d}^3x j^0_i(t, \boldsymbol{x}) \tag{3.30}$$

is a constant, independent of t, since

$$\begin{aligned}\dot{Q}_i(t) &= \int d^3x\, \partial_0 j_i^0(\boldsymbol{x}, t)\\ &= \int d^3x[\partial_\mu j_i^\mu(t, \boldsymbol{x}) - \partial_r j_i^r(t, \boldsymbol{x})]\\ &= -\int dS_r j_i^r(t, \boldsymbol{x})\\ &= 0\end{aligned} \tag{3.31}$$

if we assume that j_i^r vanishes on the boundary surface. (We have used current conservation (3.29), and Gauss's theorem to relate the spatial divergence to an integral over the boundary surface with surface element dS.) It is important to note that in deriving the current conservation (3.29) we used the Euler–Lagrange equation (3.8) for φ. Thus we are free to add to j_i^μ any quantity whose divergence vanishes by virtue of the equation of motion. Of course, we can also add to j_i^μ any quantity whose divergence is identically zero, such as

$$j_i^\mu \equiv \partial_\nu T_i^{\mu\nu} \tag{3.32a}$$

where $T_i^{\mu\nu}$ is an antisymmetric tensor:

$$T_i^{\mu\nu} = -T_i^{\nu\mu}. \tag{3.32b}$$

We have proved Noether's theorem in the simplest case that $\mathscr{L}$ is a function only of a *single* field φ and its derivative $\partial_\mu\varphi$. If there are several fields φ^a associated with each point $\boldsymbol{x}$, then the total variation Φ of each field acquires an index a, and it is easy to see that the generalisation of the conserved current (3.28) is

$$j_i^\mu = \left(\frac{\partial\mathscr{L}}{\partial(\partial_\mu\varphi^a)}(\partial_\nu\varphi^a) - \mathscr{L}\delta_\nu^\mu\right)X_i^\nu - \frac{\partial\mathscr{L}}{\partial(\partial_\mu\varphi^a)}\Phi_i^a. \tag{3.33}$$

3.3 Scalar field theory

The simplest field theory that we shall study is that described by a single real scalar field $\varphi(x)$ having a Lagrangian density

$$\mathscr{L} = \frac{1}{2}(\partial_\mu\varphi)(\partial^\mu\varphi) - \frac{1}{2}\mu^2\varphi^2 - \frac{1}{4!}\lambda\varphi^4 \tag{3.34}$$

where μ^2 and λ are constants. The Euler–Lagrange equation yields

$$(\partial_\mu\partial^\mu + \mu^2)\varphi(x) = -\tfrac{1}{6}\lambda\varphi^3(x). \tag{3.35}$$

The term on the right-hand side of (3.35) is analogous to the current source j^ν

on the right of (3.13), except that it arises from the field itself (because of its self-interaction). In the case that such a term is absent ($\lambda=0$), (3.35) reduces to the Klein–Gordon equation[2]. Interpreting $\varphi(x)$ as the wave function of a particle with energy E and momentum $\boldsymbol{p}$, the Klein–Gordon equation gives

$$-E^2+\boldsymbol{p}^2+\mu^2=0 \tag{3.36}$$

showing that the particle has rest mass μ. More generally, if

$$\mathscr{L}=\tfrac{1}{2}(\partial_\mu\varphi)(\partial^\mu\varphi)-V(\varphi) \tag{3.37}$$

and if V has an absolute minimum at $\varphi=v$, then

$$\left.\frac{\mathrm{d}V}{\mathrm{d}\varphi}\right|_{\varphi=v}=0 \tag{3.38a}$$

and

$$\left.\frac{\mathrm{d}^2V}{\mathrm{d}\varphi^2}\right|_{\varphi=v}\equiv\mu^2. \tag{3.38b}$$

The statement that 'φ is a scalar field' describes its behaviour under a Poincaré transformation. Under such a transformation

$$x^\mu \to x'^\mu=\Lambda^\mu_\nu x^\nu+a^\nu \tag{3.39}$$

where Λ^μ_ν describes a proper Lorentz transformation, and a is a (space–time) translation. Then 'φ is a scalar field' means that under this transformation

$$\varphi(x)\to\varphi'(x')=\varphi(x) \tag{3.40}$$

i.e. $\varphi(x)$ is invariant. Since φ is invariant under an arbitrary Poincaré transformation it is necessarily invariant under an infinitesimal such transformation.

Consider first an infinitesimal translation

$$x^\mu\to x^\mu+\varepsilon^\mu. \tag{3.41}$$

Then the infinitesimal parameters $\delta\omega^i$ may, on this occasion, be chosen to be the quantities ε^ν and

$$X^\mu_\nu(x)=\delta^\mu_\nu. \tag{3.42}$$

Since φ is invariant $\delta\varphi$ is zero and, from (3.21),

$$\Phi_\nu(x)=0. \tag{3.43}$$

Also from (3.34) or (3.37)

$$\frac{\partial\mathscr{L}}{\partial(\partial_\mu\varphi)}=\partial^\mu\varphi. \tag{3.44}$$

Then, from (3.33), the conserved current in this case is denoted

$$\begin{aligned} T^{\mu}_{\nu} &= \frac{\partial \mathscr{L}}{\partial(\partial_{\mu}\varphi)}(\partial_{\nu}\varphi) - \delta^{\mu}_{\nu}\mathscr{L} \\ &= (\partial^{\mu}\varphi)(\partial_{\nu}\varphi) - \delta^{\mu}_{\nu}\left(\frac{1}{2}(\partial_{\lambda}\varphi)(\partial^{\lambda}\varphi) - \frac{1}{2}\mu^2\varphi^2 - \frac{1}{4!}\lambda\varphi^4\right). \end{aligned} \tag{3.45}$$

In the free field case $\lambda = 0$ and the 'charge' is

$$\begin{aligned} P_{\nu} &\equiv \int d^3x T^0_{\nu} \\ &= \int d^3x[(\partial^0\varphi)(\partial_{\nu}\varphi) + \delta^0_{\nu}\tfrac{1}{2}\varphi(\Box\varphi + \mu^2\varphi)] \\ &= \int d^3x(\partial^0\varphi)(\partial_{\nu}\varphi). \end{aligned} \tag{3.46}$$

In deriving this we have added the total divergence $\delta^0_{\nu}\frac{1}{2}\partial_{\mu}[\varphi\,\partial^{\mu}\varphi]$ to the integrand, as we may, and then used the equation of motion (3.35) to eliminate the second term. Since $\varphi(x)$ is a free field and real we may without loss of generality write

$$\varphi(x) = \int \frac{d^3k}{(2\pi)^3}\frac{1}{2k_0}(a(k)\,e^{-ikx} + a^*(k)\,e^{ikx}) \tag{3.47}$$

where $k_0 = +(\boldsymbol{k}^2 + \mu^2)^{1/2}$. Substituting into (3.46) and performing several trivial integrations gives

$$P_{\mu} = \int \frac{d^3k}{(2\pi)^3}\frac{1}{2k_0}a^*(k)a(k)k_{\mu} \tag{3.48}$$

which *is* independent of $x^0 = ct$, as promised. P_{μ} is the energy–momentum vector of the field: each mode is labelled by a vector k (with $k^2 = \mu^2$) and $a(k)$ is the amplitude for a mode having momentum k, and $(2\pi)^{-3}(2k_0)^{-1}|a(k)|^2\,d^3k$ is the number of modes having momentum in the element d^3k around $\boldsymbol{k}$.

The $\mu = 0$ component is therefore the energy of the field. In general we have

$$P_0 = \int d^3x\left(\frac{\partial \mathscr{L}}{\partial(\partial_0\varphi)}(\partial_0\varphi) - \mathscr{L}\right) \tag{3.49a}$$

$$= \int d^3x[\tfrac{1}{2}\pi^2(x) + \tfrac{1}{2}(\partial_r\varphi^2)^2 + V(\varphi)] \equiv \int d^3x\mathscr{H} \equiv H \tag{3.49b}$$

using (3.44) and (3.37). Expressed in this way, as a functional of $\varphi(x)$ and $\pi \equiv \partial\mathscr{L}/\partial(\partial_0\varphi)$, the energy is called the Hamiltonian H of the field theory, and $\mathscr{H}$ is called the Hamiltonian density. Just as in mechanics, the equations of motion may be recast in a form reminiscent of Hamilton's equations. The first follows trivially:

$$\partial_0\varphi=\pi(x)=\frac{\delta H}{\delta\pi(x)}. \tag{3.50}$$

Also

$$\partial_0\pi=\partial_0^2\varphi \tag{3.51}$$

and

$$\frac{\delta H}{\delta\varphi(x)}=-\nabla^2\varphi+\frac{\partial V}{\partial\varphi}. \tag{3.52}$$

The Euler–Lagrange equation in this case, is

$$\Box\varphi=-\frac{\partial V}{\partial\varphi} \tag{3.53}$$

which yields the second equation:

$$\partial_0\pi=-\frac{\delta H}{\delta\varphi}. \tag{3.54}$$

Next we consider an infinitesimal Lorentz transformation

$$x^\mu\to x^\mu+\varepsilon^{\mu\nu}x_\nu \tag{3.55a}$$

with

$$\varepsilon^{\mu\nu}=-\varepsilon^{\nu\mu}. \tag{3.55b}$$

The infinitesimal parameters $\delta\omega^i$ of (3.21) are now chosen to be the quantities $-\varepsilon^{\rho\sigma}$, and now

$$X^\mu_{\rho\sigma}(x)=-\delta^\mu_\rho x_\sigma+\delta^\mu_\sigma x_\rho. \tag{3.56}$$

Since φ is scalar, $\delta\varphi$ is zero and, from (3.21),

$$\Phi_{\rho\sigma}(x)=0. \tag{3.57}$$

Thus the conserved current (3.33) may be expressed in terms of the energy–momentum tensor T, defined in (3.45). We find

$$M^\mu_{\rho\sigma}=x_\rho T^\mu_\sigma-x_\sigma T^\mu_\rho \tag{3.58}$$

and the associated 'charge'

$$M_{\rho\sigma}\equiv\int d^3x M^0_{\rho\sigma} \tag{3.59}$$

is the angular momentum tensor of the field.

Noether's theorem also applies to 'internal' symmetry transformations, i.e. those not involving the space–time coordinates x. To illustrate this we consider a field theory described by a *complex* field

$$\varphi(x)=\frac{1}{\sqrt{2}}[\varphi_1(x)+\mathrm{i}\varphi_2(x)] \tag{3.60}$$

where $\varphi_i(x)$ $(i=1,2)$ are real scalar fields. The Lagrangian density is then a function of both fields φ_i and their derivatives, or equivalently of $\varphi(x)$ and its complex conjugate $\varphi^*(x)$ (and their derivatives):

$$\mathscr{L}=(\partial_\mu\varphi^*)(\partial^\mu\varphi)-\mu^2\varphi^*\varphi-\lambda(\varphi^*\varphi)^2. \tag{3.61}$$

By varying with respect to φ^* we obtain the Euler–Lagrange equation

$$\partial_\mu(\partial^\mu\varphi)=-\mu^2\varphi-2\lambda(\varphi^*\varphi)\varphi \tag{3.62}$$

and the complex conjugate equation follows by varying with respect to φ. In the absence of any interaction ($\lambda=0$), clearly φ and φ^* describe fields having modes of mass μ since

$$(\Box+\mu^2)\varphi(x)=0 \tag{3.63}$$

as in (3.36).

The internal symmetry transformation that we wish to consider is a 'global gauge transformation'. That is to say, the transformations

$$\varphi(x)\rightarrow\varphi'(x)=\mathrm{e}^{-\mathrm{i}q\Lambda}\,\varphi(x) \tag{3.64a}$$

and

$$\varphi^*(x)\rightarrow\varphi^{*\prime}(x)=\mathrm{e}^{\mathrm{i}q\Lambda}\,\varphi^*(x) \tag{3.64b}$$

where q and Λ are real and independent of x. (The transformation is 'global' because it is the same at all space–time points x, and 'gauge' because it alters the phase of the complex field $\varphi(x)$.) If Λ is infinitesimal

$$\mathrm{e}^{-\mathrm{i}q\Lambda}\approx 1-\mathrm{i}q\Lambda \tag{3.65}$$

then

$$\delta\varphi(x)=-\mathrm{i}q\Lambda\varphi(x) \tag{3.66}$$

and, from (3.21),

$$\Phi(x)=-\mathrm{i}q\varphi(x). \tag{3.67}$$

Similarly

$$\delta(\varphi^*)=(\delta\varphi)^* \tag{3.68}$$

so

$$\Phi^*(x)=[\Phi(x)]^*. \tag{3.69}$$

Since $\mathscr{L}$ *is* invariant under (3.64), so is S. Thus the Noether current

$$j^\mu = -\frac{\partial \mathscr{L}}{\partial(\partial_\mu \varphi)}\Phi - \frac{\partial \mathscr{L}}{\partial(\partial^\mu \varphi^*)}\Phi^*$$
$$= (\partial^\mu \varphi^*)\mathrm{i}q\varphi - (\partial^\mu \varphi)\mathrm{i}q\varphi^*$$
$$\equiv -\mathrm{i}q\varphi^* \overleftrightarrow{\partial}^\mu \varphi \tag{3.70}$$

is conserved in this case.

3.4 Spinor field theory

This is defined as the field theory which yields the Dirac equation[3]

$$(\mathrm{i}\gamma^\mu \partial_\mu - m\mathbf{I})\psi(x) = 0 \tag{3.71}$$

which, it is well known, describes particles with spin angular momentum 1/2. In (3.71), ψ is a four-component column vector in 'spinor-space', $\mathbf{I}$ is the unit 4×4 matrix and the matrices γ^μ ($\mu = 0, 1, 2, 3$) satisfy

$$\{\gamma^\mu, \gamma^\nu\} \equiv \gamma^\mu\gamma^\nu + \gamma^\nu\gamma^\mu = 2g^{\mu\nu}\mathbf{I}. \tag{3.72}$$

It follows from (3.71) that the Hermitian conjugate field

$$\psi(x)^\dagger \equiv \psi(x)^{*\mathrm{T}} \tag{3.73}$$

satisfies

$$\psi(x)^\dagger(-\mathrm{i}\overleftarrow{\partial}_\mu \gamma^{\mu\dagger} - m\mathbf{I}) = 0. \tag{3.74}$$

Thus

$$\bar{\psi}(x) \equiv \psi^\dagger(x)\mathbf{A} \tag{3.75}$$

satisfies

$$\bar{\psi}(x)(-\mathrm{i}\overleftarrow{\partial}_\mu \gamma^\mu - m\mathbf{I}) = 0 \tag{3.76}$$

provided the matrix $\mathbf{A}$ satisfies

$$\gamma^{\mu\dagger}\mathbf{A} = \mathbf{A}\gamma^\mu. \tag{3.77}$$

It is easy to see that the matrices

$$\gamma^0 = \begin{pmatrix} I_2 & 0 \\ 0 & -I_2 \end{pmatrix} \qquad \gamma^i = \begin{pmatrix} 0 & \sigma^i \\ -\sigma^i & 0 \end{pmatrix} \qquad (i = 1, 2, 3) \tag{3.78}$$

where σ^i are the Pauli matrices, satisfy (3.72). In this representation, but not in all representations,

$$\gamma^0 = \gamma^{0\dagger} \qquad \gamma^i = -\gamma^{i\dagger}. \tag{3.79}$$

Thus in this representation we may take

$$\mathbf{A}=\gamma^0=\mathbf{A}^\dagger. \tag{3.80}$$

Then

$$\mathscr{L}=\bar{\psi}(x)(i\gamma^\mu\partial_\mu-m)\psi(x) \tag{3.81}$$

is the required Lagrangian density. The Euler–Lagrange equation for $\bar{\psi}$ yields (3.71) immediately, since

$$\frac{\partial\mathscr{L}}{\partial(\partial_\mu\bar{\psi})}=0. \tag{3.82}$$

Alternatively, variation with respect to ψ gives

$$\frac{\partial\mathscr{L}}{\partial(\partial_\mu\psi)}=\bar{\psi}(x)\mathrm{i}\gamma^\mu \tag{3.83a}$$

and

$$\frac{\partial\mathscr{L}}{\partial\psi}=-m\bar{\psi}(x) \tag{3.83b}$$

and the Euler–Lagrange equation then gives (3.74), as required.

Under the Poincaré transformation (3.39)

$$\mathbf{x}\rightarrow\mathbf{x}'=\mathbf{\Lambda x}+\mathbf{a} \tag{3.84}$$

and the spinor field transforms according to

$$\psi(x)\rightarrow\psi'(x')=\mathbf{S}(\Lambda)\psi(x). \tag{3.85}$$

Thus under an infinitesimal translation ψ is invariant, and (as in the scalar case) the energy–momentum tensor is

$$\begin{aligned}T^\mu_\lambda&=\frac{\partial\mathscr{L}}{\partial(\partial_\mu\psi)}\partial_\lambda\psi+\partial_\lambda\bar{\psi}\frac{\partial\mathscr{L}}{\partial(\partial_\mu\bar{\psi})}-\delta^\mu_\lambda\mathscr{L}\\&=\bar{\psi}\mathrm{i}\gamma^\mu\partial_\lambda\psi\end{aligned} \tag{3.86}$$

since $\mathscr{L}$ is zero using the Dirac equation (3.71). As before, we may decompose ψ into its Fourier components. Since there are four independent solutions of the Dirac equation, two positive energy solutions $u(k,\pm s)$ and two negative energy solutions $v(k,\pm s)$, we may write, without loss of generality,

$$\psi(x)=\int\frac{\mathrm{d}^3k}{(2\pi)^3}\frac{1}{2k_0}\sum_{\pm s}[a(k,s)u(k,s)\,\mathrm{e}^{-\mathrm{i}kx}+b^*(k,s)v(k,s)\,\mathrm{e}^{\mathrm{i}kx}] \tag{3.87}$$

where u, v satisfy

$$(\not{k}-m)u(k,\pm s)=0 \tag{3.88a}$$

$$(\not{k}+m)v(k,\pm s)=0 \tag{3.88b}$$

and we are using the Feynman notation

$$\not{p} \equiv \gamma^\mu p_\mu. \tag{3.88c}$$

The vector s is the covariant spin vector satisfying

$$s^2 = -1 \qquad s \cdot k = 0 \tag{3.89}$$

so that in the rest-frame $s = (0, \boldsymbol{n})$ specifies the spin direction. Then u, v satisfy

$$\gamma_5 \not{s} u(k, \pm s) = \pm u(k, \pm s) \tag{3.90a}$$

$$\gamma_5 \not{s} v(k, \pm s) = \pm v(k, \pm s) \tag{3.90b}$$

and are normalised so that

$$\bar{u}(k, s)\gamma_\mu u(k, s) = \bar{v}(k, s)\gamma_\mu v(k, s) = 2k_\mu \tag{3.90c}$$

$$\bar{u}(k, s)\gamma_\mu \gamma_5 u(k, s) = -\bar{v}(k, s)\gamma_\mu \gamma_5 v(k, s) = 2ms_\mu \tag{3.90d}$$

with

$$\gamma_5 \equiv \mathrm{i}\gamma^0\gamma^1\gamma^2\gamma^3 \tag{3.91a}$$

and satisfying

$$\gamma_5^2 = 1 \qquad \gamma_\mu\gamma_5 + \gamma_5\gamma_\mu = 0. \tag{3.91b}$$

In the representation (3.78)

$$\gamma_5 = \begin{pmatrix} 0 & I_2 \\ I_2 & 0 \end{pmatrix}. \tag{3.92}$$

From (3.75) it follows that

$$\bar{\psi}(x) = \int \frac{\mathrm{d}^3k}{(2\pi)^3} \frac{1}{2k_0} \sum_{\pm s} \{b(k, s)\bar{v}(k, s)\,\mathrm{e}^{-\mathrm{i}kx} + a^*(k, s)\bar{u}(k, s)\,\mathrm{e}^{\mathrm{i}kx}\} \tag{3.93a}$$

with

$$\bar{v} \equiv v^\dagger\gamma^0 \qquad \bar{u} \equiv u^\dagger\gamma^0 \tag{3.93b}$$

using (3.80). Then, using (3.33), we find the energy–momentum vector

$$\begin{aligned} P_\mu &= \int \mathrm{d}^3x\, T_\mu^0 \\ &= \int \frac{\mathrm{d}^3k}{2k_0} \frac{1}{(2\pi)^3} \sum_{\pm s} [a^*(k, s)a(k, s)k_\mu - b(k, s)b^*(k, s)k_\mu] \end{aligned} \tag{3.94}$$

It is at this point that the features characteristic of the spinor field theory begin to emerge. We might interpret $(2\pi)^{-3}(2k_0)^{-1}a^*(k, s)a(k, s)\,\mathrm{d}^3k$ as the number of a modes having spin s and momentum in the element d^3k around $\boldsymbol{k}$, just as in the scalar case. The trouble is that the second term of (3.94) then describes b modes having negative energy. In (traditional) canonical field theory the resolution of this difficulty is made by identifying the negative-energy modes as positive-energy antiparticle states. We, however, must eschew this escape. Following Berezin[4] we instead take the quantities a, a^*, b, b^* to be *anticommuting* (c-numbers). That is to say, we take all pairs of these to anticommute. Using the anticommutator notation of (3.72), this means that

$$\begin{aligned}\{a, a'\} &= \{a, a^{*\prime}\} = \{a, b'\} = \{a, b'^*\} \\ &= \{a^*, a'^*\} = \{a^*, b'\} = \{a^*, b'^*\} \\ &= \{b, b'\} = \{b, b'^*\} = \{b^*, b'^*\} = 0 \end{aligned} \tag{3.95a}$$

where

$$a = a(k, s) \qquad a' = a(k', s') \qquad \text{etc.} \tag{3.95b}$$

Mathematicians call such anticommuting variables elements of a Grassmann algebra. In the scalar field case we tacitly assumed that a and a^* were ordinary commuting c-numbers, and the reason for the different treatment of the spinor field stems ultimately, of course, from the spin and statistics theorem[5]; the scalar field describes spin zero and therefore boson particles, while the spinor field describes spin 1/2 and therefore fermion particles. This device therefore evades the positivity problem by making the energy a Grassmann variable, rather than a real number, and consequently not something whose positivity, or lack of it, can be enquired about. Thus it is clear that in this approach spin fields are essentially non-classical. We know that in quantum mechanics we may only ask what the expectation value of the energy is and *this*, when we have learned how to make the transition to quantum mechanics, had better be a real positive number.

We shall see in Chapter 4 that the formulation of the quantum version of scalar field theory is achieved by means of a functional integral over the classical scalar field configurations. We therefore anticipate that generalisation of spinor field theory to a quantum field theory will require functional integration over classical spinor field configurations. Since we have just decreed that the classical spinor field is a Grassmann variable, functional integrals over them involve certain peculiar aspects of which it is as well to be forewarned. We therefore digress briefly on differentiation and integration with respect to Grassmann variables. We start with two such variables a, b satisfying

$$\{a, a\} = \{a, b\} = \{b, b\} = 0 \tag{3.96}$$

so that $a^2 = b^2 = 0$. Without loss of generality any function $f(a, b)$ may therefore be written

$$f(a, b) = f_0 + f_1 a + \tilde{f}_1 b + f_2 ab \tag{3.97a}$$

$$= f_0 + f_1 a + \tilde{f}_1 b - f_2 ba \tag{3.97b}$$

where f_0, f_1, $\tilde{f}_1$, f_2 are ordinary c-numbers. Differentiation is defined in an obvious way:

$$\frac{\partial f}{\partial a} = f_1 + f_2 b \tag{3.98}$$

except that we must be careful to respect the anticommutativity of a, b. Thus

$$\frac{\partial f}{\partial b} = \tilde{f}_1 - f_2 a \tag{3.99}$$

using the second version (3.97b) of f. It follows that

$$\frac{\partial^2 f}{\partial a\, \partial b} = -\frac{\partial^2 f}{\partial b\, \partial a} = -f_2. \tag{3.100}$$

We may also define integration with respect to Grassmann variables. Obviously we shall want it to be a linear operation, and the 'infinitesimals' da, db will also be Grassmann variables, so that

$$\{a, \mathrm{d}a\} = \{a, \mathrm{d}b\} = \{\mathrm{d}a, b\} = \{\mathrm{d}a, \mathrm{d}b\} = 0. \tag{3.101}$$

Multiple integrals will be interpreted as iterated integrals, so that we may compute

$$\int \mathrm{d}a\, \mathrm{d}b f(a, b) \equiv \int \mathrm{d}a \left(\int \mathrm{d}b f(a, b) \right)$$

provided we know the basic single integrals $\int \mathrm{d}a$ and $\int \mathrm{d}a\, a$. Now

$$\left(\int \mathrm{d}a\right)^2 = \left(\int \mathrm{d}a\right)\left(\int \mathrm{d}b\right) = \int \mathrm{d}a\, \mathrm{d}b = -\int \mathrm{d}b\, \mathrm{d}a = -\left(\int \mathrm{d}a\right)^2 \tag{3.102}$$

so

$$\int \mathrm{d}a = 0. \tag{3.103}$$

Then, as there is no other scale to Grassmann variables, we are free to define

$$\int \mathrm{d}a\, a = 1. \tag{3.104}$$

From (3.103) and (3.104) we see that integration is the same as differentiation for Grassmann variables. With the parametrisation (3.97) we find

$$\int \mathrm{d}a\, \mathrm{d}b f(a, b) = -f_2 = \frac{\partial^2 f}{\partial a\, \partial b}. \tag{3.105}$$

The generalisation of these results to countable numbers of Grassmann variables is straightforward, and even the generalisation to Grassmann functional differentiation and integration is not especially difficult, once the ordinary functional calculus has been mastered. We defer discussion of the functional aspects to Chapter 8. Finally we note that there is no distinction between definite and indefinite integration with respect to a Grassmann variable.

Let us now return to the main problem in hand: the formulation of classical spinor field theory. We consider next the behaviour of the field under an infinitesimal Lorentz transformation (3.55)

$$x^{\mu} \rightarrow x^{\mu} + \varepsilon^{\mu\nu} x_{\nu} \tag{3.55a}$$

with the infinitesimal satisfying

$$\varepsilon^{\mu\nu} = -\varepsilon^{\nu\mu}. \tag{3.55b}$$

Under such a transformation the matrix $\mathbf{S}(\Lambda)$ of (3.85) is given by

$$\mathbf{S}(\Lambda) = \mathbf{I} - \tfrac{1}{4}\mathrm{i}\varepsilon^{\mu\nu}\sigma_{\mu\nu} \tag{3.106a}$$

where

$$\sigma_{\mu\nu} = \frac{\mathrm{i}}{2}(\gamma_{\mu}\gamma_{\nu} - \gamma_{\nu}\gamma_{\mu}). \tag{3.106b}$$

Thus, using the notation (3.20)

$$\delta\psi(x) = -\frac{\mathrm{i}}{4}\varepsilon^{\mu\nu}\sigma_{\mu\nu}\psi(x) \tag{3.107}$$

and it follows from (3.21) (taking the infinitesimal parameter to be $-\varepsilon_{\rho\sigma}$) that

$$\Phi_{\rho\sigma} = \frac{\mathrm{i}}{2}\sigma_{\rho\sigma}\psi(x). \tag{3.108}$$

Thus the conserved current (3.33)—the angular momentum density tensor $M^{\mu}_{\rho\sigma}$—is given by

$$M^{\mu}_{\rho\sigma} = x_{\rho}T^{\mu}_{\sigma} - x_{\sigma}T^{\mu}_{\rho} + \bar{\psi}(x)\gamma^{\mu}\tfrac{1}{2}\sigma_{\rho\sigma}\psi(x) \tag{3.109a}$$

and the angular momentum of the field is

$$M_{\rho\sigma} = \int \mathrm{d}^3x M^{0}_{\rho\sigma}. \tag{3.109b}$$

Clearly the first two terms describe the orbital angular momentum in the field, while the last term is its spin angular momentum.

The Lagrangian (3.81) is also invariant under a global gauge transformation analogous to (3.64)

$$\psi(x) = \mathrm{e}^{-\mathrm{i}q\Lambda}\,\psi(x) \tag{3.110a}$$

$$\bar{\psi}(x) = \mathrm{e}^{\mathrm{i}q\Lambda}\,\bar{\psi}(x). \tag{3.110b}$$

In this case, since

$$\Phi(x) = -\mathrm{i}q\psi(x) \tag{3.111}$$

and $\partial\mathscr{L}/\partial(\partial_\mu\psi)$ is given by (3.83), the conserved Noether current is

$$j^\mu = -q\bar{\psi}\gamma^\mu\psi. \tag{3.112}$$

We shall see in Chapter 8 that this is indeed (minus) the electromagnetic current which is coupled to the electromagnetic field.

The four-component (Dirac) spinor field theory with which we have so far concerned ourselves is certainly the most economical description of spin 1/2 particles with *non-zero* mass. Indeed we shall see in Chapters 8 and 9 that the electron field in quantum electrodynamics (QED) and the quark fields in quantum chromodynamics (QCD) are all described by Dirac fields. The reason is that the four-component field provides a representation of the 'parity' transformation, i.e. space reversal $\boldsymbol{x} \to -\boldsymbol{x}$. To see this we note that the parity-transformed field $\psi^{\mathrm{P}}(x^0, \boldsymbol{x})$ may be written as

$$\psi^{\mathrm{P}}(x^0, \boldsymbol{x}) = \varepsilon\gamma^0\psi(x^0, -\boldsymbol{x}) \tag{3.113}$$

(where ε is a phase factor) since ψ^{P} then satisfies

$$(\mathrm{i}\gamma^\mu\partial_\mu^{\mathrm{P}} - m)\psi^{\mathrm{P}}(x^{\mathrm{P}}) = 0 \tag{3.114a}$$

where

$$x^{\mathrm{P}} = (x^0, -\boldsymbol{x}) \qquad \partial_\mu^{\mathrm{P}} = \frac{\partial}{\partial x^{\mu\mathrm{P}}}. \tag{3.114b}$$

Since parity is conserved in QED and QCD, the use of Dirac fields provides an economic realisation of this symmetry.

It is well known that weak interactions do *not* conserve parity. A particular manifestation of this is provided by processes involving neutrinos and antineutrinos. Neutrinos are observed only in a left-handed helicity state; the right-handed neutrino state, even if it exists, is not observed in weak processes. Similarly the antineutrino is observed only in a right-handed helicity state, never in the left-handed state. Helicity is a Lorentz-*in*variant quantity only for massless particles, whereas a massive particle having right-handed helicity (say) in one inertial frame will have left-handed helicity in an inertial frame in which its momentum has opposite direction. Now, it may be that (some) neutrinos and antineutrinos are indeed massless. Thus for spin 1/2 neutrinos with *zero* mass, the use of a four-component Dirac field is at best a luxury and at worst can lead to confusion if the uncoupled modes do not even exist. In any event, we need to revise the specification (3.90) of the spin eigenstates, since for a massless particle there is no rest frame. However, the helicity, defined as the component of angular momentum ($\boldsymbol{J}$) in the direction of momentum ($\hat{\boldsymbol{k}}$), always exists. For *massless* particles we can show that it is proportional to the matrix γ_5, defined in (3.91).

Let $\psi(x)$ be any solution of the massless Dirac equation. Without loss of generality we may write it in the form

$$\psi(x)=\psi_{\mathrm{L}}(x)+\psi_{\mathrm{R}}(x) \tag{3.115a}$$

where

$$\psi_{\mathrm{L}}(x)\equiv\tfrac{1}{2}(1-\gamma_5)\psi(x) \tag{3.115b}$$

$$\psi_{\mathrm{R}}(x)\equiv\tfrac{1}{2}(1+\gamma_5)\psi(x). \tag{3.115c}$$

Then ψ_{L} and ψ_{R} are eigenvectors of γ_5, since

$$\gamma_5\psi_{\mathrm{R}}=\psi_{\mathrm{R}} \qquad \gamma_5\psi_{\mathrm{L}}=-\psi_{\mathrm{L}}. \tag{3.116}$$

Now since $\psi(x)$ satisfies the massless Dirac equation, so does $\gamma_5\psi(x)$ (since γ_5 anticommutes with γ_μ), and so therefore do $\psi_{\mathrm{L}}(x)$ and $\psi_{\mathrm{R}}(x)$. These solutions ψ_{L} and ψ_{R} are called 'Weyl spinors'. They each contain two components, as can be made explicit if we use a representation of the γ-matrices different from that given in (3.78). In the Weyl representation

$$\gamma^0=\begin{pmatrix}0 & I_2\\ I_2 & 0\end{pmatrix} \qquad \gamma^i=\begin{pmatrix}0 & -\sigma^i\\ \sigma^i & 0\end{pmatrix} \tag{3.117}$$

so that

$$\gamma_5=\mathrm{i}\gamma^0\gamma^1\gamma^2\gamma^3=\begin{pmatrix}I_2 & 0\\ 0 & -I_2\end{pmatrix}. \tag{3.118}$$

In this representation the Weyl spinor $(\psi_{\mathrm{R}})_\alpha$ has (two) non-zero upper components $\alpha=1, 2$, and $(\psi_{\mathrm{L}})_\alpha$ has two non-zero lower components $\alpha=3, 4$. The (four-component) Dirac spinor field is the sum of two (two-component) Weyl spinors.

As in (3.87) we may decompose ψ_{L} and ψ_{R} into their Fourier components

$$\psi_{\mathrm{R}}(x)=\int\frac{\mathrm{d}^3k}{(2\pi)^3}\frac{1}{2k_0}\left(a(k,+)u(k,+\tfrac{1}{2})\,\mathrm{e}^{-\mathrm{i}kx}+b^*(k,+)v(k,\tfrac{1}{2})\,\mathrm{e}^{\mathrm{i}kx}\right) \tag{3.119a}$$

$$\psi_{\mathrm{L}}(x)=\int\frac{\mathrm{d}^3k}{(2\pi)^3}\frac{1}{2k_0}\left(a(k,-)u(k,-\tfrac{1}{2})\,\mathrm{e}^{-\mathrm{i}kx}+b^*(k,-)v(k,-\tfrac{1}{2})\,\mathrm{e}^{\mathrm{i}kx}\right) \tag{3.119b}$$

where $u(k,h)$ and $v(k,h)$ $(h=\pm 1/2)$ are positive and negative energy solutions of the massless Dirac equation

$$\not{k}u(k,h)=0=\not{k}v(k,h) \tag{3.120a}$$

satisfying

$$\gamma_5 u(k,h)=2hu(k,h) \tag{3.120b}$$

$$\gamma_5 v(k,h)=2hv(k,h) \tag{3.120c}$$

and normalised so that

$$\bar{u}(k,h)\gamma_\mu u(k,h)=\bar{v}(k,h)\gamma_\mu v(k,h)=2k_\mu. \tag{3.120d}$$

From (3.109) we see that the total spin of the field $\psi(x)$ is given by

$$S_{\rho\sigma}=\int \mathrm{d}^3x\bar{\psi}(x)\gamma^0\tfrac{1}{2}\sigma_{\rho\sigma}\psi(x). \tag{3.121}$$

Thus the spin vector $\boldsymbol{S}$, defined by

$$S^k=\tfrac{1}{2}\varepsilon^{ijk}S_{ij} \tag{3.122a}$$

is given by

$$\boldsymbol{S}[\psi]=\tfrac{1}{2}\int \mathrm{d}^3x\bar{\psi}(x)\boldsymbol{\gamma}\gamma_5\psi(x). \tag{3.122b}$$

For the Weyl field ψ_R this gives

$$\boldsymbol{S}[\psi_\mathrm{R}]=\tfrac{1}{2}\int\frac{\mathrm{d}^3k}{(2\pi)^3}\frac{1}{2k_0}[a^*(k,+)a(k,+)\hat{\boldsymbol{k}}+b(k,+)b^*(k,+)\hat{\boldsymbol{k}}] \tag{3.123}$$

Evidently the (positive energy modes of the) field ψ_R has *positive* (i.e. right-handed) helicity, since the orbital angular momentum is always perpendicular to $\boldsymbol{k}$. Similarly ψ_L has negative (i.e. left-handed) helicity. Hence the labels 'L' and 'R'. Notice also that (in the Weyl representation) γ^0 interchanges left and right as it should since it represents the parity transformation. As anticipated, γ_5 gives twice the helicity.

It is easy to see that when $m=0$ the Lagrangian (3.81) is the sum of two contributions, one from each Weyl spinor

$$\mathscr{L}=\bar{\psi}\mathrm{i}\gamma^\mu\partial_\mu\psi=\chi_\mathrm{R}^\dagger\mathrm{i}\sigma^\mu\partial_\mu\chi_\mathrm{R}+\chi_\mathrm{L}^\dagger\mathrm{i}\bar{\sigma}^\mu\partial^\mu\chi_\mathrm{L} \tag{3.124a}$$

where (in the Weyl representation)

$$\psi_\alpha=\begin{pmatrix}\chi_\mathrm{R}\\ \chi_\mathrm{L}\end{pmatrix} \tag{3.124b}$$

and

$$\sigma^\mu=(1,\boldsymbol{\sigma}) \tag{3.124c}$$

$$\bar{\sigma}^\mu=(1,-\boldsymbol{\sigma}). \tag{3.124d}$$

3.5 Massless vector field theory

It is clear that Maxwell's equations

$$\partial_\mu(\partial^\mu A^\nu-\partial^\nu A^\mu)=j^\nu \tag{3.125}$$

discussed in (3.12) and thereafter, do not completely specify the vector potential $A^\mu(x)$. For, if $A^\mu(x)$ satisfies (3.125), so does

$$A'^\mu(x) \equiv A^\mu(x) + \partial^\mu \Lambda(x) \tag{3.126}$$

for an arbitrary function $\Lambda(x)$. (We shall see later that this is associated with a 'local' gauge invariance related to the global gauge invariance already discussed.) It is clear also that both vector potentials yield the same $\boldsymbol{E}$ and $\boldsymbol{B}$ fields, since $\boldsymbol{E}$ and $\boldsymbol{B}$, defined in (3.14b) and (3.15b), are invariant under the substitutions

$$A_0 \to A'_0 = A_0 + \partial_0 \Lambda \tag{3.127a}$$

$$\boldsymbol{A} \to \boldsymbol{A}' = \boldsymbol{A} - \nabla\Lambda. \tag{3.127b}$$

This lack of uniqueness of the vector potential for given electric and magnetic fields generates difficulties when, for example, we have to perform functional integrals over the different field configurations. The lack of uniqueness may be reduced by imposing a further condition on A^μ, besides those required by (3.125). It is customary to impose the 'Lorentz condition':

$$\partial_\mu A^\mu(x) = 0 \tag{3.128}$$

which is clearly the unique covariant condition which is linear in A. (Occasionally non-covariant conditions, such as $t_\mu A^\mu = 0$, are also encountered in the literature but we shall not have recourse to them.) Even the imposition of the Lorentz condition does not completely fix the vector potential, since if A and A' are related as in (3.126), then they will both satisfy (3.128) if

$$\Box\Lambda \equiv \partial^\mu \partial_\mu \Lambda = 0. \tag{3.129}$$

The imposition of the Lorentz condition (3.128) is achieved in a Lagrangian formalism by the use of a Lagrange multiplier ξ. The Lagrangian of the electromagnetic field is modified by the addition of a 'gauge fixing term' $-(1/2\xi)(\partial_\mu A^\mu)^2$. Thus instead of (3.12) we have

$$\mathscr{L} = -\frac{1}{4} F_{\mu\nu} F^{\mu\nu} - j_\mu A^\mu - \frac{1}{2\xi} (\partial_\mu A^\mu)^2 \tag{3.130a}$$

where

$$F_{\mu\nu} \equiv \partial_\mu A_\nu - \partial_\nu A_\mu. \tag{3.130b}$$

Then the Euler–Lagrange equations yield

$$\partial_\mu F^{\mu\nu} + \frac{1}{\xi} \partial^\nu (\partial_\mu A^\mu) = j^\nu. \tag{3.131}$$

Provided the current j is conserved, it follows that

$$\Box(\partial_\mu A^\mu) = 0. \tag{3.132}$$

Thus if $\partial_\mu A^\mu$ and $(\partial/\partial t)(\partial_\mu A^\mu)$ vanish at *one* time t_0, it follows from (3.132) that

$$\partial_\mu A^\mu = 0 \tag{3.133}$$

for all times. Then (3.131) reduces to Maxwell's equations, which, when (3.128) *is* satisfied, may be written as

$$\Box A^\nu = j^\nu. \tag{3.134}$$

As anticipated, this equation is also satisfied by $A^\nu + \partial^\nu \Lambda$ if $\Box \Lambda = 0$. It follows from (3.130) and the Lorentz condition (3.128) that

$$\frac{\partial \mathscr{L}}{\partial(\partial_\mu A_\nu)} = -F^{\mu\nu} - \frac{1}{\xi} g^{\mu\nu}(\partial_\lambda A^\lambda) = -F^{\mu\nu}. \tag{3.135}$$

Under the Poincaré transformation (3.39)

$$\mathbf{x} \to \mathbf{x}' = \Lambda \mathbf{x} + \mathbf{a} \tag{3.136}$$

and the vector potential transforms as

$$A^\mu(x) \to A'^\mu(x') = \Lambda^\mu{}_\nu A^\nu(x). \tag{3.137}$$

Thus A^μ is invariant under an infinitesimal translation and the energy momentum tensor is

$$\begin{aligned} T^\mu_\lambda &= \frac{\partial \mathscr{L}}{\partial(\partial_\mu A_\nu)} \partial_\lambda A_\nu - \delta^\mu_\lambda \mathscr{L} \\ &= -F^{\mu\nu} \partial_\lambda A_\nu + \delta^\mu_\lambda \tfrac{1}{4} F_{\rho\sigma} F^{\rho\sigma} \end{aligned} \tag{3.138}$$

in the free-field case $j=0$. Decomposing A into Fourier components gives

$$A_\mu(x) = \int \frac{d^3k}{(2\pi)^3} \frac{1}{2k_0} (a_\mu(k)\, e^{-ikx} + a^*_\mu(k)\, e^{ikx}) \tag{3.139a}$$

where $k_0 = |\boldsymbol{k}|$, from (3.134), and

$$k \cdot a(k) = 0 \tag{3.139b}$$

using the Lorentz condition (3.128). Proceeding as before we find that the energy–momentum vector

$$P_\nu = \int d^3x\, T^0_\nu = \int \frac{d^3k}{(2\pi)^3} \frac{1}{2k_0} [-a^*(k) \cdot a(k) k_\nu]. \tag{3.140}$$

We cannot immediately identify $(2\pi)^{-3}(2k_0)^{-1}[-a^*(k) \cdot a(k)]\, d^3k$ as the number of modes having momentum in the element dk around $\boldsymbol{k}$, since $-a^*(k) \cdot a(k)$ is not obviously positive definite. However, the Lorentz condition (3.139b) implies that

$$a_0(k) = \hat{\boldsymbol{k}} \cdot \boldsymbol{a}(k) \tag{3.141}$$

where $\hat{\boldsymbol{k}} = \boldsymbol{k}/|\boldsymbol{k}|$. Thus the time component of a_μ equals the longitudinal component $\boldsymbol{a}\cdot\boldsymbol{k}$, so that $a^*(k)\cdot a(k)$ is given entirely by the transverse component

$$\boldsymbol{a}_\perp \equiv \boldsymbol{a} - \boldsymbol{a}\cdot\hat{\boldsymbol{k}}\,\hat{\boldsymbol{k}} \tag{3.142}$$

and

$$\begin{aligned} -a^*(k)\cdot a(k) &= -\boldsymbol{a}^*(k)\cdot\hat{\boldsymbol{k}}\,\boldsymbol{a}(k)\cdot\hat{\boldsymbol{k}} + \boldsymbol{a}^*(k)\cdot\boldsymbol{a}(k) \\ &= \boldsymbol{a}_\perp^*(k)\cdot\boldsymbol{a}_\perp(k). \end{aligned} \tag{3.143}$$

Thus the Lorentz condition ensures that the 'time-like' photons are cancelled by the 'longitudinal' modes leaving a non-negative energy, as required. Now, any function $\Lambda(x)$ satisfying (3.129) may be written as

$$\Lambda(x) = \int \frac{d^3k}{(2\pi)^3}\frac{1}{2k_0}\left(\lambda(k)\,e^{-ikx} + \lambda^*(k)\,e^{ikx}\right) \tag{3.144}$$

so the residual 'gauge invariance' allows us to replace $a_\mu(k)$ by

$$a'_\mu(k) = a_\mu(k) - ik_\mu\lambda(k). \tag{3.145}$$

Thus we can always 'choose a gauge' in which

$$a'_0(k) = \boldsymbol{a}'(k)\cdot\hat{\boldsymbol{k}} = 0 \tag{3.146}$$

and the only non-zero components are the transverse ones. In *this* gauge

$$a_\mu(k) = \sum_{\lambda=1,2} a(k,\lambda)\varepsilon_\mu(k,\lambda) \tag{3.147}$$

where $\varepsilon_\mu(k,1)$, $\varepsilon_\mu(k,2)$ are two orthonormal space-like vectors in the plane transverse to $\boldsymbol{k}$:

$$\varepsilon_0(k,\lambda) = 0 \tag{3.148a}$$

$$\boldsymbol{\varepsilon}(k,\lambda)\cdot\boldsymbol{\varepsilon}(k,\lambda') = \delta_{\lambda\lambda'} \tag{3.148b}$$

$$\boldsymbol{\varepsilon}(k,1)\wedge\boldsymbol{\varepsilon}(k,2) = \hat{\boldsymbol{k}} \tag{3.148c}$$

$$\boldsymbol{k}\cdot\boldsymbol{\varepsilon}(k,\lambda) = 0. \tag{3.148d}$$

Then substituting into (3.140) we find

$$P_\nu = \int \frac{d^3k}{(2\pi)^3}\frac{1}{2k_0}\sum_{\lambda=1,2} a^*(k,\lambda)a(k,\lambda)k_\nu. \tag{3.149}$$

Under an infinitesimal Lorentz transformation

$$x^\mu \to x'^\mu = x^\mu + \varepsilon^{\mu\nu}x_\nu \tag{3.150a}$$

with the infinitesimal satisfying

$$\varepsilon_{\mu\nu} = -\varepsilon_{\mu\nu}. \tag{3.150b}$$

Then, using the notation of (3.20) and (3.21)

$$\delta A^{\mu}(x)=\varepsilon^{\mu\nu}A_{\nu}(x) \tag{3.151}$$

and

$$\Phi^{\mu}_{\rho\sigma}=\delta^{\mu}_{\rho}A_{\sigma}-\delta^{\mu}_{\sigma}A_{\rho}. \tag{3.152}$$

The angular momentum density tensor is thus

$$M^{\mu}_{\rho\sigma}=x_{\rho}T^{\mu}_{\sigma}-x_{\sigma}T^{\mu}_{\rho}+F^{\mu\nu}(g_{\nu\rho}A_{\sigma}-g_{\nu\sigma}A_{\rho}). \tag{3.153}$$

As in (3.109) the total angular momentum is composed of both orbital and spin contributions. With the gauge choice (3.146) it is easy to see that the total spin of the field

$$\begin{aligned}S_{\rho\sigma}&\equiv\int \mathrm{d}^{3}x F^{0\nu}(g_{\nu\rho}A_{\sigma}-g_{\nu\sigma}A_{\rho})\\ &=\mathrm{i}\int\frac{\mathrm{d}^{3}k}{(2\pi)^{3}}\frac{1}{2k_{0}}[a^{*}_{\rho}(k)a_{\sigma}(k)-a^{*}_{\sigma}(k)a_{\rho}(k)].\end{aligned} \tag{3.154}$$

Then the spin vector

$$S^{\tau}=\tfrac{1}{2}\varepsilon^{\rho\sigma\tau}S_{\rho\sigma} \tag{3.155}$$

may be expressed entirely in terms of the transverse modes defined in (3.147).

$$\begin{aligned}\boldsymbol{S}&=\mathrm{i}\int\frac{\mathrm{d}^{3}k}{(2\pi)^{3}}\frac{1}{2k_{0}}[a^{*}(k,1)a(k,2)-a^{*}(k,2)a(k,1)]\hat{\boldsymbol{k}}\\ &=\int\frac{\mathrm{d}^{3}k}{(2\pi)^{3}}\frac{1}{2k_{0}}[b^{*}(k,+)b(k,+)-b^{*}(k,-)b(k,-)]\hat{\boldsymbol{k}}\end{aligned} \tag{3.156a}$$

where

$$b(k,\pm)\equiv\frac{1}{\sqrt{2}}[a(k,1)\pm\mathrm{i}a(k,2)]. \tag{3.156b}$$

Clearly the modes associated with $b(k,\pm)$ have positive (negative) helicities.

Problems

3.1 Under an infinitesimal scale transformation (dilatation) $\delta x^{\mu}=\alpha x^{\mu}$ and $\delta\varphi(x)=D\alpha\varphi(x)$, the α is infinitesimal and D is constant. Show that the action for the massless real scalar field $\varphi(x)$ is dilatation invariant (in four space–time dimensions) provided $D=-1$.

3.2 Generalise the result of problem 3.1 to the case of d space–time dimensions, and for a spinor field $\psi(x)$.

3.3 Identify the Noether current associated with the dilatation invariance of a massless real scalar field theory, and verify that it is divergenceless using the field equations.

3.4 Show that the action for a massless spinor field is invariant under the 'chiral transformation', in which $\delta x^\mu = 0$ and $\delta\psi(x) = \mathrm{i}\alpha\gamma_5\psi(x)$, with α an infinitesimal. Find the Noether current associated with this transformation.

3.5 Repeat the analysis of the massless vector field theory using the gauge condition $t_\mu A^\mu = 0$, where t is a unit time-like vector.

References

The books and review articles which we have found most useful in preparing this chapter are:

Hill E L 1951 *Rev. Mod. Phys.* **23** 253

Bogoliubov N N and Shirkov D V 1959 *Introduction to the Theory of Quantized Fields* (New York: Interscience)

Itzykson C and Zuber J-B 1980 *Quantum Field Theory* (New York: McGraw-Hill)

Ramond P 1981 *Field Theory: A Modern Primer* (Reading, Mass.: Benjamin-Cummings)

References in the text

1 Noether E 1918 *Nachr. K. Geo. Wiss. Göttingen* 235

2 Klein O 1926 *Z. Phys.* **37** 895
Gordon W 1926 *Z. Phys.* **40** 117, 121

3 Dirac P A M 1928 *Proc. R. Soc.* A **117** 610

4 Berezin F A 1966 *Method of the Second Quantization* (London: Academic)

5 Fierz M 1939 *Helv. Phys. Acta* **12** 3
Pauli W 1940 *Phys. Rev.* **58** 716
Lüders G and Zumino B 1958 *Phys. Rev.* **110** 1450
See also Streater R F and Wightman A S 1964 *PCT, Spin and Statistics and All That* (New York: W A Benjamin)

4

QUANTUM FIELD THEORY OF A SCALAR FIELD

4.1 The generating functional $W[J]$

In Chapter 2, we subtracted an external source term (or driving force) $J(t)Q$ from the Hamiltonian, and studied the probability amplitude $W[J]$ to find a system in the ground state at time $+\infty$ given that it was in the ground state at time $-\infty$. We were able to cast $W[J]$ in the form of a path integral, and to show that the expectation value in the ground state of a time-ordered product of any number of operators, $\hat{Q}_{\mathrm{H}}$, could be obtained from $W[J]$ by functional differentiations. In that chapter, the discussion was restricted to a non-relativistic system with a single generalised coordinate Q and a conjugate momentum P. In the present chapter, we wish to extrapolate the results of Chapter 2 to the case of a relativistic field theory[1], where there is a continuum of degrees of freedom $\varphi(t, \boldsymbol{x})$ at any given time t. In this way we quantise the field theory. We shall assume that in making the transition from classical field theory to quantum field theory, the classical field $\varphi(t, \boldsymbol{x})$ is replaced by an operator field $\hat{\varphi}(t, \boldsymbol{x})$ in the Heisenberg picture, just as the transition from classical mechanics to non-relativistic quantum mechanics was made by replacing the classical variable $Q(t)$ by the operator $\hat{Q}_{\mathrm{H}}(t)$. (Since creation and annihilation of particles goes on in relativistic systems, we must expect $\hat{\varphi}$ to be an operator on occupation number space. In the alternative canonical quantisation procedure, the coefficients $a(k)$ and $a^*(k)$ in (3.47) become operators which annihilate and create particles of momentum $\boldsymbol{k}$, when the free-field theory is quantised.) We may define eigenstates of $\hat{\varphi}(t, \boldsymbol{x})$ denoted by $|\varphi(\boldsymbol{x}), t\rangle$ such that

$$\hat{\varphi}(t, \boldsymbol{x})|\varphi(\boldsymbol{x}), t\rangle = \varphi(\boldsymbol{x})|\varphi(\boldsymbol{x}), t\rangle. \tag{4.1}$$

The generalisation of (2.18) to quantum field theory will be

$$\langle\varphi''(\boldsymbol{x}), t''|\varphi'(\boldsymbol{x}), t'\rangle \propto \int \mathscr{D}\varphi \int \mathscr{D}\pi \exp i\hbar^{-1} \int_{t'}^{t''} \mathrm{d}t \int \mathrm{d}^3x(\pi\partial_0\varphi - \mathscr{H}(\pi, \varphi)) \tag{4.2}$$

where $\mathscr{H}$ is the Hamiltonian density, $\pi = \partial\mathscr{L}/\partial(\partial_0\varphi)$ is the conjugate momentum (density) to φ, and $\partial_0\varphi$ denotes $\partial\varphi/\partial x_0$. (We have set $c = 1$, so that there is no distinction between x_0 and t.) The path integral is over all functions $\pi(t, \boldsymbol{x})$ and over functions $\varphi(t, \boldsymbol{x})$ satisfying the boundary conditions

$$\varphi(t'', \boldsymbol{x}) = \varphi''(\boldsymbol{x}) \qquad \varphi(t', \boldsymbol{x}) = \varphi'(\boldsymbol{x}). \tag{4.3}$$

Notice that the path integral is over classical fields $\varphi(t, \boldsymbol{x})$ and $\pi(t, \boldsymbol{x})$, not over operator fields.

If an external source term $J(t, \boldsymbol{x})\varphi(t, \boldsymbol{x})$ is subtracted from the Hamiltonian density $\mathscr{H}(\pi, \varphi)$, then we may discuss the transition amplitude from the ground state at time $-\infty$ to the ground state at time $+\infty$, in the presence of this source. In the case of relativistic quantum field theory, we shall refer to the ground state as the vacuum state, since creation and annihilation of particles can occur in relativistic systems, and we would expect the lowest energy state of the theory to contain no particles. (Of course, this lowest energy state cannot necessarily be obtained in practice, because, for example, of baryon number conservation in fermi systems.) We denote the vacuum-to-vacuum amplitude in the presence of the source by $W[J]$. By analogy with (2.35), we assume that

$$W[J] = N \lim_{t'' \to \infty, t' \to -\infty} \int \mathscr{D}\varphi \int \mathscr{D}\pi$$

$$\times \exp i\hbar^{-1} \int_{t'}^{t''} \mathrm{d}t \int \mathrm{d}^3x (\pi\, \partial_0 \varphi - \mathscr{H}(\pi, \varphi) + J\varphi) \tag{4.4}$$

where again the path integral is over all functions $\pi(t, \boldsymbol{x})$ and over functions $\varphi(t, \boldsymbol{x})$ obeying the boundary conditions of (4.3). The normalisation factor N is chosen so that $W[J] = 1$ when $J = 0$. Also by analogy with (2.56),

$$(i\hbar^{-1})^n \langle 0 | T(\hat{\varphi}(x_1) \ldots \hat{\varphi}(x_n)) | 0 \rangle = \left. \frac{\delta^n W[J]}{\delta J(x_1) \ldots \delta J(x_n)} \right|_{J(x)=0} \tag{4.5}$$

where

$$x_i \equiv (x_i^0, \boldsymbol{x}_i) \qquad i = 1, \ldots, n \tag{4.6}$$

and $|0\rangle$ denotes the vacuum state. We shall refer to the vacuum expectation value of a time-ordered product of n field operators as an n-particle Green function, and use the notation

$$\mathscr{G}^{(n)}(x_1, \ldots, x_n) = \langle 0 | T(\hat{\varphi}(x_1) \ldots \hat{\varphi}(x_n)) | 0 \rangle. \tag{4.7}$$

Thus

$$(i\hbar^{-1})^n \mathscr{G}^{(n)}(x_1, \ldots, x_n) = \left. \frac{\delta^n W[J]}{\delta J(x_1) \ldots \delta J(x_n)} \right|_{J(x)=0}. \tag{4.8}$$

We shall see, in Chapter 5, that the Green functions are of the utmost importance, since the n-particle Green function defined above is directly related to the scattering amplitude involving a total of n incoming or outgoing spin zero particles. We can therefore calculate these scattering amplitudes by first evaluating $W[J]$ using (4.4) and then obtaining $\mathscr{G}(x_1, \ldots, x_n)$ from (4.5). Using (4.8) and the symmetry of $\mathscr{G}$ in its variables, we may make the expansion

$$W[J]=\sum_{n=0}^{\infty}\frac{(\mathrm{i}\hbar^{-1})^n}{n!}\int \mathrm{d}^4x_1\ldots\int \mathrm{d}^4x_n\mathscr{G}^{(n)}(x_1,\ldots,x_n)J(x_1)\ldots J(x_n) \quad (4.9)$$

where the $n=0$ term is 1.

As in §2.2, we may define the continuation $W_{\mathrm{E}}[J]$ of $W[J]$ to Euclidean space by introducing the variables

$$\bar{x}=(\bar{x}_0,\bar{\boldsymbol{x}})=(\mathrm{i}x_0,\boldsymbol{x}) \quad (4.10)$$

and rotating the integration contour in the complex x_0 plane. (Recall that we have set $c=1$, and consequently there is no distinction between x_0 and t.)

$$W_{\mathrm{E}}[J]=N\int\mathscr{D}\varphi\int\mathscr{D}\pi\exp\left[-\hbar^{-1}\int\mathrm{d}^4\bar{x}\left(-\mathrm{i}\pi\frac{\partial\varphi}{\partial\bar{x}_0}+\mathscr{H}-J\varphi\right)\right] \quad (4.11)$$

where the volume integration is over all four-dimensional Euclidean space, and the path integral is over all functions $\pi(\bar{x})$, and over functions $\varphi(\bar{x})$ obeying the boundary conditions

$$\lim_{\bar{x}_0\to\infty}\varphi(\bar{x})=\varphi'(\boldsymbol{x}) \qquad \lim_{\bar{x}_0\to-\infty}\varphi(\bar{x})=\varphi''(\boldsymbol{x}) \quad (4.12)$$

where $\varphi'(\boldsymbol{x})$ and $\varphi''(\boldsymbol{x})$ may be chosen arbitrarily, as in §2.2. In particular they may be chosen to be zero. The Euclidean space Green functions will be given by

$$\hbar^{-n}\mathscr{G}_{\mathrm{E}}^{(n)}(\bar{x}_1,\ldots,\bar{x}_n)=\left.\frac{\delta^n W_{\mathrm{E}}[J]}{\delta J(\bar{x}_1)\ldots\delta J(\bar{x}_n)}\right|_{J(\bar{x})=0}. \quad (4.13)$$

(For the Euclidean generating functional, a functional differentiation with respect to J pulls down a factor $\hbar^{-1}\varphi$, whereas for the Minkowski space generating functional a factor $\mathrm{i}\hbar^{-1}\varphi$ was produced.)

The corresponding functional expansion is

$$W_{\mathrm{E}}[J]=\sum_{n=0}^{\infty}\frac{\hbar^{-n}}{n!}\int\mathrm{d}^4\bar{x}_1\ldots\int\mathrm{d}^4\bar{x}_n\mathscr{G}_{\mathrm{E}}^{(n)}(\bar{x}_1,\ldots,\bar{x}_n)J(\bar{x}_1)\ldots J(\bar{x}_n) \quad (4.14)$$

where the $n=0$ term is 1.

In the special case where the Lagrangian density has the form

$$\mathscr{L}(\varphi,\partial_\mu\varphi)=\frac{\hbar^2}{2}(\partial_0\varphi)^2+F(\varphi,\nabla\varphi) \quad (4.15)$$

where F is an arbitrary function, the Hamiltonian density takes the form

$$\mathscr{H}=\frac{\hbar^{-2}}{2}\pi^2-F(\varphi,\nabla\varphi). \quad (4.16)$$

The path integral in (4.11) is then

$$W_{\mathrm{E}}[J]=N\int\mathscr{D}\varphi\int\mathscr{D}\pi\exp\left[-\hbar^{-1}\int\mathrm{d}^4\bar{x}\left(-\mathrm{i}\pi\frac{\partial\varphi}{\partial\bar{x}_0}+\frac{\hbar^{-2}\pi^2}{2}-F(\varphi,\nabla\varphi)-J\varphi\right)\right]. \quad (4.17)$$

The path integral over π may be carried out explicitly as follows. We have to evaluate

$$I \equiv \int \mathscr{D}\pi \exp\left[-\hbar^{-1}\int \mathrm{d}^4\bar{x}\left(-\mathrm{i}\pi\frac{\partial\varphi}{\partial\bar{x}_0}+\tfrac{1}{2}\hbar^{-2}\pi^2\right)\right]. \tag{4.18}$$

This is of the form

$$I = \int \mathscr{D}\pi \exp\left(-\tfrac{1}{2}\int \mathrm{d}^4\bar{x}'\int \mathrm{d}^4\bar{x}\pi(\bar{x}')A(\bar{x}',\bar{x})\pi(\bar{x})+\int \mathrm{d}^4\bar{x}\rho(\bar{x})\pi(\bar{x})\right) \tag{4.19}$$

with

$$A(\bar{x}',\bar{x}) = \hbar^{-3}\delta^4(\bar{x}'-\bar{x}) \tag{4.20}$$

and

$$\rho(\bar{x}) = \mathrm{i}\hbar^{-1}\,\partial\varphi/\partial\bar{x}_0. \tag{4.21}$$

Using (1.14) we see that

$$I \propto \exp\left[-\hbar^{-1}\int \mathrm{d}^4\bar{x}\frac{\hbar^2}{2}\left(\frac{\partial\varphi}{\partial\bar{x}_0}\right)^2\right]. \tag{4.22}$$

Returning to (4.17), the Euclidean generating functional is

$$W_{\mathrm{E}}[J] = N'\int \mathscr{D}\varphi \exp\left\{-\hbar^{-1}\int \mathrm{d}^4\bar{x}\left[\frac{\hbar^2}{2}\left(\frac{\partial\varphi}{\partial\bar{x}_0}\right)^2 - F(\varphi,\nabla\varphi) - J\varphi\right]\right\} \tag{4.23}$$

and in terms of the continuation $\mathscr{L}_{\mathrm{E}}$ of (4.15) to Euclidean space

$$W_{\mathrm{E}}[J] = N'\int \mathscr{D}\varphi \exp \hbar^{-1}\int \mathrm{d}^4\bar{x}(\mathscr{L}_{\mathrm{E}}+J\varphi). \tag{4.24}$$

Continuing back to Minkowski space,

$$W[J] = N'\int \mathscr{D}\varphi \exp \mathrm{i}\hbar^{-1}\int_{-\infty}^{\infty}\mathrm{d}t\int \mathrm{d}^3x\left(\mathscr{L}\left(\varphi,\frac{\partial\varphi}{\partial x^\mu}\right)+J\varphi\right). \tag{4.25}$$

The normalisation factor N' is to be chosen so that $W[J]=1$ when $J=0$. This is a simpler alternative to (4.4) when the Lagrangian is of the special form (4.15). Otherwise, it is necessary to perform the integral over π first. $W[J]$ may then be cast in the form of (4.25), but with $\mathscr{L}$ replaced by an effective I agrangian which in general differs from $\mathscr{L}$.

4.2 The generating functional for free-field theory

In the case of the free-field theory of a scalar field it is possible to evaluate the generating functional exactly. We shall see in Chapter 6 that in the case of the interacting scalar field, the best we can do is to evaluate the generating functional approximately by expanding in powers of the coupling constant, starting from the exact result for the free-field theory. The free-field Lagrangian is as in equation (3.34) with $\lambda=0$

$$\mathscr{L}=\frac{\hbar^2}{2}\,\partial_\nu\varphi\,\partial^\nu\varphi-\frac{\mu^2}{2}\,\varphi^2. \tag{4.26}$$

In order to perform an unambiguous calculation, we continue to Euclidean space, evaluate $W_{\mathrm{E}}[J]$ and then continue back to Minkowski space. The Lagrangian is of the form of (4.15) and equation (4.24) for $W_{\mathrm{E}}[J]$ is applicable. The continuation of the Lagrangian to Euclidean space is

$$\mathscr{L}_{\mathrm{E}}=-\frac{\hbar^2}{2}\,\bar{\partial}_\nu\varphi\,\bar{\partial}_\nu\varphi-\frac{\mu^2}{2}\,\varphi^2 \tag{4.27}$$

where

$$\bar{\partial}_\nu\varphi\,\bar{\partial}_\nu\varphi=\bar{\partial}_0\varphi\,\bar{\partial}_0\varphi+\bar{\partial}_1\varphi\,\bar{\partial}_1\varphi+\bar{\partial}_2\varphi\,\bar{\partial}_2\varphi+\bar{\partial}_3\varphi\,\bar{\partial}_3\varphi. \tag{4.28}$$

Thus

$$W_{\mathrm{E}}[J]=N'\int\mathscr{D}\varphi\,\exp\hbar^{-1}\int\mathrm{d}^4\bar{x}\left(-\frac{\hbar^2}{2}\,\bar{\partial}_\nu\varphi\,\bar{\partial}_\nu\varphi-\frac{\mu^2}{2}\,\varphi^2+J\varphi\right). \tag{4.29}$$

This is of the form

$$W_{\mathrm{E}}[J]=N'\int\mathscr{D}\varphi\,\exp\left(-\tfrac{1}{2}\int\mathrm{d}^4\bar{x}'\int\mathrm{d}^4\bar{x}\,\varphi(\bar{x}')A(\bar{x}',\bar{x})\varphi(\bar{x})+\int\mathrm{d}^4\bar{x}\,\rho(\bar{x})\varphi(\bar{x})\right) \tag{4.30}$$

with

$$A(\bar{x}',\bar{x})=\hbar^{-1}\left(\hbar^2\frac{\partial}{\partial\bar{x}'_\nu}\frac{\partial}{\partial\bar{x}_\nu}+\mu^2\right)\delta(\bar{x}'-\bar{x}) \tag{4.31}$$

and

$$\rho(\bar{x})=\hbar^{-1}J(\bar{x}). \tag{4.32}$$

Using (1.14)

$$W_{\mathrm{E}}[J]=\exp\frac{\hbar^{-2}}{2}\int\mathrm{d}^4\bar{x}'\int\mathrm{d}^4\bar{x}\,J(\bar{x}')A^{-1}(\bar{x}',\bar{x})J(\bar{x}) \tag{4.33}$$

where the constant of proportionality has been chosen so that $W_{\mathrm{E}}[J]=1$ when $J=0$.

We adopt the notation

$$\Delta_{\mathrm{F}}^{\mathrm{E}}(\bar{x}'-\bar{x})=\hbar^{-1}A^{-1}(\bar{x}',\bar{x}) \tag{4.34}$$

and refer to Δ_F^E as the Feynman propagator in Euclidean space for the scalar field. (We see shortly that it is a function of $\bar{x}'-\bar{x}$ alone.) Thus, we write

$$W_E[J]=\exp\frac{\hbar^{-1}}{2}\int d^4\bar{x}'\int d^4\bar{x}J(\bar{x}')\,\Delta_F^E(\bar{x}'-\bar{x})J(\bar{x}). \tag{4.35}$$

To evaluate the inverse of A^{-1} we Fourier transform by using the representation for the Dirac δ function

$$\delta(\bar{x}'-\bar{x})=\int\frac{d^4\bar{p}}{(2\pi\hbar)^4}\exp[i\hbar^{-1}\bar{p}\cdot(\bar{x}'-\bar{x})] \tag{4.36}$$

where the scalar product in Euclidean space is defined by

$$\bar{p}\cdot\bar{x}=\bar{p}_0\bar{x}_0+\bar{p}_1\bar{x}_1+\bar{p}_2\bar{x}_2+\bar{p}_3\bar{x}_3. \tag{4.37}$$

Then

$$A(\bar{x}',\bar{x})=\int\frac{d^4\bar{p}}{(2\pi\hbar)^4}\exp[i\hbar^{-1}\bar{p}\cdot(\bar{x}'-\bar{x})]\times\hbar^{-1}(\bar{p}^2+\mu^2) \tag{4.38}$$

with

$$\bar{p}^2=\bar{p}_0^2+\bar{p}_1^2+\bar{p}_2^2+\bar{p}_3^2. \tag{4.39}$$

The inverse is

$$A^{-1}(\bar{x}',\bar{x})=\int\frac{d^4\bar{p}}{(2\pi\hbar)^4}\exp[i\hbar^{-1}\bar{p}\cdot(\bar{x}'-\bar{x})]\times\hbar(\bar{p}^2+\mu^2)^{-1}. \tag{4.40}$$

Returning to (4.34) we may write

$$\Delta_F^E(\bar{x}'-\bar{x})=\int\frac{d^4\bar{p}}{(2\pi\hbar)^4}\exp[i\hbar^{-1}\bar{p}\cdot(\bar{x}'-\bar{x})]\times\tilde{\Delta}_F^E(\bar{p}) \tag{4.41}$$

with

$$\tilde{\Delta}_F^E(\bar{p})=(\bar{p}^2+\mu^2)^{-1}. \tag{4.42}$$

Now that we have evaluated $W_E[J]$ we may continue back to Minkowski space to obtain

$$W[J]=\exp\left(-\frac{i\hbar^{-1}}{2}\int d^4x'\int d^4xJ(x')\Delta_F(x'-x)J(x)\right) \tag{4.43}$$

with

$$\Delta_F(x'-x)=\int\frac{d^4p}{(2\pi\hbar)^4}\exp[-i\hbar^{-1}p\cdot(x'-x)]\times\tilde{\Delta}_F(p) \tag{4.44}$$

and

$$\tilde{\Delta}_F(p)=(p^2-\mu^2+i\varepsilon)^{-1}. \tag{4.45}$$

The analytic continuation in $\bar{x}$ and $\bar{x}'$ has been made as in (4.10), and the continuation in $\bar{p}$ has been made by writing

$$\bar{p}=(\bar{p}_0,\bar{\boldsymbol{p}})=(-\mathrm{i}p_0,\boldsymbol{p}). \tag{4.46}$$

All scalar products are now defined as appropriate to Minkowski space. The x_0, x_0' and p_0 integrations have been rotated from the imaginary axis to the real axis, and, in rotating the contour for the p_0 integration, the $\mathrm{i}\varepsilon$ of (4.45), with $\varepsilon \to 0^+$, has been introduced. This is necessary to avoid correctly the poles in p_0 when $p^2=\mu^2$.

An alternative way of deriving the generating functional for the free-field theory is to add to the Lagrangian in (4.25) a term $\frac{1}{2}\mathrm{i}\varepsilon\varphi^2$ (with $\varepsilon\to 0^+$) to guarantee convergence of the path integral. Thus,

$$W[J]=N'\int \mathscr{D}\varphi \exp \mathrm{i}\hbar^{-1}\int \mathrm{d}^4x\left(\mathscr{L}\left(\varphi,\frac{\partial\varphi}{\partial x^\mu}\right)+J\varphi+\tfrac{1}{2}\mathrm{i}\varepsilon\varphi^2\right). \tag{4.47}$$

With $\mathscr{L}$ given by (4.26), the free-field generating functional is derived by Fourier transforming. We shall not pursue this approach here since it does not generalise to the fermion field and gauge field cases.

4.3 Green functions for free-field theory

Now that we have evaluated the generating functional for free-field theory (equations (4.43) to (4.45)), the corresponding Green functions may be calculated by functional differentiation as in (4.8). Thus

$$\mathscr{G}^{(2)}(x_1,x_2)=\mathrm{i}\hbar\,\Delta_{\mathrm{F}}(x_1-x_2) \tag{4.48}$$

$$\begin{aligned}\mathscr{G}^{(4)}(x_1,x_2,x_3,x_4)=(\mathrm{i}\hbar)^2[&\Delta_{\mathrm{F}}(x_1-x_2)\,\Delta_{\mathrm{F}}(x_3-x_4)\\ &+\Delta_{\mathrm{F}}(x_1-x_3)\,\Delta_{\mathrm{F}}(x_2-x_4)\\ &+\Delta_{\mathrm{F}}(x_1-x_4)\,\Delta_{\mathrm{F}}(x_2-x_3)]\end{aligned} \tag{4.49}$$

and so on, with $\mathscr{G}^{(n)}=0$ for n odd.

We may represent these results diagramatically by using a line with end points x and y to symbolise $\mathrm{i}\hbar\,\Delta_{\mathrm{F}}(y-x)$.

$\mathrm{i}\hbar\,\Delta_{\mathrm{F}}(y-x)=$ 

(4.50)

Then

$\mathscr{G}^{(2)}(x_1,x_2)=$ 

(4.51)

and

$\mathscr{G}^{(4)}(x_1,x_2,x_3,x_4)=$ 

$+ (x_3 \leftrightarrow x_2) + (x_4 \leftrightarrow x_2)$ (4.52)

From (4.44) it may be deduced directly that

$$(\hbar^2\,\partial_\nu\partial^\nu+\mu^2)\,\Delta_{\mathrm{F}}(x-x')=-\delta(x-x'). \tag{4.53}$$

Thus, $\Delta_{\mathrm{F}}(x-x')$ is the Green function for the operator $(\hbar^2\,\partial_\nu\partial^\nu+\mu^2)$ with the boundary conditions implied by the iε in (4.45). It is therefore associated with the propagation of solutions of the Klein–Gordon wave equation

$$(\hbar^2\,\partial_\nu\partial^\nu+\mu^2)\varphi(x)=0. \tag{4.54}$$

Since this is the classical field equation for a neutral free scalar field, we conclude that $\Delta_{\mathrm{F}}(x-x')$ is associated with propagation of neutral scalar particles from x' to x. We should therefore expect $\mathscr{G}^{(4)}(x_1, x_2, x_3, x_4)$ to be intimately connected with the scattering amplitude for a process involving a total of four incoming or outgoing neutral scalar particles. We shall see in Chapter 5 that this is indeed the case.

We shall find it convenient from now on to work for the most part in units where $\hbar=c=1$. (We are already working with $c=1$.) Fourier transforming the Green functions to momentum space we write

$$\tilde{\mathscr{G}}^{(n)}(p_1,\ldots,p_n)(2\pi)^4\,\delta(p_1+\ldots+p_n)$$
$$=\int \mathrm{d}^4x_1\ldots\int \mathrm{d}^4x_n\,\exp[\mathrm{i}(p_1\cdot x_1+\ldots+p_n\cdot x_n)]\times\mathscr{G}^{(n)}(x_1,\ldots,x_n). \tag{4.55}$$

The Dirac δ function factor occurs because translation invariance implies that $\mathscr{G}^{(n)}(x_1,\ldots,x_n)$ depends only on the differences of the x_i. From (4.48), (4.44) and (4.45),

$$\tilde{\mathscr{G}}^{(2)}(p,-p)=\mathrm{i}\tilde{\Delta}_{\mathrm{F}}(p)=\mathrm{i}(p^2-\mu^2+\mathrm{i}\varepsilon)^{-1}. \tag{4.56}$$

(Because of the Dirac δ function in (4.55), $\tilde{\mathscr{G}}^{(2)}(p_1, p_2)$ is defined only for $p_1+p_2=0$.)

Diagramatically we symbolise $\mathrm{i}\tilde{\Delta}_{\mathrm{F}}(p)$ by a line with an associated momentum.

$$\tilde{\mathscr{G}}^{(2)}(p,-p)=\mathrm{i}\,\tilde{\Delta}_{\mathrm{F}}(p)= \tag{4.57}$$

The free-field Green function symbolised in (4.52) is a disconnected object. It is often convenient to study instead the so-called connected Green functions. These are constructed by first introducing a generating functional $X[J]$ for connected Green functions, defined by writing

$$W[J]=\mathrm{e}^{\mathrm{i}X[J]}. \tag{4.58}$$

The connected Green functions $G^{(n)}(x_1,\ldots,x_n)$ are then defined through the functional expansion

$$\mathrm{i}X[J]=\sum_{n=1}^{\infty}\frac{\mathrm{i}^n}{n!}\int \mathrm{d}^4x_1\ldots\int \mathrm{d}^4x_n\,G^{(n)}(x_1,\ldots,x_n)J(x_1)\ldots J(x_n). \tag{4.59}$$

Thus

$$\mathrm{i}^n G^{(n)}(x_1,\ldots,x_n)=\mathrm{i}\left.\frac{\delta^n X[J]}{\delta J(x_1)\ldots\delta J(x_n)}\right|_{J(x)=0}. \tag{4.60}$$

In the free-field case (4.43) means that

$$\mathrm{i}X[J]=-\frac{\mathrm{i}}{2}\int \mathrm{d}^4x' \int \mathrm{d}^4x\, J(x')\,\Delta_{\mathrm{F}}(x'-x)J(x) \tag{4.61}$$

and the only non-zero connected Green function is $G^{(2)}$. Thus for the free-field theory

$$G^{(2)}(x_1,x_2)=\mathrm{i}\Delta_{\mathrm{F}}(x_1-x_2)=\mathscr{G}^{(2)}(x_1,x_2). \tag{4.62}$$

When we include interactions in Chapter 6 we shall see that there are connected Green functions involving more than two scalar particles. We may Fourier transform the connected Green functions to momentum space in exact analogy with (4.55)

$$\begin{aligned}&\mathscr{G}^{(n)}(p_1,\ldots,p_n)(2\pi)^4\,\delta(p_1+\ldots+p_n)\\&\quad=\int \mathrm{d}^4x_1\ldots\int \mathrm{d}^4x_n \exp[\mathrm{i}(p_1\cdot x_1+\ldots+p_n\cdot x_n)]\times G^{(n)}(x_1,\ldots,x_n).\end{aligned} \tag{4.63}$$

4.4 The effective action and one-particle-irreducible Green functions

The developments of this section[2,3] require the introduction of the quantity $\varphi_{\mathrm{c}}(x)$, referred to as the classical field, and defined by

$$\varphi_{\mathrm{c}}(x)=\frac{\delta X[J]}{\delta J(x)}. \tag{4.64}$$

(It should be borne in mind that $\varphi_{\mathrm{c}}(x)$ is a functional of J.)

From (4.5),

$$\frac{\delta W[J]}{\delta J(x)}=\mathrm{i}\langle 0|\hat{\varphi}(x)|0\rangle_J \tag{4.65}$$

where $\langle 0|\hat{\varphi}(x)|0\rangle_J$ is the vacuum expectation value of the field operator in the presence of the source J. Thus, using (4.58),

$$\varphi_{\mathrm{c}}(x)=\langle 0|\hat{\varphi}(x)|0\rangle_J/\langle 0|0\rangle_J \tag{4.66}$$

where we have used the notation

$$W[J]=\langle 0|0\rangle_J \tag{4.67}$$

for the vacuum-to-vacuum amplitude in the presence of the source J. In the absence of the source, $\varphi_{\mathrm{c}}(x)$ is just the vacuum expectation value of the field operator.

The effective action $\Gamma[\varphi_c]$ is then defined by

$$\Gamma[\varphi_c]=X[J]-\int d^4x\, J(x)\varphi_c(x). \tag{4.68}$$

(This equation is completely analogous to the thermodynamic equation $E=F+TS$ which expresses the energy E regarded as a function of S in terms of the free energy F regarded as a function of T.)

By making a functional differentiation with respect to $J(x)$ and using (4.64), we see immediately that $\Gamma[\varphi_c]$ depends only on φ_c, as the notation implies. Just as φ_c can be obtained as in (4.64) as a functional derivative of $X[J]$, so J may be obtained as a functional derivative of $\Gamma[\varphi_c]$.

$$J(x)=-\frac{\delta\Gamma[\varphi_c]}{\delta\varphi_c(x)}. \tag{4.69}$$

In the case of free-field theory, we may obtain $\varphi_c(x)$ from (4.61). Thus

$$\varphi_c(x)=-\int d^4x'\, \Delta_F(x-x')J(x'). \tag{4.70}$$

Since $\Delta_F(x-x')$ is the Green function for the operator $\partial_\nu\partial^\nu+\mu^2$ (see (4.53)), we have

$$(\partial_\nu\partial^\nu+\mu^2)\varphi_c(x)=J(x). \tag{4.71}$$

This is identical to the classical field equation in the presence of a source J, and it is therefore appropriate to refer to $\varphi_c(x)$ as the classical field.

We may now calculate $\Gamma[\varphi_c]$ explicitly for the free-field case, because we already have an exact calculation of $X[J]$ in equation (4.61), and because (4.71) allows us to eliminate J in favour of φ_c. Thus, integrating by parts and using (4.53) we find

$$\begin{aligned}\Gamma[\varphi_c]&=-\tfrac{1}{2}\int d^4x\, \varphi_c(x)(\partial_\nu\partial^\nu+\mu^2)\varphi_c(x)\\ &=\tfrac{1}{2}\int d^4x(\partial_\nu\varphi_c\partial^\nu\varphi_c-\mu^2\varphi_c^2).\end{aligned} \tag{4.72}$$

This is exactly the action for the classical free-field theory discussed in Chapter 3, thus justifying referring to $\Gamma[\varphi_c]$ as the effective action.

In the case of an interacting field theory we will be unable to calculate $\varphi_c(x)$ and $\Gamma[\varphi_c]$ exactly, and there will be quantum corrections to (4.71) and (4.72). In general, we make the functional expansion

$$\Gamma[\varphi_c]=\sum_{n=1}^{\infty}\frac{i^n}{n!}\int d^4x_1\ldots\int d^4x_n\,\Gamma^{(n)}(x_1,\ldots,x_n)\varphi_c(x_1)\ldots\varphi_c(x_n). \tag{4.73}$$

The coefficients $\Gamma^{(n)}$ in this expansion are referred to as one-particle-irreducible (OPI) Green functions. We see later in the interacting field theory that the Feynman graphs involved in the calculation of $\Gamma^{(n)}$ are all connected, and, moreover, cannot be made disconnected by cutting a single internal line. They have also had any factors coming from external lines divided out.

In the free-field theory, we see from (4.72) that the only non-zero OPI Green function is

$$\Gamma^{(2)}(x', x) = (\partial_\nu^{x'} \partial_{x'}^\nu + \mu^2)\, \delta(x' - x). \tag{4.74}$$

We may Fourier transform the OPI Green functions to momentum space by analogy with (4.55)

$$\tilde{\Gamma}^{(n)}(p_1, \ldots, p_n)(2\pi)^4\, \delta(p_1 + \ldots + p_n)$$
$$= \int d^4x_1 \ldots \int d^4x_n \exp[i(p_1 \cdot x_1 \ldots + p_n \cdot x_n)\, \Gamma^{(n)}(x_1, \ldots, x_n). \tag{4.75}$$

Thus

$$\tilde{\Gamma}^{(2)}(p, -p) = -(p^2 - \mu^2). \tag{4.76}$$

This is consistent with the above remarks about the interpretation of OPI Green functions.

Alternatively, we may expand in powers of momentum about zero momentum. Written in position space this is an expansion of the type

$$\Gamma[\varphi_c] = \int d^4x \left(-V(\varphi_c) + \frac{A(\varphi_c)}{2}\, \partial_\nu \varphi_c\, \partial^\nu \varphi_c + \ldots \right) \tag{4.77}$$

where $V(\varphi_c)$, $A(\varphi_c)$ etc are functions of φ_c (not functionals).

The coefficient $V(\varphi_c)$ is referred to as the effective potential. In the case of a classical field $\varphi_c(x)$ which is constant in space and time, we may use (4.69) to write

$$\frac{dV}{d\varphi_c} = J. \tag{4.78}$$

In particular, if we set the source term J to zero, φ_c has the significance of the vacuum expectation value (VEV) of the field operator, and

$$\frac{dV}{d\varphi_c} = 0. \tag{4.79}$$

Thus once we know the effective potential, (4.79) is an equation which may be solved for the VEV of the field operator. In other words, the VEV of the field operator, taking account of quantum corrections, may be obtained by minimising the effective potential. (If $\varphi_c(x)$ were to vary in space or time we would instead have to calculate the VEV of the field operator from the more general equation $\delta\Gamma/\delta\varphi_c = 0$.) Strictly speaking, (4.79) only tells us that the VEV of the field operator has to be a stationary point of $V(\varphi_c)$. However, it is possible to show[3] that $V(\varphi_c)$ has the interpretation of the expectation value of the energy density in the state for which the VEV of the field is φ_c. Thus, in the true ground state, or vacuum state, φ_c must be the absolute minimum of the effective potential.

In the free-field case (4.72) means that

$$V(\varphi_c)=\frac{\mu^2}{2}\varphi_c^2. \tag{4.80}$$

In this case, minimisation of the effective potential shows that the VEV of the field operator is zero. However, in Chapter 12 we shall discuss certain interacting field theories where a non-zero VEV occurs.

It is sometimes useful to have an expansion of the effective potential in terms of Green functions. This is obtained by using the inverse of (4.75) in (4.73) (see problem 4.3) and is

$$V(\varphi_c)=-\sum_{n=1}^{\infty}\frac{\mathrm{i}^n}{n!}\tilde{\Gamma}^{(n)}(0,\ldots,0)\varphi_c^n. \tag{4.81}$$

Problems

4.1 Check in detail the analytic continuation involved in going from (4.41) to (4.44).

4.2 Carry through the alternative derivation for the free-field generating functional by adding a term $\frac{1}{2}\mathrm{i}\varepsilon\varphi^2$ to the Lagrangian.

4.3 Derive the expansion (4.81) for the effective potential.

References

1 For similar treatments see
Abars E S and Lee B W 1973 *Phys. Rep.* **9C** 1
Taylor J C 1976 *Gauge Theories of Weak Interactions* (Cambridge: Cambridge University Press)
Ramond P 1981 *Field Theory: a Modern Primer* (Reading, Mass.: Benjamin-Cummings)
2 Goldstone J, Salam A and Weinberg S 1962 *Phys. Rev.* **127** 965
Jona-Lasinio G 1964 *Nuovo Cimento* **34** 1790
3 For a treatment of the effective action with references to the original literature see
Coleman S *Lectures given at the 1973 International Summer School of Physics, Ettore Majorana* Section 3.

5

SCATTERING AMPLITUDES

5.1 Scattering amplitude in quantum mechanics

The Green functions which we have introduced in the previous chapter are not immediately measurable in experimental processes. In a scattering process a number of incident particles (usually two), which are initially widely separated, enter a region in which they interact. Subsequently a (sometimes different) number of particles emerge from the interaction region and separate. If we assume that the forces between the particles have a finite range, then at very early times ($t \to -\infty$) and at very late times ($t \to +\infty$) the particles involved are free (non-interacting). In our model field theory we have a single type of (scalar) particle with mass μ. Thus if the momenta of the incoming and outgoing particles are denoted by p_i ($i = 1, 2, \ldots$), then

$$p_i^2 = \mu^2 \tag{5.1}$$

and we say that the particles concerned are 'on the mass shell'. The generating functional $W[J]$ is defined in (4.4) as a functional integral over field configurations obeying the boundary conditions (4.3). These boundary conditions (even if we take $\varphi'(\boldsymbol{x}) = \varphi''(\boldsymbol{x}) = 0$) are not sufficient to ensure that the particles involved are free as $t \to \pm\infty$. Evidently $W[J]$ includes contributions from amplitudes having external particles with momenta p_i which do not necessarily satisfy (5.1). Thus it is the task of this chapter to find the prescription for writing down the transition amplitude (S-matrix element) for states involving particles which *do* satisfy (5.1).

We do this first for the case of non-relativistic one-dimensional quantum mechanics, considered in Chapter 2. So far we have only considered the probability amplitude (2.5) for a transition between the eigenstates $|q', t'\rangle$ and $|q'', t''\rangle$ defined in (2.2). The generalisation to an arbitrary initial state $|\mathrm{i}\rangle$ at t', and an arbitrary final state $|\mathrm{f}\rangle$ at t'', is immediate, using the completeness of the above sets of eigenstates. Denoting the required transition amplitude by $U_{\mathrm{fi}}(t'', t')$, we have

$$U_{\mathrm{fi}}(t'', t') = \int \mathrm{d}q'\, \mathrm{d}q'' \langle \mathrm{f}|q'', t''\rangle \langle q'', t''|q', t'\rangle \langle q', t'|\mathrm{i}\rangle. \tag{5.2}$$

The quantities

$$\psi_a(q, t) \equiv \langle q, t|a\rangle \qquad (a = \mathrm{i}, \mathrm{f}) \tag{5.3}$$

are, of course, the wave functions corresponding to the initial and final states. We can expand them in terms of the momentum eigenstates $|p\rangle_{\rm S}$. Then

$$\psi_a(q,t)=\int \mathrm{d}p\langle q,t|p\rangle_{\rm S}\,{}_{\rm S}\langle p|a\rangle=\int \mathrm{d}p\;{}_{\rm S}\langle q|\mathrm{e}^{-\mathrm{i}\hat{H}t}|p\rangle_{\rm S}\,C_a(p) \qquad (5.4\mathrm{a})$$

where we have used (2.4) and denoted

$$\langle p|a\rangle\equiv C_a(p) \qquad (a=\mathrm{i},\mathrm{f}). \qquad (5.4\mathrm{b})$$

We now assume that the potential $V(q)$ has a short range and is negligible when $|q|>R_0$. We are concerned with the scattering from an initial state $|\mathrm{i}\rangle$, which describes a particle well outside the range R_0 of the potential, to another such state $|\mathrm{f}\rangle$. The S-matrix element S_{fi} is defined as

$$\begin{aligned}S_{\mathrm{fi}}&=U_{\mathrm{fi}}(+\infty,-\infty)\\ &=\lim_{\substack{t'\to-\infty\\ t''\to+\infty}}\int \mathrm{d}q'\,\mathrm{d}q''\psi_{\mathrm{f}}^{*}(q'',t'')\langle q'',t''|q',t'\rangle\psi_{\mathrm{i}}(q',t').\end{aligned} \qquad (5.5)$$

Since only the behaviour of $\psi_{\mathrm{i}}(q,t')$ for $t'\to-\infty$ and of $\psi_{\mathrm{f}}(q,t'')$ for $t''\to+\infty$ is involved in calculating S_{fi}, and since in these limits ψ_{i} and ψ_{f} describe a particle well outside the range R_0 of $V(q)$, it follows that we may ignore the contribution of $\hat{V}$ to $\hat{H}$ in the calculation of $\psi_a(q,t)$ $(a=\mathrm{i},\mathrm{f})$ and retain only the kinetic energy term. In these circumstances as $t\to-\infty$

$$\begin{aligned}\psi_{i}(q,t)&\approx\int \mathrm{d}p\;C_{\mathrm{i}}(p)\;{}_{\rm S}\!\left\langle q\left|\exp\left(\frac{-\mathrm{i}}{2\mu}\hat{P}^2t\right)\right|p\right\rangle_{\rm S}\\ &=\int \mathrm{d}p\;C_{\mathrm{i}}(p)\;\mathrm{e}^{-\mathrm{i}p^2t/2\mu}\;{}_{\rm S}\langle q|p\rangle_{\rm S}\\ &=(2\pi)^{-1/2}\int \mathrm{d}p\;C_{\mathrm{i}}(p)\;\mathrm{e}^{-\mathrm{i}p^2t/2\mu}\;\mathrm{e}^{\mathrm{i}pq}\end{aligned} \qquad (5.6)$$

and a precisely similar expression holds for $\psi_{\mathrm{f}}(q,t)$ as $t\to+\infty$. Since only large values of $|t|$ (and $|q|$) are involved in these expressions, we may make an asymptotic expansion of the wave functions[1]. This may be obtained as follows. First we shift the integration variable p in (5.6) to a symmetric one. Since

$$\frac{p^2}{2\mu}t-pq=\frac{t}{2\mu}\left(p-\frac{\mu q}{t}\right)^2-\frac{\mu q^2}{2t} \qquad (5.7)$$

we define

$$k\equiv\left(\frac{|t|}{2\mu}\right)^{1/2}\left(p-\frac{\mu q}{t}\right) \qquad (5.8)$$

and then for $|t| \to \infty$

$$\psi_a(q,t) = \left(\frac{\mu}{\pi|t|}\right)^{1/2} e^{i\mu q^2/2t} \int dk\, C_a\left[\frac{\mu q}{t} + \left(\frac{2\mu}{|t|}\right)^{1/2} k\right] e^{-\varepsilon(t)ik^2} \tag{5.9a}$$

where

$$\varepsilon(t) = \begin{cases} 1 & t>0 \\ -1 & t<0. \end{cases} \tag{5.9b}$$

For $q/t = O(1)$ we may expand C_a:

$$C_a\left[\frac{\mu q}{t} + \left(\frac{2\mu}{|t|}\right)^{1/2} k\right] \sim C_a\left(\frac{\mu q}{t}\right)[1 + O(|t|^{-1/2})] \quad \text{for } |t| \to \infty \tag{5.10}$$

and then from (5.9a) we obtain

$$\psi_a(q,t) \sim \left(\frac{\mu}{|t|}\right)^{1/2} e^{i\mu q^2/2t}\, C_a\left(\frac{\mu q}{t}\right)[1 + O(|t|^{-1/2})]\, e^{-i\varepsilon(t)\pi/4} \qquad \text{as } |t| \to \infty \tag{5.11}$$

after performing the k integration. Now we substitute the leading terms back into (5.5), since the non-leading terms vanish in the limit $|t'|$, $|t''| \to \infty$. Then

$$S_{fi} = \lim_{\substack{t' \to -\infty \\ t'' \to +\infty}} i \int dq'\, dq''\, \mu |t't''|^{-1/2} C_f^*\left(\frac{\mu q''}{t''}\right) e^{-i\mu q''^2/2t''} \times \langle q'', t'' | q', t' \rangle\, e^{i\mu q'^2/2t'}\, C_i\left(\frac{\mu q'}{t'}\right). \tag{5.12}$$

Finally we may change integration variables according to

$$\frac{\mu q'}{t'} = p' \qquad \frac{\mu q''}{t''} = p''. \tag{5.13}$$

Then

$$S_{fi} = \lim_{\substack{t' \to -\infty \\ t'' \to +\infty}} \int dp'\, dp''\, \mu^{-1} |t't''|^{1/2} C_f^*(p'')\, e^{-ip''^2 t''/2\mu} \times \left\langle \frac{p''t''}{\mu}, t'' \middle| \frac{p't'}{\mu}, t' \right\rangle e^{ip'^2 t'/2\mu}\, C_i(p'). \tag{5.14}$$

Thus to calculate the S-matrix element (in the momentum representation) using the path integral method developed in Chapter 2, we must evaluate the functional integral given in (2.23) by integrating over all functions $q(t)$ defined on $-\infty<t<\infty$ and having the *asymptotic* behaviour

$$q(t)\sim\frac{p't}{\mu} \qquad \text{as} \quad t\to-\infty \tag{5.15a}$$

and

$$q(t)\sim\frac{p''t}{\mu} \qquad \text{as} \quad t\to+\infty. \tag{5.15b}$$

This should be contrasted with the *fixed* boundary conditions (2.19) which are appropriate for the calculation of $\langle q'',t''|q',t'\rangle$. The asymptotic behaviour (5.15) corresponds, of course, to the free motion of a non-relativisitic particle of mass μ, since for $|t|\to\infty$ (and therefore $|q|\to\infty$) the contribution from the potential energy is negligible.

These considerations make it apparent why the Green functions generated by $W[J]$ in Chapter 4 are not those appropriate for a physical scattering process. The boundary conditions (4.3) are the field theory analogues of the *fixed* boundary conditions (2.19). Evidently to generate physical scattering amplitudes we have to find a generating functional $S[J]$ in which the path integral (4.4) is performed over all field configurations having an *asymptotic* behaviour analogous to (5.15). This is the task to which we now turn.

5.2 Scattering amplitude in quantum field theory

We now wish to generalise the considerations of the previous section to the case where we have a relativistic field theory. We have seen in Chapter 4 how to calculate the quantum transition amplitude $\langle\varphi''(\boldsymbol{x})t''|\varphi'(\boldsymbol{x})t'\rangle^J$, when we have a real scalar field $\varphi(\boldsymbol{x},t)$ which achieves the configurations $\varphi'(\boldsymbol{x})$ at $t=t'$ and $\varphi''(\boldsymbol{x})$ at $t=t''$, in the presence of a source $J(x)$. The moral of the previous section, and of (5.14) in particular, is that to calculate the S-matrix elements we need to calculate the transition amplitude by performing the functional integral over all (classical) field configurations $\varphi(\boldsymbol{x},t)$ having free-field *asymptotic* behaviour. That is to say

$$\varphi(\boldsymbol{x},t)\sim\varphi^{\text{in}}(\boldsymbol{x},t) \qquad \text{as} \quad t\to-\infty \tag{5.16a}$$

and

$$\varphi(\boldsymbol{x},t)\sim\varphi^{\text{out}}(\boldsymbol{x},t) \qquad \text{as} \quad t\to+\infty \tag{5.16b}$$

where $\varphi^{\text{in,out}}(\boldsymbol{x},t)$ satisfy the free-field equation which follows from the Lagrangian (4.26), namely

$$(\partial^\nu\partial_\nu+\mu^2)\varphi^{\text{in,out}}=0 \tag{5.16c}$$

(which is of course (3.35) when $\lambda=0$). Now we have seen that *any* free field φ_0

may be cast in the form (3.47)

$$\varphi_0(x)=\int\frac{d^3k}{(2\pi)^3}\frac{1}{2k_0}(a(k)\,e^{-ikx}+a^*(k)e^{ikx}) \tag{5.17}$$

and $\varphi^{in}(x)$ and $\varphi^{out}(x)$ are therefore special cases of this. As we can see from (5.6), the momentum space wave function $C_i(p)$ of the initial state is associated with a factor e^{-iEt} where $E=p^2/2\mu$ is the energy of a (free) particle having momentum p. Thus the initial state ($t\to-\infty$) is associated with the 'positive frequency' piece of φ_0, namely

$$\varphi^{in}(x)=\int\frac{d^3k}{(2\pi)^3}\frac{1}{2k_0}a(k)\,e^{-ikx}. \tag{5.18}$$

Similarly, the final outgoing state ($t\to+\infty$) is associated with the 'negative frequency' part of φ_0

$$\varphi^{out}(x)=\int\frac{d^3k}{(2\pi)^3}\frac{1}{2k_0}a^*(k)\,e^{ikx}. \tag{5.19}$$

This required separation may be achieved by assigning the zeroth component of the momentum integration variable k_0 a small *negative* imaginary part

$$k_0\to k_0-i\varepsilon \qquad (\varepsilon>0). \tag{5.20}$$

This has the effect

$$e^{ik_0t}\to e^{ik_0t}\,e^{\varepsilon t}\to 0 \qquad \text{as} \quad t\to-\infty \tag{5.21}$$

and only e^{-ikx} survives as $t\to-\infty$, as required. Similarly only the part associated with e^{ikx} survives as $t\to\infty$. Thus the assignment (5.20) of a small negative imaginary part to k_0 ensures that

$$\varphi_0(x)\sim\varphi^{in}(x) \qquad \text{as} \quad x^0\to-\infty \tag{5.22}$$

and

$$\varphi_0(x)\sim\varphi^{out}(x) \qquad \text{as} \quad x^0\to+\infty. \tag{5.23}$$

It follows that the asymptotic conditions (5.16) are equivalent to the conditions

$$\varphi(x)\sim\varphi_0(x) \qquad x^0\to\pm\infty. \tag{5.24}$$

We now wish to determine the functional which generates S-matrix elements. For the reasons already given it is clear that we must perform the path integral (4.25). Thus we define

$$S[J,\varphi_0]\equiv\int\mathscr{D}\varphi\,\exp i\int dx(\mathscr{L}+J\varphi) \tag{5.25}$$

but now the integration is over all field configurations having the asymptotic behaviour (5.24). (Without ambiguity, we now denote the space–time volume element d^4x by dx.) We shall see in §6.4 how S-matrix elements emerge from this generating functional.

To determine $S[J, \varphi_0]$ we first split $\mathscr{L}$ into two pieces:

$$\mathscr{L} = \mathscr{L}_0 + \mathscr{L}_I(\varphi) \tag{5.26}$$

where

$$\mathscr{L}_0 \equiv \tfrac{1}{2}(\partial^\nu \varphi)(\partial_\nu \varphi) - \tfrac{1}{2}\mu^2 \varphi^2 \tag{5.27}$$

is the *free-field* Lagrangian density, and $\mathscr{L}_I(\varphi)$ specifies the interaction piece of $\mathscr{L}$. The only case we shall consider in detail is the example given in (3.34), where

$$\mathscr{L}_I = -\frac{\lambda}{4!}\varphi^4(x). \tag{5.28}$$

For the present, however, we shall take $\mathscr{L}_I$ to be an arbitrary function of φ. Secondly, we write the integrand in (5.25) in the form

$$\exp i \int dx(\mathscr{L} + J\varphi) = \left(\exp i \int dx \mathscr{L}_I(\varphi)\right)\left(\exp i \int dy(\mathscr{L}_0 + J\varphi)\right). \tag{5.29}$$

We can expand the first exponential

$$\exp i \int dx \mathscr{L}_I(\varphi) = 1 + i \int dx \mathscr{L}_I(\varphi) + \frac{i^2}{2!}\int dx\, dy \mathscr{L}_I(\varphi(x))\mathscr{L}_I(\varphi(y)) + \ldots. \tag{5.30}$$

Now, since

$$i\varphi(x) \exp i \int dz(\mathscr{L}_0 + J\varphi) = \frac{\delta}{\delta J(x)} \exp i \int dz(\mathscr{L}_0 + J\varphi) \tag{5.31}$$

it follows that

$$\left(\int dx \mathscr{L}_I(\varphi)\right)\left(\exp i \int dz(\mathscr{L}_0 + J\varphi)\right)$$

$$= \int dx \mathscr{L}_I\left(-i\frac{\delta}{\delta J(x)}\right)\left(\exp i \int dz(\mathscr{L}_0 + J\varphi)\right). \tag{5.32}$$

Since the operator $\int dx\, \mathscr{L}_I(-i\,\delta/\delta J(x))$ is independent of $\varphi(x)$, it can be taken outside the functional integral (5.25). Then we have

$$S[J, \varphi_0] = \exp i \int dx\, \mathscr{L}_I\left(-i\frac{\delta}{\delta J(x)}\right) S_0[J, \varphi_0] \tag{5.33}$$

where

$$S_0[J, \varphi_0] \equiv \int \mathscr{D}\varphi \exp i \int dy(\mathscr{L}_0 + J\varphi) \tag{5.34}$$

is the generating functional with only the free-field Lagrangian (5.27) in its integrand. The integration, of course, is still over fields having free-field asymptotic behaviour. Next we change the functional integration variable from φ to $\tilde{\varphi}$, where $\tilde{\varphi}$ satisfies

$$\varphi(x) \equiv \tilde{\varphi}(x) + \varphi_0(x). \tag{5.35}$$

Then (5.24) shows that

$$\tilde{\varphi}(x) \to 0 \qquad \text{as} \quad x^0 \to \pm\infty. \tag{5.36}$$

Substituting (5.35) into the integrand of (5.34) gives

$$\int \mathrm{d}y(\mathscr{L}_0 + J\varphi) = \int \mathrm{d}y[\tfrac{1}{2}(\partial^\nu\tilde{\varphi})(\partial_\nu\tilde{\varphi}) - \tfrac{1}{2}\mu^2\tilde{\varphi}^2 + J(\tilde{\varphi}+\varphi_0)$$
$$- \tilde{\varphi}\partial^\nu\partial_\nu\varphi_0 - \mu^2\tilde{\varphi}\varphi_0 - \tfrac{1}{2}\varphi_0\partial^\nu\partial_\nu\varphi_0 - \tfrac{1}{2}\mu^2\varphi_0^2] \tag{5.37}$$

after integrating by parts and dropping surface terms. Now since φ_0 is a *free* field

$$(\partial^\nu\partial_\nu + \mu^2)\varphi_0 = 0 \tag{5.38}$$

and a number of terms drop out of (5.37), which then gives

$$S_0[J, \varphi_0] = \int \mathscr{D}\tilde{\varphi} \exp \mathrm{i} \int \mathrm{d}y(\tfrac{1}{2}\partial^\nu\tilde{\varphi}\partial_\nu\tilde{\varphi} - \tfrac{1}{2}\mu^2\tilde{\varphi}^2 + J\tilde{\varphi})$$
$$\times \exp \mathrm{i} \int \mathrm{d}z\, J(z)\varphi_0(z). \tag{5.39}$$

Remember the functional integral is now over paths which satisfy (5.36). However this is a special case of the *fixed* boundary condition considered in the previous chapter. In the present case we have

$$\varphi'(\boldsymbol{x}) = \varphi''(\boldsymbol{x}) = 0. \tag{5.40}$$

So the first factor in (5.39) is just the functional $W_0[J]$ which generates the free-field Green functions $\mathscr{G}_0^{(n)}(x_1, \ldots, x_n)$. We saw in (4.43) that

$$W_0[J] = N \exp\left(-\tfrac{1}{2}\int \mathrm{d}x\, \mathrm{d}y\, J(x)\mathrm{i}\Delta_{\mathrm{F}}(x-y)J(y)\right) \tag{5.41a}$$

where

$$\Delta_{\mathrm{F}}(x-y) = \int \frac{\mathrm{d}^4p}{(2\pi)^4} \frac{\mathrm{e}^{\mathrm{i}p(x-y)}}{p^2 - \mu^2 + \mathrm{i}\varepsilon}. \tag{5.41b}$$

Now differentiating $W_0[J]$ gives

$$\frac{\delta W_0[J]}{\delta J(x)} = -\int \mathrm{d}y\, \mathrm{i}\, \Delta_{\mathrm{F}}(x-y)J(y)W_0[J]. \tag{5.42}$$

It follows from (5.41b) that

$$(\partial_x^2\partial_{x\nu}+\mu^2)\,\Delta_{\mathrm{F}}(x-y)=-\int\frac{\mathrm{d}^4p}{(2\pi)^4}\,\mathrm{e}^{\mathrm{i}p(x-y)}=-\delta(x-y). \tag{5.43}$$

So

$$(\partial_x^\nu\partial_{x\nu}+\mu^2)\frac{\delta W_0[J]}{\delta J(x)}=\mathrm{i}J(x)W_0[J]. \tag{5.44}$$

This means that we can recast (5.39) in the form

$$S_0[J,\varphi_0]=\exp\left(\int \mathrm{d}x\;\varphi_0(x)(\partial_x^\nu\partial_{x\nu}+\mu^2)\frac{\delta}{\delta J(x)}\right)W_0[J]. \tag{5.45}$$

We substitute this back into (5.33) and interchange the order of the two functional differentiations. This gives

$$S[J,\varphi_0]=\exp\left(\int \mathrm{d}x\;\varphi_0(x)(\partial_x^\nu\partial_{x\nu}+\mu^2)\frac{\delta}{\delta J(x)}\right)$$
$$\times\exp\left[\mathrm{i}\int\mathrm{d}y\;\mathscr{L}_1\left(-\mathrm{i}\frac{\delta}{\delta J(y)}\right)\right]W_0[J]. \qquad 5.46)$$

The last two terms may be combined, if we choose, to give the functional $W[J]$ which generates the Green functions $\mathscr{G}^{(n)}(x_1,\ldots,x_n)$ of the interacting field theory. This is because we can reverse the steps from (5.25) to (5.33) for the functionals $W[J]$ and $W_0[J]$, since the only difference is in the boundary conditions on the integration variable, and this was not used in the derivation. Thus finally we obtain[2]

$$S[J,\varphi_0]=\exp\left(\int \mathrm{d}x\;\varphi_0(x)(\partial_x^\nu\partial_{x\nu}+\mu^2)\frac{\delta}{\delta J(x)}\right)W[J]. \tag{5.47}$$

This gives the (asymptotic-state-to-asymptotic-state) transition amplitude in the presence of a source $J(x)$. In the actual physical processes with which we are concerned there is no source, so the quantity of physical interest is obtained by setting $J(x)$ to zero:

$$S[\varphi_0]\equiv S[0,\varphi_0]. \tag{5.48}$$

(Of course, if there really were a *physical* source $J_0(x)$ in this system, we should want to evaluate $S[J_0,\varphi_0]$.)

We shall address the task of calculating $W[J]$ and hence $S[\varphi_0]$ in the following chapter. For the moment we shall content ourselves with observing how the first factor in (5.47) fulfils the task assigned to it. Recall that $\varphi_0(x)$ is a free field, so its Fourier components k in (5.17) all satisfy the mass shell condition $k^2=\mu^2$. If we perform the x integration in (5.47) by parts, we can pull the Klein–Gordon operator $\partial_x^\nu\partial_{x\nu}+\mu^2$ back on to $\varphi_0(x)$ and hence obtain a factor $\mu^2 - k^2$ from each Fourier component k. Unless the contribution from $W[J]$ has a compensating factor $(k^2 - \mu^2)^{-1}$, its contribution to $S[J, \varphi_0]$ will vanish, since $\mu^2 - k^2 = 0$ for all Fourier components in φ_0, as we have said. Thus the effect of the pre-factor in (5.47) is to project off-mass-shell contributions to zero, and to retain the on-shell contribution, as required. Setting J zero, as in (5.48), ensures that all external lines of $W[J]$ are subjected to this treatment.

References

The non-canonical treatment of the S-matrix is rarely presented in the literature. The reference which we have used most extensively in preparing this chapter is the 1975 Les Houches Lectures of:

Fadeev L D 1976 in *Methods in Field Theory* ed R Balian and J Zinn-Justin (Amsterdam: North-Holland) p. 3

For an alternative non-canonical treatment see

Ramond P 1981 *Field Theory: A Modern Primer* (Reading, Mass.: Benjamin-Cummings) p 204

References in the text

1 See e.g. Murray J D 1974 *Asymptotic Analysis* (Oxford: Oxford University Press)
2 Cf. Itzykson C and Zuber J B 1980 *Quantum Field Theory* (New York: McGraw-Hill)

6

FEYNMAN RULES FOR $\lambda\varphi^4$ THEORY

6.1 Perturbation theory

We have seen in Chapter 4 that the functional $W[J]$ generates the Green functions of the theory, and we have seen in Chapter 5 how $W[J]$ may be related to the functional $S[\varphi_0]$ which generates the S-matrix elements of the theory. Thus all that is required for the evaluation of the scattering amplitudes is the evaluation of $W[J]$. Unfortunately this is more easily said than done. The path or functional integral which we are required to evaluate is

$$W[J]=N\int \mathscr{D}\varphi \exp \mathrm{i}\int \mathrm{d}^4x(\mathscr{L}(\varphi, \partial_\mu\varphi)+J\varphi) \tag{6.1}$$

where we have set $\hbar=1$, N is the undetermined constant of proportionality, and it is understood that the space–time integral is over all space with the time integrations along the real axis from $-\infty$ to $+\infty$. $W[J]$ can be evaluated exactly only in the trivial case of a free-field theory. In this case

$$\mathscr{L}=\mathscr{L}_0\equiv\tfrac{1}{2}(\partial_\mu\varphi)(\partial^\mu\varphi)-\tfrac{1}{2}\mu^2\varphi^2 \tag{6.2}$$

for a free scalar field φ. Then, as we saw in (4.43),

$$W[J]=W_0[J]=N\exp\left(-\frac{\mathrm{i}}{2}\int \mathrm{d}x\,\mathrm{d}y\,J(x)\,\Delta_\mathrm{F}(x-y)J(y)\right) \tag{6.3}$$

where (without ambiguity) we now denote the space–time volume element d^4x by $\mathrm{d}x$, and

$$\Delta_\mathrm{F}(x-y)=\int\frac{\mathrm{d}^4p}{(2\pi)^4}\frac{\mathrm{e}^{\mathrm{i}p(x-y)}}{p^2-\mu^2+\mathrm{i}\varepsilon} \tag{6.4}$$

is the Feynman propagator for the scalar field.

The best that can be done for interacting fields is to develop a perturbation series for $W[J]$. If

$$\mathscr{L}=\mathscr{L}_0+\mathscr{L}_\mathrm{I}(\varphi) \tag{6.5}$$

and $\mathscr{L}_\mathrm{I}$ is proportional to a parameter λ, we can expand $W[J]$ as a power series in λ. To do this we recall the result (5.33) from the previous chapter which expressed the generating functional $S[J]$ for the interacting field theory, in terms of $S_0[J]$, the generating functional of the free-field theory. Clearly a precisely similar relationship may be derived between the functionals $W[J]$

and $W_0[J]$:

$$W[J]=\exp\left[\mathrm{i}\int \mathrm{d}x\,\mathscr{L}_1\left(-\mathrm{i}\frac{\delta}{\delta J(x)}\right)\right]W_0[J] \tag{6.6}$$

since the only difference between the functionals $W[J]$ and $S[J]$ derives from the different classes of field configurations which are integrated over, and this feature was not used in the derivation of (5.33). The perturbation series comes from expanding the exponential operator as a power series:

$$\exp \mathrm{i}\int \mathrm{d}x\,\mathscr{L}_{\mathrm{I}}\left(-\mathrm{i}\frac{\delta}{\delta J(x)}\right)=1+\mathrm{i}\int \mathrm{d}x\,\mathscr{L}_{\mathrm{I}}\left(-\mathrm{i}\frac{\delta}{\delta J(x)}\right)$$
$$+\frac{\mathrm{i}^2}{2!}\int \mathrm{d}x\,\mathrm{d}y\,\mathscr{L}_{\mathrm{I}}\left(-\mathrm{i}\frac{\delta}{\delta J(x)}\right)\mathscr{L}\left(-\mathrm{i}\frac{\delta}{\delta J(y)}\right)+\dots \tag{6.7}$$

Since $\mathscr{L}_{\mathrm{I}}$ is proportional to λ, the above expression generates an expansion of $W[J]$ as a power series in λ. We illustrate this by consideration of the special case

$$\mathscr{L}_{\mathrm{I}}(\varphi)=-\frac{\lambda}{4!}\varphi^4(x). \tag{6.8}$$

In this case, substituting (6.7) into (6.6) gives

$$W[J]=W_0[J]-\frac{\mathrm{i}\lambda}{4!}\int \mathrm{d}x\frac{\delta^4 W_0}{\delta J(x)^4}+\mathrm{O}(\lambda^2). \tag{6.9}$$

From (6.3) and (6.4) we find

$$\frac{\delta W_0}{\delta J(x)}=-\int \mathrm{d}y\,\mathrm{i}\,\Delta_{\mathrm{F}}(x-y)J(y)W_0[J] \tag{6.10}$$

and repeated differentiation gives

$$\begin{aligned}W[J]=\Bigg[1-\frac{\mathrm{i}\lambda}{4!}\Bigg(&\int \mathrm{d}x\,3(\mathrm{i}\Delta_{\mathrm{F}}(0))^2\\ &-6\mathrm{i}\Delta_{\mathrm{F}}(0)\int \mathrm{d}y_1\,\mathrm{d}y_2\,\mathrm{i}\Delta_{\mathrm{F}}(x-y_1)\mathrm{i}\Delta_{\mathrm{F}}(x-y_2)J(y_1)J(y_2)\\ &+\int \mathrm{d}y_1\,\mathrm{d}y_2\,\mathrm{d}y_3\,\mathrm{d}y_4\,\mathrm{i}\,\Delta_{\mathrm{F}}(x-y_1)\mathrm{i}\Delta_{\mathrm{F}}(x-y_2)\mathrm{i}\Delta_{\mathrm{F}}(x-y_3)\mathrm{i}\Delta_{\mathrm{F}}(x-y_4)\\ &\times J(y_1)J(y_2)J(y_3)J(y_4)\Bigg)+\mathrm{O}(\lambda^2)\Bigg]W_0[J].\end{aligned} \tag{6.11}$$

The Green functions generated by $W[J]$ are defined as in (4.8) by

$$\mathrm{i}^n\mathscr{G}^{(n)}(x_1,\dots,x_n)=\left.\frac{\delta^n W[J]}{\delta J(x_1)\dots\delta J(x_n)}\right|_{J=0} \tag{6.12}$$

and it is clear that the additional terms in (6.13), generated by the interaction, change the Green functions from the values derived in Chapter 4. If the normalisation N is chosen so that

$$W_0[0]=1 \tag{6.13}$$

as in Chapter 4, then

$$\mathscr{G}^{(0)}=W[0]=1-\tfrac{1}{8}\mathrm{i}\lambda\int \mathrm{d}x\mathrm{i}\Delta_{\mathrm{F}}(x-x)\mathrm{i}\Delta_{\mathrm{F}}(x-x)+\mathrm{O}(\lambda^2). \tag{6.14}$$

Also

$$\mathscr{G}^{(2)}(x_1,x_2)=\mathscr{G}_0^{(2)}(x_1,x_2)-\tfrac{1}{2}\mathrm{i}\lambda\int \mathrm{d}x\mathrm{i}\Delta_{\mathrm{F}}(x_1-x)\mathrm{i}\Delta_{\mathrm{F}}(x-x)\mathrm{i}\Delta_{\mathrm{F}}(x-x_2)+\mathrm{O}(\lambda^2) \tag{6.15}$$

where

$$\mathscr{G}_0^{(2)}(x_1,x_2)=\mathrm{i}\Delta_{\mathrm{F}}(x_1-x_2) \tag{6.16}$$

is the free-field Green function derived in (4.48). Similarly

$$\begin{aligned}
\mathscr{G}^{(4)}(x_1,x_2,x_3,x_4) &= \mathscr{G}_0^{(4)}(x_1,x_2,x_3,x_4)\\
&-\tfrac{1}{2}\mathrm{i}\lambda\Big(\mathscr{G}_0^{(2)}(x_1,x_2)\int \mathrm{d}x\mathrm{i}\Delta_{\mathrm{F}}(x_3-x)\mathrm{i}\Delta_{\mathrm{F}}(x-x)\mathrm{i}\Delta_{\mathrm{F}}(x-x_4)\\
&+\mathscr{G}_0^{(2)}(x_3,x_4)\int \mathrm{d}x\mathrm{i}\Delta_{\mathrm{F}}(x_1-x)\mathrm{i}\Delta_{\mathrm{F}}(x-x)\mathrm{i}\Delta_{\mathrm{F}}(x-x_2)\\
&+\mathscr{G}_0^{(2)}(x_1,x_3)\int \mathrm{d}x\mathrm{i}\Delta_{\mathrm{F}}(x_2-x)\mathrm{i}\Delta_{\mathrm{F}}(x-x)\mathrm{i}\Delta_{\mathrm{F}}(x-x_4)\\
&+\mathscr{G}_0^{(2)}(x_2,x_4)\int \mathrm{d}x\mathrm{i}\Delta_{\mathrm{F}}(x_1-x)\mathrm{i}\Delta_{\mathrm{F}}(x-x)\mathrm{i}\Delta_{\mathrm{F}}(x-x_3)\\
&+\mathscr{G}_0^{(2)}(x_1,x_4)\int \mathrm{d}x\mathrm{i}\Delta_{\mathrm{F}}(x_2-x)\mathrm{i}\Delta_{\mathrm{F}}(x-x)\mathrm{i}\Delta_{\mathrm{F}}(x-x_3)\\
&+\mathscr{G}_0^{(2)}(x_2,x_3)\int \mathrm{d}x\mathrm{i}\Delta_{\mathrm{F}}(x_1-x)\mathrm{i}\Delta_{\mathrm{F}}(x-x)\mathrm{i}\Delta_{\mathrm{F}}(x-x_4)\Big)\\
&-\mathrm{i}\lambda\int \mathrm{d}x\mathrm{i}\Delta_{\mathrm{F}}(x_1-x)\mathrm{i}\Delta_{\mathrm{F}}(x_2-x)\mathrm{i}\Delta_{\mathrm{F}}(x_3-x)\mathrm{i}\Delta_{\mathrm{F}}(x_4-x)+\mathrm{O}(\lambda^2)
\end{aligned} \tag{6.17}$$

where $\mathscr{G}_0^{(4)}(x_1,x_2,x_3,x_4)$ is the free-field Green function given in (4.49). Evidently we may extend the diagrammatic notation developed in Chapter 4 to include the additional terms generated by the interaction. We write

$\mathscr{G}^{(2)}(x_1,x_2)=$ 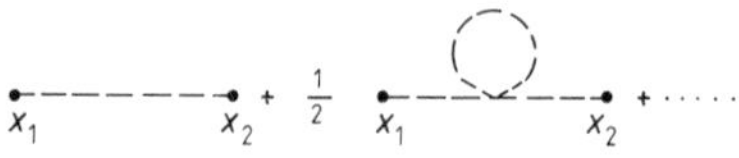 (6.18)

where the first term is just the free-field propagator given in (6.16), and the second term is the $O(\lambda)$ part of (6.15). If the internal vertex is at x the three propagators $i\Delta_F(x_1-x)i\Delta_F(x-x)i\Delta_F(x-x_2)$ arise from the rule already given associating a line with a propagator. In addition, it is apparent from (6.15) that we must associate a factor $-i\lambda$ with the vertex:

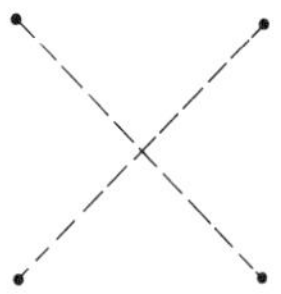

$$: \quad -i\lambda \tag{6.19}$$

and we are to integrate over the coordinates x of any internal vertex. Using this notation

$\mathscr{G}^{(4)}(x_1, x_2, x_3, x_4)=$

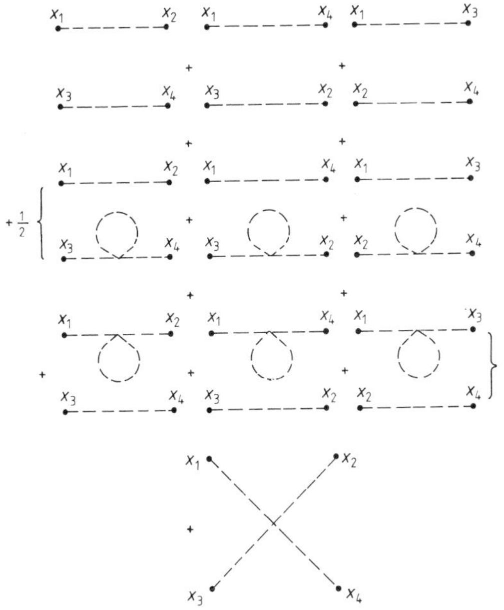

(6.20)

As before, it is easier to work with the *connected* Green functions $G^{(n)}(x_1,\ldots,x_n)$. These are generated by

$$iX[J]\equiv \ln W[J]. \tag{6.21}$$

Thus from (6.11) we find

$$
\begin{aligned}
\mathrm{i}X[J] = {} & \ln N - \tfrac{1}{2}\int \mathrm{d}y_1\,\mathrm{d}y_2\,\mathrm{i}\Delta_{\mathrm{F}}(y_1-y_2)J(y_1)J(y_2) \\
& -\frac{\mathrm{i}\lambda}{4!}\int \mathrm{d}x\bigg(3[\mathrm{i}\Delta_{\mathrm{F}}(0)]^2 - 6\int \mathrm{d}y_1\,\mathrm{d}y_2\,\mathrm{i}\Delta_{\mathrm{F}}(y_1-x) \\
& \times \mathrm{i}\Delta_{\mathrm{F}}(x-x)\mathrm{i}\Delta_{\mathrm{F}}(x-y_2)J(y_1)J(y_2) \\
& + \int \mathrm{d}y_1\,\mathrm{d}y_2\,\mathrm{d}y_3\,\mathrm{d}y_4\,\mathrm{i}\Delta_{\mathrm{F}}(y_1-x)\mathrm{i}\Delta_{\mathrm{F}}(y_2-x) \\
& \times \mathrm{i}\Delta_{\mathrm{F}}(y_3-x)\mathrm{i}\Delta_{\mathrm{F}}(y_4-x)J(y_1)J(y_2)J(y_3)J(y_4)\bigg) + \mathrm{O}(\lambda^2).
\end{aligned}
\tag{6.22}
$$

With the connected Green functions defined by (4.60):

$$
\mathrm{i}^n G^{(n)}(x_1,\ldots,x_n) = \frac{\mathrm{i}\delta^n X[J]}{\delta J(x_1)\ldots\delta J(x_n)}\bigg|_{J=0}
\tag{6.23}
$$

we find

$G^{(2)}(x_1,x_2) =$ 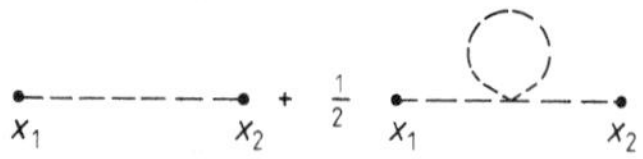 (6.24)

$G^{(4)}(x_1, x_2, x_3, x_4) =$ 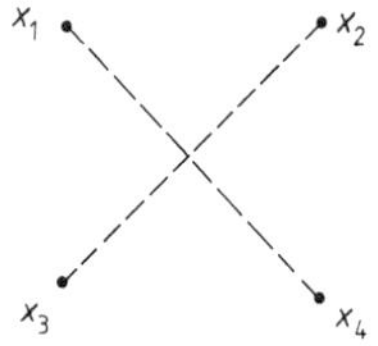 (6.25)

Thus, as anticipated in Chapter 4, the interaction generates additional terms in $G^{(2)}$ and makes $G^{(n)}$ non-zero even when $n > 2$.

6.2 Momentum space Feynman rules

It is apparent from (6.4) that the propagator has its simplest form in momentum space. For this reason it is the custom to evaluate the Fourier transforms of the Green functions with which we have so far been concerned. Thus, as in (4.55), we define

$$
\mathscr{G}^{(n)}(p_1,\ldots,p_n)(2\pi)^4\delta(p_1+\ldots+p_n) = \int \mathrm{d}x_1\ldots\mathrm{d}x_n\,\mathscr{G}^{(n)}(x_1,\ldots,x_n)\,\mathrm{e}^{-\mathrm{i}(p_1x_1+\ldots+p_nx_n)}
\tag{6.26}
$$

with a similar definition to (4.63) for the connected Green functions. Thus using (6.16) and (6.4) we find the Fourier-transformed free-field Green function

$$\mathscr{G}_0^{(2)}(p, -p) = \frac{\mathrm{i}}{p^2 - \mu^2 + \mathrm{i}\varepsilon} \tag{6.27}$$

while the connected Green function in the presence of interactions is

$$\tilde{G}^{(2)}(p, -p) = \frac{\mathrm{i}}{p^2 - \mu^2 + \mathrm{i}\varepsilon} - \frac{\mathrm{i}\lambda}{2}\int \frac{\mathrm{d}^4k}{(2\pi)^4} \frac{\mathrm{i}}{p^2 - \mu^2 + \mathrm{i}\varepsilon}\frac{\mathrm{i}}{k^2 - \mu^2 + \mathrm{i}\varepsilon}\frac{\mathrm{i}}{p^2 - \mu^2 + \mathrm{i}\varepsilon} + \mathrm{O}(\lambda^2). \tag{6.28}$$

These too may be represented diagrammatically. For example

$\tilde{G}^{(2)}(p, -p) =$ 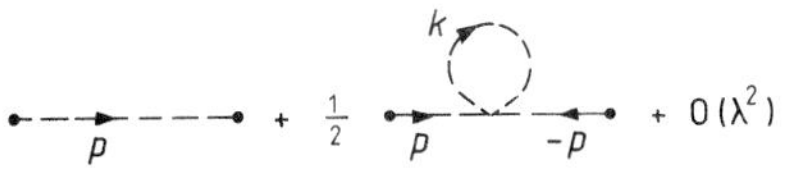

(6.29)

and the Feynman rules are as follows:

1 With each line carrying momentum p we are to associate a factor $\mathrm{i}\,(p^2 - \mu^2 + \mathrm{i}\varepsilon)^{-1}$

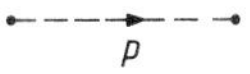

$: \quad \dfrac{\mathrm{i}}{p^2 - \mu^2 + \mathrm{i}\varepsilon}.$ (6.30)

2 With each vertex of four lines carrying momenta p_1, p_2, p_3, p_4 we associate a factor $-\mathrm{i}\lambda$, constraining the momenta so that there is overall conservation:

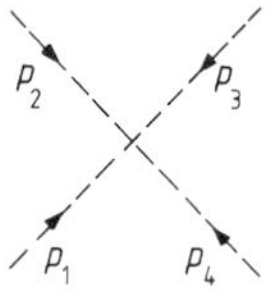

$: \quad -\mathrm{i}\lambda \qquad (p_1 + p_2 + p_3 + p_4 = 0).$ (6.31)

3 Integrate over each independent internal loop momentum k with weight $\mathrm{d}^4k(2\pi)^{-4}$.

Using this notation it is readily verified that

$\tilde{G}^{(4)}(p_1, p_2, p_3, p_4) =$ 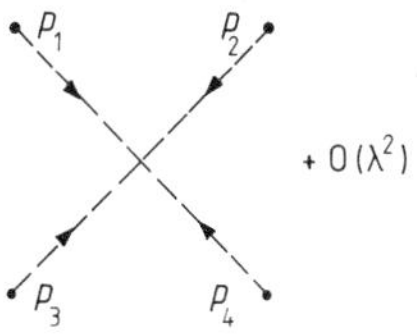

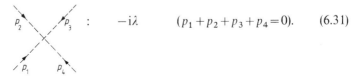

(6.32)

Notice that the Green functions have propagator factors on every external line.

6.3 One-particle-irreducible Green functions

The notation and techniques of the previous two sections may also be used to develop a perturbative expansion of the OPI Green functions in the presence of interactions. It follows from (6.22), and the definition (4.64) of the classical field $\varphi_c(x)$, that

$$\begin{aligned}\varphi_c(x) = &-\int \mathrm{d}y\, \Delta_F(x-y)J(y)\\ &+\tfrac{1}{2}\lambda \int \mathrm{d}y\, \mathrm{d}z\, \mathrm{i}\,\Delta_F(x-y)\mathrm{i}\,\Delta_F(y-y)\mathrm{i}\,\Delta_F(y-z)J(z)\\ &-\tfrac{1}{6}\lambda \int \mathrm{d}y\, \mathrm{d}z_1\, \mathrm{d}z_2\, \mathrm{d}z_3\, \mathrm{i}\,\Delta_F(x-y)\mathrm{i}\,\Delta_F(z_1-y)\mathrm{i}\,\Delta_F(z_2-y)\\ &\times \mathrm{i}\,\Delta_F(z_3-y)J(z_1)J(z_2)J(z_3)+\mathrm{O}(\lambda^2).\end{aligned} \tag{6.33}$$

Thus the generating functional $\Gamma[\varphi_c]$ (or the effective action) of the OPI Green functions

$$\begin{aligned}\Gamma[\varphi_c] \equiv{} & X[J]-\int \mathrm{d}x J(x)\varphi_c(x)\\ ={} & -\mathrm{i}\ln N+\tfrac{1}{2}\int \mathrm{d}y_1\, \mathrm{d}y_2\, \Delta_F(y_1-y_2)J(y_1)J(y_2)-\tfrac{1}{8}\lambda \int \mathrm{d}x[\mathrm{i}\,\Delta_F(0)]^2\\ & -\tfrac{1}{4}\lambda \int \mathrm{d}x\, \mathrm{d}y_1\, \mathrm{d}y_2\, \mathrm{i}\,\Delta_F(y_1-x)\,\mathrm{i}\,\Delta_F(x-x)\,\mathrm{i}\,\Delta_F(x-y_2)J(y_1)J(y_2)\\ & +\tfrac{1}{8}\lambda \int \mathrm{d}x\, \mathrm{d}y_1\, \mathrm{d}y_2\, \mathrm{d}y_3\, \mathrm{d}y_4\, \mathrm{i}\,\Delta_F(y_1-x)\,\mathrm{i}\,\Delta_F(y_2-x)\,\mathrm{i}\,\Delta_F(y_3-x)\\ & \times \mathrm{i}\,\Delta_F(y_4-x)J(y_1)J(y_2)J(y_3)J(y_4)+\mathrm{O}(\lambda^2).\end{aligned} \tag{6.34}$$

We may solve (6.33) for $J(x)$ perturbatively, and after performing integrations by parts using (4.53) we find

$$J(x)=(\partial_\mu\partial^\mu+\mu^2)\varphi_c(x)+\tfrac{1}{2}\lambda\,\mathrm{i}\,\Delta_F(0)\varphi_c(x)+\tfrac{1}{6}\lambda[\varphi_c(x)]^3+\mathrm{O}(\lambda^2). \tag{6.35}$$

Note that $\varphi_c(x)$ satisfies an equation similar to (3.35) derived in classical field theory. The differences are first the source term $J(x)$, which would have appeared in (3.35) had we included such an interaction, and second a term involving $\mathrm{i}\,\Delta_F(0)$ which is a quantum correction. If we were to restore the factors of $\hbar$ which we have set to unity it would be apparent that this term is proportional to $\hbar$. Substituting (6.35) into (6.34) gives Γ explicitly as a functional of $\varphi_c(x)$:

$$\Gamma[\varphi_c] = -\mathrm{i}\ln N - \tfrac{1}{8}\lambda \int \mathrm{d}x[\mathrm{i}\,\Delta_F(0)]^2$$
$$-\tfrac{1}{2}\int \mathrm{d}x\,\varphi_c(x)(\partial^\mu\partial_\mu + \mu^2)\varphi_c(x) - \tfrac{1}{4}\lambda\mathrm{i}\,\Delta_F(0)\int \mathrm{d}x[\varphi_c(x)]^2$$
$$-\tfrac{1}{24}\lambda\int \mathrm{d}x[\varphi_c(x)]^4 + \mathrm{O}(\lambda^2). \tag{6.36}$$

Thus the momentum space OPI Green functions defined in (4.75) satisfy

$$\mathrm{i}\tilde{\Gamma}^{(2)}(p, -p) = \mathrm{i}(p^2 - \mu^2) - \tfrac{1}{2}\mathrm{i}\lambda\mathrm{i}\Delta_F(0) + \mathrm{O}(\lambda^2)$$

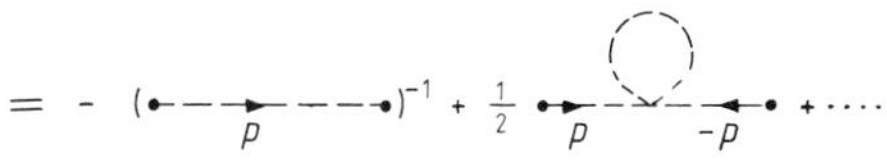

(6.37)

and

$$\mathrm{i}\tilde{\Gamma}^{(4)}(p_1, p_2, p_3, p_4) =$$ 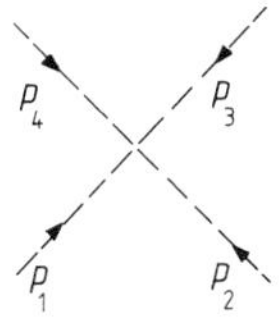

(6.38)

using the Feynman rule given in (6.30), (6.31), (6.32). Note that unlike the connected Green functions, given in (6.29), (6.32) for example, the OPI Green functions have the propagators associated with the external legs divided out.

6.4 Scattering amplitudes

In §6.1 we have seen how to evaluate the Green functions $\mathscr{G}^{(n)}(x_1, \ldots, x_n)$ of the interacting field theory, at least as a perturbation series in λ. The Green functions are generated by the functional $W[J]$ using the formula (4.8)

$$\mathrm{i}^n\mathscr{G}^{(n)}(x_1, \ldots, x_n) = \left.\frac{\delta^n W[J]}{\delta J(x_1)\ldots\delta J(x_n)}\right|_{J=0}. \tag{6.39}$$

We have also seen in §5.2 how the functional $W[J]$ is related to the functional $S[\varphi_0]$ which generates scattering amplitudes. Assuming there is no external source in the system, we find from (5.47) and (5.48) that

$$S[\varphi_0] = \exp\left(\int \mathrm{d}x\varphi_0(x)(\partial_x^\nu\partial_{x\nu} + \mu^2)\frac{\delta}{\delta J(x)}\right)W[J]\bigg|_{J=0}. \tag{6.40}$$

Thus by expanding the exponential,

$$\exp\int dx\varphi_0(x)(\partial_x^\nu\partial_{x\nu}+\mu^2)\frac{\delta}{\delta J(x)}$$

$$=\sum_{n=0}^{\infty}\frac{1}{n!}\int dx_1\ldots dx_n\varphi_0(x_1)\ldots\varphi_0(x_n)K_{x_1}\ldots K_{x_n}\frac{\delta^n}{\delta J(x_1)\ldots\delta J(x_n)} \quad (6.41)$$

we can express $S[\varphi_0]$ in terms of the Green functions $\mathscr{G}^{(n)}(x_1,\ldots x_n)$.

In (6.41) we are using the notation

$$K_{x_j}\equiv\partial_{x_j}^\nu\partial_{x_j\nu}+\mu^2 \quad (6.42)$$

to denote the Klein–Gordon operator. Substituting (6.41) into (6.40) gives

$$S[\varphi_0]=\sum_{n=0}^{\infty}\frac{i^n}{n!}\int dx_1\ldots dx_n\varphi_0(x_1)\ldots\varphi_0(x_n)K_{x_1}\ldots K_{x_n}\mathscr{G}^{(n)}(x_1,\ldots,x_n). \quad (6.43)$$

Thus, as anticipated at the end of Chapter 5, every external line of $\mathscr{G}^{(n)}(x_1,\ldots,x_n)$ has attached to it a Klein–Gordon operator K_{x_j} and a free field $\varphi_0(x_j)$. It is therefore natural to associate the contribution from the term involving $\mathscr{G}^{(n)}$ with the physical process involving a total of n incoming or outgoing free particles. In physical scattering processes these free particles are usually momentum eigenstates. It is therefore useful to have the momentum space representation of the right-hand side of (6.43). To find this we substitute for $\mathscr{G}^{(n)}$ in terms of its Fourier transform $\tilde{\mathscr{G}}^{(n)}$ defined in (4.55). Inverting (4.55) gives

$$\mathscr{G}^{(n)}(x_1,\ldots,x_n)$$

$$=\int\frac{dp_1}{(2\pi)^4}\cdots\frac{dp_n}{(2\pi)^4}e^{i(p_1x_1+\ldots+p_nx_n)}\tilde{\mathscr{G}}^{(n)}(p_1,\ldots,p_n)(2\pi)^4\delta(p_1+\ldots+p_n). \quad (6.44)$$

Each Klein–Gordon operator K_{x_j} in (6.43) acts upon $\mathscr{G}^{(n)}$ and produces a factor $\mu^2-p_j^2$, and we find

$$S[\varphi_0]=\sum_{n=0}^{\infty}\frac{i^n}{n!}\int\frac{dp_1}{(2\pi)^4}\cdots\frac{dp_n}{(2\pi)^4}(2\pi)^4\delta(p_1+\ldots+p_n)$$

$$\times(\mu^2-p_1^2)\ldots(\mu^2-p_n^2)\tilde{\mathscr{G}}^{(n)}(p_1,\ldots,p_n)\int dx_1\ldots dx_n$$

$$\times e^{i(p_1x_1+\ldots+p_nx_n)}\varphi_0(x_1)\ldots\varphi_0(x_n). \quad (6.45)$$

$\varphi_0(x)$ has already been given in (5.17) in terms of its Fourier components. Using this we find

$$\int dx\, e^{ipx}\varphi_0(x)=\int d^3k\frac{2\pi}{2k_0}[a(k)\delta(p-k)+a^*(k)\delta(p+k)] \quad (6.46a)$$

where

$$k_0 = (\boldsymbol{k}^2 + \mu^2)^{1/2}. \tag{6.46b}$$

It therefore follows that the only momenta p_j which can contribute to $S[\varphi_0]$ must satisfy $p_j = \pm k$, which, since $k^2 = \mu^2$, ensures

$$p_j^2 = \mu^2. \tag{6.47}$$

Thus the only Green functions, $\mathscr{G}^{(n)}(p_1, \ldots, p_n)$ which can contribute to $S[\varphi_0]$ are those with all external lines 'on the mass shell', as required of a physical scattering process. Also, even when (6.47) is satisfied only one of the two terms in (6.46a) can contribute; if p_{j0} is positive (and therefore equal to $+k_0$) only the first term can contribute, while if p_{j0} is negative only the second term contributes. This reflects the fact that in any such physical process some of the momenta are associated with incoming particles while others are associated with outgoing particles. As explained in (5.18) and (5.19), we associate the 'positive frequency' part of φ_0 (involving $a(k)$) with an incoming particle, while the negative frequency part (involving $a^*(k)$) is associated with an outgoing particle. (The overall energy–momentum conservation, ensured by the δ function in (6.44), guarantees that not all p_{j0} can have the same sign.) For these reasons the term involving $a(k)$ is associated with an incoming particle having momentum k, and the term involving $a^*(k)$ is associated with an outgoing particle of momentum k.

Now consider a scattering process with m particles in the initial state (i) having incoming momenta $q_1, \ldots, q_m$, and $n-m$ particles in the final state (f) having outgoing momentum $q_{m+1}, \ldots, q_n$. Since all momenta q_r are momenta of physical particles, they satisfy

$$q_r^2 = \mu^2 \qquad (r = 1, \ldots, n) \tag{6.48}$$

and also

$$q_1 + \ldots + q_m = q_{m+1} + \ldots + q_n \tag{6.49}$$

because of overall energy–momentum conservation. Then the scattering amplitude (S-matrix element) S_{fi} for this process is given by the part of $S[\varphi_0]$ involving the m factors $a(q_1) \ldots a(q_m)$ and the $n-m$ factors $a^*(q_{m+1}) \ldots a^*(q_n)$.

Thus

$$S_{\mathrm{fi}} = [\rho(q_1) \ldots \rho(q_n)]^{-1} \left. \frac{\delta^n S[\varphi_0]}{\delta a(q_1) \ldots \delta a(q_m)\, \delta a^*(q_{m+1}) \ldots \delta a^*(q_n)} \right|_{a = a^* = 0} \tag{6.50a}$$

where

$$\rho(q) \equiv (2\pi)^{-3}(2q_0)^{-1} \tag{6.50b}$$

is the (covariant) momentum integration weight function. (The reason for putting $a = a^* = 0$ after the functional differentation is to ensure that only $\mathscr{G}^{(n)}$

contributes to n-particle processes.) Substituting (6.46) into (6.45) and performing the differentiation gives

$$S_{\rm fi}=(2\pi)^4\,\delta(q_1+\ldots+q_m-q_{m+1}-\ldots-q_n)M_{\rm fi} \tag{6.51a}$$

where

$$M_{\rm fi}=(-{\rm i})^n(q_1^2-\mu^2)\ldots(q_n^2-\mu^2)\mathscr{G}^{(n)}(q_1,\ldots,q_m,-q_{m+1},\ldots,-q_n). \tag{6.51b}$$

In deriving this we have used the invariance of $\mathscr{G}^{(n)}(p_1,\ldots,p_n)$ with respect to the interchange of any p_i and p_j.

Thus to obtain the (Lorentz invariant) amplitude $M_{\rm fi}$ from the associated Green function $\mathscr{G}^{(n)}$ one merely has to multiply each leg, carrying momentum q_r, by the factor $-{\rm i}(q_r^2-\mu^2)$, which is just the inverse of the propagator (6.30) for a line having momentum q_r. Having done this there is the further multiplication by $(2\pi)^4$ times an overall energy–momentum conserving δ function to get the S-matrix element. We have observed earlier that the connected Green functions $\tilde{G}^{(n)}$ all have a propagator factor associated with each external line, and the same is clearly true of the full Green functions $\mathscr{G}^{(n)}$. Thus the multiplication by $-{\rm i}(q_r^2-\mu^2)$ cancels each of these propagators leaving a finite limit as $q_r^2\to\mu^2$. So the S-matrix elements are obtained from the *complete* Green functions by deleting the external line propagators and supplying an overall $(2\pi)^4\delta(\ldots)$. It follows that the Feynman rules for S-matrix elements are similar to those for the OPI Green functions, discussed in §6.3, in having the propagators associated with the external lines divided out. However, the diagrams to be considered differ in two respects from the OPI Green functions. First, the external lines in S-matrix elements are constrained to be on the mass shell, and second we have to include all contributing diagrams, disconnected and one-particle-reducible ones as well as the connected and OPI ones.

In certain special cases (some of) the disconnected diagrams do not contribute to the actual scattering process. For example, in the case of two incoming and two outgoing particles, energy–momentum conservation ensures that (in this case, all of) the disconnected diagrams only contribute when the initial and final states are identical, which is the case only if there is no actual scattering. The scattering cross section (by definition) measures processes in which the initial and final states differ. For this reason the disconnected diagrams do not contribute to the two-particle cross section, as we shall see in the following section.

6.5 Calculation of the scattering cross section

Finally we show how the results of the foregoing sections may be used to calculate the scattering of two incoming particles having momenta q_1, q_2 to a final state of two particles having momenta q_3, q_4. So energy–momentum conservation (6.49) gives

$$q_1+q_2=q_3+q_4 \tag{6.52}$$

and then from (6.51a) the S-matrix element is

$$S_{\mathrm{fi}}=(2\pi)^4\delta(q_1+q_2-q_3-q_4)M_{\mathrm{fi}} \tag{6.53}$$

where M_{fi} is obtained from $\mathscr{G}^{(4)}(q_1, q_2, -q_3, -q_4)$ in the manner described in the previous section. The contributing diagrams are

$M_{\mathrm{fi}}=$

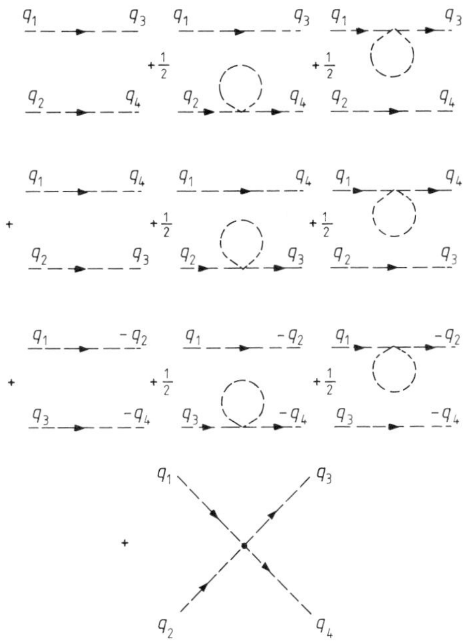

(6.54)

Since q_{10} and q_{20} are both positive, $q_1 \neq -q_2$, and the first six (disconnected) diagrams can only contribute to M_{fi} if $q_1=q_3$ or q_4, which corresponds to no actual scattering. By definition the scattering cross section is only concerned with final states which are different from the initial one. Thus with $q_1 \neq q_3$ or q_4, M_{fi} is given by the last diagram

$$M_{\mathrm{fi}}=-\mathrm{i}\lambda. \tag{6.55}$$

Now S_{fi} gives the transition amplitude, so the probability is

$$p_{\mathrm{fi}}=|S_{\mathrm{fi}}|^2=(2\pi)^8\delta(0)\delta(q_1+q_2-q_3-q_4)\lambda^2. \tag{6.56}$$

The appearance of the (infinite) $\delta(0)$ stems from the fact that we are integrating plane wave functions over an infinite space–time volume. Since

$$(2\pi)^4\delta(P)=\int \mathrm{d}^4x\ \mathrm{e}^{\mathrm{i}Px} \tag{6.57}$$

putting $P=0$ gives

$$(2\pi)^4\delta(0)=\int d^4x=VT \tag{6.58}$$

where V is the (infinite) spatial volume and T the (infinite) extent of time. Dividing p_{fi} by VT therefore gives the transition probability per unit volume per unit time:

$$W_{fi}=(2\pi)^4\delta(q_1+q_2-q_3-q_4)\lambda^2. \tag{6.59}$$

The (infinitesimal) cross section $d\sigma(i\to f)$ is defined by

$$d\sigma(i\to f)=\frac{dN(i\to f)}{F} \tag{6.60}$$

where dN = the number of particles scattered into a particular element $d\Omega$ of solid angle by a *single* target particle, and F = the flux of projectiles incident upon the target. The covariant momentum integration measure $(2\pi)^3(2k_0)^{-1}\,d^3k$ which we are using requires the state normalisation

$$\langle \boldsymbol{k}|\boldsymbol{k}'\rangle=(2\pi)^3 2k_0\,\delta(\boldsymbol{k}-\boldsymbol{k}') \tag{6.61}$$

which corresponds to $2k_0$ particles per unit volume. Thus if we regard particle 1 as the target particle, we have to divide W_{fi} by $2q_{10}$ to obtain the transition probability per unit time for scattering from a *single* target particle. Then

$$dN(i\to f)=\frac{1}{2q_{10}}W_{fi}\frac{d^3q_3}{(2\pi)^3(2q_{30})}\frac{d^3q_4}{(2\pi)^3(2q_{40})}. \tag{6.62}$$

The incident flux F is given by the density of projectiles multiplied by their velocity relative to the target (at least when the target is at rest or moving along the same line as the projectiles). So

$$F=2q_{20}v_{12}$$

where v_{12} is the relative velocity. Putting all this together gives

$$d\sigma(i\to f)=(2\pi)^{-2}\frac{1}{(2q_{10})v_{12}(2q_{20})}\lambda^2\delta(q_1+q_2-q_3-q_4)\frac{d^3q_3\,d^3q_4}{2q_{30}\,2q_{40}}. \tag{6.63}$$

The contribution $q_{10}q_{20}v_{12}$ may be written in a 'covariant' form due to Møller:

$$q_{10}v_{12}q_{20}=[(q_1\cdot q_2)^2-\mu^4]^{1/2} \tag{6.64}$$

provided $\boldsymbol{q}_1$ and $\boldsymbol{q}_2$ are collinear; and we may perform the q_4 integration trivially, using the δ function, if we write

$$\frac{d^3q_4}{2q_{40}}=\delta(q_4^2-\mu^2)\theta(q_{40})\,d^4q_4. \tag{6.65}$$

Thus

$$\mathrm{d}\sigma(\mathrm{i}\to\mathrm{f})=\frac{\lambda^2}{32\pi^2}[(q_1\cdot q_2)^2-\mu^2]^{-1/2}\,\delta[(q_1+q_2-q_3)^2-\mu^2]\frac{\mathrm{d}^3q_3}{q_{30}}. \quad (6.66)$$

From now on it is easier to work in this centre-of-mass frame, in which

$$q_1=(E,\boldsymbol{p}) \quad (6.67\mathrm{a})$$

$$q_2=(E,-\boldsymbol{p}). \quad (6.67\mathrm{b})$$

We write

$$\frac{\mathrm{d}^3q_3}{q_{30}}=q_{30}^{-1}|\boldsymbol{q}_3|^2\,\mathrm{d}|\boldsymbol{q}_3|\,\mathrm{d}\Omega_3=|\boldsymbol{q}_3|\,\mathrm{d}q_{30}\,\mathrm{d}\Omega_3 \quad (6.68)$$

and use the δ function

$$\delta[(q_1+q_2-q_3)^2-\mu^2]=\delta(4E^2-4Eq_{30}) \quad (6.69)$$

to perform the q_{30} integration. Finally we obtain the differential cross section

$$\frac{\mathrm{d}\sigma(\mathrm{i}\to\mathrm{f})}{\mathrm{d}\Omega_3}=\frac{\lambda^2}{64\pi^2s} \quad (6.70\mathrm{a})$$

where

$$s\equiv(q_1+q_2)^2=4E^2 \quad (6.70\mathrm{b})$$

is the (Lorentz invariant) total centre-of-mass energy squared.

Problems

6.1 Derive the $\mathrm{O}(\lambda^2)$ contributions to $W[J]$ in (6.11).

6.2 Derive the $\mathrm{O}(\lambda^2)$ contributions to $X[J]$ in (6.22).

6.3 Derive the $\mathrm{O}(\lambda^2)$ contributions to φ_c in (6.33), and hence determine the $\mathrm{O}(\lambda^2)$ contributions to $\Gamma(\varphi_c)$ in (6.36).

6.4 Draw the $\mathrm{O}(\lambda^2)$ contributions to $\tilde{\Gamma}^{(2)}(p,-p)$ and $\tilde{\Gamma}^{(4)}(p_1,p_2,p_3,p_4)$ and determine their weight factors.

References

The books and review articles which we have found most useful in preparing this chapter are

Iliopoulos J, Itzykson C and Martin A 1975 *Rev. Mod. Phys.* **47** 165

Ramond P 1981 *Field Theory: A Modern Primer* (Reading, Mass.: Benjamin-Cummings) Chapter IV

7

RENORMALISATION OF $\lambda\varphi^4$ THEORY

7.1 Physical motivation for renormalisation

We have already noted that the classical field $\varphi_c(x)$ satisfies an equation (6.35) which, although similar to that derived in classical field theory (3.35), differs from it in one important respect. When the source J is put to zero (6.35) may be cast in the form

$$[\partial_\mu \partial^\mu + \mu^2 + \tfrac{1}{2}\mathrm{i}\lambda\Delta_F(0)]\varphi_c(x) = -\tfrac{1}{6}\lambda[\varphi_c(x)]^3 + \mathrm{O}(\lambda^2) \tag{7.1}$$

where the extra term $\mathrm{i}\lambda\Delta_F(0)\varphi_c$ is a quantum correction arising when the field interacts. Writing it in this way makes it apparent that the interactions have had the effect of shifting ('renormalising') the mass squared from its value μ^2, when there are no interactions, to the 'renormalised' value

$$\mu_R^2 \equiv \mu^2 + \tfrac{1}{2}\mathrm{i}\lambda\Delta_F(0) + \mathrm{O}(\lambda^2). \tag{7.2}$$

Thus the parameter μ^2 which appears in the Lagrangian is only the physical mass squared in the classical (i.e. non-quantum limit). The same is true of the coupling constant λ. As would have become apparent had we calculated the $\mathrm{O}(\lambda^2)$ on the right of (7.1), the interactions have the effect of generating (quantum) corrections which shift the coupling constant from its 'bare' value λ to a renormalised value λ_R. Now the only quantities which we can measure are, of course, the renormalised quantities, since we are not able to turn the interaction on and off at will. For example, we might measure the four-point function $\tilde{\Gamma}^{(4)}(p_1, p_2, p_3, p_4)$ for a particular choice of momenta p_i, and define this to be the renormalised coupling constant λ_R. We could then, in principle, calculate $\tilde{\Gamma}^{(4)}$, or indeed any Green function $\tilde{\Gamma}^{(n)}$, using the Feynman rules, for arbitrary momenta as a function of the (input measured) renormalised parameters λ_R, μ_R^2. Obviously we should expect any perturbative approximations we make to be more successful in predicting $\tilde{\Gamma}^{(n)}$ at (say) GeV values of the momentum variables if we use renormalised parameters defined at (particular) GeV scale momenta, rather than those defined at eV or TeV momentum scales, for example. This notion, that we may vary the energy scale at which we choose to define our renormalised parameters λ_R, μ_R^2, is the key to the renormalisation group equation which will be derived in Chapter 12.

The essence of the above discussion is that the Lagrangian we have been using is given in terms of the 'bare' or 'unrenormalised' quantities, hereinafter denoted φ_B, μ_B^2, λ_B. The objective is to calculate cross sections, and other observable quantities, as functions of the renormalised quantities, which henceforth will be denoted μ^2, λ. Thus the Lagrangian of the scalar field theory with which we have been concerned hitherto is

$$\mathscr{L}_B = \frac{1}{2}(\partial_\mu \varphi_B)(\partial^\mu \varphi_B) - \frac{1}{2}\mu_B^2 \varphi_B^2 - \frac{1}{4!}\lambda_B \varphi_B^4 \qquad (7.3)$$

and the Green functions calculated in Chapter 6 are 'bare' ones, denoted G_B, $\tilde{\Gamma}_B$, etc. The choice of the coefficient $\frac{1}{2}$ of the derivative term amounts to an arbitrary choice of field strength normalisation, and we shall also introduce a renormalised field denoted φ differing from the bare field by an overall multiplicative constant.

We have seen how renormalisation of the parameters μ_B^2, λ_B occurs as a (quantum) effect of the interactions, and why it may be desirable for computational reasons to express any cross sections, for example, which we calculate in terms of suitably chosen renormalised parameters μ^2, λ. However, there is an additional reason why renormalisation is necessary, and not merely desirable. This is because the shifts of μ_B^2, λ_B caused by the interactions are actually infinite, as we shall see shortly. Since any (measured) renormalised parameters are, by definition, finite, it follows that the bare parameters are infinite. Thus the use of renormalised quantities is necessary if we are to avoid the appearance of infinities in our calculations of cross sections etc. This raises the question of whether all infinities may be expunged from the theory by the use of renormalised mass squared μ^2, coupling constant λ, and field φ. In fact, for the theory (7.3) the renormalisation of mass, coupling constant and field *is* sufficient to render the cross sections, Green functions etc of the theory finite. A quantum field theory is said to be *renormalisable* if it is rendered finite by the renormalisation of only the parameters and fields appearing in the bare Lagrangian. Not all field theories are renormalisable, but in this book we shall be concerned only with those which are. The central problem facing particle physicists is whether all interactions occurring in nature may be described by renormalisable quantum field theories. At present we have renormalisable theories which are candidates to describe the strong, weak and electromagnetic interactions, and we shall discuss all of them in the succeeding chapters. To date, there is no such theory which could provide a quantum theory of gravitation.

The main object of this chapter is to show how the theory (7.3) is renormalised. Before embarking on that enterprise let us first verify that the bare theory is indeed infinite, as we have claimed. The infinity associated with mass renormalisatiion occurs in (7.2) because $\Delta_F(0)$ is infinite, which in turn derives from the divergence of the internal momentum integration in the second diagram contributing to $\tilde{\Gamma}_B^{(2)}$ in (6.37):

$\lambda_B \Delta_F(0) =$ 

$$= \lambda_B \int \frac{d^4k}{(2\pi)^4} (k^2 - \mu^2{}_B + i\varepsilon)^{-1}. \tag{7.4a}$$

The iε in the propagator makes the poles in the k_0 integration lie just above and below the real axis of the complex k_0 plane at $\pm[(\boldsymbol{k}^2 + \mu_B^2)^{1/2} - i\varepsilon]$ as shown in figure 7.1.

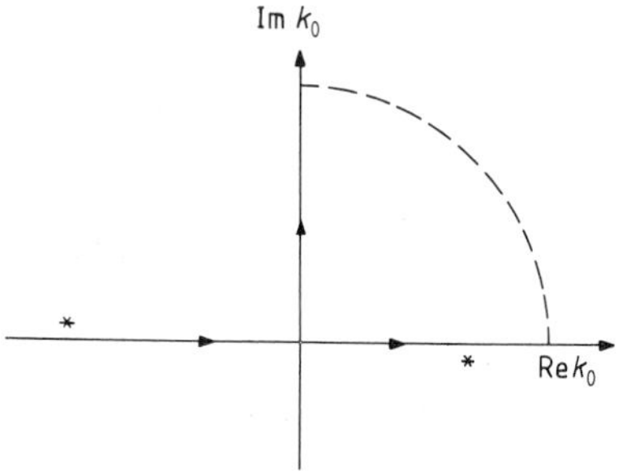

Figure 7.1 Poles at $\pm[(\boldsymbol{k}^2 + \mu_B^2)^{1/2} - i\varepsilon]$.

Since there are no other singularities in the k_0 plane, we may perform the (Wick) rotation of the contour in an *anti*clockwise direction to lie along the imaginary k_0 axis, running from $-i\infty$ to $+i\infty$. Next we change integration variables according to

$$k_0 = i\mathfrak{k}_4$$

$$\boldsymbol{k} = \boldsymbol{\mathfrak{k}}$$

so that the k_4 integration runs from $-\infty$ to $+\infty$. Thus

$$\lambda_B \Delta_F(0) = i\lambda_B \int \frac{d^4\mathfrak{k}}{(2\pi)^4} (-\mathfrak{k}^2 - \mu_B^2)^{-1} \tag{7.4b}$$

where

$$\mathfrak{k}^2 \equiv \boldsymbol{\mathfrak{k}}^2 + \mathfrak{k}_4^2 \tag{7.4c}$$

and $d^4\mathfrak{k}$ is the Euclidean volume element. (Effectively all we have done is to perform the continuation back to the Euclidean form from which we obtained $\Delta_F(x)$ in the first place.) Since $d^4\mathfrak{k}$ is proportional to $|\mathfrak{k}|^3 d|\mathfrak{k}|$, it is clear that the integral diverges for large values of $|\mathfrak{k}|$—we say that it is 'ultraviolet divergent'.

Similarly, had we evaluated $\tilde{\Gamma}_B^{(4)}$ to order λ_B^2, we would have encountered the diagram

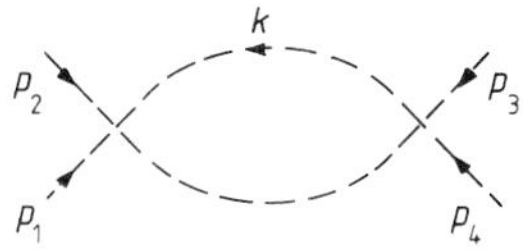

(7.5a)

According to the Feynman rules derived in Chapter 6, this represents the expression

$$\lambda_B^2 \int \frac{d^4k}{(2\pi)^4} (k^2 - \mu_B^2 + i\varepsilon)^{-1} [(p_1 + p_2 + k)^2 - \mu_B^2 + i\varepsilon]^{-1} \tag{7.5b}$$

and it is clear that it too is ultraviolet divergent. For large (Euclidean) values of k each propagator behaves as $|\not{k}|^{-2}$, while the volume element contributes $|\not{k}|^3 d|\not{k}|$. We say that this contribution to $\tilde{\Gamma}_B^{(4)}$ diverges 'logarithmically', because if we cut off the $|\not{k}|$ integration at a value $|\not{k}| = \Lambda$, then the integral (7.5) has a dominant contribution proportional to $\ln \Lambda$. Similarly the integral (7.4), and $\tilde{\Gamma}_B^{(2)}$ are said to be 'quadratically' divergent.

Having considered these two simple cases it is clear how to generalise the argument to identify which Green functions are divergent. The naive superficial 'degree of divergence' D of a diagram is given by

$$D = 4L - 2I \tag{7.6}$$

where L is the number of independent loop momenta, and I is the number of internal lines. This is because each loop momentum k has a volume element d^4k associated with it, while each (scalar) internal line is associated with a propagator which for large $|\not{k}|$ behaves like $|\not{k}|^{-2}$. If $D = 0$ the diagram is logarithmically divergent, while if $D = 2$ the diagram is quadratically divergent. The number of independent loop momenta L is less than the number of internal lines I because of momentum conservation at each vertex. In fact for a connected diagram

$$L = I - V + 1 \tag{7.7}$$

where V is the number of vertices. (Only $V - 1$ of these conservation requirements constrain L because of overall momentum conservation.) Substituting into (7.6) gives

$$D = 2I - 4V - 4. \tag{7.8}$$

The combination $4V - 2I$ is just the number of external lines E, since each vertex has four lines emerging and each internal line removes two of these. Thus

$$E = 4V - 2I. \tag{7.9}$$

Hence

$$D=4-E \tag{7.10}$$

and we see that the degree of divergence is independent of how many vertices there are in the diagram. Since in this theory E must be even, the only Green functions which are superficially divergent are $\tilde{\Gamma}^{(0)}$, $\tilde{\Gamma}^{(2)}$ and $\tilde{\Gamma}^{(4)}$.

This does not of course prove that the diagrams with $E>4$ are convergent—this is why we called D the 'superficial' degree of divergence. In fact, it is rather easy to see that there are many contributions to Green functions with more than four external lines which are indeed divergent. Consider the following two-loop contribution to $\tilde{\Gamma}^{(6)}$:

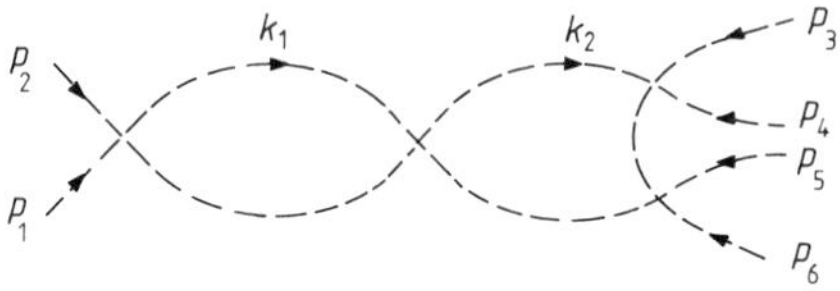

Although it has six external lines ($E=6$), it is evidently divergent, because the integration over the loop momentum k_1 diverges. But this divergence of the k_1 sub-integration, while k_2 is fixed, is just the divergence already encountered in $\tilde{\Gamma}_B^{(4)}$ and written down in (7.5). So this divergence is not a 'new' divergence as it stems from one which we might have anticipated, since $\tilde{\Gamma}_B^{(4)}$ has $D=0$. Clearly there will be a divergent contribution to every Green function arising (at least) whenever in a particular diagram we make the replacement:

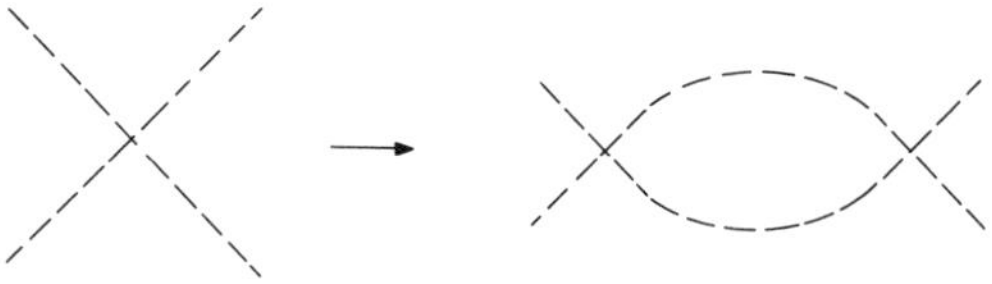

What we *can* say is that if the overall degree of divergence D of a diagram is negative, and if the degree of divergence of all of its subgraphs is also negative, then the Feynman diagram is convergent. This is Weinberg's theorem[1]. It holds for any field theory, not just the $\lambda\varphi^4$ field theory with which we are concerned in this chapter. The above conditions are sufficient for the convergence of a diagram, but not always necessary; it may happen that some invariance of the theory makes a diagram more convergent than the degree of divergence would lead one to suppose. This does not happen in the $\lambda\varphi^4$ theory, but it does occur in QED, for example, because of gauge invariance or charge conjugation invariance. If the theory is to be renormalisable, the only divergences which occur are those which arise from mass, coupling constant and field renormalisations.

7.2 Dimensional regularisation

To study the renormalisability of the $\lambda\varphi^4$ theory which we are considering we must be able to manipulate the divergences which, we have seen, certainly occur. Thus we are required to devise some procedure which renders the divergent momentum integrations finite but leaves the convergent diagrams unaffected. The simplest method is to cut off the radial Euclidean momentum integration at $|\mathscr{k}|=\Lambda$, as discussed in connection with (7.5). If Λ is large enough the convergent diagrams receive a negligible contribution from the region with $|\mathscr{k}|$ larger than Λ, so they will be unaffected, while diagrams which are quadratically or logarithmically divergent will have dominant terms proportional to Λ^2 or $\ln\Lambda$ respectively. Alternatively, and until 1972 commonly, we may use a 'covariant cut-off'. In a loop integration over k we insert one (or more) factors

$$\frac{-\Lambda^2}{k^2-\Lambda^2+\mathrm{i}\varepsilon} \tag{7.11}$$

where, again, Λ is large. Then for values of $|\mathscr{k}|$ small compared with Λ the integrand is unaffected. Since convergent diagrams receive a negligible contribution from the region where $|\mathscr{k}|$ is large, they are again unaffected by the cut-off procedure. On the other hand, diagrams which were divergent are now convergent because of the extra $|\mathscr{k}|^{-2}$ supplied by the factor for $|\mathscr{k}|\gg\Lambda$. However, both of the above methods have a drawback if they are used in the gauge theories which are the leading candidates to describe the known interactions. They both destroy the local gauge invariance from which the theories were deduced. Thus although they may be used quite satisfactorily in the renormalisation of the $\lambda\varphi^4$ theory, they are inadequate for the theories with which we are ultimately concerned.

The insertion of the cut-off factor (7.11) into the integrand of (7.5), for example, makes the integral finite ('regularises' it) by increasing the powers of $|\mathscr{k}|$ in the denominator of the integrand. Another way to regularise the integral is to decrease the powers of $|\mathscr{k}|$ deriving from the (Euclidean) volume element. In (7.5) we have d^4k because of the four-dimensional space–time continuum. To decrease the contribution from the volume element thus requires us to work in a space–time continuum of 2ω dimensions with

$$\omega<2. \tag{7.12}$$

So in the method of 'dimensional regularisation' we consider the whole theory in 2ω-dimensional space–time[2]. This makes the Green functions depend upon ω. Those corresponding to divergent Green functions in four dimensions typically have poles in $\omega-2$. Consider, for example, the divergent integral

$$I_4(\mu_{\mathrm{B}})=\int\frac{\mathrm{d}^4k}{(2\pi)^4}(k^2-\mu_{\mathrm{B}}^2+\mathrm{i}\varepsilon)^{-1} \tag{7.13}$$

contributing to $\tilde{\Gamma}_B^{(2)}$ and appearing in (7.4). In the dimensionally regularised theory we evaluate instead

$$I(\omega, \mu_B) = \int \frac{d^{2\omega}k}{(2\pi)^{2\omega}} (k^2 - \mu_B^2 + i\varepsilon)^{-1}$$

$$= -i \int \frac{d^{2\omega}\not{k}}{(2\pi)^{2\omega}} (\not{k}^2 + \mu_B^2)^{-1} \tag{7.14}$$

performing the Euclidean continuation/Wick rotation as before. Since the integral is spherically symmetric we may write

$$d^{2\omega}\not{k} = 2\omega V(2\omega)|\not{k}|^{2\omega-1}\, d|\not{k}| \tag{7.15a}$$

where

$$V(2\omega) = \frac{\pi^\omega}{\Gamma(\omega+1)} \tag{7.15b}$$

is the volume of the 2ω-dimensional unit sphere. The radial integration may be cast into standard form by the substitution

$$|\not{k}|^2 = \mu_B^2 x \tag{7.16}$$

and for $\omega < 1$ it reduces to a Beta function. Then

$$I(\omega, \mu_B) = -i\mu_B^{2\omega-2}(4\pi)^{-\omega}\Gamma(1-\omega) \tag{7.17}$$

for $\omega < 1$. This expression may be used to define $I(\omega, \mu_B)$, by analytic continuation in ω, in regions where the original definition (7.14) does not exist. The $\Gamma(1-\omega)$ has poles at $\omega = 1, 2, 3, \ldots$ and we are obviously interested in the neighbourhood of the pole at $\omega = 2$. We may not expand the factor $\mu_B^{2\omega-2}$ in powers of $\omega - 2$, as it stands, because it is dimensionful. To be general we express the dimensions of $I(\omega, \mu_B)$ in terms of an arbitrary mass scale M, which may be chosen according to convenience, taste, or, more likely, the renormalisation scheme which is adopted (see §7.5). Thus we write

$$\mu_B^{2\omega-2} = \mu_B^2 (M^2)^{\omega-2} \left(\frac{\mu_B^2}{M^2}\right)^{\omega-2}$$

and expand only the last (dimensionless) factor. This gives

$$I(\omega, \mu_B) = \frac{i\mu_B^2}{16\pi^2}(M^2)^{\omega-2}\left(\frac{1}{2-\omega} + \Gamma'(1) + 1 - \ln\frac{\mu_B^2}{4\pi M^2} + O(\omega-2)\right) \tag{7.18}$$

showing that $I(\omega, \mu_B)$ has the anticipated pole at $\omega = 2$, together with a piece which remains finite as $\omega \to 2$.

A similar treatment may be applied to other integrals. By differentiating (7.14) with respect to μ_B^2, or else directly (for non-integral n), we can show that

$$\int \frac{d^{2\omega}k}{(2\pi)^{2\omega}} (k^2 - \mu_B^2 + i\varepsilon)^{-n} = i(-1)^n \frac{\mu_B^{2\omega-2n}}{(4\pi)^\omega} \frac{\Gamma(n-\omega)}{\Gamma(n)}. \tag{7.19}$$

If $n > 2$ the right-hand side is regular at $\omega = 2$ and the left-hand side is also convergent in four dimensions. Thus dimensional regularisation has the desired property of regularising the divergent integrals while leaving convergent integrals unaffected when $\omega \to 2$.

For future reference we shall need some further integrals which may be derived easily from (7.19). By changing the integration variable to

$$k' = k + p$$

it is clear that

$$\int \frac{d^{2\omega}k}{(2\pi)^{2\omega}} (k^2 + 2p\cdot k - \mu_B^2 + i\varepsilon)^{-n} = i(-1)^n \frac{\Gamma(n-\omega)}{\Gamma(n)} \frac{(\mu_B^2 + p^2)^{\omega-n}}{(4\pi)^\omega}. \tag{7.20}$$

Differentiating with respect to p_μ, $p_\nu \ldots$ yields more formulae:

$$\int \frac{d^{2\omega}k}{(2\pi)^{2\omega}} k_\mu (k^2 + 2p\cdot k - \mu_B^2 + i\varepsilon)^{-n} = i(-1)^n \frac{\Gamma(n-\omega)}{\Gamma(n)} \frac{(\mu_B^2 + p^2)^{\omega-n}}{(4\pi)^\omega} (-p_\mu) \tag{7.21a}$$

$$\int \frac{d^{2\omega}k}{(2\pi)^{2\omega}} k_\mu k_\nu (k^2 + 2p\cdot k - \mu_B^2 + i\varepsilon)^{-n}$$

$$= i(-1)^n \frac{\Gamma(n-\omega-1)}{\Gamma(n)} \frac{(\mu_B^2 + p^2)^{\omega-n}}{(4\pi)^\omega} [p_\mu p_\nu (n-\omega-1) - \tfrac{1}{2} g_{\mu\nu} (\mu_B^2 + p^2)]. \tag{7.21b}$$

Contracting both sides with $g^{\mu\nu}$, and remembering

$$g^{\mu\nu} g_{\mu\nu} = 2\omega \tag{7.22}$$

gives

$$\int \frac{d^{2\omega}k}{(2\pi)^{2\omega}} k^2 (k^2 + 2p\cdot k - \mu_B^2 + i\varepsilon)^{-n}$$

$$= i(-1)^n \frac{\Gamma(n-\omega-1)}{\Gamma(n)} \frac{(\mu_B^2 + p^2)^{\omega-n}}{(4\pi)^\omega} [(n - 2\omega - 1)p^2 - \omega \mu_B^2]. \tag{7.23}$$

7.3 Evaluation of Feynman integrals

In the previous section we showed how to calculate the Feynman diagram (7.4a) which contributes to $\tilde{\Gamma}_B^{(2)}$. All other single-loop diagrams involve more than one propagator, and additional technology is needed to evaluate them.

For example, we shall need to calculate the diagram (7.5a) which contributes to $\tilde{\Gamma}_B^{(4)}$

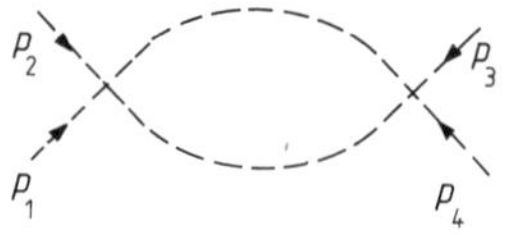

(7.5a)

In 2ω-dimensions this is proportional to the integral

$$J \equiv \int \frac{d^{2\omega}k}{(2\pi)^{2\omega}} (k^2 - \mu_B^2 + i\varepsilon)^{-1} [(P+k)^2 - \mu_B^2 + i\varepsilon]^{-1} \tag{7.24a}$$

where

$$P \equiv p_1 + p_2 = -p_3 - p_4 \tag{7.24b}$$

The technique we follow is that proposed by Feynman[3] which combines the denominators in (7.24a) into a single quadratic using the identity

$$\frac{1}{ab} = \int_0^1 dx [ax + b(1-x)]^{-2}. \tag{7.25}$$

Taking

$$a = (P+k)^2 - \mu_B^2 + i\varepsilon \tag{7.26a}$$

$$b = k^2 - \mu_B^2 + i\varepsilon \tag{7.26b}$$

and substituting into (7.24a) gives

$$J = \int \frac{d^{2\omega}k}{(2\pi)^{2\omega}} \int_0^1 dx [(k+Px)^2 + P^2 x(1-x) - \mu_B^2 + i\varepsilon]^{-2}. \tag{7.27}$$

Next we interchange the loop momentum (k) integration and the 'Feynman parameter' (x) integration. For $\omega < 2$ the k integration is convergent and we may legally shift the integration variable by the substitution

$$k' = k + Px. \tag{7.28}$$

Then the k integration is performed easily, using (7.19), to give

$$J = \int_0^1 dx \frac{i}{(4\pi)^\omega} \frac{\Gamma(2-\omega)}{\Gamma(2)} [\mu_B^2 - P^2 x(1-x)]^{\omega-2}. \tag{7.29}$$

As before, we expand in the neighbourhood of the pole at $\omega = 2$ using the mass M to carry the overall dimensions. This gives

$$J = \frac{i}{16\pi^2} (M^2)^{\omega-2} \int_0^1 dx \left[\frac{1}{2-\omega} + \Gamma'(1) - \ln \frac{\mu_B^2}{4\pi M^2} - \ln\left(1 - \frac{P^2}{\mu_B^2} x(1-x)\right) + O(\omega-2) \right. \tag{7.30}$$

The maximum value of $x(1-x)$ is 1/4, so provided $P^2 < 4\mu_B^2$ the argument of the logarithm is always positive. Then it is easy to perform the final integration, and we obtain finally:

$$J = \frac{i}{16\pi^2}(M^2)^{\omega-2}\left[\frac{1}{2-\omega} + \Gamma'(1) - \ln\frac{\mu_B^2}{4\pi M^2} - 2\left(\frac{4\mu_B^2 - P^2}{P^2}\right)^{1/2}\tan^{-1}\left(\frac{P^2}{4\mu_B^2 - P^2}\right) + 2\right] + O(\omega - 2). \quad (7.31)$$

We leave it as an exercise (problem 7.2) to find the appropriate expression for $P^2 > 4\mu_B^2$.

The loop integral we have just considered has the property that after the ultraviolet divergence has been regulated by dimensional regularisation, the remaining Feynman parameter integration is convergent when $\omega \to 2$. This is always the case in any diagrams encountered in $\lambda\varphi^4$ theory. However, it is not a property which is invariably true in the gauge theories which are our principal concern. The gauge fields characteristic of these theories are associated with particles having zero mass; for example, the photon, which is the gauge field of quantum electrodynamics, has zero mass. The presence of zero-mass particles generates additional divergences in the Feynman integrals, which stem from the small k (infrared) behaviour of the integrand. This can be illustrated in the $\lambda\varphi^4$ theory when we set $\mu_B = 0$. If we consider (7.24a), for example, and set $\omega = 2$ (and $\mu_B = 0$), then the integral is ultraviolet divergent for all P^2. It is *also* infrared divergent when $P^2 = 0$. The ultraviolet divergence is the origin of the $\Gamma(2-\omega)$ in the dimensionally regularised integral (7.29), while the infrared divergence is apparent in the remaining factor of the integrand; since $\mu_B^2 = 0$, the remaining factor is proportional to $(P^2)^{\omega-2}$ which is divergent (when $\omega < 2$) as $P^2 \to 0$. These infrared divergences are less serious than the ultraviolet ones, since they do not contribute to observable quantities[4], at least in quantum electrodynamics. (The situation is not so clear-cut in quantum chromodynamics[5], but we shall not pursue this topic in this text.) Nevertheless it is desirable that they are regulated, since, if they are not, certain (on-shell) renormalisation schemes are not well defined. One option is to introduce a mass λ for the massless particle. (In our illustration this would mean restoring $\mu_B \neq 0$.) Then everything is well defined and, since observables are independent of λ, the limit $\lambda \to 0$ can be taken at the end with impunity. This is a less fashionable option than it used to be, since in gauge theories a massive gauge field is often forbidden by gauge invariance and certain (Ward) identities will no longer be valid. The preferred option these days is to maintain $\omega \neq 2$ in the Feynman parameter integration. After the ultraviolet divergences have been removed by renormalisation (in a manner to be detailed in the following section), we are free to continue from $\omega < 2$ to $\omega > 2$, which regulates the remaining infrared divergences[6]; after the parameter integration they appear, for certain values of the external momenta, as poles in $\omega - 2$.

Integrations involving more propagators and/or more loops require more general identities than (7.25). These follow from an extension of (7.25):

$$\frac{1}{a^\alpha b^\beta}=\frac{\Gamma(\alpha+\beta)}{\Gamma(\alpha)\Gamma(\beta)}\int_0^1 \mathrm{d}x\,\frac{x^{\alpha-1}(1-x)^{\beta-1}}{[ax+b(1-x)]^{\alpha+\beta}} \qquad (\alpha,\beta>0). \quad (7.32)$$

This follows from a standard form of the Beta function after we change the integration variable to $t=(1-x)/x$. Using this we can prove

$$\frac{1}{a_1\ldots a_n}=\Gamma(n)\int_0^1 \mathrm{d}x_1\int_0^{x_1}\mathrm{d}x_2\ldots\int_0^{x_{n-2}}\mathrm{d}x_{n-1}$$

$$\times[a_1(1-x_1)+a_2(x_1-x_2)+\ldots+a_nx_{n-1}]^{-n} \quad (7.33)$$

by induction. For, assuming that (7.33) is true for some n, it follows from (7.32) with $\alpha=n$ and $\beta=1$ that

$$\frac{1}{a_1(a_2\ldots a_n)}=\Gamma(n+1)\int_0^1\mathrm{d}y\int_0^1\mathrm{d}x_1\int_0^{x_1}\mathrm{d}x_2\ldots$$

$$\int_0^{x_{n-2}}\mathrm{d}x_{n-1}y^{n-1}[a_1(1-y)+D_y]^{-(n+1)} \quad (7.34a)$$

where

$$D\equiv a_2(1-x_1)+a_3(x_1-x_2)+\ldots+a_{n+1}x_{n-1}. \quad (7.34b)$$

Now change integration variables to

$$y_1=y \quad (7.35a)$$

$$y_{i+1}=yx_i \qquad (i=1,\ldots,n-1). \quad (7.35b)$$

The Jacobian gives

$$\mathrm{d}y_1\ldots\mathrm{d}y_n=y^{n-1}\,\mathrm{d}y\,\mathrm{d}x_1\ldots\mathrm{d}x_{n-1} \quad (7.36)$$

and the integration region is clearly

$$1>y_1>y_2>\ldots>y_{n-1}>y_n>0. \quad (7.37)$$

Thus

$$\frac{1}{a_1a_2\ldots a_{n+1}}=\Gamma(n+1)\int_0^1\mathrm{d}y_1\int_0^{y_1}\mathrm{d}y_2\ldots$$

$$\int_0^{y_{n-1}}\mathrm{d}y_n[a_1(1-y_1)+a_2(y_1-y_2)+\ldots+a_{n+1}y_n]^{-(n+1)} \quad (7.38)$$

which is just (7.33) with n replaced by $n+1$. The result for $n=2$ is just (7.25), which is trivially verified. A similar treatment generalises (7.32) to

$$\frac{1}{a_1^{\alpha_1}a_2^{\alpha_2}\dots a_n^{\alpha_n}}=\frac{\Gamma(\alpha_1+\dots+\alpha_n)}{\Gamma(\alpha_1)\dots\Gamma(\alpha_n)}\int_0^1 dx_1\int_0^{x_1}dx_2\dots\int_0^{x_{n-2}}dx_{n-1}$$
$$\times(1-x_1)^{\alpha_1-1}(x_1-x_2)^{\alpha_2-1}\dots x_{n-1}^{\alpha_n-1}$$
$$\times[a_1(1-x_1)+a_2(x_1-x_2)+\dots+a_nx_{n-1}]^{-(\alpha_1+\dots+\alpha_n)}. \tag{7.39}$$

We leave the proof as an exercise (Problem 7.3). This may be used, together with the identities derived in the previous section, to prove the general results contained in Appendix A.

7.4 Renormalisation of $\lambda\varphi^4$ theory at one-loop order

We have seen in §7.2 how the use of dimensional regularisation enables us to define our Green functions in 2ω dimensions, and that, when $\omega\to 2$, $\tilde{\Gamma}_B^{(2)}$ and $\tilde{\Gamma}_B^{(4)}$ diverge. In this section we shall show how renormalisation, which is necessary in any interacting field theory, enables us to remove the infinites by absorbing them into the renormalisation constants.

We start from the Lagrangian $\mathscr{L}_B$ given in (7.3) and define a renormalised field $\varphi(x)$ by

$$\varphi_B(x)=Z^{1/2}\varphi(x) \tag{7.40}$$

where the 'wave function renormalisation constant' Z differs from unity because of quantum corrections. Thus we write

$$Z=1+\delta Z. \tag{7.41}$$

Similarly quantum effects lead to mass and coupling constant renormalisation because of the interactions. We define the renormalised mass μ by

$$Z\mu_B^2=\mu^2+\delta\mu^2 \tag{7.42}$$

where $\delta\mu^2$ is a quantum effect, and the renormalised coupling constant λ by

$$Z^2\lambda_B=\lambda+\delta\lambda \tag{7.43}$$

with $\delta\lambda$ arising from quantum effects. In terms of these parameters

$$\mathscr{L}_B=\mathscr{L}+\delta\mathscr{L} \tag{7.44a}$$

where

$$\mathscr{L}=\frac{1}{2}(\partial_\mu\varphi)(\partial^\mu\varphi)-\frac{1}{2}\mu^2\varphi^2-\frac{1}{4!}\lambda\varphi^4 \tag{7.44b}$$

and the 'counter term' Lagrangian

$$\delta\mathscr{L}=\frac{1}{2}\delta Z(\partial^\mu\varphi)(\partial^\mu\varphi)-\frac{1}{2}\delta\mu^2\varphi^2-\frac{1}{4!}\delta\lambda\varphi^4. \tag{7.44c}$$

The Feynman rules deriving from $\mathscr{L}$ are precisely those written down in §6.2. However the additional piece $\delta\mathscr{L}$ generates extra vertices. To see this we note that

$$\mathscr{L}+\delta\mathscr{L}=\mathscr{L}_0+\mathscr{L}_{\mathrm{I}} \tag{7.45}$$

where $\mathscr{L}_0$ is given in (6.2) and now the interaction is given by

$$\mathscr{L}_{\mathrm{I}}=-\frac{1}{4!}\lambda\varphi^4+\delta\mathscr{L}. \tag{7.46}$$

Then following the analysis given in §§6.1 and 6.2 for this interaction rather than that given in (6.8), we find that the Feynman rules are those given in §6.2 augmented by the following additional rules:

4 There is a vertex involving two lines. Momentum is conserved and the vertex is associated with a factor $\mathrm{i}(\delta Z p^2-\delta\mu^2)$:

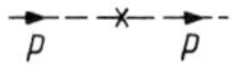

$$: \quad \mathrm{i}(\delta Z p^2-\delta\mu^2). \tag{7.47}$$

5 There is an additional vertex involving four lines, conserving momenta, which is associated with a factor $-\mathrm{i}\delta\lambda$

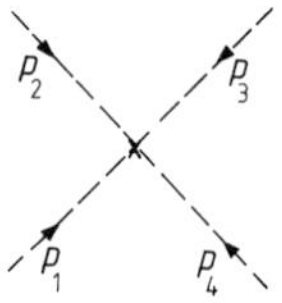

$$: \quad -\mathrm{i}\delta\lambda \qquad (p_1+p_2+p_3+p_4=0). \tag{7.48}$$

Rule 5 in any case follows immediately from rule 2, and we can check that rule 4 is correct by setting $\lambda=\delta\lambda=0$ temporarily. Then we have a non-interacting field theory in which clearly

$$\mathscr{G}_0^{(2)}(p,-p)=\mathrm{i}[p^2(1+\delta Z)-(\mu^2+\delta\mu^2)+\mathrm{i}\varepsilon]^{-1}. \tag{7.49}$$

Expanding the right-hand side to lowest order in δZ and $\delta\mu^2$ gives

$$\mathscr{G}_0^{(2)}(p,-p)=\frac{\mathrm{i}}{p^2-\mu^2+\mathrm{i}\varepsilon}+\frac{\mathrm{i}}{p^2-\mu^2+\mathrm{i}\varepsilon}\mathrm{i}(\delta Z p^2-\delta\mu^2)\frac{\mathrm{i}}{p^2-\mu^2+\mathrm{i}\varepsilon}+\dots$$

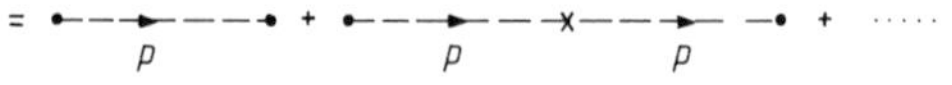

(7.50)

as required by rule 4. The quantities $\delta\mu^2$ and δZ are non-zero because of quantum effects due to the interactions. Thus we may expand them as power series in λ with leading terms of order λ:

$$\delta\mu^2=\sum_{i=1}^{\infty}\delta\mu_i^2 \tag{7.51a}$$

$$\delta Z = \sum_{i=1}^{\infty} \delta Z_i \tag{7.51b}$$

where $\delta\mu_i^2$ and δZ_i are proportional to λ^i. Similarly since $\delta\lambda/\lambda$ is a quantum effect caused by the interactions, we may expand $\delta\lambda$ as a power series starting with a term proportional to λ^2:

$$\delta\lambda = \sum_{i=2}^{\infty} \delta\lambda_i . \tag{7.52}$$

We may now calculate the (OPI) Green functions of the renormalised theory $\tilde{\Gamma}^{(n)}$, as functions of the external momenta, the renormalised parameters μ^2, λ, and the counter term parameters δZ, $\delta\mu^2$, $\delta\lambda$. For example $\tilde{\Gamma}^{(2)}$ is given in leading order by

$i\tilde{\Gamma}^{(2)}(p, -p) =$

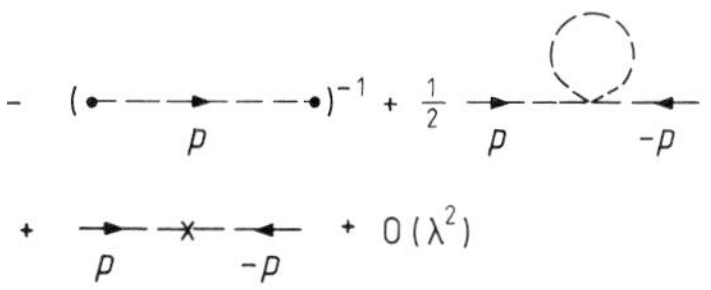

(7.53)

The first two diagrams are given in (6.37) for the bare theory (μ^2, λ were subsequently relabelled μ_B^2, λ_B), while the additional term derives from the counter term Lagrangian. Hence

$$\tilde{\Gamma}^{(2)}(p, -p) = p^2(1 + \delta Z_1) - (\mu^2 + \tfrac{1}{2}\lambda i\Delta_F(0) + \delta\mu_1^2) + O(\lambda^2). \tag{7.54}$$

The quantity $i\Delta_F(0)$ is given in our dimensionally regularised theory by

$$\lambda\Delta_F(0) = \lambda M^{2\omega-4} \frac{i\mu^2}{16\pi^2}\left(\frac{1}{2-\omega} + \Gamma'(1) + 1 - \ln\frac{\mu^2}{4\pi M^2} + O(\omega - 2)\right) \tag{7.55}$$

using (7.13) and (7.18). It looks as though our dimensions have gone awry. However it must be borne in mind that λ is only dimensionless in four dimensions. Since the action

$$S = \int d^{2\omega}x \mathscr{L} \tag{7.56}$$

is dimensionless, the field $\varphi(x)$ and the coupling constant λ have (mass) dimensions given by

$$[\varphi] = M^{\omega-1} \tag{7.57a}$$

$$[\lambda] = M^{4-2\omega}. \tag{7.57b}$$

So $\lambda M^{2\omega-4}$ is dimensionless and we define

$$\lambda M^{2\omega-4} \equiv \hat{\lambda} \tag{7.58}$$

where $\hat{\lambda}$ depends implicitly on the mass scale M. Thus

$$\tilde{\Gamma}^{(2)}(p, -p) = p^2(1+\delta Z_1) - \mu^2 - \delta\mu_1^2$$
$$+\frac{\hat{\lambda}\mu^2}{32\pi^2}\left(\frac{1}{2-\omega} + \Gamma'(1) + 1 - \ln\frac{\mu^2}{4\pi M^2} + \mathrm{O}(\omega-2)\right). \qquad (7.59)$$

The quantities $\delta\mu_1^2$ and δZ_1 are fixed by a 'renormalisation scheme', which is essentially a boundary condition on $\tilde{\Gamma}^{(2)}$. We shall discuss the most commonly used schemes in the next section. However, the essential point, shared by *all* schemes, is that, since $\tilde{\Gamma}^{(2)}$ is an observable quantity (at least in principle), it must be *finite* in four dimensions. So as $\omega \to 2$

$$\delta\mu_1^2 - \frac{\hat{\lambda}\mu^2}{32\pi^2}\frac{1}{2-\omega} \to \text{constant} \qquad (7.60a)$$

and

$$\delta Z_1 \to \text{constant}. \qquad (7.60b)$$

Clearly, since the constants have not (yet) been fixed, the finite parts of $\delta\mu_1^2$ and δZ_1 are unconstrained.

Similarly we may evaluate $\tilde{\Gamma}^{(4)}$ to order λ^2:

$\mathrm{i}\tilde{\Gamma}^{(4)}(p_1, p_2, p_3, p_4) =$

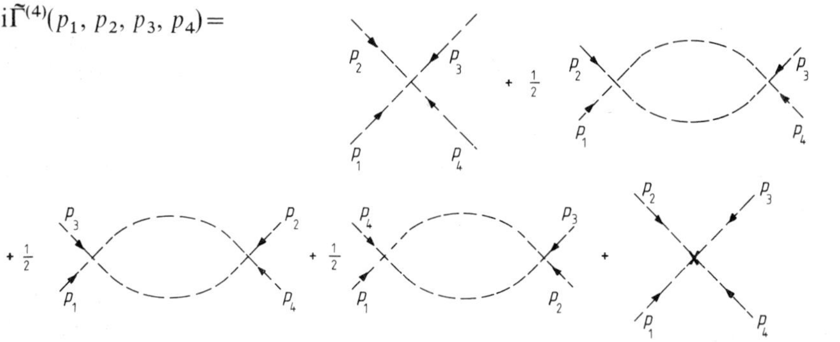

(7.61)

The integration needed to calculate the second, third and fourth diagrams was performed in §7.3 and the result given in (7.31). Then

$$\tilde{\Gamma}^{(4)}(p_1, p_2, p_3, p_4) = -\lambda + \frac{\lambda^2 M^{2\omega-4}}{32\pi^2}\left(\frac{3}{2-\omega} + 3\Gamma'(1) - 3\ln\frac{\mu^2}{4\pi M^2}\right.$$
$$\left. + A(s, \mu^2) + A(t, \mu^2) + A(u, \mu^2)\right) - \delta\lambda_2 + \mathrm{O}(\lambda^3) \qquad (7.62a)$$

where

$$A(P^2, \mu^2) \equiv 2 - 2\left(\frac{4\mu^2}{P^2} - 1\right)^{1/2} \tan^{-1}\left(\frac{4\mu^2}{P^2} - 1\right)^{-1/2} \qquad (7.62b)$$

and

$$s=(p_1+p_2)^2 \qquad t=(p_1+p_3)^2 \qquad u=(p_1+p_4)^2. \tag{7.62c}$$

Since $\tilde{\Gamma}^{(4)}$ is finite in any renormalisation scheme

$$\frac{3\lambda\hat{\lambda}}{32\pi^2}\frac{1}{2-\omega}-\delta\lambda_2 \rightarrow \text{constant} \tag{7.63}$$

as $\omega \rightarrow 2$. As before, the finite part of $\delta\lambda_2$ is arbitrary. It too is fixed by the particular renormalisation scheme which is adopted. Evidently different schemes correspond to different choices of the finite parts of the counter terms, and therefore to different choice of the renormalised parameters.

We saw in §7.1 that the only bare OPI Green functions which are divergent (in one loop order) are $\tilde{\Gamma}_B^{(2)}$ and $\tilde{\Gamma}_B^{(4)}$, so it is of some interest to relate these to the renormalised Green functions $\tilde{\Gamma}^{(2)}$ and $\tilde{\Gamma}^{(4)}$, which we have shown in (7.60) and (7.63) are finite in one-loop order. To do this we note that (7.40) and (7.44) imply that

$$\mathscr{L}+\delta\mathscr{L}+J(x)\varphi(x)=\mathscr{L}_B+J_B(x)\varphi_B(x) \tag{7.64a}$$

where

$$J_B(x)=Z^{-1/2}J(x). \tag{7.64b}$$

We may use this to relate the generating functionals of the bare and renormalised theories. We denote by $W_B[J_B]$ the generating functional defined in (4.25) where now $\mathscr{L}$ is $\mathscr{L}_B$, $J(x)$ is $J_B(x)$ and the functional integration variable is $\varphi_B(x)$. We *now* denote by $W[J]$ the generating functional when $\mathscr{L}$ is $\mathscr{L}+\delta\mathscr{L}$, J is J and the functional integration variable is $\varphi(x)$. Then (7.64) shows that

$$W[J]=W_B[J_B]=W_B[Z^{-1/2}J]. \tag{7.65}$$

This may be used to relate the Green functions of the bare and renormalised theories. Let us denote by $G^{(n)}$ the Green functions (ordinary or connected) generated by $W[J]$, and by $G_B^{(n)}$ those generated by $W_B[J_B]$. Then, using (7.65) and (4.8), we find that

$$\tilde{G}^{(n)}(p_1,\ldots,p_n)=(Z^{-1/2})^n\tilde{G}_B^{(n)}(p_1,\ldots,p_n) \tag{7.66}$$

gives the relationship between (the Fourier transforms of) the Green functions.

The generating functionals of the OPI Green functions may be related similarly. In an obvious notation, it follows from (7.65) that

$$X[J]=X_B[J_B] \tag{7.67a}$$

so that

$$\varphi_c(x)\equiv\frac{\delta X[J]}{\delta J(x)}=Z^{-1/2}\frac{\delta X_B}{\delta J_B(x)}=Z^{-1/2}\varphi_{cB}(x) \tag{7.67b}$$

and hence that

$$\Gamma[\varphi_c] = \Gamma_B[\varphi_{cB}] = \Gamma_B[Z^{1/2}\varphi_c]. \tag{7.67c}$$

Using (4.73) and (4.75) it follows that the relation between the Green functions is

$$\tilde{\Gamma}^{(n)}(p_1, \ldots, p_n) = Z^{n/2}\tilde{\Gamma}_B^{(n)}(p_1, \ldots, p_n). \tag{7.68}$$

For $n=2$ the right-hand side is

$$Z\left[p^2 - \mu_B^2 + \frac{\hat{\lambda}_B\mu_B^2}{32\pi^2}\left(\frac{1}{2-\omega} + \Gamma'(1) + 1 - \ln\frac{\mu_B^2}{4\pi M^2} + O(\omega-2)\right) + O(\lambda_B^2)\right] \tag{7.69}$$

which is easily seen to equal $\tilde{\Gamma}^{(2)}$, as calculated in (7.59), using (7.42) and remembering that δZ is of order λ.

Thus the divergence in $\tilde{\Gamma}_B^{(2)}$ has been absorbed into the renormalisation constants, and the same is true of $\tilde{\Gamma}^{(4)}$. We have thereby verified that $\lambda\varphi^4$ theory *is* renormalisable at one loop order, as claimed. The proof that this is true in all orders is more difficult, and therefore beyond the scope of this book. The interested reader is referred to reference 7.

7.5 Renormalisation schemes

The precise way in which the parameters $\delta\lambda$, $\delta\mu^2$ and δZ of the counter term Lagrangian are fixed is called a 'renormalisation scheme'. The simplest such scheme, which is also particularly suited to gauge theories, is one which emerges naturally from the dimensional regularisation which we are using. It is called the 'minimal subtraction' (MS) scheme[8]. In it the counter terms remove just the divergence and no more.

Let us state this more precisely. With dimensional regularisation the bare Green functions develop poles and higher order singularities in $\omega-2$, as we have seen. In *any* renormalisation scheme these singularities are removed by the counter terms, which have the form

$$\delta\lambda = M^{4-2\omega}\left(a_0(\hat{\lambda}, M/\mu, \omega) + \sum_{\nu=1}^{\infty}\frac{a_\nu(\hat{\lambda}, M/\mu)}{(2-\omega)^\nu}\right) \tag{7.70a}$$

$$\delta\mu^2 = \mu^2\left(b_0(\hat{\lambda}, M/\mu, \omega) + \sum_{\nu=1}^{\infty}\frac{b_\nu(\hat{\lambda}, M/\mu)}{(2-\omega)^\nu}\right) \tag{7.70b}$$

$$\delta Z = c_0(\hat{\lambda}, M/\mu, \omega) + \sum_{\nu=1}^{\infty}\frac{c_\nu(\hat{\lambda}, M/\mu)}{(2-\omega)^\nu} \tag{7.70c}$$

where a_0, b_0 and c_0 are regular as $\omega \to 2$. Since a_ν, b_ν and c_ν are dimensionless, they can only depend on the dimensionless parameters M/μ and $\hat{\lambda} \equiv \lambda M^{2\omega-4}$, as indicated. In the MS scheme these counter terms remove only the singularities in $\omega-2$, so

$$a_0^{\mathrm{MS}}=b_0^{\mathrm{MS}}=c_0^{\mathrm{MS}}\equiv 0. \tag{7.71}$$

The beauty of the MS scheme is that the remaining coefficients a_ν, b_ν, c_ν turn out to be mass-independent. That is to say

$$a_\nu^{\mathrm{MS}}(\hat\lambda, M/\mu)=a_\nu(\hat\lambda) \qquad \text{etc.} \tag{7.72}$$

We can verify this in lowest order using the results derived in §7.4. Comparing (7.63) with (7.70a) and using (7.71) we find

$$\delta\lambda^{\mathrm{MS}}=\frac{3\lambda\hat\lambda}{32\pi^2}\frac{1}{2-\omega}+\mathrm{O}(\lambda^3) \tag{7.73a}$$

and

$$a_1^{\mathrm{MS}}=\frac{3\hat\lambda^2}{32\pi^2}+\mathrm{O}(\lambda^3) \tag{7.73b}$$

so that a_1 is mass-independent to order $\hat\lambda^2$, as claimed. Similarly, we find

$$\delta\mu^{2\,\mathrm{MS}}=\frac{\hat\lambda\mu^2}{32\pi^2}\frac{1}{2-\omega}+\mathrm{O}(\lambda^2) \qquad b_1^{\mathrm{MS}}=\frac{\hat\lambda}{32\pi^2}+\mathrm{O}(\lambda^2) \tag{7.74a}$$

$$\delta Z^{\mathrm{MS}}=\mathrm{O}(\lambda^2) \qquad c_1^{\mathrm{MS}}=\mathrm{O}(\lambda^2). \tag{7.74b}$$

This mass-independence of the counter terms permits a ready solution of the 'renormalisation group equation' in this scheme, as we shall see in Chapter 12.

Having fixed the counter terms, the Green functions are now finite unambiguous functions of the renormalised parameters, and we may take the limit $\omega\to 2$. Then in the MS scheme

$$\tilde\Gamma^{(4)}(p_1, p_2, p_3, p_4)=-\lambda+\frac{\lambda^2}{32\pi^2}\left(3\Gamma'(1)-3\ln\frac{\mu^2}{4\pi M^2}\right.$$
$$\left.+A(s,\mu^2)+A(t,\mu^2)+A(u,\mu^2)\right)+\mathrm{O}(\lambda^3) \tag{7.75a}$$

$$\tilde\Gamma^{(2)}(p,-p)=p^2-\mu^2+\frac{\lambda\mu^2}{32\pi^2}\left(\Gamma'(1)+1-\ln\frac{\mu^2}{4\pi M^2}\right)+\mathrm{O}(\lambda^2) \tag{7.75b}$$

using (7.59), (7.62), (7.73a), (7.74). In (7.75) there is an implicit dependence of the parameters λ, μ^2 on the renormalisation scheme.

The MS scheme is just one of a class of mass-independent schemes, in all of which a_ν etc are independent of M/μ, but in which a_0 etc are not necessarily zero, as in (7.71). One such scheme, the $\overline{\mathrm{MS}}$ scheme[9], is clearly related to MS, as the name suggests. It derives from the observation that the factors $\Gamma'(1)$ and $\ln 4\pi$, appearing in (7.19), (7.31) etc, appear as a result of expanding the quantity $(4\pi)^{2-\omega}\Gamma(2-\omega)$ which occurs *naturally* in the dimensionally regularised theory. In the $\overline{\text{MS}}$ scheme these (finite mass-independent) quantities are *also* subtracted by the counter terms. So, for example,

$$\delta\lambda^{\overline{\text{MS}}} = \frac{3\lambda\hat{\lambda}}{32\pi^2}\left(\frac{1}{2-\omega} + \Gamma'(1) + \ln 4\pi\right) + \text{O}(\lambda^3) \tag{7.76}$$

instead of (7.73a); and

$$a_1^{\overline{\text{MS}}} = \frac{3\hat{\lambda}^2}{32\pi^2} + \text{O}(\lambda^3) \tag{7.77a}$$

$$a_0^{\overline{\text{MS}}} = \frac{3\hat{\lambda}^2}{32\pi^2}[\Gamma'(1) + \ln 4\pi] + \text{O}(\lambda^3). \tag{7.77b}$$

In the $\overline{\text{MS}}$ scheme then, taking $\omega \to 2$,

$$\tilde{\Gamma}^{(4)}(p_1, p_2, p_3, p_4) = -\lambda + \frac{\lambda^2}{32\pi^2}\left(-3\ln\frac{\mu^2}{M^2} + A(s, \mu^2) + A(t, \mu^2) + A(u, \mu^2)\right) + \text{O}(\lambda^3) \tag{7.78a}$$

$$\tilde{\Gamma}^{(2)}(p_1, p_2, p_3, p_4) = p^2 - \mu^2 + \frac{\lambda\mu^2}{32\pi^2}\left(1 - \ln\frac{\mu^2}{M^2}\right) + \text{O}(\lambda^2). \tag{7.78b}$$

In the two remaining schemes which we shall mention the counter terms are determined incidentally, by imposing boundary conditions on $\tilde{\Gamma}^{(2)}$ and $\tilde{\Gamma}^{(4)}$. Clearly three conditions are needed to fix the three quantities $\delta\lambda$, $\delta\mu^2$, δZ. In the 'momentum scheme'[10] the boundary conditions are imposed at a point where the external momenta are characterised by a single scale m. The conditions are that at the (Euclidean) point $p^2 = -m^2$

$$\tilde{\Gamma}^{(2)}(p, -p) = -\mu^2 - m^2 \tag{7.79a}$$

and

$$\frac{\partial}{\partial p^2}\tilde{\Gamma}^{(2)}(p, -p) = 1. \tag{7.79b}$$

In other words, the conditions require that

$$\tilde{\Gamma}^{(2)}(p, -p) = p^2 - \mu^2 + \text{O}[(p^2 + m^2)^2]. \tag{7.80}$$

Referring back to (7.59) we see that these require

$$\delta Z^{\text{mom}} = \text{O}(\lambda^2) \tag{7.81}$$

and

$$\delta\mu^{2\,\text{mom}} = \frac{\hat{\lambda}\mu^2}{32\pi^2}\left(\frac{1}{2-\omega} + \Gamma'(1) + 1 - \ln\frac{\mu^2}{4\pi M^2}\right) + \text{O}(\lambda^2). \tag{7.82}$$

$\delta\lambda$ is fixed by a similar boundary condition on $\tilde{\Gamma}^{(4)}$, namely

$$\tilde{\Gamma}^{(4)}(p_1, p_2, p_3, p_4) = -\lambda \tag{7.83}$$

at the 'symmetric point'

$$p_i \cdot p_j = m^2(\tfrac{1}{3} - \tfrac{4}{3}\delta_{ij}) \qquad (i, j = 1, \ldots, 4). \tag{7.84}$$

At this point

$$s = t = u = -\tfrac{4}{3}m^2 \tag{7.85}$$

so from (7.62) we find

$$\delta\lambda^{\text{mom}} = \frac{\lambda\hat{\lambda}}{32\pi^2}\left(\frac{3}{2-\omega} + 3\Gamma'(1) - 3\ln\frac{\mu^2}{4\pi M^2} + 3A(-\tfrac{4}{3}m^2, \mu^2)\right) + \mathrm{O}(\lambda^3). \tag{7.86}$$

Thus in the momentum scheme, taking $\omega \to 2$,

$$\tilde{\Gamma}^{(2)}(p, -p) = p^2 - \mu^2 + \mathrm{O}(\lambda^2) \tag{7.87a}$$

$$\tilde{\Gamma}^{(4)}(p_1, p_2, p_3, p_4) = -\lambda + \frac{\lambda^2}{32\pi^2}[A(s, \mu^2) + A(t, \mu^2) + A(u, \mu^2) - 3A(-\tfrac{4}{3}m^2, \mu^2)] + \mathrm{O}(\lambda^3). \tag{7.87b}$$

In this scheme, therefore, the mass scale M has disappeared from the Green functions, but the counter terms $\delta\mu^2$, $\delta\lambda$ are manifestly *not* mass-independent.

In all of the schemes so far discussed the quantities μ and λ, which are called the 'renormalised mass' and the 'renormalised coupling constant', have been merely the parameters which we chose to characterise the Green functions. In the 'on-shell' or 'physical' scheme, μ and λ are 'the' mass and 'the' coupling constant. That is to say, they have the values which are actually measured. The physical mass is defined as the position of the pole in $\tilde{G}^{(2)}$, or equivalently of the zero in $\tilde{\Gamma}^{(2)}$. Thus the boundary conditions on $\tilde{\Gamma}^{(2)}$ are that at $p^2 = \mu^2$

$$\tilde{\Gamma}^{(2)}(p, -p) = 0 \tag{7.88a}$$

$$\frac{\partial}{\partial p^2}\tilde{\Gamma}^{(2)}(p, -p) = 1. \tag{7.88b}$$

Comparing these with (7.79) we see that the on-shell scheme is merely the special case of the momentum scheme in which $m^2 = -\mu^2$. The Green functions in this scheme are therefore trivially obtained from (7.87) by making this substitution. The only mass parameter appearing in the Green functions is μ, but the counter terms remain mass-dependent.

Finally, we remark that any Green function has a unique *value*, no matter which scheme has been chosen to define the parameters. This means that the values of the parameters differ by finite amounts which depend upon the scheme adopted. For example, comparing (7.77) with (7.87) we find

$$\mu^{2\,\mathrm{mom}}=\mu^{2\,\overline{\mathrm{MS}}}\left[1+\frac{\lambda^{\overline{\mathrm{MS}}}}{32\pi^4}\left(\ln\frac{\mu^{2\,\overline{\mathrm{MS}}}}{M^2}-1\right)\right]+\mathrm{O}(\lambda^{2\,\overline{\mathrm{MS}}}) \tag{7.89a}$$

$$\lambda^{\mathrm{mom}}=\lambda^{\overline{\mathrm{MS}}}+\frac{3\lambda^{2\,\overline{\mathrm{MS}}}}{32\pi^2}\left(\ln\frac{\mu^{2\,\overline{\mathrm{MS}}}}{M^2}-A(-\tfrac{4}{3}m^2,\mu^{2\,\overline{\mathrm{MS}}})\right)+\mathrm{O}(\lambda^{3\,\overline{\mathrm{MS}}}). \tag{7.89b}$$

Problems

7.1 Identify the Green functions $\tilde{\Gamma}^{(n)}$ which are superficially divergent in a $\lambda\varphi^3$ field theory.

7.2 Evaluate the Feynman integral J, defined in (7.24), when $P^2 > 4\mu_B^2$.

7.3 Prove (7.39).

7.4 Verify the formulae quoted in Appendix A.

References

The books and review articles which we have found most useful in preparing this chapter are:

Bogoliubov N N and Shirkov D V 1959 *Introduction to the Theory of Quantised Fields* (New York: Interscience)
Coleman S 1971 Renormalization: a Review for Non-Specialists, in *Properties of the Fundamental Interactions* ed A Zichichi (Bologna: Editrice Compositori) p. 604
Itzykson C and Zuber J B 1980 *Quantum Field Theory* (New York: McGraw-Hill)
Ramond P 1981 *Field Theory: A Modern Primer* (Reading, Mass.: Benjamin-Cummings)
Liebbrandt G 1975 *Rev. Mod. Phys.* **47** 849

References in the text

1 Weinberg S 1960 *Phys. Rev.* **118** 838
2 Bollini C G 1972 *Nuov. Cim.* B **12** 20
Ashmore J F 1972 *Nuov. Cim. Lett.* **4** 289
't Hooft G and Veltman M J G 1972 *Nucl. Phys.* B **44** 189
3 Feynman R P 1949 *Phys. Rev.* **76** 769
4 Bloch F and Nordsieck A 1937 *Phys. Rev.* **52** 54
5 Frenkel J and Taylor J C 1976 *Nucl. Phys.* B **116** 185
—— 1977 *Nucl. Phys.* B **124** 268
6 Gastmans R and Meuldermans R 1973 *Nucl. Phys.* B **63** 277
7 See, for example, the proof given by Callan C G 1976 Introduction to

Renormalization Theory, in *Methods in Field Theory* ed R Balian and J Zinn-Justin (Amsterdam: North-Holland) p 41
8 't Hooft G 1973 *Nucl. Phys.* B **61** 455
9 Bardeen W A *et al.* 1978 *Phys. Rev.* D **18** 3998
10 Celmaster W and Gonsalves R J 1979 *Phys. Rev. Lett.* **42** 1435

8

QUANTUM FIELD THEORY WITH FERMIONS

8.1 Path integrals over Grassmann variables

As discussed in Chapter 3, it is necessary to interpret the classical field theory of a spinor field in terms of fields which are Grassmann (anticommuting) variables rather than ordinary commuting variables. It will therefore be necessary when we quantise the field theory of a spinor field to introduce path (or functional) integrals over Grassmann variables[1,2]. In this section, we shall introduce Gaussian path integrals over Grassmann variables by generalising integrals over a finite number of (Grassmann) degrees of freedom (much as we did for ordinary variables in Chapter 1).

Let us consider first the case where the variables are real Grassmann numbers. When there is only a single Grassmann variable θ, then as discussed in Chapter 3,

$$\{\theta, \theta\}=0 \tag{8.1}$$

so that $\theta^2=0$. Suppose we want to evaluate a Gaussian integral involving θ. Then, expanding the exponential

$$\int d\theta \exp(-\tfrac{1}{2}a\theta^2)=\int d\theta=0. \tag{8.2}$$

To obtain a non-trivial integral we must go to more than one variable.

For n Grassmann variables $\theta_1, \ldots, \theta_n$, we have

$$\{\theta_i, \theta_j\}=0 \qquad i, j=1, \ldots, n \tag{8.3}$$

so that $\theta_i^2=0$, and consequently any function $f(\theta_1, \ldots, \theta_n)$ may be expanded in a power series in the θ_i which terminates when there are at most n factors, $\theta_1\theta_2\ldots\theta_n$. As discussed in Chapter 3, integration has the properties

$$\int d\theta_i=0 \tag{8.4}$$

$$\int d\theta_i\theta_i=1 \tag{8.5}$$

(where *no* summation on i is implied) and multiple integrals are interpreted as iterated integrals. We now consider a Gaussian integral over n Grassmann variables

$$I_n \equiv \int d\theta_1 \ldots d\theta_n \exp(-\tfrac{1}{2}\boldsymbol{\Theta}^{\mathrm{T}}\mathbf{A}\boldsymbol{\Theta}) \tag{8.6}$$

where $\mathbf{A}$ is a real antisymmetric matrix, and $\boldsymbol{\Theta}$ is the column vector with components $(\theta_1, \ldots, \theta_n)$. We must *not* take $\mathbf{A}$ to be a symmetric matrix here as we did in Chapter 1, otherwise (8.3) will immediately imply that the integral is zero. Each non-zero term in the expansion of the exponential in (8.6) involves a even number of factors of θ_i which must all differ because of (8.3). On the other hand, when n is odd, there is an odd number of factors $d\theta$. There must therefore be at least one factor $\int d\theta_i$ where the integrand is 1. Thus, using (8.4),

$$I_n = 0 \qquad \text{for } n \text{ odd.} \tag{8.7}$$

When n is even, the only term which need be retained in the expansion of the exponential in (8.6) is the one which involves n factors of θ. Terms with more than n factors of θ are immediately zero because of (8.3). Terms with less than n factors of θ give zero upon integration because there is at least one factor $\int d\theta_i$ where the integrand is 1. Thus for n even,

$$\begin{aligned} I_n &= \int d\theta_1 \ldots d\theta_n \frac{1}{(n/2)!}\left(-\frac{1}{2}\boldsymbol{\Theta}^{\mathrm{T}}\mathbf{A}\boldsymbol{\Theta}\right)^{n/2} \\ &= (\det \mathbf{A})^{1/2}. \end{aligned} \tag{8.8}$$

One easily convinces oneself of the correctness of this last step (see Berezin[1]) by starting with the simple cases $n=2$ and $n=4$. We may also use (8.8) for n odd consistently with (8.7), because the determinant of an antisymmetric $n \times n$ matrix with n odd is zero. It is worth noting at this stage that a positive power of the determinant occurs in (8.8) whereas in (1.2) it was a negative power that arose.

Generalising to the case where the integration is over the continuous infinity of components of a (Grassmann) function $\psi(x)$, instead of over the finite number of components of the column vector $\boldsymbol{\Theta}$, we obtain the path (or functional) integral

$$\int \mathscr{D}\psi \exp\left(-\tfrac{1}{2}\int dx' \int dx\, \psi(x')A(x',x)\psi(x)\right) = (\det \mathbf{A})^{1/2} = \exp(\tfrac{1}{2}\operatorname{Tr}\ln \mathbf{A}). \tag{8.9}$$

A useful extension of (8.8) is to include a linear term in the exponent in (8.6). We then obtain

$$\int d\theta_1 \ldots d\theta_n \exp(-\tfrac{1}{2}\boldsymbol{\Theta}^{\mathrm{T}}\mathbf{A}\boldsymbol{\Theta} + \rho^{\mathrm{T}}\boldsymbol{\Theta}) = \exp(\tfrac{1}{2}\operatorname{Tr}\ln \mathbf{A})\exp(-\tfrac{1}{2}\rho^{\mathrm{T}}\mathbf{A}^{-1}\rho) \tag{8.10}$$

where ρ is a given column vector consisting of Grassmann variables

$(\rho_1, \ldots, \rho_n)$. Thus, in addition to (8.3), we have

$$\{\rho_i, \rho_j\} = 0 = \{\rho_i, \theta_j\}. \tag{8.11}$$

Equation (8.10) may be derived from (8.8) by completing the square and changing variables. Thus

$$\mathbf{\Theta}^{\mathrm{T}}\mathbf{A}\mathbf{\Theta} - 2\rho^{\mathrm{T}}\mathbf{\Theta} = (\mathbf{\Theta} + \mathbf{A}^{-1}\rho)^{\mathrm{T}}\mathbf{A}(\mathbf{\Theta} + \mathbf{A}^{-1}\rho) + \rho^{\mathrm{T}}\mathbf{A}^{-1}\rho \tag{8.12}$$

where we have used the antisymmetry of the matrix, $\mathbf{A}$, and we then make the change of variables

$$\mathbf{\Theta}' = \mathbf{\Theta} + \mathbf{A}^{-1}\rho. \tag{8.13}$$

$$\int \mathscr{D}\psi \exp\left(-\tfrac{1}{2} \int \mathrm{d}x' \int \mathrm{d}x \psi(x') A(x', x) \psi(x) + \int \mathrm{d}x \rho(x) \psi(x) \right)$$
$$= \exp(\tfrac{1}{2} \operatorname{Tr} \ln \mathbf{A}) \exp\left(-\tfrac{1}{2} \int \mathrm{d}x' \int \mathrm{d}x \rho(x') A^{-1}(x', x) \rho(x) \right) \tag{8.14}$$

where $\rho(x)$ is a given (Grassmann) function.

So far we have been discussing real Grassmann variables $\theta_1, \ldots, \theta_n$ or a real Grassmann function $\psi(x)$. In the case of complex Grassmann variables, the generalisation of (8.6) and (8.8) is

$$\int \mathrm{d}\theta_1^* \, \mathrm{d}\theta_1 \ldots \int \mathrm{d}\theta_n^* \, \mathrm{d}\theta_n \exp(-\mathbf{\Theta}^\dagger \mathbf{A} \mathbf{\Theta}) = \det \mathbf{A} \tag{8.15}$$

where $\mathbf{A}$ is a skew Hermitian matrix and we define integration over complex variables by

$$\int \mathrm{d}\theta_i^* \, \mathrm{d}\theta_i = 2 \int \mathrm{d}(\operatorname{Re} \theta_i) \, \mathrm{d}(\operatorname{Im} \theta_i). \tag{8.16}$$

The corresponding path integral is

$$\int \mathscr{D}\psi^* \mathscr{D}\psi \exp\left(-\int \mathrm{d}x' \int \mathrm{d}x \psi^*(x') A(x', x) \psi(x) \right) = \det \mathbf{A} = \exp(\operatorname{Tr} \ln \mathbf{A}). \tag{8.17}$$

If a linear term is included in the exponent, the generalisation of (8.14) is

$$\int \mathscr{D}\psi^* \mathscr{D}\psi \exp\left(-\int \mathrm{d}x' \int \mathrm{d}x \psi^*(x') A(x', x) \psi(x) + \int \mathrm{d}x [\rho^*(x)\psi(x) - \psi^*(x)\rho(x)] \right)$$
$$= \exp(\operatorname{Tr} \ln \mathbf{A}) \exp\left(-\int \mathrm{d}x' \int \mathrm{d}x \rho^*(x') A^{-1}(x', x) \rho(x) \right). \tag{8.18}$$

In the next sections, we shall find it convenient to work in four-dimensional Minkowski space. In that case, (8.17) is replaced by

$$\int \mathscr{D}\psi^* \mathscr{D}\psi \exp\left(\mathrm{i} \int \mathrm{d}^4x' \int \mathrm{d}^4x \psi^*(x') B(x', x) \psi(x) \right) = \det(\mathrm{i}\mathbf{B}) = \exp \operatorname{Tr} \ln(\mathrm{i}\mathbf{B}) \tag{8.19}$$

where **B** is Hermitian, and (8.18) is replaced by

$$\int \mathscr{D}\psi^* \mathscr{D}\psi \exp\left(\mathrm{i} \int \mathrm{d}^4x' \int \mathrm{d}^4x \psi^*(x')B(x',x)\psi(x) \right.$$
$$\left. + \mathrm{i} \int \mathrm{d}^4x [\sigma^*(x)\psi(x) + \psi^*(x)\sigma(x)] \right)$$
$$= \exp \mathrm{Tr} \ln(\mathrm{i}\mathbf{B}) \exp\left(-\mathrm{i} \int \mathrm{d}^4x' \int \mathrm{d}^4x \sigma^*(x') B^{-1}(x',x)\sigma(x) \right). \quad (8.20)$$

8.2 The generating functional for spinor field theories

In analogy with (4.47) we write the generating functional

$$W[\sigma, \bar{\sigma}] = N' \int \mathscr{D}\bar{\psi} \mathscr{D}\psi \exp \mathrm{i} \int \mathrm{d}^4x (\mathscr{L} + \bar{\psi}\sigma + \bar{\sigma}\psi) \quad (8.21)$$

where $\sigma(x)$ is an external source which is a Grassmann variable, and the normalisation factor N' is to be chosen so that $W=1$ when $\sigma=0$. We have not added a small term quadratic in the field as in (4.47), because for integration over Grassmann variables this does not provide a convergence factor. The ambiguity arising from the definition of the path integral will be discussed later.

Green functions may be defined by

$$\mathscr{G}^{(2n)}(x_1, \ldots, x_n; y_1, \ldots, y_n)$$
$$= \langle 0 | T(\hat{\psi}(x_1) \ldots \hat{\psi}(x_n) \hat{\bar{\psi}}(y_1) \ldots \hat{\bar{\psi}}(y_n)) | 0 \rangle \quad (8.22)$$

where $\hat{\psi}$ denotes a field operator, and the time ordering operation T is defined for Dirac fields so that it not only reorders the fields in chronological order (as for scalar fields), but also introduces a minus sign each time two Dirac fields have to be transposed in the reordering. (This is necessary because of the anticommuting nature of Grassmann variables.) The Green functions are related to the generating functional $W[\sigma, \bar{\sigma}]$ by

$$\mathrm{i}^{2n} \mathscr{G}^{(2n)}(x_1, \ldots, x_n; y_1, \ldots, y_n) = \frac{\delta^{(2n)} W[\sigma, \bar{\sigma}]}{\delta\bar{\sigma}(x_1) \ldots \delta\bar{\sigma}(x_n)\, \delta\sigma(y_1) \ldots \delta\sigma(y_n)}. \quad (8.23)$$

$\mathscr{G}^{(2n)}$ is antisymmetric in the indices x_i, and in the indices y_i (as appropriate for fermions) because

$$\frac{\delta^2}{\delta\sigma(x_i)\, \delta\sigma(x_j)} = -\frac{\delta^2}{\delta\sigma(x_j)\, \delta\sigma(x_i)}. \quad (8.24)$$

8.3 Propagator for the Dirac field

It is possible to evaluate exactly the generating functional for the free-field theory of a Dirac field, much as for the free-field theory of a scalar field. From (3.81), the appropriate free-field Lagrangian is

$$\mathscr{L}=\bar{\psi}(x)(\mathrm{i}\gamma^{\mu}\partial_{\mu}-m)\psi(x). \tag{8.25}$$

This Lagrangian must be substituted in (8.21) to obtain the generating functional. This is of the form

$$W[\sigma,\bar{\sigma}]=N'\int\mathscr{D}\bar{\psi}\mathscr{D}\psi\exp\left(\mathrm{i}\int\mathrm{d}^4x'\int\mathrm{d}^4x\bar{\psi}(x')B(x',x)\psi(x)\right.$$
$$\left.+\mathrm{i}\int\mathrm{d}^4x[\bar{\psi}(x)\sigma(x)+\bar{\sigma}(x)\psi(x)\right) \tag{8.26}$$

with

$$B(x',x)=(-\mathrm{i}\gamma^{\mu}\partial_{\mu}^{x}-m)\delta^4(x'-x) \tag{8.27}$$

where ∂_{μ}^{x} signifies that the differentiation is on x rather than x'. Using (8.20),

$$W[\sigma,\bar{\sigma}]=\exp-\mathrm{i}\int\mathrm{d}^4x'\int\mathrm{d}^4x\bar{\sigma}(x')B^{-1}(x',x)\sigma(x) \tag{8.28}$$

where we have chosen the constant of proportionality so that $W[\sigma,\bar{\sigma}]=1$ when $\sigma=0$. (It is easy to check that replacing ψ^* by $\psi^{\dagger}\gamma^0$ and σ^* by $\bar{\sigma}=\sigma^{\dagger}\gamma^0$ in (8.20) does not affect the result.) Thus

$$W[\sigma,\bar{\sigma}]=\exp-\mathrm{i}\int\mathrm{d}^4x'\int\mathrm{d}^4x\bar{\sigma}(x')S_{\mathrm{F}}(x'-x)\sigma(x) \tag{8.29}$$

with

$$S_{\mathrm{F}}(x'-x)=B^{-1}(x',x) \tag{8.30}$$

the propagator for the Dirac field. We may construct the inverse $B^{-1}(x',x)$ by Fourier transforming:

$$B(x',x)=\int\frac{\mathrm{d}^4p}{(2\pi)^4}\,\mathrm{e}^{-\mathrm{i}p\cdot(x'-x)}\,(\not{p}-m) \tag{8.31}$$

implies that

$$B^{-1}(x',x)=\int\frac{\mathrm{d}^4p}{(2\pi)^4}\,\mathrm{e}^{-\mathrm{i}p\cdot(x'-x)}\,\frac{(\not{p}+m)}{p^2-m^2}. \tag{8.32}$$

Thus,

$$S_{\mathrm{F}}(x'-x)=\int\frac{\mathrm{d}^4p}{(2\pi)^4}\,\mathrm{e}^{-\mathrm{i}p\cdot(x'-x)}\,\frac{(\not{p}+m)}{p^2-m^2}. \tag{8.33}$$

We notice that an ambiguity has arisen because of the pole at $p^2 = m^2$. This ambiguity may be resolved by analogy with (4.45) for the scalar case by introducing iε. Thus we write

$$S_F(x'-x) = \int \frac{d^4p}{(2\pi)^4}\, e^{-ip\cdot(x'-x)}\, \tilde{S}_F(p) \tag{8.34}$$

with

$$\tilde{S}_F(p) = \frac{(\not{p}+m)}{p^2-m^2+i\varepsilon} = (\not{p}-m+i\varepsilon)^{-1}. \tag{8.35}$$

It is possible to justify this prescription by continuing to Euclidean space to carry out the evaluation (as for the scalar case).

8.4 Renormalisable theories of Dirac fields and scalar fields

As is discussed in Chapter 7, for a field theory to be renormalisable it is necessary that there should be only a finite number of primitively divergent diagrams. When this criterion was applied to scalar field theories in four dimensions we were restricted to terms involving not more than four powers of φ in the Lagrangian. We shall now apply the criterion to theories of a Dirac spinor field ψ and a scalar field φ.

Consider a theory in which there are interaction vertices (*not* involving derivatives of fields) with various numbers of fermion and scalar boson lines. In general, let an interaction vertex have N_B boson lines and N_F fermion lines. The (superficial) degree of divergence D of a Feynman diagram with I_B internal boson lines, I_F internal fermion lines, and L loops is

$$D = 4L - 2I_B - I_F \tag{8.36}$$

(with the fermion propagator as in (8.35)). We shall express D in terms of the number of external lines and the number of vertices. Let a given Feynman diagram have $V(N_B, N_F)$ vertices with N_B boson lines and N_F fermion lines attached, let there be E_B external boson lines and E_F external fermion lines, let the total number of internal lines be I, the total number of external lines be E, and the total number of vertices be V. Then the following relations hold:

$$E_B + 2I_B = \sum_{N_B, N_F} V(N_B, N_F) N_B \tag{8.37}$$

$$E_F + 2I_F = \sum_{N_B, N_F} V(N_B, N_F) N_F \tag{8.38}$$

$$V = \sum_{N_B, N_F} V(N_B, N_F) \tag{8.39}$$

$$I = I_B + I_F \tag{8.40}$$

$$E = E_B + E_F \tag{8.41}$$

and as before (Chapter 7)

$$L = I - V + 1. \tag{8.42}$$

Using (8.37)–(8.42) we see that

$$D = 4 - E_B - \tfrac{3}{2}E_F + \sum_{N_B, N_F} V(N_B, N_F)(\tfrac{3}{2}N_F + N_B - 4). \tag{8.43}$$

For a renormalisable theory we want to avoid interaction vertices which lead to the degree of divergence of a diagram growing with the number of vertices. We must therefore restrict ourselves to interactions for which

$$\tfrac{3}{2}N_F + N_B - 4 \leqslant 0. \tag{8.44}$$

The solutions of (8.44) are

$$N_F = 0 \qquad N_B \leqslant 4 \tag{8.45}$$

as in Chapter 7,

$$N_B = 0 \qquad N_F = 2 \tag{8.46}$$

and

$$N_B = 1 \qquad N_F = 2. \tag{8.47}$$

(The apparent solution $N_F = 1, N_B = 2$ is inconsistent with angular momentum conservation.) There are thus *no* renormalisable pure fermion interactions. (The solution (8.46) is just a mass term.) The only renormalisable interactions of fermions with scalars are given by (8.47). For a Dirac spinor field ψ, and a scalar field φ, the explicit interactions allowed are Yukawa interactions of the form $\bar{\psi}\psi\varphi$ and $\bar{\psi}\gamma_5\psi\varphi$. In either case we represent the interaction vertex as in figure 8.1.

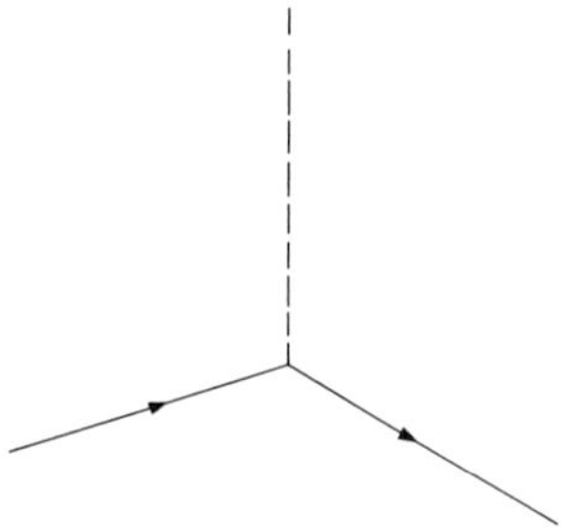

Figure 8.1 Yukawa interaction vertex.

If we were to allow derivatives of fields in the interaction vertices, this would only lead to a degree of divergence growing more rapidly with the number of vertices, and there are *no* renormalisable derivative interactions of Dirac fields and scalar fields.

8.5 Feynman rules for Yukawa interactions

In §6.1, Feynman rules have been developed for a theory involving only scalar fields. Now that the theory involves both Dirac and scalar fields, a slight generalisation of the formalism of Chapter 6 is required. The generating functional depends on both scalar and spinor sources

$$W[J,\sigma,\bar{\sigma}]=N'\int\mathscr{D}\varphi\mathscr{D}\bar{\psi}\mathscr{D}\psi\,\exp\mathrm{i}\int\mathrm{d}x(\mathscr{L}+J\varphi+\bar{\psi}\sigma+\bar{\sigma}\psi) \qquad (8.48)$$

where $\mathscr{L}$ depends on φ and ψ. The normalisation factor N' is to be chosen so that $W=1$ when $J=0$, $\sigma=0$. In the free-field case, it follows from (4.43) and (8.29) that

$$W[J,\sigma,\bar{\sigma}]=W_0[J,\sigma,\bar{\sigma}]=\exp\left(-\frac{\mathrm{i}}{2}\int\mathrm{d}x\,\mathrm{d}y J(x)\Delta_{\mathrm{F}}(x-y)J(y)\right)$$

$$\times\exp\left(-\mathrm{i}\int\mathrm{d}x\,\mathrm{d}y\bar{\sigma}(x)S_{\mathrm{F}}(x-y)\sigma(y)\right). \qquad (8.49)$$

In (8.48), we separate $\mathscr{L}$ into a free-field part and an interaction part:

$$\mathscr{L}=\mathscr{L}_0+\mathscr{L}_{\mathrm{I}}(\varphi,\psi,\bar{\psi}). \qquad (8.50)$$

Then

$$\exp\mathrm{i}\int\mathrm{d}x(\mathscr{L}+J\varphi+\bar{\psi}\sigma+\bar{\sigma}\psi)$$

$$=\exp\left(\mathrm{i}\int\mathrm{d}x\mathscr{L}_{\mathrm{I}}(\varphi,\psi,\bar{\psi})\right)\exp\left(\mathrm{i}\int\mathrm{d}x(\mathscr{L}_0+J\varphi+\bar{\psi}\sigma+\bar{\sigma}\psi)\right). \qquad (8.51)$$

The expansion in powers of the interaction may conveniently be made by observing that

$$\frac{\delta}{\delta\sigma(y)}\exp\mathrm{i}\int\mathrm{d}x(\mathscr{L}_0+\sigma\varphi+\bar{\psi}\sigma+\bar{\sigma}\psi)$$

$$=-\mathrm{i}\bar{\psi}(y)\exp\mathrm{i}\int\mathrm{d}x(\mathscr{L}_0+J\varphi+\bar{\psi}\sigma+\bar{\sigma}\psi) \qquad (8.52)$$

and

$$\frac{\delta}{\delta\bar{\sigma}(y)}\exp \mathrm{i}\int \mathrm{d}x(\mathscr{L}_0+J\varphi+\bar{\psi}\sigma+\bar{\sigma}\psi)$$

$$=\mathrm{i}\psi(y)\exp \mathrm{i}\int \mathrm{d}x(\mathscr{L}_0+J\varphi+\bar{\psi}\sigma+\bar{\sigma}\psi). \quad (8.53)$$

The minus sign in (8.53) relative to (8.52) arises because $\delta/\delta\sigma$ and $\bar{\psi}$ anticommute, being Grassmann variables. Taking (8.52) and (8.53) together with (6.7), we have

$$\left(\int \mathrm{d}x\mathscr{L}_{\mathrm{I}}(\varphi,\psi,\bar{\psi})\right)\left(\exp \mathrm{i}\int \mathrm{d}x(\mathscr{L}_0+J\varphi+\bar{\psi}\sigma+\bar{\sigma}\psi)\right)$$

$$=\int \mathrm{d}x\mathscr{L}_{\mathrm{I}}\left(-\mathrm{i}\frac{\delta}{\delta J},-\mathrm{i}\frac{\delta}{\delta\bar{\sigma}},\mathrm{i}\frac{\delta}{\delta\sigma}\right)\left(\exp \mathrm{i}\int \mathrm{d}x(\mathscr{L}_0+J\varphi+\bar{\psi}\sigma+\bar{\sigma}\psi)\right). \quad (8.54)$$

From (8.54), (8.51) and (8.48) it follows that

$$W[J,\sigma,\bar{\sigma}]=\exp\left[\mathrm{i}\int \mathrm{d}x\mathscr{L}_{\mathrm{I}}\left(-\mathrm{i}\frac{\delta}{\delta J},-\mathrm{i}\frac{\delta}{\delta\bar{\sigma}},\mathrm{i}\frac{\delta}{\delta\sigma}\right)\right]W_0[J,\sigma,\bar{\sigma}]. \quad (8.55)$$

The perturbation series is now obtained as in Chapter 6, by expanding the exponential. We need to know the functional derivatives of W_0 with respect to the sources. These are obtained from (8.49). Thus

$$\frac{\delta W_0}{\delta\bar{\sigma}(x)}=\left(-\int \mathrm{d}y\mathrm{i}S_{\mathrm{F}}(x-y)\sigma(y)\right)W_0 \quad (8.56)$$

$$\frac{\delta W_0}{\delta\sigma(x)}=\left(\int \mathrm{d}y\bar{\sigma}(y)\mathrm{i}S_{\mathrm{F}}(y-x)\right)W_0 \quad (8.57)$$

and $\delta W_0/\delta J(x)$ is given by (6.12). It is important to notice that there is an extra minus sign in (8.57) arising because $\delta/\delta\sigma$ and $\bar{\sigma}$ anticommute, being Grassmann variables. Proceeding in strict analogy with Chapter 6, we obtain the additional momentum space Feynman rules for diagrams involving spin 1/2 particles with Yukawa interaction

$$\mathscr{L}_{\mathrm{I}}=g\bar{\psi}\gamma\psi\varphi \quad (8.58)$$

with $\gamma=\mathbf{I}$ or γ_5.

1 With each fermion line carrying momentum p there is associated a factor $\mathrm{i}\tilde{S}_{\mathrm{F}}(p)$

: $\mathrm{i}(\not{p}+m)/(p^2-m^2+\mathrm{i}\varepsilon)$.

2 With each vertex there is associated a factor $-ig\gamma$

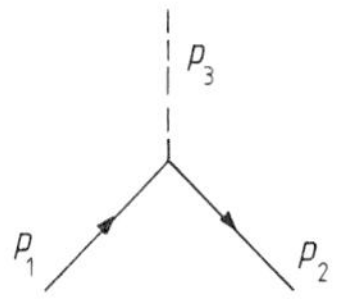

$: \quad -ig\gamma$

where $p_3 = p_2 - p_1$.

3 For each closed fermion loop there is a factor -1.

Because the fermion propagators and vertices are matrices we must be careful to maintain the order of the lines and vertices in the diagram. The rules for scalar lines and vertices involving only scalar particles are as in Chapter 6. It should be noted that a theory of scalar and spin 1/2 particles interacting through a Yukawa interaction will also necessarily involve a φ^4 interaction. This is because there are diagrams like figure 8.2 contributing to the renormalised φ^4 vertex.

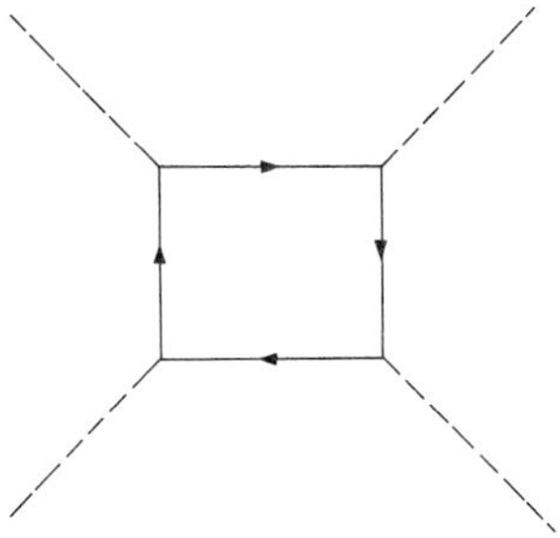

Figure 8.2 Contribution to renormalised φ^4 vertex.

The way in which the Feynman rule assigning a minus sign to closed fermion loops arises may be illustrated by considering the scalar meson two-point function, $G^{(2)}(p, -p)$. This has a contribution from the diagram of figure 8.3. The relevant terms in $W[J, \sigma, \bar{\sigma}]$ for the derivation of this diagram are given by

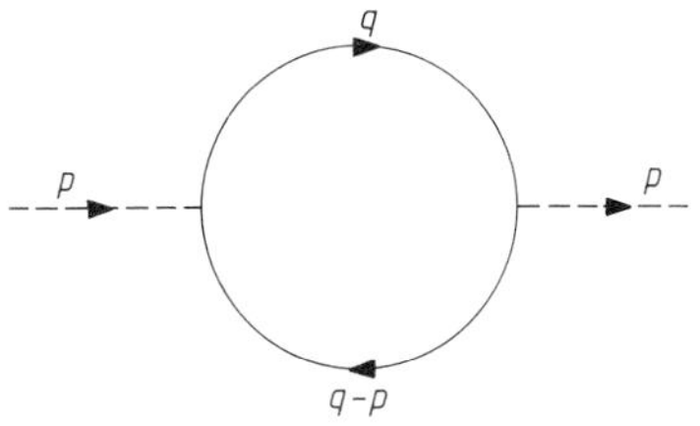

Figure 8.3 Contribution to scalar meson two-point function.

(8.55) and (8.58) as

$$W[J,\sigma,\bar\sigma]=-\frac{g^2}{2}\int \mathrm{d}x\,\mathrm{d}y\left(\frac{\delta}{\delta\sigma(x)}\gamma\frac{\delta}{\delta\bar\sigma(x)}\frac{\delta}{\delta J(y)}\right)$$
$$\times\left(\frac{\delta}{\delta\sigma(y)}\gamma\frac{\delta}{\delta\bar\sigma(y)}\frac{\delta}{\delta J(y)}\right)W_0[J,\sigma,\bar\sigma]. \tag{8.59}$$

Using (8.56), (8.57) and (6.12), we obtain

$$W[J,\sigma,\bar\sigma]=\frac{g^2}{2}\int \mathrm{d}x\left(\frac{\delta}{\delta\sigma(x)}\gamma\frac{\delta}{\delta\bar\sigma(x)}\frac{\delta}{\delta J(y)}\right)$$
$$\times\int \mathrm{d}y\int \mathrm{d}y_1\,\mathrm{d}y_2\,\mathrm{d}y_3\bar\sigma(y_1)[-\mathrm{i}S_\mathrm{F}(y_1-y)]\gamma[-\mathrm{i}S_\mathrm{F}(y-y_2)]\sigma(y_2)$$
$$\times[-\mathrm{i}\Delta_\mathrm{F}(y-y_3)]J(y_3)W_0[J,\sigma,\bar\sigma]. \tag{8.60}$$

A crucial extra factor of -1 arises because of the difference in sign between (8.56) and (8.57). The diagram in which we are interested corresponds to two external boson lines, but no external fermion lines. We are therefore interested in the terms in $W[J,\sigma,\bar\sigma]$ with two factors of J and no factors of $\sigma,\bar\sigma$. The derivatives $\delta/\delta\sigma(x)$ and $\delta/\delta\bar\sigma(x)$ must therefore be allowed to act on the fermion sources $\sigma(y_2)$ and $\bar\sigma(y_1)$ rather than on $W_0[J,\sigma,\bar\sigma]$ in (8.60). The term we want is therefore

$$W[J,\sigma,\bar\sigma]=\frac{g^2}{2}\int \mathrm{d}y_3\,\mathrm{d}y_4[-\mathrm{i}\Delta_\mathrm{F}(x-y_4)][-\mathrm{i}\Delta_\mathrm{F}(y-y_3)]J(y_4)J(y_3)$$
$$\times[\mathrm{i}S_\mathrm{F}(x-y)]_{\alpha\beta}\gamma_{\beta\gamma}[-\mathrm{i}S_\mathrm{F}(y-x)]_{\gamma\delta}\gamma_{\delta\alpha}W_0[J,\sigma,\bar\sigma]+\ldots. \tag{8.61}$$

No further minus signs arise at this stage from $\delta/\delta\sigma(x)$ and $\delta/\delta\bar\sigma(x)$ because σ and $\bar\sigma$ are already in the correct order for immediate differentiation. To proceed from (8.61) to the momentum space Feynman rule for this diagram is now exactly along the lines of Chapter 6. The effect for this diagram of the anticommuting nature of the Grassmann fields has been an overall minus sign, as required by the third Feynman rule above.

8.6 Massless fermions

As we have seen in Chapter 3, massless fermions such as the neutrino which possess only a single helicity state may be described by two-component Weyl spinor fields. For a left-handed Weyl spinor field χ_L we may write the generating functional

$$W[\sigma_\mathrm{R},\sigma_\mathrm{R}^\dagger]=N'\int \mathscr{D}\chi_\mathrm{L}^\dagger\mathscr{D}\chi_\mathrm{L}\exp\mathrm{i}\int \mathrm{d}^4x(\mathscr{L}+\chi_\mathrm{L}^\dagger\sigma_\mathrm{R}+\sigma_\mathrm{R}^\dagger\chi_\mathrm{L}) \tag{8.62}$$

where the source $\sigma_R(x)$ is a right-handed Weyl spinor. In the free-field case the Lagrangian is

$$\mathscr{L} = i\chi_L^\dagger \bar{\sigma}^\mu \partial_\mu \chi_L \tag{8.63}$$

where

$$\bar{\sigma}^\mu = (\mathbf{I}, -\boldsymbol{\sigma}). \tag{8.64}$$

Following steps exactly analogous to those of §8.3 we obtain the momentum space propagator

$$S_L(p) = \frac{\sigma^\mu p_\mu}{(p^2 + i\varepsilon)} = (\bar{\sigma}^\mu p_\mu + i\varepsilon)^{-1} \tag{8.65}$$

where

$$\sigma^\mu = (\mathbf{I}, \boldsymbol{\sigma}). \tag{8.66}$$

For a right-handed Weyl spinor field χ_R the free-field Lagrangian is

$$\mathscr{L} = i\chi_R^\dagger \sigma^\mu \partial_\mu \chi_R \tag{8.67}$$

and the momentum space propagator is

$$S_R(p) = \frac{\bar{\sigma}^\mu p_\mu}{(p^2 + i\varepsilon)} = (\sigma^\mu p_\mu + i\varepsilon)^{-1}. \tag{8.68}$$

An alternative procedure for massless fermions is to use the propagator of (8.35) in the zero mass limit. For $m \to 0$,

$$S_F(p) \to \frac{\not{p}}{(p^2 + i\varepsilon)} = (\not{p} + i\varepsilon)^{-1}. \tag{8.69}$$

For calculations at zero temperature and density this procedure is satisfactory because the vertices involving neutrinos in electroweak theory always involve a factor $\frac{1}{2}(1 - \gamma_5)$ which projects out the left-handed part of the propagator in any neutrino loop. However, at finite temperature (or density) the use of a massless Dirac propagator for the neutrino can lead to errors corresponding to the thermal (or Fermi) energy of unphysical right-handed neutrinos. In such cases, it may be safer to use Weyl spinors for massless fields.

8.7 Scattering amplitudes with fermions

By analogy with §6.4 for scalar fields, we obtain a (Lorentz invariant) scattering amplitude involving fermions by calculating a (momentum space) Green function with the appropriate external legs. (See problem 8.3.) The propagators associated with external lines should be divided out as for OPI Green functions, but all connected diagrams, not just one-particle irreducible ones, should be included. The only essential difference from the scalar case is that incoming external particle (antiparticle) lines for fermions will have associated factors $u(p, s)(\bar{v}(p, s))$ and outgoing lines will have associated factors $\bar{u}(p, s)(v(p, s))$. These arise because in the analogue of (6.45) and (6.46), the free fermion fields contain factors of $u(p, s)$, $v(p, s)$, $\bar{u}(p, s)$ and $\bar{v}(p, s)$ from the expansions in Fourier components (3.87) and (3.93). These factors remain after carrying out differentiations with respect to $a(p, s)$, $b^*(p, s)$, $a^*(p, s)$ and $b(p, s)$ in the analogue of (6.50). Because of the antisymmetry of Green functions under interchange of identical fermions, the scattering amplitudes have this antisymmetry when external fermion lines are interchanged, as one should expect.

As an example of a scattering amplitude calculation, consider the scattering of two identical fermions with a Yukawa interaction given by (8.58). At lowest order, the two contributing diagrams are as in figure 8.4, where 1, 2, 3, 4 label the momentum and spin degrees of freedom of the fermions. The contributions of these two diagrams to $M_{p_3p_4,p_1p_2}$ differ by a sign because of the antisymmetry of the amplitude under interchange of the two final fermions.

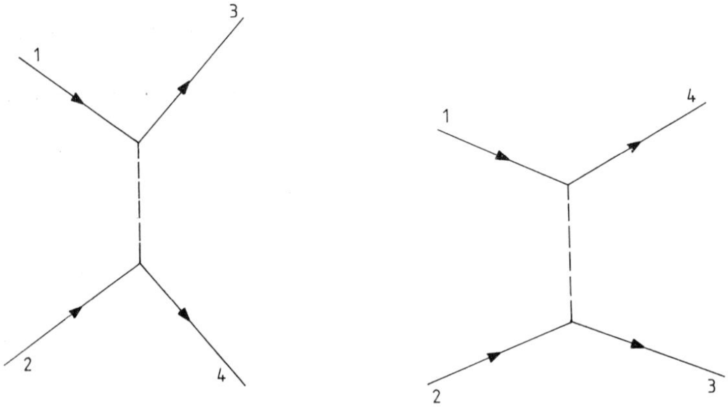

Figure 8.4 Contributions to fermion–fermion scattering.

The invariant amplitude is

$$M_{p_3p_4,p_1p_2} = -ig^2\{\bar{u}(p_3, s_3)\gamma u(p_1, s_1)\bar{u}(p_4, s_4)\gamma u(p_2, s_2)[(p_3-p_1)^2-\mu^2]^{-1}$$
$$-\bar{u}(p_4, s_4)\gamma u(p_1, s_1)\bar{u}(p_3, s_3)\gamma u(p_2, s_2)[(p_4-p_1)^2-\mu^2]^{-1}\} \quad (8.70)$$

As in §6.5, the differential cross section may now be calculated (see problem 8.4) using the usual gamma matrix trace theorems of relativisitic quantum mechanics[3], and being careful to include a factor of $\frac{1}{2}$ to allow for identical fermions in the final state.

Problems

8.1 Derive (8.8) for $n=2$ and $n=4$.

8.2 By analogy with Chapter 6, derive the Feynman rules for fermion Green functions, listed after (8.58).

8.3 Carry out in detail the derivation of the Feynman rules for scattering amplitudes involving fermions described in §8.7.

8.4 Calculate the differential cross section in the centre-of-mass frame for scattering of identical fermions by a Yukawa interaction (with $\gamma=\mathbf{I}$ or γ_5), taking the incident fermions to be unpolarised.

References

1 Berezin F A 1966 *The Method of Second Quantization* (New York: Academic)
2 For a similar treatment see
Ramond P 1981 *Field Theory: a Modern Primer* (Reading, Mass.: Benjamin-Cummings)
3 See, for example,
Bjorken J S and Drell S D 1964 *Relativistic Quantum Mechanics* (New York: McGraw-Hill)
Aitchison I J R and Hey A J G 1982 *Gauge Theories in Particle Physics* (Bristol: Adam Hilger)

9

GAUGE FIELD THEORIES

9.1 Abelian gauge field theory

So far we have been dealing with field theories involving only spin zero and spin 1/2 particles. We now wish to include spin 1 particles. We shall restrict attention to massless spin 1 particles. The reason for this is that it turns out for the non-Abelian case (Yang–Mills theories[1]) that renormalisability of the field theory requires the vector fields to be massless in the first instance. The masses of the vector fields are generated at a later stage through certain scalar fields (Higgs scalars) acquiring vacuum expectation values, as we shall see in Chapter 12.

The earliest field theory involving massless vector fields is quantum electrodynamics (QED), the theory of the electromagnetic interactions of particles. For the interaction of a Dirac field with the electromagnetic field, the Lagrangian density takes the form

$$\mathscr{L}=\bar{\psi}(\mathrm{i}\gamma^{\mu}\partial_{\mu}-m)\psi-\tfrac{1}{4}F_{\mu\nu}F^{\mu\nu}-q\bar{\psi}\gamma^{\mu}\psi A_{\mu} \tag{9.1}$$

where we have used (3.81) and (3.118), and have inserted the electromagnetic current (3.112) for a spin 1/2 particle of charge q.

It is possible to arrive at this form of Lagrangian using a principle called gauge invariance which is fruitful in suggesting generalisations. Suppose we start from the Lagrangian for the free Dirac field

$$\mathscr{L}_1=\bar{\psi}(\mathrm{i}\gamma^{\mu}\partial_{\mu}-m)\psi. \tag{9.2}$$

As discussed in §3.4, this Lagrangian is invariant under the phase transformation

$$\psi(x)\rightarrow \mathrm{e}^{-\mathrm{i}q\Lambda}\,\psi(x) \tag{9.3}$$

where Λ is an arbitrary real number, the same for all values of x. We might generalise this global symmetry of the Lagrangian to a local symmetry, if we could arrange invariance under the transformation

$$\psi(x)\rightarrow \mathrm{e}^{-\mathrm{i}q\Lambda(x)}\,\psi(x) \tag{9.4}$$

where Λ may now depend on x, i.e. we have an independent phase transformation for each point of (four-) space. This is referred to as Abelian gauge invariance. (Abelian because there is no non-trivial group theory involved.) The Lagrangian of (9.2) is not, however, invariant under this local symmetry. In fact under the transformation (9.4),

$$\mathscr{L}_1 \to \mathscr{L}_1 + q\bar{\psi}\gamma^\mu\psi\,\partial_\mu\Lambda. \tag{9.5}$$

If we wish to insist on gauge invariance it is necessary to add more terms to the Lagrangian. This may be done as follows. The derivative in (9.2) is replaced by a covariant derivative D_μ defined by

$$\mathrm{D}_\mu\psi = (\partial_\mu + \mathrm{i}qA_\mu)\psi \tag{9.6}$$

where A_μ is a vector field referred to as the gauge field. The vector field is then required to have the transformation property

$$A_\mu \to A_\mu + \partial_\mu\Lambda \tag{9.7}$$

under the gauge transformation which acts on ψ according to (9.4). We then see that $\mathrm{D}_\mu\psi$ transforms in the same way as ψ

$$\mathrm{D}_\mu\psi \to \mathrm{e}^{-\mathrm{i}q\Lambda}\,\mathrm{D}_\mu\psi. \tag{9.8}$$

Thus, the Lagrangian

$$\mathscr{L}_2 = \bar{\psi}(\mathrm{i}\gamma^\mu\mathrm{D}_\mu - m)\psi \tag{9.9}$$

is gauge invariant.

It remains to include gauge invariant terms for the vector field A_μ. It follows immediately that

$$F_{\mu\nu} = \partial_\mu A_\nu - \partial_\nu A_\mu \tag{9.10}$$

is invariant under the gauge transformation (9.7). Thus, we may write the final gauge invariant Lagrangian

$$\mathscr{L} = \bar{\psi}(\mathrm{i}\gamma^\mu\mathrm{D}_\mu - m)\psi - \tfrac{1}{4}F_{\mu\nu}F^{\mu\nu} \tag{9.11}$$

where we have build a Lorentz scalar from $F_{\mu\nu}$, and included a factor of $\frac{1}{4}$ to give a conventional normalisation of the field. With the definition of the (gauge) covariant derivative given by (9.6), this is just the Lagrangian density for the interaction of a Dirac field with the electromagnetic field, as in (9.1).

This Lagrangian density describes a massless vector field. We might ask whether it is possible to give the vector field a mass, while maintaining the gauge invariance. The answer is 'no', because the mass term $A_\mu A^\mu$ is not invariant under the gauge transformation of (9.7). Thus a gauge invariant Lagrangian describing the interaction of a vector field with a spinor field necessarily means a massless vector field. However, we shall see later that there is a way round this, by introducing scalar fields, some of which develop vacuum expectation values.

9.2 Non-Abelian gauge field theories

Suppose we want to construct a theory of a number of Dirac spinor fields ψ_i, $i=1,\ldots,p$ interacting with a number of vector fields A^μ_a, $a=1,\ldots,r$. It will be convenient to assemble the Dirac fields into a column vector which we shall denote by $\boldsymbol{\psi}$. A gauge invariant theory may be set up by giving each of the Dirac fields a 'charge' for its coupling to each vector field A^μ_a, and proceeding exactly as in the last section. This is then a theory of r Abelian gauge fields.

It is natural to ask whether there might not be generalisations[1,2] of the principle of gauge invariance which differ from simply having r distinct Abelian gauge field theories, as above. To explore this conjecture, we first generalise the gauge transformation of (9.3) to

$$\boldsymbol{\psi}(x) \to \mathrm{e}^{-ig\mathbf{T}_a\Lambda_a(x)}\,\boldsymbol{\psi}(x) \tag{9.12}$$

where the $\mathbf{T}_a$ are $p\times p$ matrices which act on the column vector $\boldsymbol{\psi}(x)$, the $\Lambda_a(x)$ are arbitrary functions of x, a sum over a is understood, and g is eventually going to be a coupling constant. (If the matrices $\mathbf{T}_a$ are all multiples of the identity, then we will simply have a succession of Abelian gauge transformations like that of (9.4).) More succinctly, we write

$$\boldsymbol{\psi}(x) \to \mathrm{e}^{-ig\boldsymbol{t}\cdot\Lambda(x)}\,\boldsymbol{\psi}(x) \tag{9.13}$$

where

$$\mathbf{T}\cdot\Lambda(x) = \mathbf{T}_a\Lambda_a(x). \tag{9.14}$$

By analogy with (9.6), we write

$$\mathrm{D}^\mu\boldsymbol{\psi} = (\partial^\mu + ig\mathbf{T}\cdot A^\mu)\boldsymbol{\psi}. \tag{9.15}$$

(As before, each function $\Lambda_a(x)$ is replaced by a vector field A^μ_a, to define the covariant derivative.) The development is easier if we now consider an infinitesimal gauge transformation

$$\boldsymbol{\psi}(x) \to (\mathbf{I} - ig\mathbf{T}\cdot\Lambda)\boldsymbol{\psi}(x). \tag{9.16}$$

Under this infinitesimal transformation

$$\partial^\mu\boldsymbol{\psi} \to (\mathbf{I} - ig\mathbf{T}\cdot\Lambda)\partial^\mu\boldsymbol{\psi} - ig(\mathbf{T}\cdot\partial\Lambda)\boldsymbol{\psi}. \tag{9.17}$$

We now adopt a gauge transformation property for the gauge fields

$$A^\mu_a \to A^\mu_a + \partial^\mu\Lambda_a + gf_{abc}\Lambda_b A^\mu_c \tag{9.18}$$

where f_{abc} are some constants. This is analogous to (9.7) except for the last term, which has been introduced to give the gauge fields an opportunity to fulfil their role of cancelling out the unwanted terms in (9.17). We want the covariant derivative of $\boldsymbol{\psi}$ to transform in the same way as $\boldsymbol{\psi}$.

$$\mathrm{D}^\mu\boldsymbol{\psi} \to (\mathbf{I} - ig\mathbf{T}\cdot\Lambda)\mathrm{D}^\mu\boldsymbol{\psi}. \tag{9.19}$$

This will occur provided

$$[\mathbf{T}\cdot\Lambda, \mathbf{T}\cdot A^{\mu}] = \mathrm{i}f_{abc}\mathbf{T}_a\Lambda_b A_c^{\mu}. \tag{9.20}$$

Consequently,

$$[\mathbf{T}_b, \mathbf{T}_c] = \mathrm{i}f_{abc}\mathbf{T}_a. \tag{9.21}$$

If we assume that the coefficients f_{abc} are antisymmetric in all indices, then this may be written as

$$[\mathbf{T}_b, \mathbf{T}_c] = \mathrm{i}f_{bca}\mathbf{T}_a. \tag{9.22}$$

Thus, the matrices $\mathbf{T}_a$ give a representation of the Lie algebra with structure constants f_{abc}. (The assumption of antisymmetry of f_{abc} in its indices is necessary to obtain invariant terms for the gauge fields. Antisymmetry in the indices b and c is evident.) If we take a gauge transformation with constant $\Lambda_a(x)$ in (9.18), we see that the gauge fields transform as the adjoint (or regular) representation of the Lie group.

It is sometimes convenient to use the finite gauge transformation of (9.13)

$$\psi(x) \to \mathbf{U}(x)\psi(x) \tag{9.23}$$

where

$$\mathbf{U}(x) = \mathrm{e}^{-\mathrm{i}g\mathbf{T}\cdot\Lambda(x)}. \tag{9.24}$$

The corresponding finite gauge transformation on the gauge fields may be developed from the infinitesimal transformation of (9.18). We first introduce the $p \times p$ matrix

$$\mathbf{A}^{\mu} = \mathbf{A}_a^{\mu}\mathbf{T}_a = \mathbf{A}^{\mu}\cdot\mathbf{T}. \tag{9.25}$$

Multiplying (9.18) by $\mathbf{T}_a$, and using (9.21), the infinitesimal transformation becomes

$$\mathbf{A}^{\mu} \to \mathbf{A}^{\mu} + \mathbf{T}\cdot\partial^{\mu}\Lambda - \mathrm{i}g[\mathbf{T}\cdot\Lambda, \mathbf{A}^{\mu}]. \tag{9.26}$$

This corresponds to the finite transformation

$$\mathbf{A}^{\mu}(x) \to \mathbf{U}(x)(\mathbf{A}^{\mu}(x) - \mathrm{i}g^{-1}\partial^{\mu})\mathbf{U}^{-1}(x) \tag{9.27}$$

taken to linear order in Λ_a.

Next we construct a gauge invariant Lagrangian for the gauge fields themselves. To do this we shall need an object $F_a^{\mu\nu}$ with two Lorentz indices which transforms in a covariant way under the gauge group. This may be constructed directly from the covariant derivative of (9.15) by defining

$$F^{\mu\nu} = F_a^{\mu\nu}\mathbf{T}_a = -\mathrm{i}g^{-1}[\mathrm{D}^{\mu}, \mathrm{D}^{\nu}]. \tag{9.28}$$

Thus

$$F^{\mu\nu} = \partial^{\mu}A^{\nu} - \partial^{\nu}A^{\mu} + \mathrm{i}g[A^{\mu}, A^{\nu}] \tag{9.29}$$

where we have dropped a total derivative, or equivalently

$$F_a^{\mu\nu} = \partial^\mu A_a^\nu - \partial^\nu A_a^\mu - g f_{abc} A_b^\mu A_c^\nu \tag{9.30}$$

where we have used (9.21). We notice that (9.30) is independent of the fermion representation chosen in (9.25).

The transformation property of $F^{\mu\nu}$ under the gauge group derived from (9.27) is

$$F^{\mu\nu}(x) \to U(x) F^{\mu\nu}(x) U^{-1}(x). \tag{9.31}$$

A gauge invariant Lagrangian $\mathscr{L}_{\text{YM}}$ for the gauge (or Yang–Mills) fields may now be written down. It is usual for this purpose to use the generators t_a for the fundamental representation of the gauge group in (9.25). Then, correctly normalised,

$$\mathscr{L}_{\text{YM}} = -\tfrac{1}{2}\,\text{Tr}(F_{\mu\nu} F^{\mu\nu}) \tag{9.32}$$

or equivalently

$$\mathscr{L}_{\text{YM}} = -\tfrac{1}{4} F_a^{\mu\nu} F_{\mu\nu}^a. \tag{9.33}$$

The equivalence of these two forms follows from the conventional normalisation of the generators of the gauge group, which gives for the fundamental representation

$$\text{Tr}(t_a t_b) = \tfrac{1}{2}\delta_{ab}. \tag{9.34}$$

Summarising, we may write down gauge invariant Lagrangians, for Dirac spinor fields interacting with vector gauge fields, of the form

$$\mathscr{L} = \bar{\psi}(\text{i}\gamma^\mu \text{D}_\mu - m)\psi - \tfrac{1}{2}\,\text{Tr}(F_{\mu\nu} F^{\mu\nu}) \tag{9.35}$$

or equivalently

$$\mathscr{L} = \bar{\psi}(\text{i}\gamma^\mu \text{D}_\mu - m)\psi - \tfrac{1}{4} F_{\mu\nu}^a F_a^{\mu\nu}. \tag{9.36}$$

Here, the covariant derivative D_μ is given by (9.15), and the covariant curl $F_{\mu\nu}$ (or $F_{\mu\nu}^a$) by (9.29) and (9.25), with $\mathbf{T}_a$ replaced by t_a, the generators of the fundamental representation, (or (9.30)). The gauge fields transform as the adjoint representation of some Lie group (referred to as the gauge group), and the spinor fields transform as some representation of the gauge group with matrix generators $\mathbf{T}_a$. We may, of course, include further Dirac spinor fields, transforming as chosen representations of the gauge group, in the same fashion. As for the Abelian case, it is not possible to construct gauge invariant mass terms, and the gauge fields are necessarily massless. However, we shall discuss in Chapter 12 a method for generating masses for the gauge fields, if required, by using scalar fields some of which have non-zero vacuum expectation values. We shall also discuss situations where the left- and right-handed components of Dirac fields transform as different representations of the gauge group.

If the gauge group is a simple Lie group (e.g. SU(N)) then there is a single gauge coupling constant g. However, if the gauge group is a semi-simple one, which can be written as a product of simple factors (e.g. $\mathrm{SU}(2)\times\mathrm{SU}(2)$) then there are independent gauge coupling constants for the various simple factors.

9.3 Field equations for gauge field theories

The Euler–Lagrange equations corresponding to the Lagrangian (9.36) are

$$\partial^\nu \frac{\partial \mathscr{L}}{\partial(\partial^\nu A^\mu_a)} = \frac{\partial \mathscr{L}}{\partial A^\mu_a} \tag{9.37}$$

and

$$\partial^\nu \frac{\partial \mathscr{L}}{\partial(\partial^\nu \bar{\psi})} = \frac{\partial \mathscr{L}}{\partial \bar{\psi}}. \tag{9.38}$$

It is easy to check that (9.37) leads to

$$\partial^\mu F^a_{\mu\nu} - g f_{abc} A^\mu_b F^c_{\mu\nu} = g\bar{\psi}\gamma^\mu \mathbf{T}_a \psi. \tag{9.39}$$

This can be written neatly in terms of the covariant derivative of a gauge field,

$$\mathrm{D}^\mu A^\nu{}_a = \partial^\mu A^\nu_a - g f_{abc} A^\mu_b A^\nu_c \tag{9.40}$$

which arises from (9.15) with the replacement

$$(\mathbf{T}_a)_{bc} \to -\mathrm{i} f_{abc} \tag{9.41}$$

for the adjoint representation to which the gauge fields belong. Thus, (9.39) is

$$\mathrm{D}^\mu F^a_{\mu\nu} = g\bar{\psi}\gamma^\mu \mathbf{T}_a \psi. \tag{9.42}$$

The other Euler–Lagrange equation (9.38) leads immediately to

$$(\mathrm{i}\gamma^\mu \mathrm{D}_\mu - m)\psi = 0 \tag{9.43}$$

where D_μ for the fermion fields is as in (9.15).

References

1 Yang C N and Mills R L 1954 *Phys. Rev.* **96** 191
2 Gell-Mann M and Glashow S L 1961 *Ann. Phys., NY* **15** 437

10

FEYNMAN RULES FOR QUANTUM CHROMODYNAMICS AND QUANTUM ELECTRODYNAMICS

10.1 Quantum chromodynamics

We shall see later that quantum electrodynamics (QED) can be unified with the weak interactions leading to a gauge field theory (electroweak theory) based on the non-Abelian group SU(2) ⊗ U(1), with independent coupling constants for the SU(2) and U(1) factors. For electroweak theory, it is necessary to generate masses for the gauge fields, through scalar fields with non-zero VEVS. This is discussed in Chapter 12.

The theory of strong interactions, quantum chromodynamics (QCD), is based on the colour SU(3) group. This is a group which acts on the so-called colour indices of the quarks. Each flavour of quark u, d, s, c, b, ... comes in three 'colours' labelled 1, 2, 3 which form a basis for the three-dimensional representation of colour SU(3). The colour degrees of freedom were introduced, in the first instance, to allow three quarks to be in an s-wave ground state, consistently with Fermi statistics, by having a colour singlet wave function antisymmetric in the colour indices. It was later observed that if this group were gauged, it might provide a theory of the strong interactions, while the flavour degrees of freedom are more closely related to the gauge group of the weak and electromagnetic interactions. In the case of QCD, it has been thought possible to live with massless gauge fields. This possibility rests on the hypothesis of colour confinement, namely that colour degrees of freedom are never observed, and all observed particles are colour singlets (at least at low density and temperature). We shall have more to say about this later. In what follows in this chapter we shall discuss a general simple gauge group with the example of QCD, where the gauge group is colour SU(3), in mind.

In Chapter 15, the possibility will be discussed of combining electroweak theory and QCD into a single gauge theory, referred to as a grand unified theory.

10.2 Problems in quantising gauge field theories

The simplest guess to quantise a (non-Abelian) gauge field theory would be to write for the generating functional

$$W[J_a^\mu] \propto \int \mathscr{D}A^\mu \exp \mathrm{i}\hbar^{-1} \int \mathrm{d}^4x[\mathscr{L}_{\mathrm{YM}}(A_a^\mu) + J_a^\mu A_{a\mu}] \tag{10.1}$$

where the x^0 integration is from $-\infty$ to ∞, and a continuation to Euclidean space is expected to make the path integral well defined. The Lagrangian $\mathscr{L}_{\mathrm{YM}}$ is as in (9.33), and $\int \mathscr{D}A^\mu$ is used as shorthand for $\prod_a \int \mathscr{D}A_a^\mu$. Unfortunately (10.1) is unsatisfactory for several reasons.

First, if we proceed with this generating functional in the free-field case, $g \to 0$, we arrive at

$$W[J_a^\mu] = \exp\left(-\tfrac{1}{2}\mathrm{i}\hbar^{-1} \int \mathrm{d}^4x' \int \mathrm{d}^4x J_{a\mu}(x') D_{\mathrm{F}ab}^{\mu\nu}(x', x) J_{b\nu}(x)\right) \tag{10.2}$$

where

$$D_{\mathrm{F}ab}^{\mu\nu}(x', x) = \delta_{ab} D_{\mathrm{F}}^{\mu\nu}(x', x) \tag{10.3}$$

with

$$D_{\mathrm{F}}^{\mu\nu}(x', x) = \int \frac{\mathrm{d}^4p}{(2\pi)^4} \mathrm{e}^{-\mathrm{i}p(x'-x)} \tilde{D}_{\mathrm{F}}^{\mu\nu}(p) \tag{10.4}$$

and

$$[\tilde{D}_{\mathrm{F}}^{\mu\nu}(p)]^{-1} = -p^2 g^{\mu\nu} + p^\mu p^\nu. \tag{10.5}$$

In order to invert (10.5) we have to separate in terms of the transverse and longitudinal projection operators

$$P_{\mathrm{T}}^{\mu\nu}(p) = g^{\mu\nu} - p^\mu p^\nu / p^2 \tag{10.6}$$

and

$$P_{\mathrm{L}}^{\mu\nu}(p) = p^\mu p^\nu / p^2. \tag{10.7}$$

Thus

$$[\tilde{D}_{\mathrm{F}}^{\mu\nu}(p)]^{-1} = -p^2 P_{\mathrm{T}}^{\mu\nu}(p) + \mathrm{O} P_{\mathrm{L}}^{\mu\nu}(p). \tag{10.8}$$

Inverting gives

$$\tilde{D}_{\mathrm{F}}^{\mu\nu}(p) = -(p^2)^{-1} P_{\mathrm{T}}^{\mu\nu}(p) + \mathrm{O}^{-1} P_{\mathrm{L}}^{\mu\nu}(p) \tag{10.9}$$

where it is understood we will have to introduce the usual $\mathrm{i}\varepsilon$ in the denominator. The second term in (10.9) is infinite, and so the Feynman propagator for the gauge field theory makes no sense.

Second, we should not really have expected a sensible result because (10.1) is overcounting degrees of freedom. In (10.1) we are integrating over all $A_a^\mu(x)$ including those that are connected by a gauge transformation. Consider the case $J_a^\mu = 0$ for all a. Then because $\mathscr{L}_{\mathrm{YM}}$ is invariant under gauge transformations, the integral is constant over the infinite surface in gauge field space obtained from a given A_a^μ by applying all possible gauge transformations. The contribution to the path integral corresponding to integrating over this surface is therefore infinite. We must find some way of separating out this infinity.

Another closely related way of looking at the problem is to say that the action does not depend on all the components of the gauge field. In the free-field case $g \to 0$, we may Fourier transform the gauge fields to obtain

$$\int \mathrm{d}^4x \mathscr{L}_{\mathrm{YM}} = \tfrac{1}{2} \int \frac{\mathrm{d}^4p}{(2\pi)^4} \tilde{A}_a^\mu(p)(-p^2 g_{\mu\nu} + p_\mu p_\nu)\tilde{A}_a^\nu(-p) \tag{10.10}$$

where we have written

$$A_a^\mu(x) = \int \frac{\mathrm{d}^4p}{(2\pi)^4} \mathrm{e}^{-\mathrm{i}p\cdot x} \tilde{A}_a^\mu(p). \tag{10.11}$$

In terms of the transverse and longitudinal projection operators of (10.6) and (10.7)

$$\int \mathrm{d}^4x \mathscr{L}_{\mathrm{YM}} = -\tfrac{1}{2} \int \frac{\mathrm{d}^4p}{(2\pi)^4} \tilde{A}_a^\mu(p) P_{\mu\nu}^{\mathrm{T}}(p) \tilde{A}_a^\nu(-p). \tag{10.12}$$

Thus the action depends only on the transverse components of the gauge field

$$\tilde{A}_{a\mu}^{\mathrm{T}}(p) \equiv P_{\mu\nu}^{\mathrm{T}}(p)\tilde{A}_a^\nu(p) \tag{10.13}$$

and not on the longitudinal components

$$\tilde{A}_{a\mu}^{\mathrm{L}}(p) \equiv P_{\mu\nu}^{\mathrm{L}}(p)\tilde{A}_a^\nu(p). \tag{10.14}$$

The path integral integrates over all the components of A_a^μ. Since the action does not depend on some of these components, an infinity is bound to arise.

10.3 An analogy with ordinary integrals

Before showing how to overcome the problems discussed in §10.2, we discuss an analogous situation which arises in connection with ordinary integrals rather than path integrals. This illuminating analogy has been particularly emphasised by Coleman[1]. Suppose that S is given as a function of $m+n$ real variables x_i $(i=1,\ldots,m+n)$ but that S depends on the last n variables $x_{m+1},\ldots,x_{m+n}$ but not on the first m variables $x_1,\ldots,x_m$. The variables $x_{m+1},\ldots, x_{m+n}$ model the transverse components of the gauge fields, the variables $x_1,\ldots,x_m$ model the longitudinal components, and S models the action. Write

$$W = \int \mathrm{d}x_1 \ldots \int \mathrm{d}x_{m+n}\, \mathrm{e}^{\mathrm{i}S}. \tag{10.15}$$

This would model the generating functional of (10.1) for zero sources, and would be infinite because S does not depend on all the variables of integration. (The integrations are understood to be from $-\infty$ to ∞.) Suppose we consider instead

$$\tilde{W} = \int dx_{m+1} \cdots \int dx_{m+n}\, e^{iS} \tag{10.16}$$

which integrates over only the variables on which S depends, and which we assume is finite (at least after continuation to Euclidean space in some sense.) This can be recast as an integral over all the variables $x_1, \ldots, x_{m+n}$ by using Dirac δ functions.

Let

$$x_i = f_i(x_{m+1}, \ldots, x_{m+n}) \qquad (i = 1, \ldots, m) \tag{10.17}$$

define an arbitrary surface. Then we may rewrite (10.16) as

$$\tilde{W} = \int dx_1 \ldots \int dx_{m+n}\, e^{iS} \prod_{i=1}^{m} \delta[x_i - f_i(x_{m+1}, \ldots, x_{m+n})]. \tag{10.18}$$

If instead we are given (10.17) in the implicit form

$$F_i(x_1, \ldots, x_m, x_{m+1}, \ldots, x_{m+n}) = 0 \qquad (i = 1, \ldots, m) \tag{10.19}$$

for more appropriate functions F_i then we can recast (10.16) in terms of the F_i as follows:

$$\tilde{W} = \int dx_{m+1}, \ldots \int dx_{m+n} \int dF_1 \ldots \int dF_m\, e^{iS} \prod_{i=1}^{m} \delta(F_i) \tag{10.20}$$

changing variables from $F_1, \ldots, F_m$ to $x_1, \ldots, x_m$ gives

$$\tilde{W} = \int dx_1 \ldots \int dx_{m+n}\, e^{iS} \det\left(\frac{\partial F_j}{\partial x_k}\right) \prod_{i=1}^{m} \delta(F_i). \tag{10.21}$$

The determinant is for derivatives of $F_1, \ldots, F_m$ with respect to $x_1, \ldots, x_m$ and because of the Dirac δ functions we need evaluate the determinant only on the surface defined by (10.19). The Faddeev–Popov quantisation procedure for gauge fields is the analogue of (10.21).

10.4 Quantisation of gauge field theory

A procedure which overcomes the difficulties discussed in §10.2 has been devised by Faddeev and Popov[2]. We shall be considering how to amend (10.1) so as to make it well defined and, for simplicity, we shall discuss the path integral for zero source terms. There are some additional difficulties when the sources are included which we shall mention at the end. As discussed in

Chapter 9, the action

$$S[A_{a\mu}]=\int \mathrm{d}^4x\mathscr{L}_{\mathrm{YM}}(A_{a\mu}) \tag{10.22}$$

is gauge invariant, where $\mathscr{L}_{\mathrm{YM}}$ is as in (9.33). Let us denote by $A^{\mathbf{U}}_{a\mu}$ the gauge fields obtained from $A_{a\mu}$ by the finite gauge transformation specified in (9.24), (9.25) and (9.27). The gauge invariance of the action means that

$$S[A^{\mathbf{U}}_{a\mu}]=S[A_{a\mu}]. \tag{10.23}$$

The difficulties discussed earlier arise because we overcount degrees of freedom by integrating over gauge fields which are connected by a gauge transformation. If we could factor out, from the path integral, a path integral over gauge transformations, then we might resolve the problem. The general guage transformation is associated with

$$\mathbf{U}(x)=\mathrm{e}^{-ig\mathbf{T}\cdot\Lambda(x)} \tag{10.24}$$

as in (9.24) and is determined by the gauge parameters $\Lambda_a(x)$. We shall adopt the notation $\int \mathscr{D}\mathbf{U}$ to denote an integration over the gauge group elements in some sense, and we shall try to factor out this path integral. It is possible[3] to define $\int \mathscr{D}\mathbf{U}$ in such a way that

$$\int \mathscr{D}\mathbf{U}f[\mathbf{U}]=\int \mathscr{D}\mathbf{U}f[\mathbf{U}\mathbf{U}'] \tag{10.25}$$

where $f[\mathbf{U}]$ is any functional of $\mathbf{U}(x)$, and $\mathbf{U}'(x)$ is a fixed gauge transformation. In the case where the integration is restricted to an infinitesimal region we may take

$$\mathscr{D}\mathbf{U}=\prod_a \mathscr{D}\Lambda_a. \tag{10.26}$$

To check the correctness of (10.25) is to check that

$$\int \mathscr{D}\mathbf{U}''f[\mathbf{U}'']=\int \mathscr{D}\mathbf{U}f[\mathbf{U}''] \tag{10.27}$$

where

$$\mathbf{U}''=\mathbf{U}\mathbf{U}'. \tag{10.28}$$

This is indeed the case for infinitesimal regions with the $\mathscr{D}\mathbf{U}$ of (10.26) because

$$\det\left(\frac{\partial\Lambda''_a}{\partial\Lambda_b}\right)=1+\mathrm{O}(\Lambda'^2). \tag{10.29}$$

(See problem 10.1.)

In the end we want to replace (10.1) by a path integral that does not integrate over gauge fields which are connected by a gauge transformation. In other words, we want to have a path integral which remains in a definite gauge. Thus, we want to be able to introduce a gauge fixing term as in (3.130a) for the Abelian case. Generalising (3.128) a little, we may think of a choice of gauge as a set of conditions

$$F_a(A_b^\mu)=0 \tag{10.30}$$

which may involve the derivatives of the gauge fields. The gauge choices in which we shall be particularly interested are given by

$$F_a(A_b^\mu)\equiv\partial_\mu A_a^\mu - f_a(x)=0 \tag{10.31}$$

where the $f_a(x)$ are some given functions of x. With a view to importing the required gauge fixing term, consider the functional

$$\Delta[A_{a\mu}]=\int \mathscr{D}\mathbf{U}\,\delta[F_a(A_{b\mu}^{\mathbf{U}})] \tag{10.32}$$

where a functional delta function has been introduced, and for compactness of notation we write

$$\delta[F_a]\equiv\prod_a \delta[F_a]. \tag{10.33}$$

The functional Δ is invariant under gauge transformations:

$$\Delta[A_{a\mu}^{\mathbf{U}'}]=\Delta[A_{a\mu}]. \tag{10.34}$$

The proof is brief:

$$\Delta[A_{a\mu}^{\mathbf{U}'}]=\int \mathscr{D}\mathbf{U}\,\delta[F_a(A_{b\mu}^{\mathbf{U}\mathbf{U}'})]. \tag{10.35}$$

Using (10.25) the result follows. The inverse of Δ will have the defining property

$$\Delta^{-1}[A_{a\mu}]\Delta[A_{a\mu}]=1. \tag{10.36}$$

With the aid of (10.36) and (10.32) the path integral we want may be recast as

$$\int \mathscr{D}A^\mu \exp\left(\mathrm{i}\int \mathrm{d}^4x\,\mathscr{L}_{\mathrm{YM}}(A_{a\mu})\right)=\int \mathscr{D}A^\mu\,\mathrm{e}^{\mathrm{i}S[A_{a\mu}]}$$

$$=\int \mathscr{D}A^\mu \Delta^{-1}[A_{a\mu}]\int \mathscr{D}\mathbf{U}\,\delta[F_a(A_{b\mu}^{\mathbf{U}})]\,\mathrm{e}^{\mathrm{i}S[A_{a\mu}]}. \tag{10.37}$$

Using the gauge invariance of the action and of Δ, as in (10.23) and (10.34),

$$\int \mathscr{D}A^\mu\,\mathrm{e}^{\mathrm{i}S[A_{a\mu}]}=\int \mathscr{D}A^\mu \Delta^{-1}[A_{a\mu}]\int \mathscr{D}\mathbf{U}\,\delta[F_a(A_{b\mu}^{\mathbf{U}})]\,\mathrm{e}^{\mathrm{i}S[A_{a\mu}^{\mathbf{U}}]}. \tag{10.38}$$

Observing that $\int \mathscr{D}A_\mu^{\mathbf{U}}$ is the same path integral as $\int \mathscr{D}A_\mu$, we now see that

$$\int \mathscr{D}A^\mu \, \mathrm{e}^{\mathrm{i}S[A_{a\mu}]} = \int \mathscr{D}\mathbf{U} \int \mathscr{D}A^\mu \Delta^{-1}[A_{a\mu}]\delta[F_a(A_{b\mu})] \, \mathrm{e}^{\mathrm{i}S[A_{a\mu}]}. \tag{10.39}$$

The integral over gauge transformations $\int \mathscr{D}\mathbf{U}$ has now factored out, as we had hoped, and the gauge is being fixed by the functional δ function, as we had also intended. Thus if we write

$$\int \mathscr{D}A^\mu \, \mathrm{e}^{\mathrm{i}S[A_{a\mu}]} \propto \int \mathscr{D}A^\mu \Delta^{-1}[A_{a\mu}]\delta(F_a(A_{b\mu})] \, \mathrm{e}^{\mathrm{i}S[A_{a\mu}]} \tag{10.40}$$

then the difficulties of §10.2 should have been removed.

It is necessary to be able to evaluate Δ^{-1} in (10.40). This is done by changing the integration variables in (10.32) from $\mathscr{D}\mathbf{U}$ to $\prod_a \mathscr{D}F_a$. We may use the $\mathscr{D}\mathbf{U}$ of (10.26) because the functional δ functions in (10.40) and (10.32) restrict $\mathbf{U}$ to an infinitesimal region around the identity. Thus

$$\int \mathscr{D}\mathbf{U} = \int \prod_a \mathscr{D}\Lambda_a = \int \prod_a \mathscr{D}F_a \det\left(\frac{\delta\Lambda_b(x)}{\delta F_a(x')}\right) \tag{10.41}$$

where the functional differentiation is in the sense of (1.18).

Now

$$\begin{aligned} \Delta[A_{a\mu}] &= \int \prod_a \mathscr{D}F_a \det\left(\frac{\delta\Lambda_b(x)}{\delta F_a(x')}\right)\delta[F_a] \\ &= \det\left(\frac{\delta\Lambda_b(x)}{\delta F_a(x')}\right)\bigg|_{F_a=0}. \end{aligned} \tag{10.42}$$

The inverse is

$$\Delta^{-1}[A_{a\mu}] = \det\left(\frac{\delta F_a(x')}{\delta\Lambda_b(x)}\right)\bigg|_{F_a=0}. \tag{10.43}$$

Thus, we have the explicit expression (analogous to (10.21) for the case of ordinary integrals),

$$\int \mathscr{D}A^\mu \exp\left(\mathrm{i}\int \mathrm{d}^4x \mathscr{L}_{\mathrm{YM}}(A_{a\mu})\right) = \int \mathscr{D}A^\mu \, \mathrm{e}^{\mathrm{i}S[A_{a\mu}]} \propto \int \mathscr{D}A^\mu \det\left(\frac{\delta F_a(x')}{\delta\Lambda_b(x)}\right)\delta[F_a] \, \mathrm{e}^{\mathrm{i}S[A_{a\mu}]} \tag{10.44}$$

where

$$F_a(x) = F_a(A_{b\mu}^{\mathbf{U}}(x)) \tag{10.45}$$

and $A_{a\mu}^{\mathbf{U}}$ is the gauge field obtained from $A_{a\mu}$ by the gauge transformation defined in (9.24), (9.25) and (9.27). The generalisation when the source terms are non-zero is

$$W[J_a^\mu] \propto \int \mathscr{D}A^\mu \det\left(\frac{\delta F_a(x')}{\delta \Lambda_b(x)}\right)\delta[F_a]$$
$$\times \exp\left[\mathrm{i}\int \mathrm{d}^4x\left(\mathscr{L}_{\mathrm{YM}}(A_a^\mu)+J_a^\mu A_{a\mu}\right)\right]. \tag{10.46}$$

That this generalisation is correct is by no means obvious because the derivation we have given depends on the gauge invariance of the action, and this is broken by the source terms. However, it can be shown that (10.46) is indeed correct provided we are in the end going to use the generating functional to calculate S-matrix elements. This point is discussed in Taylor[3].

10.5 Gauge fixing terms and Faddeev–Popov ghosts

If we are to use the generating functional for gauge field theory to develop a perturbation theory, we need to convert the functional δ function and the functional determinant in (10.46) into exponentials.

First, we use the functional δ function to obtain a conventional gauge fixing term in the exponent. With the choice of F_a of (10.31),

$$\delta[F_a]=\delta[\partial_\mu A_a^\mu - f_a(x)]. \tag{10.47}$$

Multiplying the generating functional by

$$\int\left(\prod_c \mathscr{D}f_c\right)\exp\left(-\frac{\mathrm{i}}{2\xi}\int \mathrm{d}^4x f_a^2(x)\right)$$

simply multiplies by a constant which can be absorbed into the overall normalisation. But

$$\int\prod_a \mathscr{D}f_a \exp\left(\frac{-\mathrm{i}}{2\xi}\int \mathrm{d}^4x f_a^2(x)\right)\delta[F_a]=\exp\left(\frac{-\mathrm{i}}{2\xi}\int \mathrm{d}^4x(\partial_\mu A_a^\mu)^2\right). \tag{10.48}$$

Consequently, the generating functional of (10.46) becomes

$$W[J_a^\mu] \propto \int \mathscr{D}A^\mu \det\left(\frac{\delta F_a(x')}{\delta \Lambda_b(x)}\right)$$
$$\times \exp\left[\mathrm{i}\int \mathrm{d}^4x\left(\mathscr{L}_{\mathrm{YM}}(A_a^\mu)+J_a^\mu A_{a\mu}-\frac{1}{2\xi}(\partial_\mu A_a^\mu)^2\right)\right]. \tag{10.49}$$

Secondly, we convert the determinant in (10.49) into an exponential by introducing the Faddeev–Popov ghost fields. These do not correspond to physical particles but are simply a mathematical device to enable a perturbation series to be developed. Using (8.19) for complex Grassmann variables $\eta_a(x)$ we have

$$\int \mathscr{D}\eta^* \mathscr{D}\eta \exp\left(\mathrm{i} \int \mathrm{d}^4x \int \mathrm{d}^4x' \eta_a^*(x') B_{ab}(x', x) \eta_b(x) \right) \propto \det(-\mathbf{B}) \quad (10.50)$$

where

$$\mathscr{D}\eta^* \mathscr{D}\eta \equiv \prod_a \mathscr{D}\eta_a^* \mathscr{D}\eta_a. \quad (10.51)$$

(We have absorbed $\det(-\mathrm{i}\mathbf{I})$ into the constant of proportionality for later convenience.)

Taking

$$B_{ab}(x', x) = -\frac{\delta F_a(x')}{\delta \Lambda_b(x)} \quad (10.52)$$

gives

$$\det\left(\frac{\delta F_a(x')}{\delta \Lambda_b(x)}\right) \propto \int \mathscr{D}\eta^* \mathscr{D}\eta \exp\left(-\mathrm{i} \int \mathrm{d}^4x \int \mathrm{d}^4x' \eta_a^*(x') \frac{\delta F_a(x')}{\delta \Lambda_b(x)} \eta_b(x) \right). \quad (10.53)$$

Using the infinitesimal gauge transformation specified in (9.18) we have

$$\frac{\delta A_a^\mu(x')}{\delta \Lambda_b(x)} = \delta_{ab} \partial_{x'}^\mu \delta^4(x'-x) + g f_{abc} A_c^\mu(x) \delta^4(x'-x). \quad (10.54)$$

The functional differentiation is in the sense of (1.18). Thus, with the F_a of (10.31)

$$\frac{\delta F_a(x')}{\delta \Lambda_b(x)} = \partial_{\mu x'} \{[\delta_{ab} \partial_{x'}^\mu + g f_{abc} A_c^\mu(x')] \delta^4(x'-x)\}. \quad (10.55)$$

The required determinant (10.53) now becomes (after integration by parts)

$$\det\left(\frac{\delta F_a(x')}{\delta \Lambda_b(x)}\right) \propto \int \mathscr{D}\eta^* \mathscr{D}\eta \exp\left(\mathrm{i} \int \mathrm{d}^4x \partial_\mu \eta_a^* (\partial^\mu \eta_a + g f_{abc} \eta_b A_c^\mu) \right). \quad (10.56)$$

In the covariant gauge of (10.31), the final generating functional for a general gauge field theory is

$$W[J_a^\mu] \propto \int \mathscr{D}A^\mu \int \mathscr{D}\eta^* \mathscr{D}\eta \exp\left[\mathrm{i} \int \mathrm{d}^4x \left(\mathscr{L}_{\mathrm{YM}} - \frac{1}{2\xi} (\partial_\mu A_a^\mu)^2 + \mathscr{L}_{\mathrm{FP}} + J_a^\mu (A_\mu)_a \right) \right]. \quad (10.57)$$

where

$$\mathscr{L}_{\mathrm{FP}} = \partial_\mu \eta_a^* (\partial^\mu \eta_a + g f_{abc} \eta_b A_c^\mu) \quad (10.58)$$

and, as in (9.33),

$$\mathscr{L}_{\mathrm{YM}} = -\tfrac{1}{4} F_a^{\mu\nu} F_{a\mu\nu}. \tag{10.59}$$

A gauge fixing term has appeared, and, additionally, Faddeev–Popov ghost terms. In the Abelian case of QED, the Faddeev–Popov ghost Lagrangian reduces to $\partial_\mu \eta^* \partial^\mu \eta$. There is no coupling of the ghost fields to the gauge field, and the ghosts simply contribute a multiplicative constant, which may be absorbed into the normalisation of the generating functional. Thus, in the covariant gauges used here, there is no need to introduce Faddeev–Popov ghosts into QED. However, the ghosts play an important role in the non-Abelian case of QCD.

Other choices of gauge are possible. For example, we may choose a gauge by writing

$$F_a(A_b^\mu) \equiv t_\mu A_a^\mu - f_a(x) = 0 \tag{10.60}$$

where t_μ is a four-vector with

$$t_\mu t^\mu = 1. \tag{10.61}$$

These non-covariant gauges are referred to as axial gauges. They have the advantage that Faddeev–Popov ghosts decouple from gauge fields even in the non-Abelian case. However, there is the (more than) compensating disadvantage that the gauge field propagator turns out to be very complicated in these gauges. (See problem 10.3.) We shall not use these gauges here.

Finally, we may include fermion fields ψ transforming as an arbitrary representation of the gauge group, by adding to $\mathscr{L}$ the fermion Lagrangian

$$\mathscr{L}_{\mathrm{F}} = \bar{\psi}(\mathrm{i}\gamma^\mu \mathrm{D}_\mu - m)\psi \tag{10.62}$$

with the covariant derivative D_μ as in (9.15):

$$\mathrm{D}_\mu \psi \equiv (\partial_\mu + \mathrm{i} g \mathbf{T} \cdot A_\mu)\psi. \tag{10.63}$$

10.6 Feynman rules for gauge field theories

Our experience in Chapter 6 of deriving Feynman rules from a Lagrangian will allow us to read off the Feynman rules for gauge field theory. We shall want the Feynman rules in momentum space, so we first Fourier transform the fields as in (10.11):

$$A_a^\mu(x) = \int \frac{\mathrm{d}^4 p}{(2\pi)^4} \, \mathrm{e}^{-\mathrm{i}p \cdot x} \, \tilde{A}_a^\mu(p) \tag{10.64}$$

and similarly for $\eta_a(x)$ and $\psi(x)$. The terms quadratic in the fields in the action of (10.57) yield

$$\int d^4x\mathscr{L}\text{ (quadratic)}$$

$$=\frac{1}{2}\int\frac{d^4p}{(2\pi)^4}\tilde{A}_a^\mu(p)[-p^2g_{\mu\nu}+(1-\xi^{-1})p_\mu p_\nu]\tilde{A}_a^\nu(-p)+\text{ghost terms.}$$

(10.65)

Corresponding to (10.5) we now have the inverse propagator

$$[\tilde{D}_F^{\mu\nu}(p)]^{-1}=-p^2g^{\mu\nu}+(1-\xi^{-1})p^\mu p^\nu \quad (10.66)$$

and in terms of the transverse and longitudinal projection operators of (10.6) and (10.7)

$$[\tilde{D}_F^{\mu\nu}(p)]^{-1}=-p^2P_T^{\mu\nu}(p)-\xi^{-1}p^2P_L^{\mu\nu}(p). \quad (10.67)$$

Now that we have the gauge fixing term, it is possible to invert (10.67) to obtain a sensible answer in contrast to (10.9). Thus

$$\tilde{D}_F^{\mu\nu}(p)=(p^2+i\varepsilon)^{-1}[-g^{\mu\nu}+(1-\xi)p^\mu p^\nu/p^2] \quad (10.68)$$

where we have introduced the usual iε in the denominator, to resolve ambiguity in the meaning of this expression. The two most frequently used gauges are Feynman gauge, $\xi=1$ and Landau gauge, $\xi=0$. The propagator takes its simplest form in Feynman gauge, but in Landau gauge the propagator is purely transverse and this can sometimes have calculational advantages. As we shall see later, the Landau gauge also has advantages in theories with Higgs scalar mesons, because the scalars decouple from the Faddeev–Popov ghosts in this gauge. (This is related to the transverse nature of the gauge field.) For Feynman diagrams we shall use a wiggly line

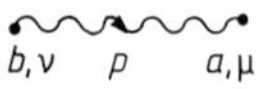

: $i\delta_{ab}\tilde{D}_F^{\mu\nu}(p)$. (10.69)

The quadratic terms for the Faddeev–Popov ghosts are those of a massless complex scalar field (though the Faddeev–Popov ghost fields are Grassmann variables). Correspondingly, the propagator is (see problem 4.1)

$$\tilde{\Delta}_{Fab}^{\text{ghost}}(p)=\delta_{ab}(p^2+i\varepsilon)^{-1}. \quad (10.70)$$

For Feynman diagrams we shall use a dotted line.

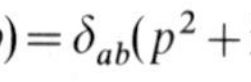

: $i\tilde{\Delta}_{Fab}^{\text{ghost}}(p)$. (10.71)

Because Faddeev–Popov ghost fields are Grassmann variables, the closed loops in Feynman diagrams involving them will each attract a minus sign. To obtain the Feynman rules for the various interaction vertices in the theory, we next

Fourier transform the remaining terms in (10.57) and (10.62). For the gauge fields this leads to

$$\int \mathrm{d}^4x \mathscr{L} \text{ (interaction)}$$
$$= -\mathrm{i}gf_{abc}\left(\int \frac{\mathrm{d}^4p}{(2\pi)^4}\frac{\mathrm{d}^4q}{(2\pi)^4}\frac{\mathrm{d}^4r}{(2\pi)^4}\, p_\mu g_{\lambda\nu} \times \delta(p+q+r)\tilde{A}^\lambda_a(p)\tilde{A}^\mu_b(q)\tilde{A}^\nu_c(r)\right)$$
$$-\tfrac{1}{4}g^2 f_{abc}f_{ade}\left(\int \frac{\mathrm{d}^4p}{(2\pi)^4}\int \frac{\mathrm{d}^4q}{(2\pi)^4}\int \frac{\mathrm{d}^4r}{(2\pi)^4}\int \frac{\mathrm{d}^4s}{(2\pi)^4}\right.$$
$$\left.\times\, \delta(p+q+r+s)g_{\lambda\nu}g_{\mu\rho}\tilde{A}^\lambda_b(p)\tilde{A}^\mu_c(q)\tilde{A}^\nu_d(r)\tilde{A}^\rho_e(s)\right). \tag{10.72}$$

The trilinear term is antisymmetric under interchanges of (p, λ), (q, μ), (r, ν) in the coefficient of $\tilde{A}^\lambda_a(p)\tilde{A}^\mu_b(q)\tilde{A}^\nu_c(r)$ as a result of the antisymmetry of f_{abc} in its indices. We may make this antisymmetry explicit by making the replacement

$$-\mathrm{i}gf_{abc}p_\mu g_{\lambda\nu} \rightarrow -\frac{1}{3!}\mathrm{i}gf_{abc}[(r-q)_\lambda g_{\mu\nu}+(q-p)_\nu g_{\lambda\mu}+(p-r)_\mu g_{\nu\lambda}]. \tag{10.73}$$

It is easily checked that each of the six terms on the right-hand side of (10.73) is identical to the original term on the left-hand side. Thus the replacement does not change the value of the integral. (See problem 10.4.) Since we now have an expression which treats $\tilde{A}^\lambda_a(p)$, $\tilde{A}^\mu_b(q)$ and $\tilde{A}^\nu_c(r)$ on the same footing, we may read off the Feynman rule (including the usual factor of i for vertices).

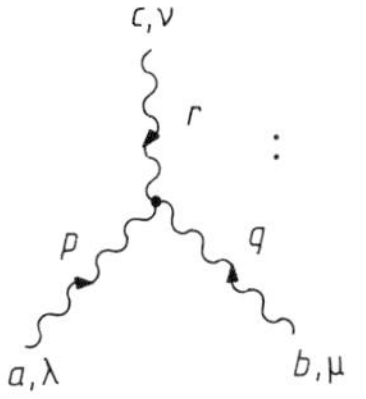

$$: \quad gf_{abc}[(r-q)_\lambda g_{\mu\nu}+(q-p)_\nu g_{\lambda\mu}+(q-p)_\nu g_{\lambda\mu}+(p-r)_\mu g_{\nu\lambda}] \tag{10.74}$$

with

$$p+q+r=0. \tag{10.75}$$

The quadrilinear term is symmetric under interchanges of (b, λ), (c, μ), (d, ν), (e, ρ) in the coefficient of $\tilde{A}^\lambda_b(p)\tilde{A}^\mu_c(q)\tilde{A}^\nu_d(r)\tilde{A}^\rho_e(s)$. The symmetry becomes explicit on making the replacement

$$-\frac{1}{4}g^2 f_{abc}f_{ade}g_{\lambda\nu}g_{\mu\rho} \rightarrow -\frac{1}{4!}g^2[f_{abc}f_{ade}(g_{\lambda\nu}g_{\mu\rho}-g_{\mu\nu}g_{\lambda\rho})$$
$$+f_{adc}f_{abe}(g_{\nu\lambda}g_{\mu\rho}-g_{\mu\lambda}g_{\nu\rho})$$
$$+f_{abd}f_{ace}(g_{\lambda\mu}g_{\nu\rho}-g_{\mu\nu}g_{\lambda\rho})]. \tag{10.76}$$

Again each term on the right-hand side of the equation is identical to the original term on the left-hand side. We now have an expression which treats $\tilde{A}_b^{\lambda}(p)$, $\tilde{A}_c^{\mu}(q)$, $\tilde{A}_d^{\nu}(r)$ and $\tilde{A}_e^{\rho}(s)$ on the same footing, and we may read off the Feynman rule:

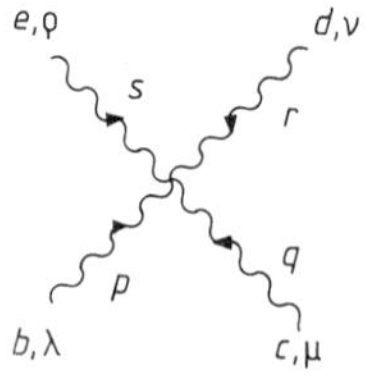

$$-ig^2[f_{abc}f_{ade}(g_{\lambda\nu}g_{\mu\rho}-g_{\mu\nu}g_{\lambda\rho}) \\ +f_{adc}f_{abe}(g_{\nu\lambda}g_{\mu\rho}-g_{\mu\lambda}g_{\nu\rho}) \\ +f_{abd}f_{ace}(g_{\lambda\mu}g_{\nu\rho}-g_{\nu\mu}g_{\lambda\rho})] \tag{10.77}$$

with four-momentum conservation at the vertex

$$p+q+r+s=0. \tag{10.78}$$

In the case of QED, where the gauge group is Abelian, $f_{abc}=0$, neither the trilinear nor the quadrilinear vertex in the gauge field occurs.

For the (ghost)–(gauge field) interaction, the relevant Fourier transform is

$$\int d^4x \mathscr{L}_{FP}\text{ (interaction)}$$

$$=igf_{abc}\int\frac{d^4p}{(2\pi)^4}\int\frac{d^4q}{(2\pi)^4}\int\frac{d^4r}{(2\pi)^4}\,p_{\mu}\delta^4(q+r-p)\tilde{\eta}_a^*(p)\tilde{\eta}_b(q)\tilde{A}_c^{\mu}(r) \tag{10.79}$$

giving the Feynman rule

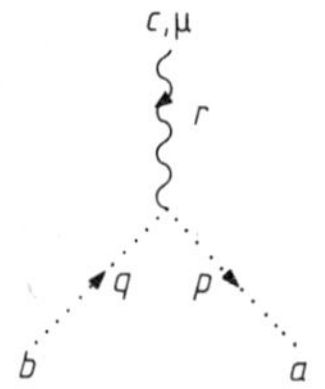

$$: \quad -gf_{abc}p_{\mu} \tag{10.80}$$

with

$$q+r-p=0. \tag{10.81}$$

Again this vertex does not occur for QED.

As in Chapter 8, we denote Fermion propagators by a solid line.

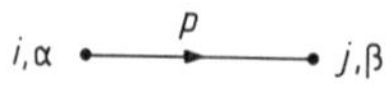

$$: \quad i\delta_{ji}[\tilde{S}_F(p)]_{\beta\alpha} \tag{10.82}$$

where α and β are spinor indices. The (gauge field)–(fermion) interaction may be read straight off from (10.62) and (10.63):

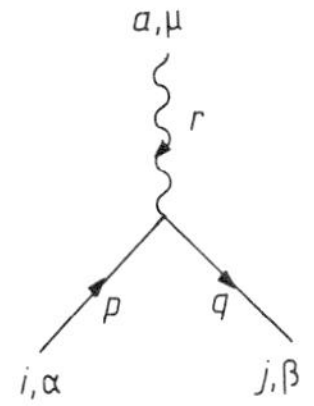

$$: \quad -\mathrm{i}g(\gamma_\mu)_{\beta\alpha}(T_a)_{ji} \tag{10.83}$$

with

$$p+r-q=0. \tag{10.84}$$

For QED, $\mathrm{i}gT_a$ is simply replaced by $\mathrm{i}q$, where q is the charge of the fermion and the Feynman rule is

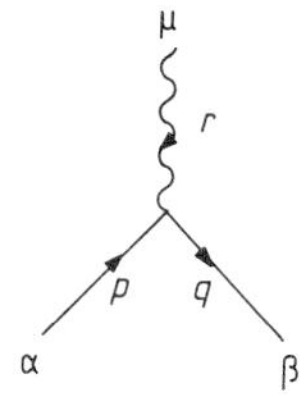

$$: \quad -\mathrm{i}q(\gamma_\mu)_{\alpha\beta}. \tag{10.85}$$

10.7 Scattering amplitudes with gauge fields

By analogy with §6.4 for scalar fields, we obtain a (Lorentz invariant) scattering amplitude when there are gauge field external legs, by calculating a (momentum space) Green function with the appropriate external legs. Again the propagators associated with external lines should be divided out, as for OPI Green functions, but all connected diagrams, not just one-particle irreducible ones, should be included. The only difference from the scalar case is that incoming (or outgoing) external gauge field lines will have associated factors $\varepsilon^\mu(k, \lambda)$ (or $\varepsilon^{\mu *}(k, \lambda)$, as defined in §3.4, where λ specifies the helicity of the vector particle, and μ is a Lorentz index. These factors arise from the expansion in Fourier components of the free massless vector field, (3.127) and (3.135), substituted in the analogue of (6.45) and (6.46). After carrying out differentiations with respect to $a(k, \lambda)$ in the analogue of (6.50), these factors remain.

Problems

10.1 $\mathbf{U}$, $\mathbf{U}'$, $\mathbf{U}''$ are infinitesimal gauge transformations satisfying $\mathbf{U}\mathbf{U}'=\mathbf{U}''$ as in (10.28), with gauge parameters Λ_a, Λ'_a, Λ''_a respectively. Obtain an expression for Λ''_a correct to second order Λ_a, Λ'_a and check (10.29).

10.2 Show that the right-hand side of (10.44) is independent of the choice of gauge. (One way of doing this is to follow the analogous step in §10.3.)

10.3(*a*) Derive the gauge field propagator for the axial gauges specified by (10.60) and (10.61). (*b*) Show that in axial gauges there is no coupling of the Faddeev–Popov ghosts to the gauge fields.

10.4 Check directly that each of the terms on the right-hand side of (10.73) and (10.76) is identical to the original term of the left-hand side, and that the various claims made about symmetry and antisymmetry for these two vertices are correct.

10.5 Derive the Feynman rules for the Lagrangian

$$\mathscr{L}=\tfrac{1}{2}(\partial_\mu\varphi)(\partial^\mu\varphi)(1+g\varphi)^2-\tfrac{1}{2}\mu^2\varphi^2(1+\tfrac{1}{2}g\varphi)^2+J\varphi(1+\tfrac{1}{2}g\varphi)$$

(Coleman p. 48).

References

1 Coleman S 1973 *Lectures given at 1973 International Summer School of Physics, Ettore Majorana* Sections 4 and 5
2 Faddeev L D and Popov V N 1967 *Phys. Lett.* **25B** 29
3 Taylor J C 1976 *Gauge Theories of Weak Interactions* (Cambridge: Cambridge University Press) Chapter 11

11

RENORMALISATION OF QCD AND QED AT ONE-LOOP ORDER

11.1 Counter terms for gauge field theories

In the covariant gauges discussed in Chapter 10, the gauge field propagator of (10.68) has a high momentum behaviour $\sim(p^2)^{-1}$ just like the scalar propagator. Thus, the (superficial) degree of divergence of a Feynman diagram involving gauge fields will be given by a slight generalisation to include derivative interactions of the expressions derived in §8.4. (See problem 11.1.) The gauge field interactions in (10.58), (10.59), (10.62) and (10.63) all satisfy the criterion to avoid the degree of divergence increasing with the number of vertices

$$\tfrac{3}{2}N_{\mathrm{F}}+N_{\mathrm{B}}+N_{\mathrm{D}}-4\leqslant 0 \tag{11.1}$$

where N_{B} is the number of boson lines, N_{F} is the number of fermion lines, and N_{D} is the number of derivatives for a particular type of vertex. Consequently, we expect a renormalisable theory, and we should be able to generate all necessary counter terms for the renormalisation from a bare Lagrangian which involves the same types of vertices as the renormalised Lagrangian. However, at first sight there could be rather a lot of counter terms because there are two different vertices involving only gauge fields, one involving gauge fields and ghost fields, and one involving fermion fields and a gauge field for each type of fermion field. Fortunately, the situation is simpler than this, because it is possible to prove[1] that if the bare Lagrangian is gauge invariant so is the renormalised Lagrangian. There is thus just a single bare coupling constant *and* a single renormalised coupling constant for a simple gauge group. We shall not prove that this is true to all orders of perturbation theory here, but shall simply assume that it is true. However, even if we were not to assume this general result, we would be able to prove its correctness at one-loop order by computing all the vertices. (See problem 11.2.)

The renormalised Lagrangian for a general simple gauge group is

$$\begin{aligned}\mathscr{L}=&-\frac{1}{4}F_a^{\mu\nu}F^a_{\mu\nu}-\frac{1}{2\xi}(\partial_\mu A^\mu_a)^2\\&+\partial_\mu\eta_a^*(\partial^\mu\eta_a+gf_{abc}\eta_b A^\mu_c)\\&+\bar{\psi}(\mathrm{i}\gamma^\mu \mathrm{D}_\mu-m)\psi\end{aligned} \tag{11.2}$$

where

$$F_a^{\mu\nu} = \partial^\mu A_a^\nu - \partial^\nu A_a^\mu - g f_{abc} A_b^\mu A_c^\nu \tag{11.3}$$

and

$$\mathrm{D}_\mu \psi = (\partial_\mu + \mathrm{i} g \mathbf{T} \cdot A_\mu)\psi \tag{11.4}$$

for a fermion field transforming according to the representation $\mathbf{T}_a$ of the gauge group. We write for the bare Lagrangian

$$\mathscr{L}_\mathrm{B} = \mathscr{L} + \Delta\mathscr{L} \tag{11.5}$$

where $\Delta\mathscr{L}$ is the counter term Lagrangian.

We assume that the bare Lagrangian takes exactly the same form as the renormalised Lagrangian but with the bare quantities ψ_B, $(A_a^\mu)_\mathrm{B}$, $(\eta_a)_\mathrm{B}$, ξ_B, m_B and g_B replacing the corresponding renormalised quantities. Then the counter term Lagrangian is of the form

$$\begin{aligned} \Delta\mathscr{L} = &-\frac{1}{4}\Delta Z_A F_a^{\mu\nu} F_{\mu\nu}^a - \frac{K_\xi}{2\xi}(\partial_\mu A_a^\mu)^2 \\ &+ \Delta Z_\eta \partial_\mu \eta_a^* \partial^\mu \eta_a + \Delta Z_\psi \bar{\psi} \mathrm{i} \gamma^\mu \partial_\mu \psi \\ &- m K_m \bar{\psi}\psi + g K_1 \partial_\mu \eta_a^* \eta_b A_c^\mu f_{abc} \\ &- g K_2 \bar{\psi} \gamma^\mu \mathbf{T}_a \psi A_a^\mu + g K_3 f_{abc} A_b^\mu A_c^\nu \partial^\mu (A_a)_\nu \\ &- \frac{g^2}{4} K_4 f_{abc} f_{ade} A_b^\mu A_c^\nu (A_d)_\mu (A_e)_\nu . \end{aligned} \tag{11.6}$$

The relationships between bare and renormalised fields and masses are

$$(A_a^\mu)_\mathrm{B} = (1 + \Delta Z_A)^{1/2} A_a^\mu = Z_A^{1/2} A_a^\mu \tag{11.7}$$

$$(\eta_a)_\mathrm{B} = (1 + \Delta Z_\eta)^{1/2} \eta_a = Z_\eta^{1/2} \eta_a \tag{11.8}$$

$$\psi_\mathrm{B} = (1 + \Delta Z_\psi)^{1/2} \psi = Z_\psi^{1/2} \psi \tag{11.9}$$

$$\xi_\mathrm{B}^{-1} = \xi^{-1}(1 + K_\xi) Z_A^{-1} = \xi^{-1} Z_\xi Z_A^{-1} \tag{11.10}$$

$$m_\mathrm{B} = m(1 + K_m) Z_\psi^{-1} = m Z_m Z_\psi^{-1} \tag{11.11}$$

$$g_\mathrm{B} = g(1 + K_1) Z_\eta^{-1} Z_A^{-1/2} = g Z_1 Z_\eta^{-1} Z_A^{-1/2} \tag{11.12}$$

$$g_\mathrm{B} = g(1 + K_2) Z_\psi^{-1} Z_A^{-1/2} = g Z_2 Z_\psi^{-1} Z_A^{-1/2} \tag{11.13}$$

$$g_\mathrm{B} = g(1 + K_3) Z_A^{-3/2} = g Z_3 Z_A^{-1/3} \tag{11.14}$$

$$g_\mathrm{B}^2 = g^2(1 + K_4) Z_A^{-2} = g^2 Z_4 Z_A^{-2} \tag{11.15}$$

where we have defined

$$Z_A = 1 + \Delta Z_A \tag{11.16}$$

$$Z_\eta = 1 + \Delta Z_\eta \tag{11.17}$$

$$Z_\psi = 1 + \Delta Z_\psi \tag{11.18}$$

$$Z_\xi = 1 + K_\xi \tag{11.19}$$

$$Z_m = 1 + K_m \tag{11.20}$$

$$Z_i = 1 + K_i \qquad i = 1, \ldots, 4. \tag{11.21}$$

There are four different expressions for the bare coupling constant (11.12) to (11.15) depending on which vertex we consider. Thus, there are the relationships amongst renormalisation constants

$$Z_1 Z_\eta^{-1} = Z_2 Z_\eta^{-1} = Z_3 Z_A^{-1} = Z_4^{1/2} Z_A^{-1/2}. \tag{11.22}$$

11.2 Calculation of renormalisation constants

We now evaluate to one-loop order the renormalisation constants defined in the last section. We shall regularise the divergent loop integrations using dimensional regularisation, and shall adopt the MS renormalisation scheme described in §7.5. As in Chapter 7, in 2ω spatial dimensions, the dimensionless coupling constant

$$\hat{g} = gM^{\omega - 2} = gM^{-\varepsilon/2} \tag{11.23}$$

is introduced, where M is an arbitrary mass scale.

To compute Z_A we must consider the OPI Green function for two gauge fields $\tilde{\Gamma}^{(2)}_{ba,\nu\mu}(-p, p)$ which we denote by

$\tilde{\Gamma}^{(2)}_{ba,\nu\mu}(-p, p)$: 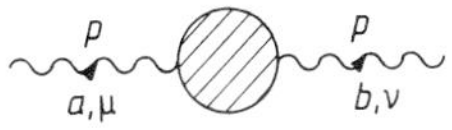

(11.24)

(This is often referred to as calculating the vacuum polarisation diagrams.) The contributions to this Green function to one-loop order are

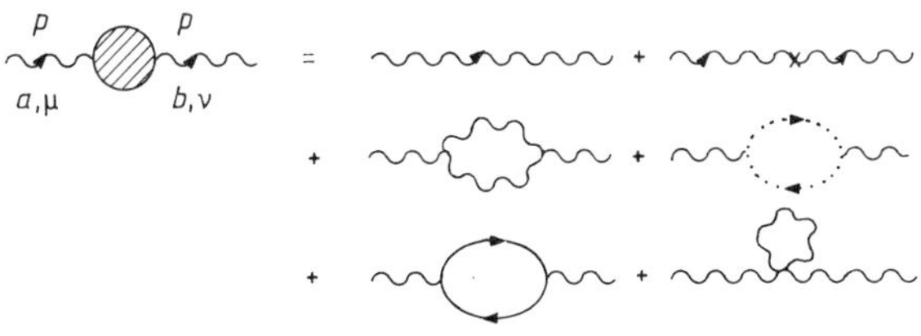

(11.25)

Here we have used a cross to denote a counter term as in Chapter 7. The last diagram is quadratically divergent with no momentum flowing into the loop and vanishes in dimensional regularisation. This is a good thing because this diagram has no dependence on p and could only contribute to a gauge boson mass. The assumed gauge invariance of both the renormalised and unrenormalised Lagrangian means that there can be no bare or renormalised gauge boson mass, and that there can be no mass counter term for the gauge bosons. We evaluate the remaining three one-loop diagrams.

Consider first

(diagram 1)= 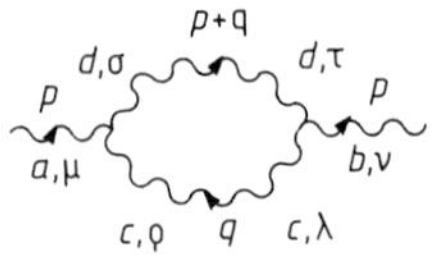

(11.26)

where we have taken advantage of the Kronecker delta in the colour indices in (10.69) to write the same colour index at each end of an internal gauge boson line. Using the Feynman rules of Chapter 10, including a symmetry factor of $\frac{1}{2}$ for the two identical internal gauge boson lines,

$$(\text{diagram 1}) = \tfrac{1}{2} M^{\varepsilon} \hat{g}^2 f_{acd} f_{cbd} \int \frac{\mathrm{d}^{2\omega} q}{(2\pi)^{2\omega}} J_{\mu\nu}(p, q) \tag{11.27}$$

where

$$\begin{aligned} J_{\mu\nu}(p, q) = {} & [-(p+2q)_\mu g_{\rho\sigma} + (q-p)_\sigma g_{\rho\mu} + (2p+q)_\rho g_{\mu\sigma}] \\ & \times [-(p+2q)_\nu g_{\lambda\tau} + (q-p)_\tau g_{\lambda\nu} + (2p+q)_\lambda g_{\tau\nu}] \\ & \times \tilde{D}_{\mathrm{F}}^{\sigma\tau}(p+q) \tilde{D}_{\mathrm{F}}^{\lambda\rho}(q). \end{aligned} \tag{11.28}$$

Taking the Feynman propagator from (10.68), we obtain (after some labour)

$$\begin{aligned} & J_{\mu\nu}(p, q)[(p+q)^2 + \mathrm{i}\varepsilon]^2 (q^2 + \mathrm{i}\varepsilon)^2 \\ & \quad = [2(4\omega - 3) q_\mu q_\nu + (4\omega - 3)(q_\mu p_\nu + p_\mu q_\nu) + (2\omega - 6) p_\mu p_\nu \\ & \qquad + (5p^2 + 2p\cdot q + 2q^2) g_{\mu\nu}] q^2 (p+q)^2 \\ & \qquad + \eta^2 [p^4 q_\mu q_\nu + (p\cdot q)^2 p_\mu p_\nu - (p\cdot q) p^2 (p_\mu q_\nu + q_\mu p_\nu)] \\ & \qquad + \eta (p+q)^2 [(q^2 + 2p\cdot q - p^2) q_\mu q_\nu + (q^2 + 3p\cdot q)(p_\mu q_\nu + q_\mu p_\nu) \\ & \qquad\qquad [(q^2 + 2p - q - p^2) q_\mu q_\nu + (q^2 + 3p\cdot q)(p_\mu q_\nu + q_\mu p_\nu) \\ & \qquad\qquad - q^2 p_\mu p_\nu - (q^2 + 2p\cdot q)^2 g_{\mu\nu}] \\ & \qquad + \eta q^2 [(q^2 - 2p^2) q_\mu q_\nu + (p\cdot q)(q_\mu p_\nu + p_\mu q_\nu) \\ & \qquad\qquad + (p^2 - 2q^2) p_\mu p_\nu - (q^2 - p^2)^2 g_{\mu\nu}] \end{aligned} \tag{11.29}$$

where

$$\eta = 1 - \xi. \tag{11.30}$$

With the aid of the integrals of Appendix A, we find the pole in $\varepsilon = 4 - 2\omega$ to be

$$(\text{diagram 1}) = \frac{\mathrm{i}\hat{g}^2 c_1 \delta_{ab}}{16\pi^2 \varepsilon} \left[\left(-\frac{11}{3} - 2\eta \right) p_\mu p_\nu + \left(\frac{19}{6} + \eta \right) p^2 g_{\mu\nu} \right] + \dots \tag{11.31}$$

where the group theory factor is defined by

$$c_1\delta_{ab}=f_{acd}f_{bcd} \tag{11.32}$$

Consider next

(diagram 2)= 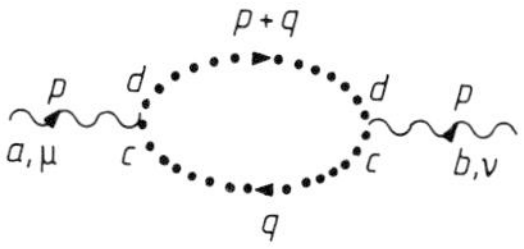

(11.33)

With the Faddeev–Popov ghost propagator of (10.70) and the vertex of (10.80),

$$(\text{diagram } 2)=-M^{\varepsilon}\hat{g}^2c_1\delta_{ab}\int\frac{\mathrm{d}^{2\omega}q}{(2\pi)^{2\omega}}\frac{(p+q)_\mu q_\nu}{[(p+q)^2+\mathrm{i}\varepsilon](q^2+\mathrm{i}\varepsilon)} \tag{11.34}$$

including a minus sign for the closed loop of Grassmann fields. Performing the integral with the aid of Appendix A we find the pole term

$$(\text{diagram } 2)=\frac{\mathrm{i}\hat{g}^2c_1\delta_{ab}}{16\pi^2\varepsilon}(\tfrac{1}{3}p_\mu p_\nu+\tfrac{1}{6}p^2g_{\mu\nu})+\ldots. \tag{11.35}$$

Consider finally

(diagram 3)= 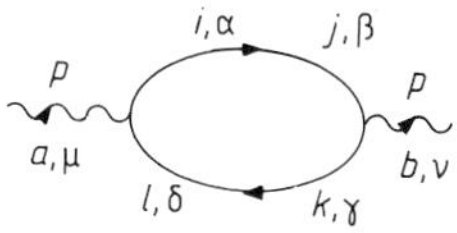

(11.36)

Using the fermion propagator of (10.82) and the vertex of (10.83), and including a minus sign for the closed fermion loop,

$$(\text{diagram } 3)=-\hat{g}^2M^{\varepsilon}c_2\delta_{ab}\int\frac{\mathrm{d}^{2\omega}q}{(2\pi)^{2\omega}}\,\mathrm{Tr}(\gamma_\mu\tilde{S}_{\mathrm{F}}(q)\gamma_\nu\tilde{S}_{\mathrm{F}}(p+q) \tag{11.37}$$

where the group theory factor is defined by

$$c_2\delta_{ab}=\mathrm{Tr}(\mathbf{T}_a\mathbf{T}_b) \tag{11.38}$$

and $S_{\mathrm{F}}(q)$ is as in (8.35). Evaluating the Dirac gamma matrix trace gives

$$(\text{diagram } 3)=-\hat{g}^2M^{\varepsilon}c_2\delta_{ab}2^{\omega}\int\frac{\mathrm{d}^{2\omega}q}{(2\pi)^{2\omega}}$$

$$\times\frac{q_\mu(p+q)_\nu+(p+q)_\mu q_\nu+(m^2-q^2-p-q)g_{\mu\nu}}{(q^2-m^2+\mathrm{i}\varepsilon)[(p+q)^2-m^2+\mathrm{i}\varepsilon]}. \tag{11.39}$$

The pole term in $\varepsilon=4-2\omega$ is most easily extracted by putting $m=0$ in (11.39). Then we may use the integrals of Appendix A to obtain

$$(\text{diagram } 3)=\frac{-\mathrm{i}\hat{g}^2c_2\delta_{ab}}{16\pi^2\varepsilon}(-\tfrac{8}{3}p_\mu p_\nu+\tfrac{8}{3}p^2g_{\mu\nu})+\ldots. \tag{11.40}$$

We may now calculate the renormalisation constants ΔZ_A and K_ξ from (11.25). Thus, if we adopt the MS renormalisation of §7.5, where the counter terms exactly cancel the poles in ε, and read off the counter terms from (11.6), we have

$$i\Delta Z_A(-p^2g_{\mu\nu}+p_\mu p_\nu)-i\xi^{-1}K_\xi p_\mu p_\nu = \frac{-i\hat{g}^2}{16\pi^2\varepsilon}\{(-p^2g_{\mu\nu}+p_\mu p_\nu)[(-\tfrac{10}{3}-\eta)c_1+\tfrac{8}{3}c_2]-\eta c_1 p_\mu p_\nu\}. \quad (11.41)$$

Thus

$$\Delta Z_A = -\frac{\hat{g}^2}{16\pi^2\varepsilon}[(-\tfrac{13}{3}+\xi)c_1+\tfrac{8}{3}c_2] \quad (11.42)$$

and

$$\xi^{-1}K_\xi = \frac{-\hat{g}^2}{16\pi^2\varepsilon}(1-\xi)c_1 \quad (11.43)$$

where we have used (11.30). The renormalised OPI Green function of (11.24) may now be evaluated from (11.25) using the above counter terms. (See problem 11.3.)

Consider next the OPI Green function for two Faddeev–Popov ghost fields. The contributions to this Green function at one-loop order are

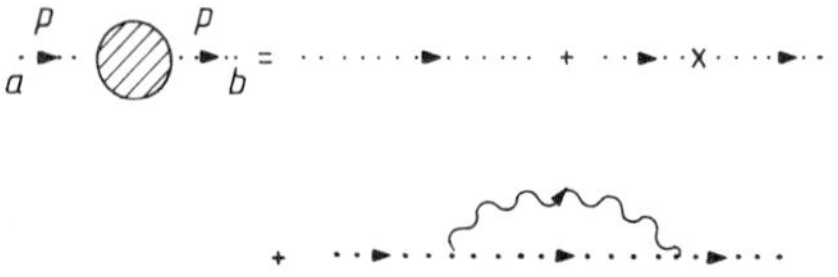

(11.44)

There is only one non-trivial diagram to consider:

(diagram 4) = 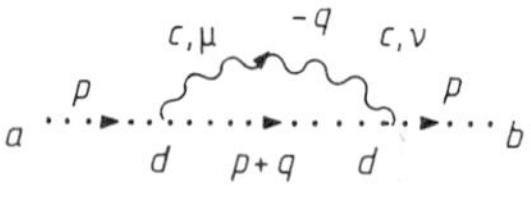

(11.45)

Using the propagators and vertices of §10.6,

$$(\text{diagram } 4) = \hat{g}^2 M^\varepsilon c_1 \delta_{ab} \int \frac{d^{2\omega}q}{(2\pi)^{2\omega}} D_F^{\mu\nu}(-q)(p+q)_\mu p_\nu[(p+q)^2+i\varepsilon]^{-1}. \quad (11.46)$$

Performing the integral with the aid of Appendix A, the pole term is

$$(\text{diagram } 4) = \frac{2i\hat{g}^2c_1\delta_{ab}}{16\pi^2\varepsilon}p^2\left(\frac{3}{4}-\frac{5\xi}{4}\right)+\ldots. \quad (11.47)$$

The counter term ΔZ_η may now be obtained from (11.44) and (11.6). In the MS renormalisation scheme we require an exact cancellation of the pole term by the counter term. Then

$$\mathrm{i}\delta_{ab}p^2\,\Delta Z_\eta = -\frac{2\mathrm{i}\hat{g}^2 c_1 \delta_{ab}}{16\pi^2\varepsilon}\,p^2\left(\frac{3}{4}-\frac{5\xi}{4}\right) \tag{11.48}$$

and so

$$\Delta Z_\eta = \frac{-2\hat{g}^2 c_1}{16\pi^2\varepsilon}\left(\frac{3}{4}-\frac{5\xi}{4}\right). \tag{11.49}$$

The fermion wave function and mass renormalisation counter terms may be obtained from the equation for the OPI Green function:

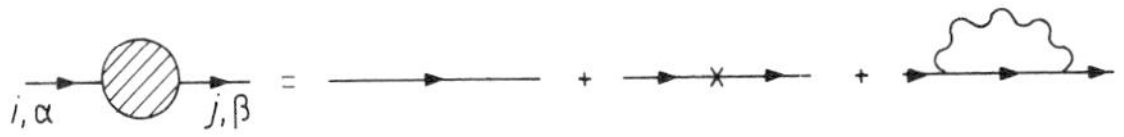

(11.50)

There is only one diagram to evaluate, namely,

(diagram 5) = 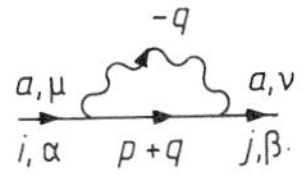

(11.51)

Using the propagators and vertices of §10.6,

$$(\text{diagram } 5) = \hat{g}^2 M^{\varepsilon} c_3 \delta_{ji} \int \frac{\mathrm{d}^{2\omega}q}{(2\pi)^{2\omega}} (\gamma_\nu S_{\mathrm{F}}(p+q)\gamma_\mu)_{\beta\alpha} \tilde{D}_{\mathrm{F}}^{\mu\nu}(-q) \tag{11.52}$$

where the group theory factor is defined by

$$(\mathbf{T}_a \mathbf{T}_a)_{ji} = c_3 \delta_{ji}. \tag{11.53}$$

It follows immediately from the definitions of c_2 and c_3 that

$$c_3 = \frac{d_{\mathrm{G}}}{d_{\mathrm{F}}} c_2 \tag{11.54}$$

where d_{G} is the dimensionality of the adjoint representation of the gauge group to which the gauge fields belong and d_{F} is the dimensionality of the irreducible representation to which the fermions belong. To isolate the pole term from diagram 5, we may put $m=0$ in the denominator of the integrand, and use the integrals of Appendix A. The result is

$$(\text{diagram } 5) = \frac{2\mathrm{i}\hat{g}^2 c_3 \delta_{ji}}{16\pi^2\varepsilon} (\xi \not{p} - (3+\xi) m \mathbf{I})_{\beta\alpha} + \ldots \tag{11.55}$$

where we have used the identities

$$\gamma^\mu \gamma_\mu = 2\omega \mathbf{I} \tag{11.56}$$

and

$$\gamma_\mu \gamma_\rho \gamma^\mu = 2(1-\omega)\gamma_\rho. \tag{11.57}$$

The fermion wave function and mass renormalisation counter terms may now be derived using (11.50) and (11.6).

Thus, in the MS renormalisation scheme,

$$\mathrm{i}(\Delta Z_\psi \not{p} - mK_m \mathbf{I})\delta_{ji} = -\frac{2\mathrm{i}\hat{g}^2 c_3 \delta_{ji}}{16\pi^2 \varepsilon}(\xi\not{p} - (3+\xi)m\mathbf{I}) \tag{11.58}$$

and so

$$\Delta Z_\psi = -\frac{2\hat{g}^2 c_3 \xi}{16\pi^2 \varepsilon} \tag{11.59}$$

and

$$K_m = -\frac{2\hat{g}^2 c_3}{16\pi^2 \varepsilon}(3+\xi). \tag{11.60}$$

Given that there is a single bare coupling constant, it will be sufficient to calculate one of the renormalisation constants K_i, $i=1,\ldots,4$. The other three are obtained from the one we choose to calculate by using (11.22), now that we have evaluated ΔZ_η and ΔZ_A. For ease of evaluation we should choose either K_1 or K_2. We select K_2. (The diagrams needed to calculate K_3 or K_4 involve more gauge field internal lines.) The contributions at one-loop order to the OPI Green function with two external fermion legs and one external gauge field leg are

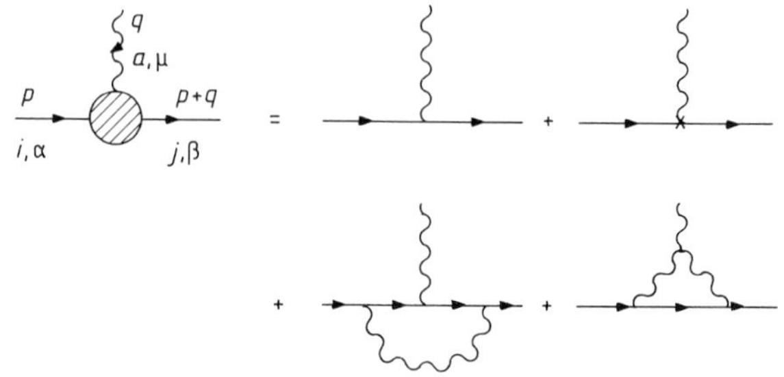

(11.61)

Consider first

(diagram 6) = 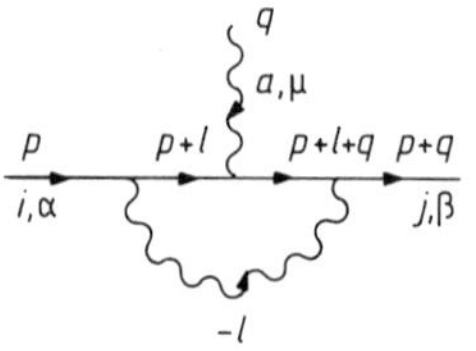

(11.62)

With the propagators and vertices of §10.6,

$$(\text{diagram } 6) = g\hat{g}^2 M^\varepsilon (\mathbf{T}_b \mathbf{T}_a \mathbf{T}_b)_{ji}$$

$$\times \int \frac{\mathrm{d}^{2\omega} l}{(2\pi)^{2\omega}} D_F^{\rho\nu}(l)[\gamma_\rho S_F(p+l+q)\gamma_\mu S_F(p+l)\gamma_\nu]_{\beta\alpha}. \tag{11.63}$$

The group theory factor is easily obtained in terms of the group theoretical constants defined in (11.32) and (11.38)

$$\mathbf{T}_b\mathbf{T}_a\mathbf{T}_b=(c_3-\tfrac{1}{2}c_1)\mathbf{T}_a. \tag{11.64}$$

To evaluate the pole term we may put $m=0$ because, as discussed in Chapter 7, the pole terms are independent of the renormalised mass. We may shorten the derivation further by noticing that the counter term for this vertex has no q dependence, and so the pole term can have no such dependence in a renormalisable theory. We may therefore set $q=0$ to calculate the pole term. We must not, however, also put $p=0$, because we would then run into trouble with infrared divergences. (An additional contribution to the pole in ε would then appear which is only present at $p=0$. This should not be cancelled by the counter term, whose function is to cancel ultraviolet divergences, which are present for all values of p. The infrared divergences should remain after renormalisation, but should cancel when physical processes are calculated.)

Carrying out the evaluation at $m=0$, $q=0$, with the aid of the integrals of Appendix A, we find the pole term,

$$(\text{diagram } 6)=-\frac{2ig\hat{g}^2(c_3-\frac{1}{2}c_1)\xi}{16\pi^2\varepsilon}(\mathbf{T}_a)_{ji}(\gamma_\mu)_{\beta\alpha}+\ldots \tag{11.65}$$

which is independent of p, as expected. In arriving at this result we have used the identities for Dirac gamma matrices (11.57) and

$$\gamma_\rho\gamma_\lambda\gamma_\mu\gamma_\nu\gamma^\rho=-2\gamma_\nu\gamma_\mu\gamma_\lambda+2(2-\omega)\gamma_\lambda\gamma_\mu\gamma_\nu. \tag{11.66}$$

Finally, consider

(diagram 7)= 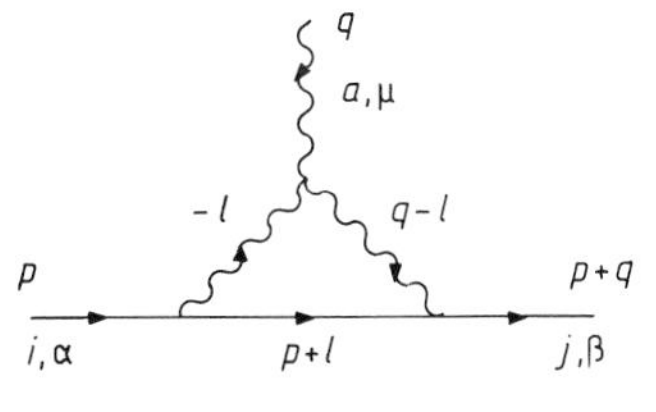

(11.67)

With the Feynman rules of §10.6,

$$(\text{diagram } 7)=ig\hat{g}^2M^\varepsilon f_{abc}(\mathbf{T}_c\mathbf{T}_b)_{ji}$$
$$\times\int\frac{d^{2\omega}l}{(2\pi)^{2\omega}}[\gamma_\tau S_F(p+l)\gamma_\sigma]_{\beta\alpha}\tilde{D}_F^{\tau\rho}(q-l)\tilde{D}_F^{\nu\sigma}(l)$$
$$\times[(2l-q)_\mu g_{\nu\rho}-(q+l)_\rho g_{\mu\nu}+(2q-l)_\nu g_{\rho\mu}]. \tag{11.68}$$

The group theory factor may be evaluated in terms of c_1 of (11.32)

$$f_{abc}\mathbf{T}_c\mathbf{T}_b=-\frac{i}{2}c_1\mathbf{T}_a. \tag{11.69}$$

Again the pole term is most easily evaluated by setting $m=0$, $q=0$. With the aid of the integrals of Appendix A we find the pole term

$$(\text{diagram } 7)=\frac{3ig\hat{g}^2c_1(1-\xi)}{32\pi^2\varepsilon}(\mathbf{T}_a)_{ji}(\gamma_\mu)_{\beta\alpha}+\ldots \tag{11.70}$$

which is again independent of p as expected. To confirm that the theory is indeed renormalisable (at one-loop order) we should calculate the q dependence of diagrams 6 and 7 by combining denominators in (11.63) and (11.68) and make sure that there is no q dependence in the pole term (problem 11.4).

We may now derive the renormalisation constant K_2. Using (11.61), (11.6), (11.65) and (11.70), we find, in the MS renormalisation scheme,

$$K_2=-\frac{\hat{g}^2}{32\pi^2\varepsilon}[3c_1+(c_1+4c_3)\xi]. \tag{11.71}$$

The renormalisation constants K_1, K_3 and K_4 are determined using (11.22). We now have at our disposal all the renormalisation counter terms necessary to renormalise the theory (to sufficient accuracy to carry out the renormalisation at one-loop order).

11.3 The electron anomalous magnetic moment

In this section, we specialise to the case of QED (Abelian gauge theory) and derive the electron anomalous magnetic moment. For convenience we shall work in Feynman gauge, $\xi=1$. To one-loop order, the contributions to the appropriate OPI Green function are

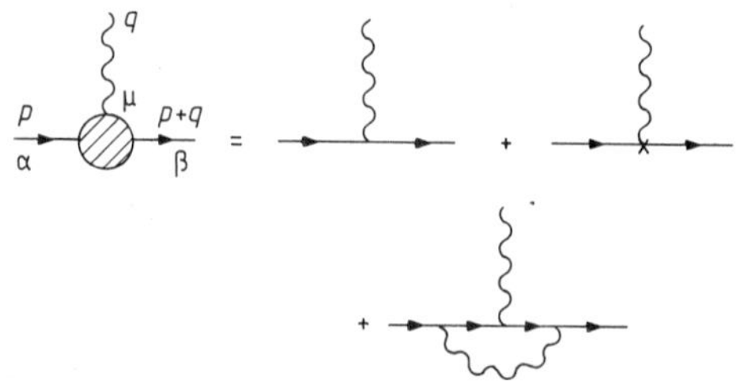

(11.72)

The final diagram of (11.61) does not contribute here because the gauge field self-interactions only occur in a non-Abelian theory. Thus, the only Feynman diagram we need to study is diagram 6 of (11.62). We make the transition to the Abelian case of QED by the substitutions

$$\mathbf{T}_a=\mathbf{I} \qquad c_1=0 \qquad c_2=c_3=1 \qquad g=e. \tag{11.73}$$

Thus, for QED,

(diagram 6) =

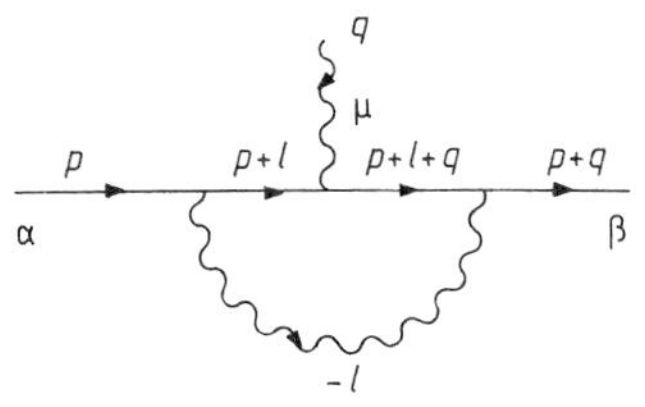

$$= e\hat{e}^2 M^\varepsilon \int \frac{d^{2\omega}l}{(2\pi)^{2\omega}} D_F^{\rho\nu}(l)[\gamma_\rho \tilde{S}_F(p+l+q)\gamma_\mu \tilde{S}_F(p+l)\gamma_\nu)_{\beta\alpha}. \quad (11.74)$$

This time we shall focus on the finite part of the Feynman integral. Combining denominators with the aid of (7.33), we have

$$(\text{diagram } 6) = -2e\hat{e}^2 M^\varepsilon \int_0^1 dx \int_0^x dy \int \frac{d^{2\omega}l}{(2\pi)^{2\omega}} \frac{N}{(l^2 + 2P\cdot l - \mu^2)^3} \quad (11.75)$$

where

$$P = (1 + y - x)p + yq \quad (11.76)$$

$$-\mu^2 = [(p+q)^2 - m^2]y + (p^2 - m^2)(1-x) \quad (11.77)$$

and

$$N = \gamma_\rho(\not{p} + \not{l} + \not{q} + m)\gamma^\mu(\not{p} + \not{l} + m)\gamma^\rho. \quad (11.78)$$

One is usually interested in situations where the electron–photon vertex finds itself sandwiched between Dirac spinors $\bar{u}(p+q)$ and $u(p)$ associated with on-mass-shell electrons. Then

$$p^2 - m^2 = (p+q)^2 - m^2 = \mu^2 = 0. \quad (11.79)$$

Also, the Dirac equation for the spinors allows us to simplify the numerator in (11.75), leading to

$$\begin{aligned} -N = {} & 2(1-\omega)m^2\gamma_\mu + 4m\omega(p+l)_\mu + 2m\omega\not{q}\gamma_\mu + 2m[\gamma_\mu, \not{q}] \\ & - 2(\not{p} + \not{l})\gamma_\mu(\not{p} + \not{l} + \not{q}) + 2(2-\omega)(\not{p} + \not{l} + \not{q})\gamma_\mu(\not{p} + \not{l}) \end{aligned} \quad (11.80)$$

where we have used (11.56), (11.57) and (11.66). The $d^{2\omega}l$ integrations may be performed with the aid of (7.20), (7.21a) and (7.21b), and simplifications may be made by noticing that

$$\int_0^1 dx \int_0^x dy\,\varphi(x, y) = \int_0^1 dx \int_0^x dy\,\varphi(1-y, 1-x) \quad (11.81)$$

for any function. Everything may be expressed in terms of the covariants γ_μ and

$\frac{1}{2}[\gamma_\mu, \gamma_\nu]$ by using the Gordon reduction for on-mass-shell electrons:

$$(p'+p)_\mu \bar{u}(p')u(p) = 2m\bar{u}(p')\gamma_\mu u(p) + \tfrac{1}{2}\bar{u}(p')[\gamma_\mu, \gamma_\nu]u(p)(p'-p)^\nu \quad (11.82)$$

together with the identity

$$\not{q}\gamma_\mu\not{q} = 2q_\mu\not{q} - q^2\gamma_\mu. \quad (11.83)$$

This leads to the anomalous magnetic moment term (associated with the coefficient of $\frac{1}{2}[\gamma_\mu, \gamma_\nu]$ at $q^2=0$):

$$(\text{diagram 6}) = \frac{2ie^3}{16\pi^2 m}[\gamma_\mu, \not{q}]\int_0^1 dx \int_0^x dy[xy-(1-x)^2](1+y-x)^{-2} + \ldots . \quad (11.84)$$

Carrying out the parameter integration,

$$(\text{diagram 6}) = \frac{\alpha}{2\pi}\frac{e}{2m}\frac{i}{2}[\gamma_\mu, \not{q}] + \ldots \quad (11.85)$$

where the omitted terms are contributions to the coefficient of $\frac{1}{2}[\gamma_\mu, \gamma_\nu]$ when $q^2 \neq 0$ and contributions to the coefficient of γ_μ, and

$$\alpha = \frac{e^2}{4\pi} \quad (11.86)$$

is the fine structure constant.

Thus, the anomalous magnetic moment of the electron μ_{AMM} is

$$\mu_{\mathrm{AMM}} = \frac{\alpha}{2\pi}\frac{e}{2m}. \quad (11.87)$$

Strictly, we should define α in terms of the 'physical' electromagnetic coupling constant e_{phys} which is the coefficient of $-i\gamma_\mu$ when $q=0$, with on-mass-shell electrons. However, e_{phys} only differs from e in order e^3, and we are working in lowest order perturbation theory. (The relation between e and e_{phys} depends on the scale of mass M used in the renormalisation of e.)

One may now proceed to calculate the coefficient of $-ie\gamma_\mu$ to order q^2. (See problem 11.5.) The result of such a calculation exhibits a pole in ε even after renormalisation. The reason for this is an infrared divergence. Such divergences, though present in individual Green functions, always cancel when physical processes are calculated, provided all relevant diagrams are included, and appropriate integrations over phase space are carried out. Thus, for example, when scattering of an electron off a Coulomb potential is studied, the infrared divergence in the electron–photon vertex mentioned here is cancelled by contributions where the electron emits a real photon. Both types of contributions are needed to describe the physical process, because any experimental apparatus has a finite energy resolution and the emission of a real photon of sufficiently low energy cannot be ruled out.

Problems

11.1 Derive the criterion (11.1) for renormalisable inreractions involving derivatives, where the bosons have propagators with the high momentum behaviour $(p^2)^{-1}$.

11.2 By studying the renormalisation of all the vertices involved in a non-Abelian gauge theory, show that at one-loop order there is a single renormalised coupling constant if the bare Lagrangian is gauge invariant.

11.3 Derive the renormalised OPI Green function for a gauge field at one-loop order using the counter terms (11.42) and (11.43).

11.4 Calculate the q dependence of diagrams 6 and 7, and check that there is *no* q dependence in the pole term in ε.

11.5 Calculate the coefficient of $-ie\gamma_\mu$ for the vertex of (11.78) to order q^2, and show that a pole in ε remains after renormalisation.

References

1 't Hooft G and Veltman M 1972a *Nucl. Phys.* B **44** 189
—— 1972b *Nucl. Phys.* B **50** 318
Lee B W 1972 *Phys. Rev.* D **5** 823
Lee B W and Zinn-Justin J 1972 *Phys. Rev.* D **5** 3137
Taylor J C 1976 *Gauge Theories of Weak Interactions* (Cambridge: Cambridge University Press) Chapter 14

12

QCD AND ASYMPTOTIC FREEDOM

12.1 The renormalisation group equation

We have seen in Chapters 7 and 11 that renormalisation of a field theory depends on some mass M, and that the Green functions depend on this mass. (Of course scattering amplitudes and other directly observable quantities must be independent of M, which has no physical significance.) A question that may be asked is how do the Green functions of a gauge field theory change as the renormalisation scale M is varied? This question may be answered by recalling the connection between the renormalised and unrenormalised OPI Green functions (in 2ω dimensions). Let $\tilde{\Gamma}^{(n)}(p_1, \ldots, p_n, \hat{g}, \xi, m, M)$ be a normalised OPI Green function with a total of n external legs, n_A of which are gauge fields and n_ψ of which are fermions

$$n = n_A + n_\psi. \tag{12.1}$$

Here, $\hat{g}$ and ξ are the dimensionless renormalised gauge coupling constant and gauge parameter, and m is a renormalised fermion mass as in §11.1. Reference to the Lorentz indices of the gauge fields, to the spinor indices of the fermion fields, and to the gauge group indices, has been suppressed since these will have no bearing on the derivation which follows. Let the corresponding unrenormalised OPI Green function be $\tilde{\Gamma}_{\mathrm{B}}^{(n)}(p_1, \ldots, p_n, g_{\mathrm{B}}, \xi_{\mathrm{B}}, m_{\mathrm{B}})$. A crucial point to notice is that $\tilde{\Gamma}_{\mathrm{B}}^{(n)}$ does not depend on M, which only enters when renormalisation is carried out. The connection between the renormalised and unrenormalised Green functions is

$$\tilde{\Gamma}^{(n)}(p_1, \ldots, p_n, \hat{g}, \xi, m, M) = Z_A^{n_A/2} Z_\psi^{n_\psi/2} \tilde{\Gamma}_{\mathrm{B}}^{(n)}(p_1, \ldots, p_n, g_{\mathrm{B}}, \xi_{\mathrm{B}}, m_{\mathrm{B}}) \tag{12.2}$$

where Z_A and Z_ψ are defined in (11.7) and (11.9). If we carry out the differentiation $M(\partial/\partial M)$ holding g_{B}, ξ_{B} and m_{B} fixed, we obtain

$$M \frac{\partial}{\partial M} \tilde{\Gamma}^{(n)} + M \frac{\partial \hat{g}}{\partial M} \frac{\partial \tilde{\Gamma}^{(n)}}{\partial \hat{g}} + M \frac{\partial \xi}{\partial M} \frac{\partial \tilde{\Gamma}^{(n)}}{\partial \xi} + M \frac{\partial m}{\partial M} \frac{\partial \tilde{\Gamma}^{(n)}}{\partial m}$$
$$= \frac{n_A}{2} Z_\psi^{n_\psi/2} Z_A^{n_A/2} Z_A^{-1} M \frac{\partial Z_A}{\partial M} \tilde{\Gamma}_{\mathrm{B}}^{(n)} + \frac{n_\psi}{2} Z_\psi^{n_A/2} Z_A^{n_A/2} Z_\psi^{-1} M \frac{\partial Z_\psi}{\partial M} \tilde{\Gamma}_{\mathrm{B}}^{(n)} \tag{12.3}$$

where $M\,\partial\tilde{\Gamma}^{(n)}/\partial M$ denotes a differentiation at constant $\hat{g}$, ξ and m, and similarly for $\partial/\partial\hat{g}$, $\partial/\partial\xi$ and $\partial/\partial m$. Thus

$$\left(M\frac{\partial}{\partial M}+\beta_{\hat{g}}\frac{\partial}{\partial\hat{g}}+\beta_{\xi}\frac{\partial}{\partial\xi}-\gamma_m m\frac{\partial}{\partial m}-n_A\gamma_A-n_\psi\gamma_\psi\right)\tilde{\Gamma}^{(n)}(p_1,\ldots,p_n,\hat{g},\xi,m,M)=0 \tag{12.4}$$

where the coefficients are defined by

$$\beta_{\hat{g}}=M\frac{\partial\hat{g}}{\partial M} \tag{12.5}$$

$$\beta_{\xi}=M\frac{\partial\xi}{\partial M} \tag{12.6}$$

$$\gamma_m=-\frac{M}{m}\frac{\partial m}{\partial M} \tag{12.7}$$

$$\gamma_A=Z_A^{-1/2}M\frac{\partial Z_A^{1/2}}{\partial M} \tag{12.8}$$

$$\gamma_\psi=Z_\psi^{-1/2}M\frac{\partial Z_\psi^{1/2}}{\partial M}. \tag{12.9}$$

In (12.5)–(12.9), differentiation is understood to be holding $g_{\rm B}$, $\xi_{\rm B}$ and $m_{\rm B}$ constant. The essence of (12.4), called the renormalisation group equation, is that when the renormalisation scale M is changed, the corresponding changes in the renormalised quantities $\hat{g}$, ξ and m are such that the unrenormalised Green function (which does not depend on M) does not change.

The dimensionless coefficients in the renormalisation group equation depend in general on $\hat{g}$ and m/M. However, if we adopt a mass-independent renormalisation scheme, such as the MS or $\overline{\text{MS}}$ scheme, then the m/M dependence drops out, and the renormalisation group equation is considerably easier to use[1,2,3]. Accordingly, we shall always assume that such a renormalisation scheme is used in what follows, so that the renormalisation group coefficients depend only on $\hat{g}$. These coefficients may be computed in the MS scheme to one-loop order from the renormalisation constants of Chapter 11 (see problem 12.1). The results are as follows:

$$\beta_{\hat{g}}=-\frac{\varepsilon}{2}\hat{g}-b\hat{g}^3 \tag{12.10}$$

$$\beta_{\xi}=\left[\left(\frac{13}{6}-\frac{\xi^2}{2}\right)C_1-\frac{4}{3}\sum_R C_2^R\right]\frac{\hat{g}^2\xi}{8\pi^2} \tag{12.11}$$

$$\gamma_m=\frac{3C_3^R\hat{g}^2}{8\pi^2}\equiv b_m\hat{g}^2 \tag{12.12}$$

$$\gamma_A = -\left[\left(\frac{13}{6}-\frac{\xi}{2}\right)C_1 - \frac{4}{3}\sum_R C_2^R\right]\frac{\hat{g}^2}{16\pi^2} \tag{12.13}$$

and

$$\gamma_\psi = \frac{\xi C_3^R \hat{g}^2}{16\pi^2} \tag{12.14}$$

where

$$b = \frac{1}{16\pi^2}\left(\frac{11}{3}C_1 - \frac{4}{3}\sum_R C_2^R\right). \tag{12.15}$$

We have assumed that the fermions belong to irreducible representations R of the gauge group, with values C_2^R of the group theory factor C_2, and C_3^R of C_3. The group theory factors C_1, C_2 and C_3 are defined in (11.32), (11.38) and (11.53). In four dimensions the linear term in (12.10) vanishes. (We are assuming here that both chiral components of a fermion belong to the same irreducible representation R. This assumption may be relaxed when necessary, as in problem 16.2.)

An important use of the renormalisation group equation is to discuss the behaviour of Green functions as the momenta of the external legs are scaled, i.e. when $p_1, \ldots, p_n$ are replaced by $sp_1, \ldots, sp_n$ where s is dimensionless. The Green function has energy dimensions (see problem 12.2)

$$d_\Gamma = 2\omega + n_A(1-\omega) + n_\psi(\tfrac{1}{2}-\omega). \tag{12.16}$$

Since $\tilde{\Gamma}^{(n)}(sp_1, \ldots, sp_n, \hat{g}, \xi, m, M)$ is homogeneous of degree d_Γ in $p_1, \ldots, p_n, m, M$, we have

$$\left(s\frac{\partial}{\partial s} + m\frac{\partial}{\partial m} + M\frac{\partial}{\partial M}\right)\tilde{\Gamma}^{(n)}(sp_1, \ldots, sp_n, \hat{g}, \xi, m, M) = d_\Gamma \tilde{\Gamma}^{(n)}(sp_1, \ldots, sp_n, \hat{g}, \xi, m, M). \tag{12.17}$$

Combining (12.17) with the renormalisation group equation (12.4), we may eliminate $M\,\partial\tilde{\Gamma}^{(n)}/\partial M$ to obtain

$$\left(-s\frac{\partial}{\partial s} + \beta_{\hat{g}}\frac{\partial}{\partial \hat{g}} + \beta_\xi \frac{\partial}{\partial \xi} - (1+\gamma_m)m\frac{\partial}{\partial m} - n_A\gamma_A - n_\psi\gamma_\psi + d_\Gamma\right) \tilde{\Gamma}^{(n)}(sp_1, \ldots, sp_n, \hat{g}, \xi, m, M) = 0. \tag{12.18}$$

This equation may be solved with the aid of running coupling constant, gauge parameter and mass $\bar{g}(s)$, $\bar{\xi}(s)$ and $\bar{m}(s)$, defined as the solutions of

$$s\frac{\partial \bar{g}(s)}{\partial s} = \beta_{\hat{g}}(\bar{g}(s)) \tag{12.19}$$

$$s\frac{\partial \bar{\xi}(s)}{\partial s} = \beta_\xi(\bar{g}(s), \bar{\xi}(s)) \tag{12.20}$$

and

$$s\frac{\partial\bar{m}(s)}{\partial s}=-[1+\gamma_m(\bar{g}(s),\bar{\xi}(s))]\bar{m}(s) \tag{12.21}$$

with the initial conditions

$$\bar{g}(1)=\hat{g}(M)\equiv\hat{g} \tag{12.22}$$

$$\bar{\xi}(1)=\xi(M)\equiv\xi \tag{12.23}$$

and

$$\bar{m}(1)=m(M)\equiv m. \tag{12.24}$$

Thus from (12.5), (12.6) and (12.7), we see that

$$\bar{g}(s)=\hat{g}(sM) \tag{12.25}$$

$$\bar{\xi}(s)=\xi(sM) \tag{12.26}$$

and

$$\bar{m}(s)=s^{-1}m(sM) \tag{12.27}$$

where $\hat{g}(sM)$, $\xi(sM)$ and $m(sM)$ are the renormalised quantities when the renormalisation scale is sM instead of M. The solution of (12.18) may now be written as

$$\tilde{\Gamma}^{(n)}(sp_1,\ldots,sp_n,\hat{g},\xi,m,M)=s^{d_\Gamma}\exp\left(-\int_1^s\frac{ds'}{s'}\,[n_A\gamma_A(\bar{g}(s'),\bar{\xi}(s'))+n_\psi\gamma_\psi(\bar{g}(s'),\bar{\xi}(s')]\right)\tilde{\Gamma}^{(n)}(p_1,\ldots,p_n,\bar{g}(s),\bar{\xi}(s),\bar{m}(s),M) \tag{12.28}$$

(see, for example, the book of Piaggio[4]).

With the explicit expressions at one-loop order of (12.10)–(12.12) we may solve for $\bar{g}(s)$, $\bar{\xi}(s)$ and $\bar{m}(s)$. In four dimensions ($\varepsilon=0$), the solution for $\bar{g}(s)$ is

$$\bar{g}^2(s)=g^2(1+2bg^2\ln s)^{-1} \tag{12.29a}$$

or equivalently

$$\bar{g}^{-2}(s)=g^{-2}+2b\ln s. \tag{12.29b}$$

Provided $b>0$, we see that $\bar{g}^2(s)$ decreases as s increases and tends to zero as $s\to\infty$. The theory approaches a free-field theory (logarithmically). This phenomenon[5] is spoken of as asymptotic freedom. Referring back to (12.28), we see that if we want to calculate Green functions at large momenta, then the solution is given in terms of a running coupling constant $\bar{g}(s)$ which is small. It should therefore be possible to use perturbation theory in $\bar{g}(s)$ for this purpose, despite the fact that in QCD we are dealing with the strong interactions, and perturbation theory in g (as opposed to $\bar{g}(s)$) is not expected to be useful.

Conversely, $\bar{g}^2(s)$ increases as s decreases and so phenomena at small momenta should be described by a running coupling constant which is large.

Thus, perturbation theory is expected to be useless for QCD when long distance behaviour is studied. In particular, we do not expect quarks, which are fields in the free Lagrangian, to be asymptotic states of the theory before and after scattering. Rather, we expect to have to calculate what the asymptotic states are in a non-perturbative fashion, and hope to find the known hadrons if we do so. Such calculations tend to rely on lattice gauge theory methods and are outside the scope of this book.

The restriction that b should be positive to obtain asymptotic freedom is not a very severe one for QCD where the gauge group is colour SU(3). With N_f flavours of quark, each belonging to the three dimensional fundamental representation of the colour group, we find from (12.15) that

$$b=(11-\tfrac{2}{3}N_f)/16\pi^2. \tag{12.30}$$

Thus, provided there are not more than 16 flavours of quark, QCD is asymptotically free.

However, for QED the situation is quite different. In that case, the gauge group is U(1) with coupling constant e, the gauge field does not couple to itself, so that $C_1=0$, and C_2 is q^2 in units of e. With N_G generations of quarks and leptons, each generation containing a quark of charge $\frac{2}{3}$, a quark of charge $-\frac{1}{3}$, a lepton of charge -1, and a lepton (neutrino) of charge 0, we have (instead of (12.30))

$$b=-\frac{4}{3}\frac{N_G}{16\pi^2}(1+0+\tfrac{4}{9}+\tfrac{1}{9})=-\frac{56}{27}\frac{N_G}{16\pi^2}. \tag{12.31}$$

In any case, $-b$ is always given by a sum of squares of quark and lepton charges, and so b is always negative. Thus, in QED $\bar{e}^2(s)$ grows as s increases and the theory is not asymptotically free. (It appears from (12.29) that $\bar{e}^2(s)$ will become infinite as $2be^2 \ln s \to -1$, but this is not the case, because lowest order perturbation theory has broken down by this stage.) On the other hand, there is no difficulty in deciding what the asymptotic states of the theory are since $\bar{e}^2(s)$ decreases as s decreases so that the theory approaches a free-field theory at large distances. (There is, of course, the usual well understood problem associated with the long range nature of the electromagnetic interaction.) Scalar field theory resembles QED in these respects. (See problem 12.3.)

Combining (12.20) and (12.11) we see that

$$s\frac{\partial\bar{\xi}(s)}{\partial s}=\left[\left(\frac{13}{6}-\frac{\bar{\xi}^2}{2}\right)C_1-\frac{4}{3}\sum_R C_2^R\right]\frac{\bar{g}^2(s)\bar{\xi}(s)}{8\pi^2}. \tag{12.32}$$

In general, $\bar{\xi}(s)$ will vary with s. This complication can be avoided by working in Landau gauge. Then the initial condition is

$$\bar{\xi}(1)=0 \tag{12.33}$$

and (12.32) shows that

$$\bar{\xi}(s)=0 \tag{12.34}$$

for all values of s. We shall always adopt Landau gauge in what follows.

Apart from the (order $\bar{g}^2(s)$) correction from γ_m, (12.21) shows that $\bar{m}(s)$ decreases like s^{-1} for large values of s. Since $\bar{m}(s)$ decreases as a power of s, whereas $\bar{g}(s)$ decreases only as a logarithm, it is often a good enough approximation to put $\bar{m}(s)=0$ in discussing the large momentum behaviour of QCD. (We shall discuss the effect of the running mass $\bar{m}(s)$ in §16.5 in the context of grand unified theories.)

At asymptotically high momenta, in Landau gauge, (12.28) gives for the behaviour of the Green functions

$$\tilde{\Gamma}^{(n)}(sp_1,\ldots,sp_n,\hat{g},\xi=0,m,M)$$
$$\approx s^{d_\Gamma}\exp\left(n_A d_A\int_1^s\frac{ds'}{s'}\bar{g}^2(s')\right)\tilde{\Gamma}^{(n)}(p_1,\ldots,p_n,\bar{g}(s),\bar{\xi}(s)=0,\bar{m}(s)=0,M) \tag{12.35}$$

where we have written

$$\gamma_A(\bar{g}(s),\bar{\xi}(s)=0)=-d_A\bar{g}^2(s) \tag{12.36}$$

with (from 12.13))

$$d_A=\left(\frac{13}{6}C_1-\frac{4}{3}\sum_R C_2^R\right)\bigg/16\pi^2. \tag{12.37}$$

Using (12.29), we have (in four dimensions),

$$\tilde{\Gamma}^{(n)}(sp_1,\ldots,sp_n,g,\xi=0,m,M)$$
$$\approx s^{d_\Gamma}(1+2bg^2\ln s)^{n_Ad_A/2b}\tilde{\Gamma}^{(n)}(p_1,\ldots,p_n,\bar{g}(s),\bar{\xi}(s)=0,\bar{m}(s)=0,M). \tag{12.38}$$

The behaviour is free-field behaviour, apart from the multiplicative power of $1+2bg^2\ln s$ and apart from an additive logarithmic correction if we expand $\tilde{\Gamma}^{(n)}(p_1,\ldots,p_n,\bar{g}(s),\bar{\xi}(s)=0,\bar{m}(s)=0,M)$ about $\bar{g}^2(s)=0$.

12.2 Deep inelastic electron–nucleon scattering

An important process where tests of QCD are possible is the inclusive process $eN\to eX$ where N denotes a nucleon, and X denotes an arbitrary unobserved final state. When the scattering is approximated by one-photon exchange (see figure 12.1) then the electron–photon vertex is just $ie\gamma_\mu$, but the γNX vertex contains effects of the strong interactions. Thus, what we have to study using QCD is the total cross section $\gamma N\to X$ or, because of the optical theorem, the absorptive part of the forward scattering amplitude for $\gamma N\to\gamma N$, where the photon is off-mass-shell ($q^2\neq 0$). In the case where we do not have a spin

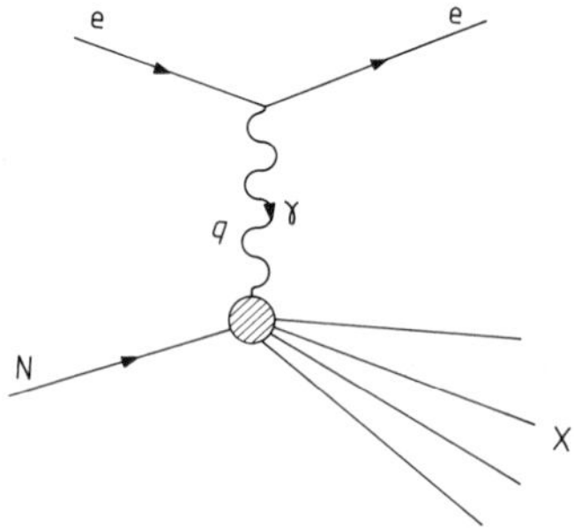

Figure 12.1 One-photon exchange contribution to eN→eX.

polarised target, the object we need to describe eN → eX is

$$
\begin{aligned}
W_{\mu\nu}(p,q) &= \frac{1}{2\pi}\sum_{|X\rangle}\int d^4x\, e^{iq\cdot x}\langle N,p|j_\mu(x)|X\rangle\langle X|j_\nu(0)|N,p\rangle \\
&= \frac{1}{2\pi}\int d^4x\, e^{iq\cdot x}\langle N,p|j_\mu(x)j_\nu(0)|N,p\rangle \\
&= \frac{1}{2\pi}\int d^4x\, e^{iq\cdot x}\langle N,p|[j_\mu(x),j_\nu(0)]|N,p\rangle \qquad (12.39)
\end{aligned}
$$

where p and q are the four-momenta of the nucleon and off-mass-shell photon, respectively, and $j_\mu(x)$ denotes the electromagnetic current operator. It is understood that the nucleon spin states are averaged over. (A proof that the product of currents may be replaced by the commutator may be found, for example, in §4.2.11 of reference 6.) The structure functions W_1 and W_2 are defined by the expansion in terms of covariants,

$$
W_{\mu\nu}(p,q) = \left(-g_{\mu\nu}+\frac{q_\mu q_\nu}{q^2}\right)W_1 + \frac{\left(p_\mu-\dfrac{p\cdot q}{q^2}q_\mu\right)\left(p_\nu-\dfrac{p\cdot q}{q^2}q_\nu\right)}{m_N^2}W_2. \qquad (12.40)
$$

The forward scattering amplitude of $\gamma N\to\gamma N$ (the forward Compton amplitude) is determined by

$$
T_{\mu\nu}(p,q) = i\int d^4x\, e^{iq\cdot x}\langle N,p|T(j_\mu(x)j_\nu(0)|N,p\rangle \qquad (12.41)
$$

with the decomposition in covariants

$$
T_{\mu\nu}(p,q) = \left(-g_{\mu\nu}+\frac{q_\mu q_\nu}{q^2}\right)T_1 + \frac{[p_\mu-(p\cdot q/q^2)q_\mu][p_\nu-(p\cdot q/q^2)q_\nu]}{m_N^2}T_2 \qquad (12.42)
$$

where again an average over the nucleon spin states is understood. Because of the optical theorem, there is the connection

$$W_\sigma = \frac{1}{\pi}\,\text{Im}\,T_\sigma \qquad \sigma = 1, 2. \tag{12.43}$$

In writing down (12.41) and (12.39), we are going somewhat beyond what we have found in Chapter 6 about scalar boson scattering amplitudes, and its generalisation to amplitudes involving photons. Inspection of (6.43) shows that scalar boson scattering amplitudes are obtained from Green functions by acting with Klein–Gordon operators, or, in the case of photon scattering, by acting with factors of $\Box_x$. Thus, essentially, scattering amplitudes are obtained from Green functions by replacing fields by currents. Such expressions are *vacuum* expectation values. What we are assuming here is that similar expressions for scattering amplitudes exist where *not* all the external particles occur as currents, but instead *some* external particles (the nucleons in (12.41)) occur as states in which the expectation value is taken.

The kinematical region in which we shall be able to apply asymptotic freedom will turn out to be the region referred to as the Bjorken limit,

$$q^2 \to -\infty,\ p\cdot q \to \infty \qquad X \equiv \frac{-q^2}{2p\cdot q} \quad \text{fixed.} \tag{12.44}$$

As we now show, the Bjorken limit corresponds to studying the light cone in coordinate space. To see this it is convenient to work in the laboratory frame in which

$$p = (m_\text{N}, 0, 0, 0) \qquad q = (q^0, 0, 0, q^0) \tag{12.45}$$

with the z axis chosen along the direction of the momentum of the (virtual) photon. We next introduce the (light cone) variables

$$q_\pm = q^0 \pm q^3 \tag{12.46}$$

and

$$x_\pm = x^0 \pm x^3. \tag{12.47}$$

In terms of these variables,

$$q^2 = q_+ q_- \tag{12.48}$$

$$p\cdot q = (m_\text{N}/2)(q_+ + q_-) \tag{12.49}$$

and

$$X = \frac{-q_+ q_-}{m_\text{N}(q_+ + q_-)}. \tag{12.50}$$

The Bjorken limit is thus the limit $q_+ \to \infty$ with q_- fixed (and negative), and consequently, $X = -q_-/m_N$. In (12.39), we may write

$$e^{iq\cdot x} = \exp\left(\frac{i}{2}(q_+x_- + q_-x_+)\right). \tag{12.51}$$

Because the exponential oscillates rapidly as $q_+ \to \infty$ the only contribution to the integral in the Bjorken limit comes from the region

$$x_- = 0. \tag{12.52}$$

Then

$$x^2 = x_+x_- - x_1^2 - x_2^2 \leqslant 0. \tag{12.53}$$

However, (microscopic) causality means that the commutator in (12.39) vanishes for $x^2 < 0$. Thus, the only contribution to the integral is from the light cone

$$x^2 = 0. \tag{12.54}$$

Consequently, to study deep inelastic electron–nucleon scattering in the Bjorken limit, we must study the product of two electromagnetic current operators on the light cone. This is a task which is facilitated by the Wilson operator product expansion.

12.3 The Wilson operator product expansion

Wilson[7] has shown that the product of two local operators $A(x)$ and $B(y)$ (for example, two electromagnetic or weak current operators) can be expanded in the form

$$A(x)B(y) = \sum_i C_i(x-y)O_i\left(\frac{x+y}{2}\right) \tag{12.55}$$

where $O_i(x)$ are the local operators of the theory with the quantum numbers of AB, and the $C_i(x-y)$ are c-number coefficients. The result is not too difficult to prove in free-field theory (see problem 12.4). The importance of the Wilson operator product expansion is that the behaviour of the product $A(x)B(y)$ at short distances, $x-y \to 0$, is controlled by those local operators O_i for which $C_i(x-y)$ is most singular as $x-y \to 0$. If there were no dimensionful parameters in the theory then the coefficients $C_i(x-y)$ would involve a number of powers of $x-y$ given by dimensional analysis and the (mass) dimensions of O_i. Then, the most singular coefficients, $C_i(x-y)$, would be those associated with operators O_i with the lowest mass dimensions, and these operators would dominate the short distance behaviour of $A(x)B(y)$. Our experience in §12.1 suggests that this will be true for asymptotically free theories like QCD, apart from logarithmic corrections which bring in the renormalisation scale. We shall see later that this is indeed the case, so that dimensional analysis enables us to isolate the leading contributions to the Wilson expansion in QCD.

The Wilson expansion may be used in applications of asymptotic freedom to decay processes where the exchange of a heavy gauge boson requires us to study a product of vector or axial current operators at short distances. Examples of this are the $\Delta I = \frac{1}{2}$ rule for hadronic decays in electroweak theory, and the baryon-number-violating decay of a proton in grand unified theory.

For present purposes, we are not so much interested in short distances, $x - y \to 0$, as in the vicinity of the light cone, $(x-y)^2 \to 0$. Let us choose local operators O_i of definite spin l_i and let us consider for definiteness the product of two electromagnetic currents, j_μ. Then the Wilson expansion (for $y=0$) for the time ordered product is of the form

$$\mathrm{i}T(j_\mu(x)j_\nu(0)) = \sum_i C^i_{\mu\nu\mu_1\ldots\mu_{l_i}}(x)O_i^{\mu_1\ldots\mu_{l_i}}(0). \tag{12.56}$$

(To reach this form we have first expanded $O_i^{\mu_1\ldots\mu_{l_i}}$ about $x=0$ and then regrouped the coefficients using the fact that the derivatives of a particular O_i are other O_i's in the series.) When $O_i^{\mu_1\ldots\mu_{l_i}}$ has definite spin l_i, we may write

$$\begin{aligned} C^i_{\mu\nu\mu_1\ldots\mu_{l_i}}(x) = &-a^i(x^2)g_{\mu\nu}x_{\mu_1}\ldots x_{\mu_{l_i}} \\ &+b^i(x^2)g_{\mu\mu_1}g_{\nu\mu_2}x_{\mu_3}\ldots x_{\mu_{l_i}} \\ &+\ldots. \end{aligned} \tag{12.57}$$

Here, we are using the fact that $O_i^{\mu_1\ldots\mu_{l_i}}$ is traceless to drop terms involving $g_{\mu_1\mu_2}$ etc, and the fact that it is symmetric to condense a collection of terms into the $b^i(x^2)$ term. There are other possible covariants of the form $x_\mu x_\nu x_{\mu_1}\ldots x_{\mu_{l_i}}$ and $(x_\mu g_{\nu\mu_1}x_{\mu_2}\ldots x_{\mu_{l_i}} + (\mu \leftrightarrow \nu))$. However, it will be sufficient to retain only those terms shown explicitly in (12.57) in order to identify the contributions to T_1 and T_2 in (12.42). A term of the type $\varepsilon_{\mu\nu\mu_1\lambda}x^\lambda x_{\mu_2}\ldots x_{\mu_{l_i}}$ is forbidden by the symmetry of (12.41) under $\mu \leftrightarrow \nu$, $x \leftrightarrow -x$. In the absence of dimensionful parameters, the coefficients $a^i(x^2)$ and $b^i(x^2)$ involve a single power of x^2 which is dictated by the mass dimensions of the operator O_i and by the number of factors $x_{\mu_1}\ldots x_{\mu_{l_i}}$ which increase the mass dimensions available to $a^i(x^2)$ and $b^i(x^2)$. Thus, the leading terms on the light cone are those for which the twist

$$t_i \equiv d_i - l_i \tag{12.58}$$

is smallest, where d_i is the mass dimension of O_i, and l_i is the spin of O_i.

We now see that asymptotic freedom is relevant to the Bjorken limit. To study the operator product of currents on the light cone, it suffices to study the behaviour of the functions $a^i(x^2)$ and $b^i(x^2)$ for $x^2 \to 0$. Since $a^i(x^2)$ and $b^i(x^2)$ are scalar functions of x^2 alone, all we need do is study them for $x \to 0$, i.e. at short distances, where asymptotic freedom will allow us to do a reliable calculation. We will be able to identify the leading contributions on the light cone as those of lowest twist. This is because our experience in §12.1 with asymptotically free QCD is that at short distances dimensionful quantities enter only through logarithmic corrections involving the renormalisation scale M. To derive the logarithmic corrections for the coefficients $a^i(x^2)$ and $b^i(x^2)$, it is necessary to derive renormalisation group equations for Green functions involving composite local operators built from the fundamental field operators. This we shall do in §12.5.

12.4 Wilson coefficients and moments of structure functions

In this section, we establish a connection between the structure functions for $eN \to eX$, defined in §12.2, and the coefficients in the Wilson operator product expansion, defined in §12.3. In the next section, we shall use asymptotic freedom to calculate the behaviour of the Wilson coefficients in QCD. First rewrite (12.57) as

$$C^i_{\mu\nu\mu_1\ldots\mu_{l_i}}(x) = -(-1)^{l_i} g_{\mu\nu}\partial_{\mu_1}\ldots\partial_{\mu_{l_i}} A^i(x^2) + (-1)^{l_i-2} g_{\mu\mu_1} g_{\nu\mu_2}\partial_{\mu_3}\ldots\partial_{\mu_{l_i}} B^i(x^2) + \ldots \tag{12.59}$$

where

$$(-1)^{l_i} 2^{l_i} \frac{d^{l_i}}{d(x^2)^{l_i}} A^i(x^2) = a^i(x^2) \tag{12.60}$$

and

$$(-1)^{l_i-2} 2^{l_i-2} \frac{d^{l_i-2}}{d(x^2)^{l_i-2}} B^i(x^2) = b^i(x^2). \tag{12.61}$$

Taking the matrix element of (12.56) between spin-averaged nucleon states, and performing the Fourier transform $\int d^4x\, e^{iq\cdot x}$, to reconstruct (12.41), we see that

$$T_{\mu\nu}(p, q) = \sum_i K_i[-g_{\mu\nu}(p\cdot q)^{l_i}\tilde{A}^i(q^2) + p_\mu p_\nu (p\cdot q)^{l_i-2}\tilde{B}^i(q^2)] + \ldots \tag{12.62}$$

where

$$\tilde{A}^i(q^2) = \int d^4x\, e^{iq\cdot x} A^i(x^2) \tag{12.63}$$

$$\tilde{B}^i(q^2) = \int d^4x\, e^{iq\cdot x} B^i(x^2) \tag{12.64}$$

and

$$\langle \mathrm{N}, p|O_i^{\mu_1 \ldots \mu_{l_i}}|\mathrm{N}, p\rangle = K_i p^{\mu_1} \ldots p^{\mu_{l_i}} + \ldots \qquad (12.65)$$

where K_i is some (in general unknown) constant. In (12.65), the omitted terms involve at least one factor of the type $g^{\mu_1\mu_2}$. As can be seen from (12.57), these terms produce at least one extra factor of x^2, and are thus less singular on the light cone, and may be dropped. On dimensional grounds, we may write,

$$\tilde{A}^i(q^2) = (-q^2)^{-l_i} \hat{A}^i(q^2) \qquad (12.66)$$

and

$$\tilde{B}^i(q^2) = (-q^2)^{-l_i+1} \hat{B}^i(q^2) \qquad (12.67)$$

where $\hat{A}^i(q^2)$ and $\hat{B}^i(q^2)$ are dimensionless functions. (In QCD, they will be logarithmic functions of q^2/M^2, where M is the renormalisation scale.) We may now write (12.62) as

$$T_{\mu\nu}(p, q) = \sum_i K_i[-g_{\mu\nu}(2X)^{-l_i} \hat{A}^i(q^2) + p_\mu p_\nu (p \cdot q)^{-1} (2X)^{-l_i+1} \hat{B}^i(q^2)] + \ldots \qquad (12.68)$$

with the variable X as defined in (12.44).

Comparing with (12.42) we see that

$$T_1(p, q) = \sum_i K_i (2X)^{-l_i} \hat{A}^i(q^2) \qquad (12.69)$$

and

$$\nu T_2(p, q) = m_\mathrm{N} \sum_i K_i (2X)^{-l_i+1} \hat{B}^i(q^2) \qquad (12.70)$$

where

$$\nu \equiv p \cdot q / m_\mathrm{N}. \qquad (12.71)$$

Thus, apart from the factors $\hat{A}^i(q^2)$ and $\hat{B}^i(q^2)$, $T_1(p, q)$ and $\nu T_2(p, q)$ are functions of the variable X, alone. (This is referred to as Bjorken scaling.) In QCD, the factors $\hat{A}^i(q^2)$ and $\hat{B}^i(q^2)$ produce slowly varying corrections[8] to Bjorken scaling, which are logarithmic in q^2/M^2 where M is the renormalisation scale. (There will also be contributions of order m_N^2/q^2 from the operators of higher twist in the operator product expansion.)

It remains to make the connection with the structure functions W_1 and W_2 for deep inelastic electroproduction. The connection is given in the physical region $0 < X < 1$ by (12.43). However, in this region for X sufficiently close to 0, the series (12.69) and (12.70) diverge. It is therefore necessary to use an analytic continuation in the variable X, and to isolate individual terms in the Laurent series. Taking a large circular counter-clockwise contour $\mathscr{C}$ in the X plane, we obtain for the coefficients in the Laurent expansion

$$\frac{1}{2\pi \mathrm{i}}\int_{\mathscr{C}} \mathrm{d}X\, X^{l-1} T_1 = \frac{1}{2^l} \sum_{i,l_i=l} K_i \hat{A}^i(q^2) \tag{12.72}$$

and

$$\frac{1}{2\pi \mathrm{i}}\int_{\mathscr{C}} \mathrm{d}X\, X^{l-2} \nu T_2 = \frac{m_{\mathrm{N}}}{2^{l-1}} \sum_{i,l_i=l} K_i \hat{B}^i(q^2). \tag{12.73}$$

(In (12.72) and (12.73), the sum over i is now a sum over only those operators which have the same spin $l_i = l$.) The contour integrals may be obtained as integrals along a branch cut running between $X = \pm 1$ with the discontinuity across the cut obtained from (12.43).

Thus

$$\frac{1}{2\pi \mathrm{i}}\int_{\mathscr{C}} \mathrm{d}X X^{l-1} T_1 = 2\int_0^1 \mathrm{d}X X^{l-1} W_1 \tag{12.74}$$

and

$$\frac{1}{2\pi \mathrm{i}}\int_{\mathscr{C}} \mathrm{d}X X^{l-2} \nu T_2 = 2\int_0^1 \mathrm{d}X X^{l-2} \nu W_2. \tag{12.75}$$

Consequently,

$$\int_0^1 \mathrm{d}X X^{l-1} W_1 = \frac{1}{2^{l+1}} \sum_{i,l_i=l} K_i \hat{A}_i(q^2) \tag{12.76}$$

and

$$\int_0^1 \mathrm{d}X X^{l-2} \nu W_2 = \frac{m_{\mathrm{N}}}{2^l} \sum_{i,l_i=l} K_i \hat{B}_i(q^2). \tag{12.77}$$

We see that it is the moments of the structure functions (integrals with powers of X) that are related to the Fourier transforms of coefficients in the operator product expansion, $\hat{A}_i(q^2)$ and $\hat{B}_i(q^2)$ defined in (12.66), (12.67), (12.63), (12.64) and (12.59). There also enter the (in general) unknown coefficients K_i, which are matrix elements between nucleon states of operators $O_i^{\mu_1 \cdots \mu_{l_i}}$ as defined in (12.65). In an asymptotically free theory, we will expect $\hat{A}_i(q^2)$ and $\hat{\zeta}_i(q^2)$ to depend logarithmically on q^2/M^2. In general (12.76) and (12.77) will be difficult to test because more than one logarithmic term will be involved, with unknown coefficients K_i, whenever there is more than one operator $O_i^{\mu_1 \cdots \mu_{l_i}}$ for a given spin $l_i = l$. We shall see later that, by studying combinations of structure functions which are non-singlet with respect to SU(3) of flavour, we shall be able to ensure that only a single logarithmic term occurs[9]. Then experimental tests are feasible. Our next step is to calculate the behaviour of the coefficients $\hat{A}_i(q^2)$ and $\hat{B}_i(q^2)$ in QCD, using renormalisation group equations for Green functions involving composite local operators.

12.5 Renormalisation group equation for Wilson coefficients

In this section, we shall derive renormalisation group equations for the coefficients in the Wilson operator product expansion by first obtaining renormalisation group equations for Green functions involving some legs which are composite operators (like j^μ and $O_i^{\mu_1\cdots\mu_{l_i}}$). For succinctness, we shall denote a composite operator $O_i^{\mu_1\cdots\mu_{l_i}}$ by O_i, for the moment. Green functions with composite operator legs may be defined formally by adding source terms $J_i(x)O_i(x)$ to the Lagrangian and setting up generating functionals, in the way discussed for ordinary Green functions in Chapter 4. In general, the relationship between bare operators $(O_i)_\mathrm{B}$ and renormalised operators O_i is the matrix one,

$$(O_i)_\mathrm{B} = Z_{ij} O_j \tag{12.78}$$

when there is more than one operator O_i in the operator product expansion with the same quantum numbers, including spin. (We shall see in §12.6 how the renormalisation constants Z_{ij} may be computed from Feynman diagrams for Green functions with composite operator legs.) In exact analogy with §12.1, we introduce the notation $\tilde{\Gamma}^{(n)}_{O_i}(p_1, \ldots, p_n, \hat{g}, \xi, m, M)$ for a renormalised OPI Green function, with n external legs which are ordinary fields, n_A of which are gauge fields, and n_ψ of which are fermions, and, additionally, one external leg with zero four momentum which is a composite operator O_i. Such Green functions are referred to as inserted Green functions. More than one composite external leg introduces no further difficulties, other than notational. The connection between the bare and renormalised inserted Green functions is

$$\tilde{\Gamma}^{(n)}_{O_i}(p_1, \ldots, p_n, p_n, \hat{g}, \xi, m, M) = Z_{ij} Z_A^{n_A/2} Z_\psi^{n_\psi/2} \tilde{\Gamma}^{(n)}_{O_j\mathrm{B}}(p_1, \ldots, p_n, g_\mathrm{B}, \xi_\mathrm{B}, m_\mathrm{B}). \tag{12.79}$$

Carrying out the differentiation $M\partial/\partial M$ with g_B, ξ_B and m_B held fixed, just as in §12.1, gives the renormalisation group equation

$$\left[\delta_{ij}\left(M\frac{\partial}{\partial M} + \beta_{\hat{g}}\frac{\partial}{\partial \hat{g}} + \beta_\xi \frac{\partial}{\partial \xi} - \gamma_m m \frac{\partial}{\partial m} - n_A\gamma_A - n_\psi \gamma_\psi\right) - \gamma_{ij}\right] \times \tilde{\Gamma}^{(n)}_{O_j}(p_1, \ldots, p_n, \hat{g}, \xi, m, M) = 0 \tag{12.80}$$

where

$$\gamma_{ij} = M \frac{\partial Z_{ik}}{\partial M} Z_{kj}^{-1} \tag{12.81}$$

and the other coefficients are as in (12.5)–(12.9).

A renormalisation group equation for the Wilson coefficients may now be derived by utilising (12.80). We first multiply the operator product expansion (12.56) by n_A factors of gauge fields and n_ψ factors of fermion fields, time order, and Fourier transform to momentum space. This gives (see problem 12.5) the relationship between inserted OPI Green functions

$$\mathrm{i}\tilde{\Gamma}^{(n)}_{j_\mu j_\nu}(q,-q,p_1,\ldots,p_n)=\sum_i C^i(q)\tilde{\Gamma}^{(n)}_{O_i}(p_1,\ldots,p_n) \tag{12.82}$$

where the left-hand side denotes an OPI Green function, as in §12.1, but with two additional external legs which are composite operators j_μ and j_ν with four-momenta q and $-q$, respectively. On the right-hand side of (12.82) we have suppressed the indices μ, ν and $\mu_1,\ldots,\mu_{l_i}$. (The relationship is, in the first instance, one between ordinary Green functions, but can be made one between OPI Green functions because the subsets of Feynman diagrams which are disconnected or one-particle-reducible are in one-to-one correspondence on the two sides of the equation.) The Green function $\tilde{\Gamma}^{(n)}_{j_\mu j_\nu}$ obeys exactly the same renormalisation group equation (12.4) as $\tilde{\Gamma}^{(n)}$, because the conserved current j_μ requires *no* renormalisation as may be checked directly at one-loop order by calculating Feynman diagrams. Substituting (12.82) into the renormalisation group equation for $\tilde{\Gamma}^{(n)}_{j_\mu j_\nu}$ and using the renormalisation group equation for $\tilde{\Gamma}^{(n)}_{O_i}$, we obtain

$$\left[\delta_{ij}\left(M\frac{\partial}{\partial M}+\beta_{\hat{g}}\frac{\partial}{\partial\hat{g}}+\beta_\xi\frac{\partial}{\partial\xi}-\gamma_m m\frac{\partial}{\partial m}\right)+\gamma_{ji}\right]C_j(q)=0. \tag{12.83}$$

This is the required renormalisation group equation for the Wilson coefficients. It involves the so-called anomalous dimensions matrix γ_{ji} defined through (12.81) and (12.78). Reintroducing the Lorentz indices in the operator product expansion, and defining the independent covariants as in §12.4, we see that $\hat{A}^i(q^2)$ and $\hat{B}^i(q^2)$ obey the renormalisation group equations

$$\left[\delta_{ij}\left(M\frac{\partial}{\partial M}+\beta_{\hat{g}}\frac{\partial}{\partial\hat{g}}+\beta_\xi\frac{\partial}{\partial\xi}-\gamma_m m\frac{\partial}{\partial m}\right)+\gamma_{ji}\right]\hat{A}^j(q^2)=0 \tag{12.84}$$

and

$$\left[\delta_{ij}\left(M\frac{\partial}{\partial M}+\beta_{\hat{g}}\frac{\partial}{\partial\hat{g}}+\beta_\xi\frac{\partial}{\partial\xi}-\gamma_m m\frac{\partial}{\partial m}\right)+\gamma_{ji}\right]\hat{B}^j(q^2)=0. \tag{12.85}$$

Here, the sum over j is over operators with some fixed value of spin $l_j=l$.

Life is particularly simple if there is only one operator $O_i^{\mu_1\cdots\mu_{l_i}}$ (of lowest twist) for the given value of spin l_i. (We see later that this happens if we take a flavour singlet combination of structure functions.) Then, we no longer need a *matrix* of anomalous dimensions γ_{ij} and we may write

$$\gamma_{ij}=\gamma_{O_i}\delta_{ij}. \tag{12.86}$$

The renormalisation group equations are now diagonal, and have a solution, when q is scaled by a factor s, exactly analogous to that of (12.28)

$$\hat{A}^i(s^2q^2,\hat{g},m,M)=\exp\left(\int_1^s\frac{\mathrm{d}s'}{s'}\gamma_{O_i}(\bar{g}(s),\bar{\xi}(s))\right)\times\hat{A}^i(q^2,\bar{g}(s),\bar{\xi}(s),\bar{m}(s),M). \tag{12.87}$$

(Since $\hat{A}^i$ is dimensionless, there is no factor like s^{d_Γ}.) At asymptotically high momentum, and in Landau gauge, we will have, in analogy with (12.38),

$$\hat{A}^i(s^2q^2, g, \xi=0, m, M) \approx (1+2bg^2 \ln s)^{-d_i/2b}\hat{A}^i(q^2, \bar{g}(s), \bar{\xi}(s)=0, \bar{m}(s)=0, M) \quad (12.88)$$

where we have defined d_i by

$$\gamma_{O_i}(g, \xi=0) = -d_i g^2 + O(g^3). \quad (12.89)$$

Thus

$$\hat{A}^i(s^2q^2) \sim (\ln s)^{-d_i/2b}. \quad (12.90)$$

(The dependence of $\hat{A}^i(q^2, \bar{g}(s), \bar{\xi}(s)=0, \bar{m}(s)=0, M)$ on $\bar{g}(s)$ is additive, and gives a non-leading term.) If $q^2 = -M^2$ is adopted as a reference momentum,

$$\hat{A}^i(q^2) \sim [(\ln(-q^2/M^2)]^{-d_i/2b} \quad (12.91)$$

and similarly for $\hat{B}^i(q^2)$. As promised in §12.4, the Wilson coefficients $\hat{A}^i(q^2)$ and $\hat{B}^i(q^2)$, and so the moments of structure functions, depend logarithmically on q^2/M^2. The final step, which is the content of the next section, is to evaluate the anomalous dimensions.

12.6 Calculation of anomalous dimensions

For QCD, when we analyse the Wilson operator product expansion for a product of two electromagnetic currents, the gauge invariant operators of lowest twist, $d-l$, have twist 2. The operators which are non-singlet under SU(3) of flavour are

$$O_{l,\alpha}^{\mu_1\ldots\mu_l} = \frac{i^{l-1}}{l!}(\bar{\psi}\lambda^\alpha\gamma^{\mu_1}D^{\mu_2}\ldots D^{\mu_l}\psi + \text{permutations}) - (\text{trace terms}) \quad (12.92)$$

for $\alpha = 1, \ldots, 8$, where ψ denotes a quark field operator which is a triplet under colour SU(3), and also under SU(3) of flavour. (The c, b and t quarks do not need to be considered for moderate energies of scattering.) The Gell-Mann matrix λ^α is a matrix in the flavour space, and D^μ is the covariant derivative

$$D^\mu\psi = \left(\partial^\mu + ig\frac{\lambda^a}{2}A_a^\mu\right)\psi \quad (12.93)$$

where λ^a is a matrix in colour space, and A_a^μ are the colour gluons.

There are also flavour singlet operators

$$O_{l,0}^{\mu_1\ldots\mu_l} = \frac{i^{l-1}}{l!}(\bar{\psi}\gamma^{\mu_1}D^{\mu_2}\ldots D^{\mu_l}\psi + \text{permutations}) - (\text{trace terms}) \quad (12.94)$$

and

$$O^{\mu_1\ldots\mu_l}_{l,A} = \frac{\mathrm{i}^{l-2}}{(2l)!}\sum_a (F_a^{\mu\mu_1}\mathrm{D}^{\mu_2}\ldots\mathrm{D}^{\mu_{l-1}}(F_a)^{\mu_l}_{\mu} + \text{permutations})$$

$$-(\text{trace terms}) \tag{12.95}$$

where the covariant derivative acting on quark fields is as in (12.93), the covariant derivative acting on gauge fields is

$$\mathrm{D}^\mu A^\nu_a = \partial^\mu A^\nu_a - g f_{abc} A^\mu_b A^\nu_c \tag{12.96}$$

and $F^{\mu\nu}_a$ is the covariant curl as in (9.30).

If we consider a combination of structure functions which is flavour non-singlet, then we need only consider an operator of the type (12.92) and the anomalous dimensions matrix will be diagonal. (The simplest case is to consider the difference of νW_2 with a proton target and νW_2 with a neutron target.) The renormalisation constant for $O_{l,\alpha}$ (suppressing the Lorentz indices on the operator) will be denoted by $Z_{l,\alpha}$. It may be determined by considering the renormalisation of the Green function $\tilde{\Gamma}^{(2)}_{O_{l,\alpha}}$ with two quark external legs and one $O_{l,\alpha}$ external leg. From (12.79),

$$\tilde{\Gamma}^{(2)}_{O_{l,\alpha}} = Z_{l,\alpha} Z_\psi \tilde{\Gamma}^{(2)}_{O_{l,\alpha}\mathrm{B}}. \tag{12.97}$$

At one loop order, this Green function is given by

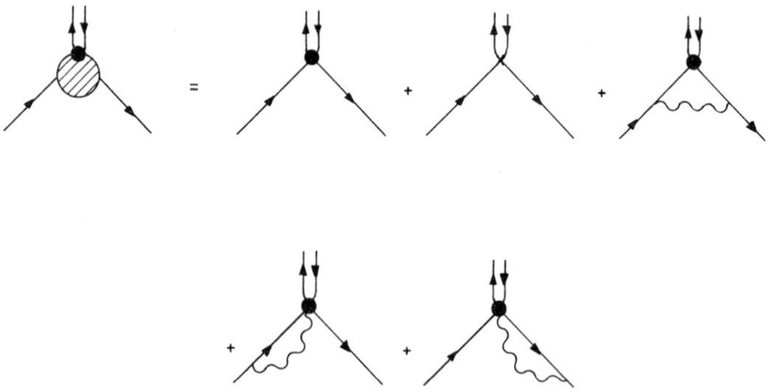

(12.98)

where the two parallel fermion lines leaving and entering a black blob are being used to denote an $O_{l,\alpha}$ external line, and a cross to denote a counter-term. The zeroth order term and counter term are

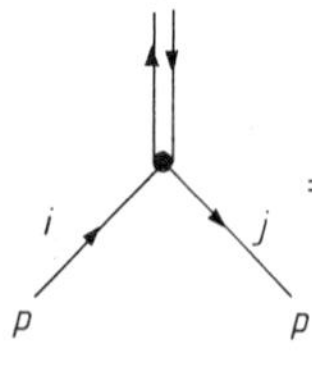

$$= -\frac{1}{l!}(\gamma^{\mu_1} p^{\mu_2}\ldots p^{\mu_l} + \text{permutations})\lambda^\alpha \delta_{ij} - (\text{trace terms})$$

(12.99)

and

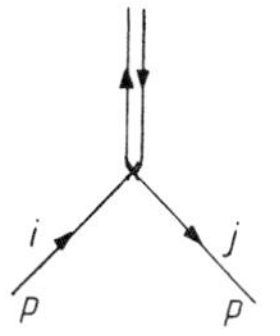

$$= \frac{-(\Delta Z_\psi + \Delta Z_{l,\alpha})}{l!} (\gamma^{\mu_1} p^{\mu_2} \ldots p^{\mu_l} + \text{permutations}) \lambda^\alpha \delta_{ij}$$

$$-(\text{trace terms}) \qquad (12.100)$$

where we have written

$$Z_{l,\alpha} = 1 + \Delta Z_{l,\alpha} \qquad (12.101)$$

and (as in §11.1)

$$Z_\psi = 1 + \Delta Z_\psi. \qquad (12.102)$$

Calculating the non-trivial diagrams in Feynman gauge (for this gauge invariant[8] object) gives

(diagram 1) =

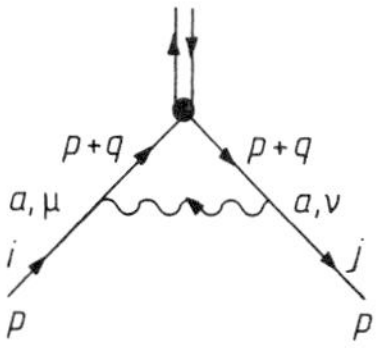

$$= -\frac{\hat{g}^2 M^\varepsilon \lambda^\alpha C_3 \delta_{ij}}{16\pi^2 \varepsilon} \frac{4}{l(l+1)} \frac{1}{l!} (\gamma^{\mu_1} p^{\mu_2} \ldots p^{\mu_l} + \text{permutations})$$

$$-(\text{trace terms}) \qquad (12.103)$$

where $\hat{g}$ is the dimensionless coupling constant and M is the renormalisation scale, as in Chapter 11, and the group theory factor C_3 is defined in (11.53).

(diagram 2) =

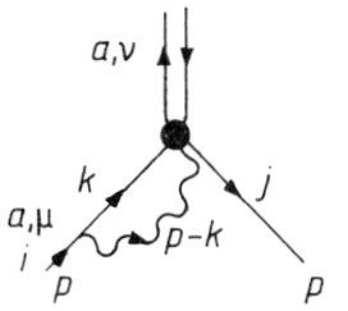

$$= \frac{\hat{g}^2 M^\varepsilon \lambda^\alpha C_3 \delta_{ij}}{16\pi^2 \varepsilon} \left(\sum_{s=2}^{l} \frac{4}{s} \right) \frac{1}{l!} (\gamma^{\mu_1} p^{\mu_2} \ldots p^{\mu_l} + \text{permutations})$$

$$-(\text{trace terms}). \qquad (12.104)$$

Also

(diagram 3) = 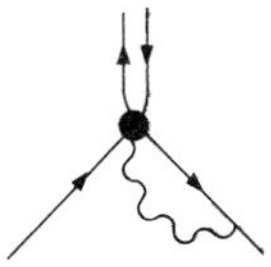 = (diagram 2). (12.105)

Returning to (12.98), and using (12.99)–(12.105), we obtain for the renormalisation constant in the MS scheme

$$\Delta Z_{\psi}+\Delta Z_{l,\alpha}=-\frac{2\hat{g}^2C_3}{16\pi^2\varepsilon}\left(\frac{2}{l(l+1)}-4\sum_{s=2}^{l}\frac{1}{s}\right). \qquad (12.106)$$

Taking the fermion wave function renormalisation counter term ΔZ_{ψ}, from (11.59), we find

$$\Delta Z_{l,\alpha}=\frac{2\hat{g}^2C_3}{16\pi^2\varepsilon}\left(1-\frac{2}{l(l+1)}+4\sum_{s=2}^{l}\frac{1}{s}\right) \qquad (12.107)$$

and

$$Z_{l,\alpha}=1+\Delta Z_{l,\alpha}. \qquad (12.108)$$

Following (12.81) and (12.86), we obtain the anomalous dimension for $O_{l,\alpha}$ from

$$\gamma_{O_{l,\alpha}}=M\frac{\partial}{\partial M}\ln(1+\Delta Z_{l,\alpha}). \qquad (12.109)$$

Thus, using (12.107), and the renormalisation mass dependence of $\hat{g}$ given in (11.23), we find

$$\gamma_{O_{l,\alpha}}=-\frac{2\hat{g}^2C_3}{16\pi^2}\left(1-\frac{2}{l(l+1)}+4\sum_{s=2}^{l}\frac{1}{s}\right). \qquad (12.110)$$

In the notation of (12.89),

$$d_{l,\alpha}=\frac{2C_3}{16\pi^2}\left(1-\frac{2}{l(l+1)}+4\sum_{s=2}^{l}\frac{1}{s}\right) \qquad (12.111)$$

and from (12.91) and (12.92) we see that

$$\int_0^1 \mathrm{d}X X^{l-2}\nu W_2\sim[\ln(-q^2/M^2)]^{-d_{l,\alpha}/2b} \qquad (12.112)$$

with b as in (12.30), and the appropriate flavour non-singlet combination of structure functions understood (e.g. the difference of νW_2 off protons and νW_2 off neutrons).

Similar calculations may be carried out for the flavour singlet structure functions[8] using the operators of (12.94) and (12.95). In that case, there is a 2×2 matrix of anomalous dimensions, and, in general, the moments of structure functions depend on two distinct powers of $\ln(-q^2/M^2)$ with unknown coefficients K_i (as in (12.76) and (12.77)). The calculations of this chapter may also be extended to deep inelastic neutrino production[9], where moments of structure functions which involve a single power of $\ln(-q^2/M^2)$ may be found, with the aid of charge conjugation invariance, without considering different targets.

12.7 Comparison with experiment, and Λ_{QCD}

To compare with experiment the prediction (12.112) for the variation with q^2 of the moments of flavour non-singlet structure functions, it is convenient to take logarithms. Thus

$$\ln M_l(q^2) = -\frac{d_{l,\alpha}}{2b}\ln(-q^2/M^2) + \text{constant} \tag{12.113}$$

where the lth moment is

$$M_l(q^2) \equiv \int_0^1 \mathrm{d}X\, X^{l-2}\nu W_2 \tag{12.114}$$

and $d_{l,\alpha}$ and b are as in (12.111) and (12.30). If we now plot $\ln M_l(q^2)$ against $\ln M_{l'}(q^2)$, where l and l' refer to two different moments of the structure function, the slope of the plot is $d_{l,\alpha}/d_{l',\alpha}$

$$\ln M_l(q^2) = \frac{d_{l,\alpha}}{d_{l',\alpha}}\ln M_{l'}(q^2) + \text{constant}. \tag{12.115}$$

Since the $d_{l,\alpha}$ are as in (12.111), there is a clean prediction, which is in quite good agreement with experiment.

The alert reader will have noticed that, following on from (12.91), we have written the prediction for the structure function moment, (12.112), in terms of a reference mass M which is entirely arbitrary. Moreover, we have not yet fixed the value of the QCD coupling constant g in (12.29), defined as the value of the renormalised coupling constant for renormalisation scale M. The reason for this is that (12.29) is only valid when $2bg^2 \ln s$ is very much greater than one (otherwise $\bar{g}^2(s)$ is not necessarily small, and expansion in powers of $\bar{g}(s)$ in $\beta_g(\bar{g}(s))$ is not valid), and then, to leading order, g^2 divides out. Correspondingly, (12.112) is only valid when $\ln(-q^2/M^2)$ is very much greater than one, at which stage $\ln(-q^2) \gg \ln M^2$ (so to speak) and the scale M cannot be determined reliably. However, by going to next-to-leading order in the coupling strength $\bar{g}^2$, in performing the QCD calculations and comparing with experiment, it is possible to determine the scale on which $\bar{g}^2$, and consequently the structure function moments, vary. We may introduce this scale in the following way. In the leading order expression (12.29) take

$$s = \tilde{M}/M \tag{12.116}$$

where $\tilde{M}$ is some new renormalisation mass. Then,

$$\bar{g}^2(\tilde{M}/M) = g^2(\tilde{M}) = g^2/[1 + 2bg^2 \ln(\tilde{M}/M)] \tag{12.117}$$

where we have used (12.25), and

$$g \equiv g(M). \tag{12.118}$$

We may rewrite (12.117) in the form

$$g^2(\tilde{M})=[2b\ln(\tilde{M}/\Lambda_{QCD})]^{-1} \tag{12.119}$$

valid for $\tilde{M} \gg \Lambda_{QCD}$ where Λ_{QCD} is defined by

$$\ln(\Lambda_{QCD}/M)=-(2bg^2)^{-1}. \tag{12.120}$$

The value of the coupling constant $g(\tilde{M})$ at the new renormalisation scale $\tilde{M}$ cannot depend on the original renormalisation scale M. Thus (12.119) must be independent of M, and Λ_{QCD} must be independent of M (the M dependence cancelling between $\ln M$ and $g \equiv g(M)$). The QCD coupling constant at the Z mass $g(m_Z)$, as determined from comparison with experiment of next-to-leading-order QCD calculations is given by

$$\alpha_s(m_Z)=0.113 \qquad m_Z=91.18\text{ GeV} \tag{12.121}$$

where

$$\alpha_s(\tilde{M}) \equiv g^2(\tilde{M})/4\pi. \tag{12.122}$$

With b as in (12.30) with 5 flavours of quark operative in the range of energy up to the Z mass, the corresponding value of the QCD scale parameter Λ_{QCD} is

$$\Lambda_{QCD}=0.065\text{ GeV}. \tag{12.123}$$

We may determine $g^2(\tilde{M})$ for any renormalisation scale $\tilde{M}$ from this value of Λ_{QCD}.

The coupling constant for QED may be treated in a similar way by writing

$$e^2(\tilde{M})=-[2b\ln(\Lambda_{QED}/\tilde{M})]^{-1} \tag{12.124}$$

valid for $\tilde{M} \ll \Lambda_{QED}$. In this case b is negative, which accounts for the slight difference in form (12.124) and (12.119), and, for $\tilde{M}$ in the range of energy up to the Z mass, we take b to be given by (12.31) with two complete generations and the top quark contribution of 4/9 omitted for the third generation. Then Λ_{QED} may be determined from the known value of $e^2_{PHYS}/4\pi$ where e_{PHYS} is the coupling constant for on-mass-shell electrons.

From §11.3, we see that $e^2(\tilde{M})$ for $\tilde{M}=m_e$, the electron mass, differs from e^2_{PHYS} by less than 1%, so we write

$$\alpha(\tilde{M}) \simeq 1/137 \simeq 7.3\times 10^{-3} \qquad \tilde{M}=m_e \tag{12.125}$$

where

$$\alpha(\tilde{M}) \equiv e^2(\tilde{M})/4\pi. \tag{12.126}$$

Using (12.124) we then find

$$\Lambda_{QED} \simeq 2.5 \times 10^{66} m_e \tag{12.127}$$

This is such an enormous number that $e^2(\tilde{M})$ grows exceedingly slowly with $\tilde{M}$. Thus, for example,

$$e^2(m_Z)/e^2(m_e) = 1.09 \tag{12.128}$$

so that

$$\alpha^{-1}(m_Z) = 126.1. \tag{12.129}$$

12.8 e^+e^- annihilation

The total cross section into hadrons for the inclusive process

$$e^+e^- \to X \tag{12.130}$$

where X is an arbitrary unobserved hadronic final state, provides another test[10] of QCD. Treating the process to lowest order in the electromagnetic interaction, what we have to study is the total cross section for a photon to produce hadrons (see figure 12.2). One way to approach this problem is to use the optical theorem to relate the required cross section to the OPI Green function for two photon fields, and then to use the renormalisation group equation with two running coupling constants, $\bar{e}$ and $\bar{g}$, because both the electromagnetic coupling constant e, and the QCD coupling constant g enter the diagrams. The variation of the electromagnetic coupling constant with renormalisation scale is extremely slow (see §12.7) and to a good approximation only the QCD coupling constant need be allowed to run. This approach is described in detail in the review of Politzer[11].

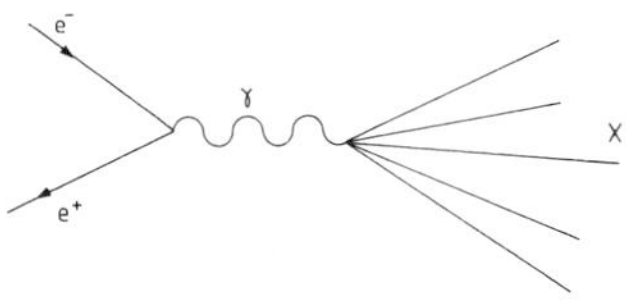

Figure 12.2 Electron–positron annihilation into hadrons.

There is an alternative less rigorous approach, which has the virtue that it can be extended to study the differential cross section for quark or gluon jets, as well as the total cross section for e^+e^- annihilation. It can also be extended to other situations where the operator product expansion is not applicable. In this approach, one calculates (zero- and) one-loop diagrams for e^+e^- annihilation into hadrons of figure 12.2. We have to include not only diagrams for $e^+e^- \to \bar{q}q$ (where q is a quark) but also diagrams for $e^+e^- \to \bar{q}qg$ (where g is a gluon), because, as a matter of principle, any apparatus used for observations has a finite energy resolution, and there is no way of excluding the possibility that a sufficiently low energy gluon has been emitted. Because of the zero mass of the gluon, there are infrared divergences in the diagrams of figure 12.2(*b*), (*c*) and (*d*) which appear in dimensional regularisation as poles in ε which remain after the counter terms have been subtracted using the diagrams of figure 12.4. However, after integrations over phase space have been made to obtain observable cross sections, these infrared divergences are cancelled by the infrared divergences in the diagrams of figures 12.3(*e*) and 12.3(*b*). This is true in particular of the total cross section for e^+e^- annihilation. Thus, when all diagrams of figures 12.3 and 12.4 are included, a finite result is obtained for the total cross section into hadrons for virtual photon four-momentum, namely,

$$\sigma(e^+e^- \to X) = \frac{4\pi\alpha^2}{3q^2}\left(3\sum_f Q_f^2\right)\left(1 + \frac{3\alpha_s(\sqrt{q^2})}{4\pi}c_3\right) \qquad (12.131)$$

where the sum over f is a sum over the charges squared of all (active) quark flavours, the preceding factor of 3 is from the three quark colours, and the group theory factor c_3 (as defined in (11.53)) is for the **3** of colour SU(3), and so has the value $\frac{4}{3}$. As observed earlier, the electromagnetic coupling constant varies only very slowly with renormalisation scale, and the fine structure

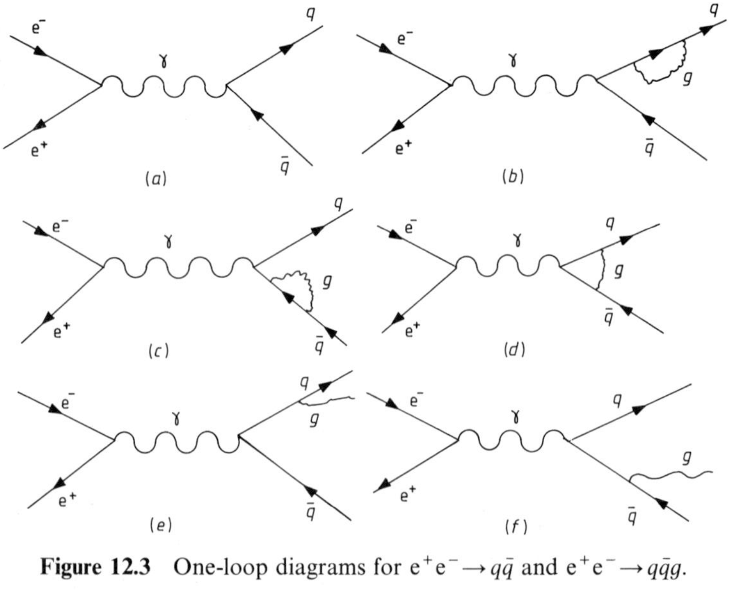

Figure 12.3 One-loop diagrams for $e^+e^- \to q\bar{q}$ and $e^+e^- \to q\bar{q}g$.

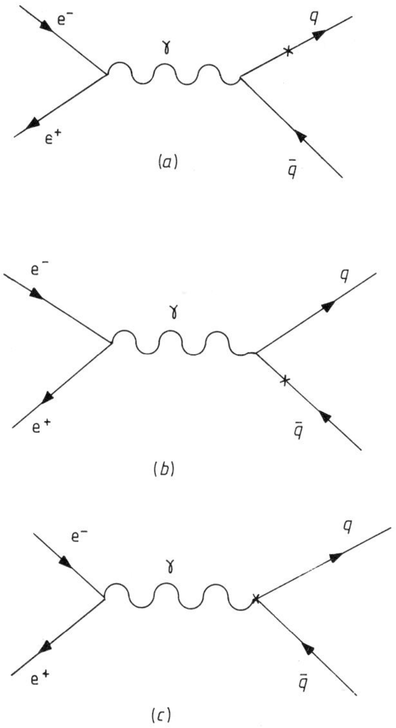

Figure 12.4 Counter term diagrams for $e^+e^- \rightarrow q\bar{q}$.

constant α in (12.131) may be taken to be 1/137. The QCD fine structure constant α_s must be allowed to run, and is given by (12.119). More detail of this calculation, together with extensions of the method to $eN \rightarrow eX$ and other processes may be found in the reviews of Pennington[12] and Sachrajda[13].

Problems

12.1 Calculate the renormalisation group coefficients of (12.10)–(12.14) from the renormalisation constants of Chapter 11.

12.2 Show that the energy dimensions of the n-point OPI Green functions are as in (12.16).

12.3 Derive the renormalisation group equation for OPI Green functions in scalar field theory with $\lambda\varphi^4$ interaction, and find the behaviour of the running coupling constant $\bar{\lambda}(s)$ as a function of s.

12.4 Derive the operator product expansion (12.55) for two electromagnetic currents in free-field theory.

12.5 Derive the relation (12.82) between Green functions with electromagnetic currents as external legs and inserted Green functions.

References

1 't Hooft G 1973 *Nucl. Phys.* B **61** 455
2 Weinberg S 1973 *Phys. Rev.* D **8** 3497
3 Collins J C and McFarlane A M 1973 *Phys. Rev.* D **10** 1201
4 Piaggio H T H 1950 *Elementary Treatise on Differential Equations and Their Applications* (London: Bell)
5 't Hooft G 1972 *Conf. Lagrangian Field Theory, Marseilles*
Politzer H D 1973 *Phys. Rev. Lett.* **26** 1346
Gross D and Wilczek F 1973 *Phys. Rev. Lett.* **26** 1343
6 Bailin D 1982 *Weak Interactions* (Bristol: Hilger)
7 Wilson K 1969 *Phys. Rev.* **179** 1499
8 Georgi H and Politzer H D 1974 *Phys. Rev.* D **9** 416
Gross D and Wilczek F 1974 *Phys. Rev.* D **9** 980
9 Bailin D, Love A and Nanopoulos D V 1974 *Lett. Nuovo Cimento* **9** 501
Gross D and Wilczek F 1974 *Phys. Rev.* D **9** 980
10 Applequist T and Georgi M 1973 *Phys. Rev.* D **8** 4000
Poggio E, Quinn H and Weinberg S 1976 *Phys. Rev.* D **13** 1958
11 Politzer H D 1974 *Phys. Rep.* **14C** 129 and references therein
12 Pennington M R 1983 *Rep. Prog. Phys.* **46** 293 and references therein
13 Sachrajda C 1982 Introduction to Perturbative Quantum Chromodynamics *Southampton Preprint* and references therein

13

SPONTANEOUS SYMMETRY BREAKING

13.1 Introduction

We have seen in Chapter 9 that the local gauge invariance of QED requires the vector field $A_\mu(x)$ – the gauge field—to be massless (since a mass term $m_A^2 A_\mu A^\mu$ is not invariant under the transformation (9.7)). A_μ determines the electromagnetic field whose quantum, the photon, is indeed massless. So in this respect, and in many others too numerous to detail here, experiment is consistent with the predictions derived from gauge invariance.

This masslessness is, of course, intimately related to the long (infinite) range of electromagnetic interactions. With the exception of gravitational interactions, which are not discussed in this book, these are the only long-range forces found in nature. In particular, the weak interactions are known to have a very short range. We shall see in Chapter 14 that the fermion currents observed in weak processes have precisely the form which follows from a non-Abelian gauge invariance based on the group $SU(2) \times U(1)$. It is therefore tempting to suppose that a gauge field theory may be responsible for both weak and electromagnetic interactions. However, the gauge invariance requires, as before, that the associated gauge fields are massless, as noted in Chapter 9, and this masslessness implies a long-range weak interaction which is not in accord with experiment. Some of the quanta of the weak fields have electric charge, as we shall see, and 'charged photons' are simply not seen. So the immediate obstacle to implementing gauge invariance in weak interactions is to reconcile it with the massive gauge particles needed to generate the short-range force actually observed. This is the objective of this chapter.

However, it is not immediately apparent that we *must* reconcile these two facets of the weak interactions. Why do we not simply add the gauge invariant Lagrangian to whatever (non-invariant) mass terms are needed to make the interaction sufficiently short-range? The answer is that if we do, the resulting field theory is not renormalisable. Renormalisability was discussed in §7.1, but the essence is that in an unrenormalisable theory the infinities which occur cannot be removed by the renormalisation of only the parameters and fields of the bare Lagrangian. Their removal requires the introduction of an infinite number of unpredicted but measurable quantities. Such theories therefore lack predictive power and we shall not discuss them further. The reason why the field theory described above is unrenormalisable is because its divergences are 'worse' than those which appear in the massless gauge invariant theory. The difference between the two stems from the form of the gauge field's propagator in the two cases.

We start by deriving the propagator of the massive vector field. If we simply add a mass term to the Lagrangian for a free vector field A_μ, given in (3.12), we obtain

$$\mathscr{L} = -\tfrac{1}{4}(\partial_\mu A_\nu - \partial_\nu A_\mu)(\partial^\mu A^\nu - \partial^\nu A^\mu) + \tfrac{1}{2}m_A^2 A_\mu A^\mu. \tag{13.1}$$

The first term is invariant under the gauge transformation (9.7), while the second term is not. The Euler–Lagrange equations (3.8) now give

$$-\partial_\mu(\partial^\mu A^\nu - \partial^\nu A^\mu) = m_A^2 A^\nu. \tag{13.2}$$

Thus

$$\partial_\nu A^\nu = 0 \tag{13.3}$$

since $m_A^2 \neq 0$, and substituting back we find

$$(\partial_\mu \partial^\mu + m_A^2)A^\nu = 0. \tag{13.4}$$

So A_μ now describes a particle of mass m_A, as anticipated, and the Lorentz condition (3.121) is a consequence of the field equations; it does not have to be imposed, as it was in the massless case. There is therefore no necessity for a gauge fixing term, as in (3.130). This is because the Lagrangian is no longer gauge invariant, so the field A_μ in this case *is* uniquely specified. The derivation of the propagator is now straightforward. We write

$$\int \mathrm{d}^4x \mathscr{L} = \int \mathrm{d}x\, \mathrm{d}x' \tfrac{1}{2} A^\rho(x') C_{\rho\sigma}(x', x) A^\sigma(x) \tag{13.5a}$$

where

$$C_{\rho\sigma}(x', x) = \int \frac{\mathrm{d}p}{(2\pi)^4}\, \mathrm{e}^{-\mathrm{i}p(x'-x)}\, [(m_A^2 - p^2)g_{\rho\sigma} + p_\rho p_\sigma]. \tag{13.5b}$$

The inverse is easily found (as in Chapter 4) to be

$$C_{\rho\sigma}^{-1}(x', x) = \int \frac{\mathrm{d}p}{(2\pi)^4}\, \tilde{\Delta}_{\mathrm{F}\rho\sigma}(p)\, \mathrm{e}^{-\mathrm{i}p(x'-x)} \tag{13.6a}$$

where $\mathrm{i}\tilde{\Delta}_{\mathrm{F}\rho\sigma}(p)$ is the massive vector boson propagator

$$\mathrm{i}\tilde{\Delta}_{\mathrm{F}\rho\sigma}(p) = \frac{-\mathrm{i}(g_{\rho\sigma} - p_\rho p_\sigma / m_A^2)}{p^2 - m_A^2 + \mathrm{i}\varepsilon}. \tag{13.6b}$$

This may be compared with the propagator derived in (10.68) for the massless (gauge-invariant) case:

$$\mathrm{i}\tilde{D}_{\mathrm{F}\rho\sigma}(p) = \frac{-\mathrm{i}[g_{\rho\sigma} + (\xi - 1)p_\rho p_\sigma / p^2]}{p^2 + \mathrm{i}\varepsilon}. \tag{13.7}$$

The divergences which arise in the integration over loop momenta are determined by the large (Euclidean) momentum behaviour of the propagators and vertices appearing in any Feynman diagram. In the massless case we see that

$$\tilde{D}_{\mathrm{F}}(p) \sim |\bar{p}|^{-2} \tag{13.8}$$

whereas in the massive case we have

$$\tilde{\Delta}_{\mathrm{F}}(p) \sim |\bar{p}|^{0}. \tag{13.9}$$

In general, therefore, we shall expect that some diagrams which are convergent with massless vector boson propagators will be divergent when massive vector boson propagators are substituted. This is why we say that the divergences are 'worse' in the massive case.

The difference is easily seen to arise from the difference between the numerators in the two cases; the $p_\rho p_\sigma/m_A^2$ term in the massive case removes the $|\bar{p}|^{-2}$ supplied by the denominator. The numerator is in fact the sum over polarisation vectors

$$\sum_{\lambda=1}^{3} \varepsilon_\rho^{(\lambda)}(p)\varepsilon_\sigma^{(\lambda)}(p) = -g_{\rho\sigma} + p_\rho p_\sigma/m_A^2 \tag{13.10}$$

where the sum is over the *three* orthonormal space-like vectors transverse to p:

$$\varepsilon^{(\lambda)} \cdot \varepsilon^{(\lambda')} = -\delta^{\lambda\lambda'} \tag{13.11a}$$

$$p \cdot \varepsilon^{(\lambda)} = 0. \tag{13.11b}$$

Recall that in §3.5 we showed that in the massless case the 'time-like' and 'longitudinal' modes cancel and that we may choose a gauge in which only the *two* 'transverse' modes appear. In the massive case a third polarisation state, with a (longitudinal) component parallel to $\boldsymbol{p}$, exists. In fact

$$\begin{aligned} \varepsilon_\rho^{(3)}(p) &= m_A^{-1}(|\boldsymbol{p}|, p_0\hat{\boldsymbol{p}}) \\ &= \frac{p_\rho}{m_A} - \frac{m_A}{p_0 + |\boldsymbol{p}|}(1, \hat{\boldsymbol{p}}) \\ &\sim p_\rho/m_A \qquad \text{as} \quad p_0, |\boldsymbol{p}| \to \infty. \end{aligned} \tag{13.12}$$

Thus the $p_\rho p_\sigma/m_A^2$ term in the numerator of $\tilde{\Delta}_{\mathrm{F}\rho\sigma}(p)$ is just the contribution from this longitudinal mode, at least in the large p limit.

Of course it may happen that this longitudinal mode is not coupled by any of the interactions in the theory, in which case the massive theory is no worse than the massless one. This happens, for example, in 'massive QED', in which we simply give the photon a mass. But in general, and for the weak interactions in particular, this prescription leads to an unrenormalisable theory.

13.2 Spontaneous symmetry breaking in a ferromagnet

In order to implement gauge invariance in weak interactions, we have to find some method of generating gauge vector boson masses without destroying the renormalisability of the gauge theory. Any such mass terms break the (gauge) symmetry, and the only known method of doing so in a renormalisable manner is called 'spontaneous' symmetry breaking, although it has been observed[1] that the symmetry is not so much 'broken' as 'secret', or 'hidden'.

The inspiration of the technique is to be found in the collective behaviour of certain many-body systems. Consider, for example, a ferromagnetic material in zero external magnetic field. Its properties are well understood in terms of the Heisenberg nearest-neighbour spin–spin interaction model with a Hamiltonian

$$H = -\tfrac{1}{2}J \sum_{(i,j)} \boldsymbol{\sigma}_i \cdot \boldsymbol{\sigma}_j \tag{13.13}$$

where the sum is over all nearest-neighbour sites (i, j) and $\boldsymbol{\sigma}_i$ is the spin on the site i. Clearly H is rotationally invariant, so the unitary operator $U(R)$ describing a rotation R commutes with H:

$$U(R)H = HU(R). \tag{13.14}$$

However, the energy eigenstates are not always rotationally invariant. In fact, it is well known that (below the Curie temperature T_C) the ground state of the system has a non-zero magnetisation $\boldsymbol{M}$, which is clearly not rotationally invariant. The invariance expressed by (13.14) merely implies that the ground state $|\boldsymbol{M}\rangle$ and the state $|\boldsymbol{M}'\rangle$, where

$$M'_i = R_{ij} M_j \tag{13.15}$$

are degenerate. That is to say that the state with magnetisation $\boldsymbol{M}$ has the same energy as that in which $\boldsymbol{M}$ has been rotated into some other direction $\boldsymbol{M}'$. Indeed, if the ferromagnet is heated up above T_C (at which point $\boldsymbol{M}$ vanishes), and is then cooled down to the original temperature, still in zero external field, then in general the ground state will have a magnetisation $\boldsymbol{M}' \neq \boldsymbol{M}$. Thus the symmetry resides in the degeneracy of the ground state; any particular ground state is not symmetric since the magnetisation points in a definite direction. This direction is selected 'spontaneously' by the system as it cools, and this is why the symmetry is said to be 'spontaneously broken'.

In Landau's mean-field theory the free energy functional F of the system has a form reminiscent of (4.77)

$$F = \int \mathrm{d}^3x [\mathscr{F}(\boldsymbol{M}) + \tfrac{1}{2}K_L(\boldsymbol{M})(\nabla \cdot \boldsymbol{M})^2 + \tfrac{1}{2}K_T(\boldsymbol{M})(\nabla \wedge \boldsymbol{M})^2 + \ldots] \tag{13.16a}$$

where

$$\mathscr{F}(\boldsymbol{M}) = N\left(\frac{T - T_C}{T_C}\boldsymbol{M}^2 + \beta(\boldsymbol{M}^2)^2 + \ldots\right) \tag{13.16b}$$

with N a (density of states) normalisation, T the temperature and β positive. In (13.16a) the dots indicate terms involving more than two derivatives, while in (13.16b) they stand for higher powers of $\boldsymbol{M}^2$. The ground state of the system has no dependence upon spatial position

$$\boldsymbol{M}(x) = \boldsymbol{M} \tag{13.17}$$

so we may drop all derivative terms, and then F is a function only of $\boldsymbol{M}^2$ (because of the rotational invariance). Clearly, therefore, *if* $\boldsymbol{M}$ is non-zero we cannot predict its direction, and the symmetry will be spontaneously broken. $|\boldsymbol{M}|$ is found by minimising

$$F = VN\left(\frac{T - T_C}{T_C}|\boldsymbol{M}|^2 + \beta|\boldsymbol{M}|^4\right) \tag{13.18}$$

where V is the volume of the system and we have dropped the higher powers ($\boldsymbol{M}$ is small). It is clear from figure 13.1, or directly, that when $T > T_C$ F has a minimum when $\boldsymbol{M} = 0$, while for $T < T_C$ the minimum is when $\boldsymbol{M}$ is non-zero. Thus in the first case the ground state is (rotationally) symmetric, but in the second case the symmetry is spontaneously broken.

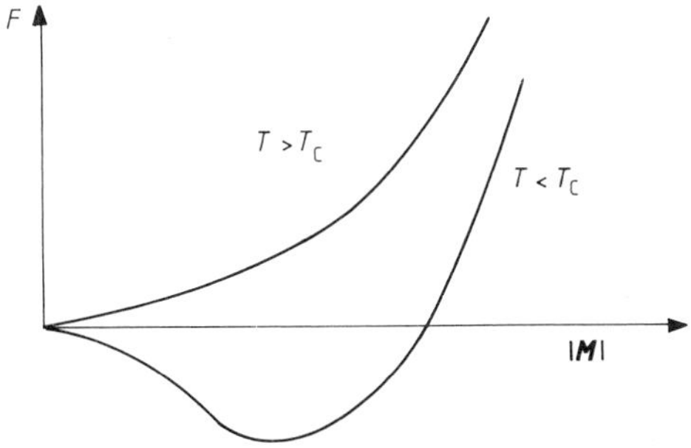

Figure 13.1 Free energy functional of a ferromagnet.

13.3 Spontaneous breaking of a discrete symmetry

The task then is to apply this technique to a (particle physics) field theory at zero temperature, and to apply it not to break rotational symmetry but some other (internal) symmetry. The analogue of the ground state of a many-body system is, of course, the vacuum in particle physics[2]. We must take the

Hamiltonian (Lagrangian) of the field theory to be invariant under the symmetry, but the vacuum to be characterised by some field which is (non-zero and) *not* invariant under the symmetry transformation. If the field in question were a spinor or vector field, for example, then the vacuum would be characterised by a non-zero angular momentum $J(=1/2$ or $1)$, and the rotational invariance would have been broken. The particle physics vacuum is observed to be rotationally invariant, so it is clear that the internal symmetry with which we are concerned must be broken by a *scalar* field having a non-zero value in the vacuum.

This scalar field is called the Higgs field, and, although it has never been measured in the way that $\boldsymbol{M}$ has, we are postulating its existence in order to break the internal symmetry spontaneously. Saying that it has a non-zero value in the vacuum means that there is a non-zero classical field *in vacuo.* Thus in the language of §4.4 we are saying that there is a scalar field operator $\hat{\varphi}(x)$ having a non-zero vacuum expectation value (VEV) in the absence of any source

$$\langle 0|\hat{\varphi}(x)|0\rangle = \varphi_c(x) \neq 0 \tag{13.19}$$

where $\varphi_c(x)$ is the field measured in the vacuum. Since the vacuum is observed to be translation invariant, when there is no source, we require $\varphi_c(x)$ to be independent of x:

$$\varphi_c(x) = \varphi_c. \tag{13.20}$$

It is clear from (4.5, 4.43) that the VEV of $\hat{\varphi}$ is zero in every order of perturbation theory, at least in the $\lambda\varphi^4$ theory considered there. So spontaneous symmetry breaking must be a non-perturbative effect. We saw in (4.79) that φ_c is determined by minimising the effective potential. Further, if we ignore quantum effects temporarily, the effective potential is simply given by *the* potential $V(\varphi)$. This is apparent from (6.35), which, as already noted, reduces to the (classical) field equation (3.35) when the source J is absent and the quantum effect $\lambda_i\Delta_F(0)$ dropped. The field theory discussed in Chapters 3 and 4 is described by the Lagrangian density

$$\mathscr{L} = \tfrac{1}{2}(\partial_\mu\varphi)(\partial^\mu\varphi) - V(\varphi) \tag{13.21a}$$

with

$$V(\varphi) = \frac{1}{2}\mu^2\varphi^2 + \frac{1}{4!}\lambda\varphi^4. \tag{13.21b}$$

The only symmetry of this simple model is the invariance under the discrete transformation

$$\varphi(x) \to \varphi'(x) \equiv -\varphi(x). \tag{13.22}$$

Obviously V will only have an absolute minimum if

$$\lambda \geqslant 0 \tag{13.23}$$

and in any case this is required to ensure the convergence of the functional integral. When μ^2 is positive V has a minimum only at $\varphi=0$ and μ is the mass of the field φ. V does have a minimum at a non-zero value of φ provided

$$\mu^2<0 \tag{13.24}$$

and then

$$\varphi_c=\pm(-6\mu^2/\lambda)^{1/2}. \tag{13.25}$$

This, of course, does not fix which sign of φ_c is actually selected by the system, because of the symmetry, but whichever one is chosen breaks the symmetry, since neither is invariant under (13.22). It is easy enough to define a new field which does have zero VEV. We let

$$\tilde{\varphi}\equiv\hat{\varphi}-\varphi_c \tag{13.26a}$$

so that, using (13.19, 13.20),

$$\langle 0|\tilde{\varphi}|0\rangle=0. \tag{13.26b}$$

When $\mathscr{L}$ is expressed as a function of $\tilde{\varphi}$ it will obviously not possess the reflection symmetry $\tilde{\varphi}\to-\tilde{\varphi}$, since $\tilde{\varphi}$ measures fluctuations about the asymmetric point $\varphi=\varphi_c$. Using (13.25) we find

$$\mathscr{L}=\frac{1}{2}[(\partial_\mu\tilde{\varphi})(\partial^\mu\tilde{\varphi})+2\mu^2\tilde{\varphi}^2]-\frac{\lambda}{4!}(\tilde{\varphi}^4+4\tilde{\varphi}^3\varphi_c)-\frac{1}{4}\mu^2\varphi_c^2. \tag{13.27}$$

The cubic term $\tilde{\varphi}^3$ shows that the symmetry is spontaneously 'broken', as expected, although since this is the same Lagrangian as the symmetric (13.21) we can see why some[1] prefer to describe the symmetry as 'secret'; it is secret because only with the particular coefficient of the $\tilde{\varphi}^3$ term given in (13.27) can the Lagrangian be recast in a symmetric form. Note that, since φ_c is proportional to $\lambda^{-1/2}$, the spontaneous symmetry breaking is indeed non-perturbative, as anticipated. Also, the mass squared of the field $\tilde{\varphi}$ is clearly $-2\mu^2$, which is just the second derivative anticipated in (3.38)

$$\left.\frac{d^2V}{d\varphi^2}\right|_{\varphi=\varphi_c}=-2\mu^2. \tag{13.28}$$

13.4 Spontaneous breaking of a continuous global symmetry

The real scalar field theory (13.21) discussed so far has only the discrete symmetry (13.22), whereas we are concerned with a continuous gauge symmetry. The spontaneous breaking of a continuous symmetry exhibits novel features which do not arise in the discrete case. For this reason we shall discuss the *complex* scalar field theory introduced in (3.61). The Lagrangian

$$\mathscr{L}=(\partial_\mu\varphi)(\partial^\mu\varphi^*)-V(\varphi,\varphi^*) \tag{13.29}$$

is invariant under a global U(1) gauge transformation

$$\varphi(x) \rightarrow \varphi'(x) \equiv e^{-iq\Lambda} \varphi(x) \tag{13.30a}$$

$$\varphi(x)^* \rightarrow \varphi'(x)^* = e^{iq\Lambda}\varphi(x)^* \tag{13.30b}$$

(with q, Λ real and constant) provided

$$V(\varphi, \varphi^*) = V(\varphi\varphi^*). \tag{13.31}$$

If we restrict our attention to renormalisable theories, then (13.31) implies that V has the form

$$V(\varphi, \varphi^*) = \mu^2\varphi\varphi^* + \tfrac{1}{4}\lambda(\varphi\varphi^*)^2 \tag{13.32}$$

analogous to (13.21b). As before we require λ to be positive, and then if μ^2 is positive V has an absolute minimum only at $\varphi = 0$. When μ^2 is negative V acquires a minimum at a non-zero value φ_c of φ which satisfies

$$|\varphi_c|^2 = -2\mu^2/\lambda. \tag{13.33}$$

However, in this case there is a circle of degenerate minima, since (13.33) obviously does not fix the phase of φ_c, because of the gauge invariance (13.30). Thus we have a situation analogous to that of the ferromagnetic system discussed in §13.2. Any particular choice of φ_c breaks the symmetry spontaneously, since under a gauge transformation (13.30) the ground state $|\varphi_c\rangle$ is transformed into a different ground state $|e^{-iq\Lambda}\varphi_c\rangle$. The novel feature which arises when we break a continuous symmetry emerges when we define a new field having zero VEV. Let the phase of φ_c be δ, so that

$$\varphi_c = \frac{1}{\sqrt{2}} v\, e^{i\delta} \tag{13.34a}$$

with

$$v = +(-4\mu^2/\lambda)^{1/2} \tag{13.34b}$$

and similarly for $\hat{\varphi}$.

Then we may express φ in terms of two real fields φ_1, φ_2 by

$$\varphi \equiv \frac{1}{\sqrt{2}} (\varphi_1 + i\varphi_2)\, e^{i\delta}. \tag{13.35}$$

Since

$$\langle 0|\hat{\varphi}|0\rangle = \varphi_c \tag{13.36}$$

it follows from (13.34) and (13.35) that only $\hat{\varphi}_1$ has a non-zero VEV:

$$\langle 0|\hat{\varphi}_i|0\rangle = v\delta_{i1} \qquad (i = 1, 2). \tag{13.37}$$

Thus we define new fields $\tilde{\varphi}_i$ having zero VEV

$$\tilde{\varphi}_i \equiv \hat{\varphi}_i - v\delta_{i1} \qquad (i = 1, 2). \tag{13.38}$$

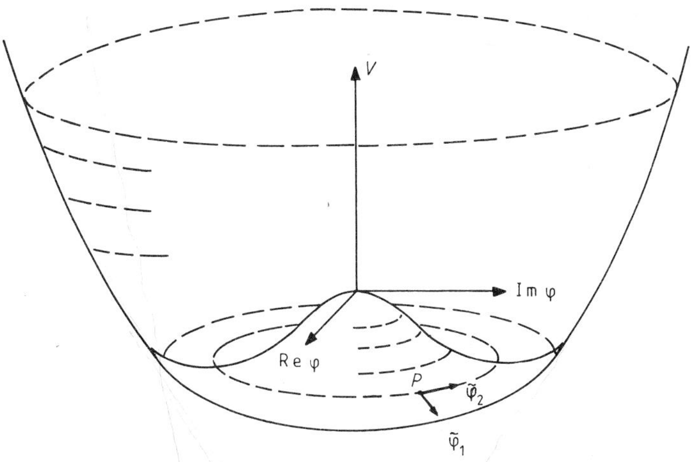

Figure 13.2 The gauge invariant potential (13.32).

Evidently $\tilde{\varphi}_1$ and $\tilde{\varphi}_2$ measure deviations from the asymmetric point P (see figure 13.2) in the directions radial and tangential to the circle of degenerate minima passing through P.

The quadratic terms of the Lagrangian (13.32) are diagonalised by these variables and we find

$$\mathscr{L}=\tfrac{1}{2}[(\partial_\mu\tilde{\varphi}_1)(\partial^\mu\tilde{\varphi}_1)+(\partial_\mu\tilde{\varphi}_2)(\partial^\mu\tilde{\varphi}_2)+2\mu^2(\tilde{\varphi}_1)^2]$$
$$-\tfrac{1}{16}\lambda[(\tilde{\varphi}_1)^2+(\tilde{\varphi}_2)^2]^2-\tfrac{1}{4}\lambda v\tilde{\varphi}_1[(\tilde{\varphi}_1)^2+(\tilde{\varphi}_2)^2]-\tfrac{1}{4}\mu^2v^2. \tag{13.39}$$

Obviously the symmetry is spontaneously broken, as expected, and the field $\tilde{\varphi}_1$ has a (positive) mass squared of $-2\mu^2$ as before. This is because V has a true minimum at P in the plane $\tilde{\varphi}_2=0$

$$\left.\frac{\partial^2 V}{\partial\varphi_1^2}\right|_{(\varphi_1,\varphi_2)=(v,0)}=-2\mu^2. \tag{13.40}$$

The novel feature is that the field $\tilde{\varphi}_2$ is *massless*. This too could have been anticipated as

$$\left.\frac{\partial^2 V}{\partial\varphi_2^2}\right|_{(\varphi_1,\varphi_2)=(v,0)}=0 \tag{13.41}$$

since $\varphi_2\,(=\tilde{\varphi}_2)$ measures deviations in the direction in which V is flat, because of the gauge symmetry. Such massless modes, which arise from the degeneracy of the ground state after spontaneous symmetry breaking, are called 'Goldstone bosons'.

In fact Goldstone bosons are a general consequence of the spontaneous breaking of a continuous global symmetry[3]. To see this consider a general non-Abelian gauge symmetry G, defined in (9.13), and some scalar fields transforming as some (possibly reducible) representation of G. Without loss of generality we may express these in terms of n (say) real scalar fields

$$\boldsymbol{\varphi}(x)=\begin{pmatrix}\varphi_1(x)\\ \varphi_2(x)\\ \vdots\\ \varphi_n(x)\end{pmatrix}. \tag{13.42}$$

Under an infinitesimal global gauge transformation

$$\boldsymbol{\varphi}(x)\rightarrow\boldsymbol{\varphi}(x)'=\boldsymbol{\varphi}(x)+\boldsymbol{\delta\varphi}(x) \tag{13.43a}$$

with

$$\boldsymbol{\delta\varphi}(x)=-\mathrm{i}g\mathbf{T}^a\Lambda^a\boldsymbol{\varphi}(x) \tag{13.43b}$$

where g, Λ^a are real, and $\mathbf{T}^a$ ($a=1,\ldots,N$) are the $n\times n$ matrices satisfying the Lie algebra (9.21); since $\mathrm{i}\mathbf{T}^a$ is real and $\mathbf{T}^a$ is Hermitian, $\mathbf{T}^a$ must be antisymmetric. If $\mathscr{L}$ is invariant under the gauge transformation (13.43), as shown in (3.33), there is a conserved Noether current

$$j_\mu^a=\boldsymbol{\pi}_\mu^{\mathrm{T}}(x)\mathrm{i}\mathbf{T}^a\boldsymbol{\varphi}(x) \qquad (a=1,\ldots,N) \tag{13.44a}$$

where

$$\pi_{\mu i}\equiv\frac{\partial\mathscr{L}}{\partial(\partial^\mu\varphi_i)} \qquad (i=1,\ldots,n). \tag{13.44b}$$

The Euler–Lagrange equations (3.8) give

$$\partial^\mu\boldsymbol{\pi}_\mu=\frac{\partial\mathscr{L}}{\partial\boldsymbol{\varphi}} \tag{13.45}$$

and current conservation then implies that

$$\left(\frac{\partial\mathscr{L}}{\partial\boldsymbol{\varphi}}\right)^{\mathrm{T}}\mathrm{i}\mathbf{T}^a\boldsymbol{\varphi}+\boldsymbol{\pi}_\mu\mathrm{i}\mathbf{T}^a\partial^\mu\boldsymbol{\varphi}=0. \tag{13.46}$$

In the field theories with which we are concerned the Lagrangian has the form

$$\mathscr{L}=\tfrac{1}{2}(\partial_\mu\boldsymbol{\varphi})^{\mathrm{T}}(\partial^\mu\boldsymbol{\varphi})-V(\boldsymbol{\varphi}) \tag{13.47}$$

so

$$\boldsymbol{\pi}_\mu=\partial_\mu\boldsymbol{\varphi} \tag{13.48a}$$

$$\frac{\partial\mathscr{L}}{\partial\boldsymbol{\varphi}}=-\frac{\partial V}{\partial\boldsymbol{\varphi}}. \tag{13.48b}$$

It follows from (13.48a) that the second term of (13.46) vanishes, since $\mathbf{T}^a$ is antisymmetric, and we deduce that V satisfies

$$\frac{\partial V^{\mathrm{T}}}{\partial \boldsymbol{\varphi}} \mathbf{T}^a \boldsymbol{\varphi} = 0 \tag{13.49}$$

for all $\boldsymbol{\varphi}$, as a consequence of the symmetry.

The masses of the various modes are controlled by the behaviour of V in the vicinity of its minimum. Since we are considering a spontaneously broken symmetry, V has its minimum at some value $\boldsymbol{v}$ of $\boldsymbol{\varphi}$ which fixes the VEV of the field operators. Thus

$$\langle 0|\hat{\boldsymbol{\varphi}}(x)|0\rangle = \boldsymbol{v} \tag{13.50a}$$

where

$$\left.\frac{\partial V}{\partial \boldsymbol{\varphi}}\right|_{\boldsymbol{\varphi}=\boldsymbol{v}} = \mathbf{0}. \tag{13.50b}$$

Further, the ground state described by $\boldsymbol{v}$ is not in general invariant under a gauge transformation, which means that

$$(1 - \mathrm{i}g\Lambda^a \mathbf{T}^a)\boldsymbol{v} \neq \boldsymbol{v} \tag{13.51}$$

for all choices of the infinitesimals Λ^a. So, for at least one a,

$$\mathrm{i}\mathbf{T}^a \boldsymbol{v} \neq \mathbf{0}. \tag{13.52}$$

We now define fields

$$\tilde{\boldsymbol{\varphi}} \equiv \hat{\boldsymbol{\varphi}} - \boldsymbol{v} \tag{13.53}$$

having zero VEV

$$\langle 0|\tilde{\boldsymbol{\varphi}}|0\rangle = \mathbf{0} \tag{13.54}$$

and express $\mathscr{L}$ in terms of $\tilde{\boldsymbol{\varphi}}$. Then using (13.47) and (13.50b)

$$\mathscr{L} = \frac{1}{2}\left((\partial_\mu \tilde{\varphi}_i)(\partial^\mu \tilde{\varphi}_i) - \tilde{\varphi}_i \tilde{\varphi}_j \left.\frac{\partial^2 V}{\partial \varphi_i \partial \varphi_j}\right|_{\boldsymbol{\varphi}=\boldsymbol{v}} \right) - V(\boldsymbol{v}) + \mathrm{O}(\tilde{\varphi}^3). \tag{13.55}$$

Clearly the masses of the fields $\tilde{\varphi}_i$ are the eigenvalues of the mass matrix

$$(\mu^2)_{ij} \equiv \left.\frac{\partial^2 V}{\partial \varphi_i \partial \varphi_j}\right|_{\boldsymbol{\varphi}=\boldsymbol{v}}. \tag{13.56}$$

Differentiating (13.49) with respect to $\boldsymbol{\varphi}$ and evaluating at $\boldsymbol{\varphi} = \boldsymbol{v}$, we find

$$(\mu^2)\mathrm{i}\mathbf{T}^a \boldsymbol{v} = \mathbf{0} \qquad (a = 1, \ldots, N) \tag{13.57}$$

using (13.50b). It follows from (13.52) that μ^2 has at least one eigenvector with zero eigenvalue, and consequently that the linear combination $\tilde{\boldsymbol{\varphi}}^{\mathrm{T}}\mathrm{i}\mathbf{T}^a\boldsymbol{v}$ is a Goldstone boson.

Now suppose that the ground state $|\boldsymbol{v}\rangle$ is left invariant under gauge transformations belonging to some (maximal) subgroup S of G. Then we may choose generators $\mathbf{T}^a$ ($a=1,\ldots,N$) of G such that $\mathbf{T}^a$ ($a=1,\ldots,M$) generate S. Since $|\mathfrak{v}\rangle$ is invariant under the transformations belonging to S,

$$\mathbf{T}^a\mathfrak{v}=\mathbf{0} \qquad (a=1,\ldots,M) \tag{13.58a}$$

but

$$\mathbf{T}^a\mathfrak{v}\neq\mathbf{0} \qquad (a=M+1,\ldots,N). \tag{13.58b}$$

The $N-M$ vectors $\mathbf{T}^a\mathfrak{v}$ ($a=M+1,\ldots,N$) are clearly linearly independent, and it follows that there are $N-M$ Goldstone bosons.

13.5 The Higgs mechanism

We are now in a position to attack the main objective of the chapter, namely the generation of masses for the gauge vector boson fields in a way which does not destroy the renormalisability of locally gauge invariant theories. In §13.1 we saw that the breaking of a local gauge invariance by the addition of gauge boson mass terms, which explicitly break the symmetry, leads in general to an unrenormalisable theory. We have also seen, in the following sections, how the global invariance of a field theory may be broken (or 'hidden') by the ground state (vacuum) spontaneously selecting one of the degenerate minima of the potential. This suggests that we study the effect of breaking a local gauge invariance spontaneously, in the hope that the breaking will induce gauge boson masses, while the (hidden) symmetry will protect the renormalisability.

In fact this is precisely what happens. We illustrate the mechanism (now called the Higgs mechanism) by applying it to the locally gauge invariant version of the model discussed in §13.4. This model has the Lagrangian of 'scalar electrodynamics', but when it is spontaneously broken it is called the 'Higgs model'[4]. Thus we start with

$$\mathscr{L}=(\mathrm{D}_\mu\varphi)(\mathrm{D}^\mu\varphi^*)-\mu^2\varphi\varphi^*-\tfrac{1}{4}\lambda(\varphi\varphi^*)^2-\tfrac{1}{4}F_{\mu\nu}F^{\mu\nu} \tag{13.59a}$$

where

$$\mathrm{D}_\mu\varphi\equiv(\partial_\mu+\mathrm{i}qA_\mu)\varphi \tag{13.59b}$$

$$\mathrm{D}_\mu\varphi^*\equiv(\partial_\mu-\mathrm{i}qA_\mu)\varphi^* \tag{13.59c}$$

are the U(1) gauge covariant derivatives defined in (9.6), and

$$F_{\mu\nu}=\partial_\mu A_\nu-\partial_\nu A_\mu \tag{13.59d}$$

is the gauge invariant field tensor, defined in (9.10). The last term is the Lagrangian density for the electromagnetic field (3.12) without an external source j_μ. When μ^2 is positive the U(1) invariance is unbroken and (13.59a) evidently describes a scalar particle of mass μ and charge q interacting with a massless electromagnetic field; hence the name scalar electrodynamics.

We are concerned with the case when $\mu^2 < 0$, so the symmetry is broken spontaneously and $\hat{\varphi}$ acquires a VEV

$$\langle 0|\hat{\varphi}(x)|0\rangle = \frac{1}{\sqrt{2}} v\,\mathrm{e}^{\mathrm{i}\delta} \tag{13.60}$$

where v is given in (13.34b) and δ is arbitrary. As before we change variables and use the fields $\tilde{\varphi}_i$ $(i=1,2)$, defined in (13.35) and (13.38), which do have zero VEVS. In terms of these variables the covariant derivative becomes:

$$\mathrm{D}_\mu\hat{\varphi} = \frac{\mathrm{e}^{\mathrm{i}\delta}}{\sqrt{2}}\,[\partial_\mu\tilde{\varphi}_1 + \mathrm{i}(\partial_\mu\tilde{\varphi}_2 + qvA_\mu) + \mathrm{i}qA_\mu(\tilde{\varphi}_1 + \mathrm{i}\tilde{\varphi}_2)]. \tag{13.61}$$

Notice that the erstwhile Goldstone boson $\tilde{\varphi}_2$ is inextricably attached to the hitherto massless gauge field A_μ. Indeed, aside from interaction terms, $\tilde{\varphi}_2$ and A_μ enter the Lagrangian only in the combination

$$A'_\mu \equiv A_\mu + \frac{1}{qv}\,\partial_\mu\tilde{\varphi}_2. \tag{13.62}$$

In other words, because of the spontaneous symmetry breaking the gauge field is mixed with the Goldstone mode $\tilde{\varphi}_2$, which in momentum space provides a longitudinal degree of freedom. From our discussion at the end of §13.1 this suggests that the field A'_μ has a non-zero mass. This is indeed the case. If we eliminate A_μ in favour of A'_μ in (13.59a), we find a mass term for the field A'_μ, as in (13.1), with

$$m(A') = qv \tag{13.63}$$

and we see that the mass requires both the spontaneous symmetry breaking $(v \neq 0)$ and the coupling of the gauge field to the scalar field $(q \neq 0)$, as anticipated. We shall not exhibit the precise form of $\mathscr{L}$ as a function of A'_μ, $\tilde{\varphi}_1$ and $\tilde{\varphi}_2$, because $\tilde{\varphi}_2$ can be eliminated from the Lagrangian. We can see this by exploiting the gauge invariance of $\mathscr{L}$. (Remember any gauge transformation leaves $\mathscr{L}$ *in*variant.) Comparing (13.62) with (9.7) we notice that A'_μ may be obtained from A_μ by a particular gauge transformation, namely one with

$$\Lambda(x) = \frac{1}{qv}\,\tilde{\varphi}_2(x). \tag{13.64}$$

This suggests that the *whole* of the dependence of $\mathscr{L}$ upon $\tilde{\varphi}_2$ might be absorbable into a (different) gauge transformation. From (13.35) and (13.38) we have that

$$\varphi = \frac{1}{\sqrt{2}}\,(v + \tilde{\varphi}_1 + \mathrm{i}\tilde{\varphi}_2)\,\mathrm{e}^{\mathrm{i}\delta}. \tag{13.65}$$

Under a gauge transformation

$$\varphi \to \varphi' = \mathrm{e}^{-\mathrm{i}q\Lambda}\,\varphi \tag{13.66}$$

$$\equiv \frac{1}{\sqrt{2}}(v + \tilde{\varphi}_1' + \mathrm{i}\tilde{\varphi}_2')\,\mathrm{e}^{\mathrm{i}\delta}. \tag{13.67}$$

So by choosing

$$q\Lambda = \arctan\frac{\tilde{\varphi}_2}{v + \tilde{\varphi}_1} \tag{13.68}$$

we can arrange that

$$\tilde{\varphi}_2' = 0. \tag{13.69}$$

In *this* gauge we denote $\tilde{\varphi}_1'$ by H, so that

$$\varphi(x) \to \varphi'(x) = \frac{1}{\sqrt{2}}[v + H(x)] \tag{13.70}$$

and using (13.61) this gives

$$\mathrm{D}_\mu\varphi \to (\mathrm{D}_\mu\varphi)' = \frac{\mathrm{e}^{\mathrm{i}\delta}}{\sqrt{2}}(\partial_\mu H + \mathrm{i}qvA_\mu' + \mathrm{i}qA_\mu' H) \tag{13.71}$$

where now A_μ' is the field A_μ gauge-transformed using (13.68). Since $\mathscr{L}$ is gauge invariant we may evaluate it in any gauge, and in this gauge we obtain from (13.59) using (13.70) and (13.71)

$$\mathscr{L} = \frac{1}{2}(\partial_\mu H)(\partial^\mu H) + \frac{1}{2}q^2 A_\mu' A'^\mu (v + H)^2$$

$$-\frac{1}{2}\mu^2(v+H)^2 - \frac{\lambda}{16}\lambda(v+H)^4 - \frac{1}{4}F_{\mu\nu}' F'^{\mu\nu} \tag{13.72a}$$

where

$$F_{\mu\nu}' = \partial_\mu A_\nu' - \partial_\nu A_\mu'. \tag{13.72b}$$

We may simplify (13.72) using the fact, expressed by (13.34b), that $v/\sqrt{2}$ minimises the potential, and finally we have

$$\mathscr{L} = \frac{1}{2}[(\partial_\mu H)(\partial^\mu H) + 2\mu^2 H^2]$$

$$-\frac{1}{4}\mu^2 v^2 - \frac{\lambda}{16}(H^4 + 4vH^3)$$

$$-\frac{1}{4}F_{\mu\nu}' F'^{\mu\nu} + \frac{1}{2}q^2 A_\mu' A'^\mu (v^2 + 2vH + H^2). \tag{13.73}$$

The Goldstone mode has been completely 'eaten' by the gauge-transformed boson A'_μ which has a mass qv, as in (13.63). There is one remaining scalar field, the Higgs field H, which is *real*, having a mass $(-2\mu^2)^{1/2}$. Thus the total number (four) of degrees of freedom is unaltered. Instead of a massless gauge boson, having two (transverse) modes, plus a complex field φ composed of two real fields, we now have a massive vector field A'_μ having three modes (two transverse and one longitudinal), plus one real scalar field H. Clearly the gauge invariance is completely broken, since A'_μ is massive and H is real. However, the renormalisability of the theory, if it has been preserved, is not manifest, because of the problems with massive vector bosons discussed in §13.1.

To verify the renormalisability we work in a different gauge from that specified in (13.68). The gauge given in (13.68) is called the 'unitary' gauge, since it demonstrates that the Goldstone mode may be eliminated, (13.69), while the surviving fields H, A'_μ are perfectly normal fields having the normal propagators for massive scalar and vector particles. In other words the only poles occurring in Green functions and Feynman diagrams are those deriving from real particles. In all other gauges, and in particular in the R_ξ gauges which we shall shortly define, there are spurious vector and scalar poles which must cancel from S-matrix elements since they are absent in the unitary gauge. (See Appendix B.) In other words, the R_ξ gauges are not manifestly unitary. However they are manifestly renormalisable; the ultraviolet divergences encountered are no 'worse' than those occurring in QED. The R_ξ gauge is specified by imposing a condition in the gauge field. For example in §3.5 we demonstrated that the addition of a gauge-fixing term to the Lagrangian

$$\mathscr{L}_{\mathrm{GF}} = -\frac{1}{2\xi}(\partial_\mu A^\mu)^2 \tag{13.74}$$

ensures that the gauge field A_μ may be made to satisfy the Lorentz condition

$$\partial_\mu A^\mu = 0. \tag{3.116}$$

In the present context it is useful to use a different gauge-fixing Lagrangian, first suggested by 't Hooft[5]:

$$\mathscr{L}_{\mathrm{GF}} = -\frac{1}{2\xi}(\partial_\mu A^\mu - \xi q v \tilde{\varphi}_2)^2. \tag{13.75}$$

This ensures that A_μ can be chosen so that

$$\partial_\mu A^\mu = \xi q v \tilde{\varphi}_2. \tag{13.76}$$

Provided qv is non-zero, we see that this reduces to (13.69) in the limit $\xi \to \infty$. Thus we shall expect the unitary gauge to be a limiting case of the R_ξ gauge. The advantage of the 't Hooft gauge fixing is that it removes the bilinear mixing of A_μ and φ_2. We recall that this derives from the $\mathrm{D}_\mu\varphi\mathrm{D}^\mu\varphi^*$ term of $\mathscr{L}$, and from (13.61) we see that this contains a quadratic term

$$\tfrac{1}{2}[(\partial_\mu\tilde{\varphi}_2)(\partial^\mu\tilde{\varphi}_2)+2qvA^\mu(\partial_\mu\tilde{\varphi}_2)+(qv)^2A_\mu A^\mu]$$
$$=\tfrac{1}{2}[(\partial_\mu\tilde{\varphi}_2)(\partial^\mu\tilde{\varphi}_2)-2qv(\partial_\mu A^\mu)\tilde{\varphi}_2+(qv)^2A_\mu A^\mu]+qv\partial_\mu(A^\mu\tilde{\varphi}_2). \quad (13.77)$$

The total divergence may be dropped, since it does not contribute to the action, and the cross term now cancels that in (13.75) precisely. As explained in §10.5, there is no necessity to introduce ghost fields into this theory since it is based on an Abelian group U(1). The full Lagrangian of the Higgs model in the R_ξ gauge is therefore $\mathscr{L}+\mathscr{L}_{\mathrm{GF}}$ which gives

$$\begin{aligned}\mathscr{L}_{\text{Higgs}}=&\frac{1}{2}[(\partial_\mu\tilde{\varphi}_1)(\partial^\mu\tilde{\varphi}_1)+2\mu^2\tilde{\varphi}_1^2]-\frac{1}{4}\mu^2v^2\\&+\frac{1}{2}[(\partial_\mu\tilde{\varphi}_2)(\partial^\mu\tilde{\varphi}_2)-\xi m_A^2\tilde{\varphi}_2^2]\\&+\frac{1}{2}[(1-\xi^{-1})(\partial_\mu A^\mu)^2-(\partial_\mu A_\nu)(\partial^\mu A^\nu)+m_A^2A_\mu A^\mu]\\&-\frac{\lambda}{16}[4v\tilde{\varphi}_1(\tilde{\varphi}_1^2+\tilde{\varphi}_2^2)+(\tilde{\varphi}_1^2+\tilde{\varphi}_2^2)^2]\\&+qA^\mu\tilde{\varphi}_1\overleftrightarrow{\partial}_\mu\tilde{\varphi}_2+\frac{1}{2}q^2A_\mu A^\mu\tilde{\varphi}_1^2+q^2vA_\mu A^\mu\tilde{\varphi}_1\end{aligned} \quad (13.78a)$$

where

$$m_A=qv. \quad (13.78b)$$

Thus the $\tilde{\varphi}_1$ field has a mass squared $\mu^2+3\lambda v^2=-2\mu^2$ and the (unphysical) erstwhile Goldstone boson mode $\tilde{\varphi}_2$ now has mass squared ξm_A^2. The propagator of the vector field may be found, as in §10.6, and we find

$$\mathrm{i}\tilde{\Delta}_{\mathrm{F}\rho\sigma}(p)=-\mathrm{i}\frac{g_{\rho\sigma}+(\xi-1)p_\rho p_\sigma(p^2-\xi m_A^2)^{-1}}{p^2-m_A^2+\mathrm{i}\varepsilon}. \quad (13.79)$$

As anticipated, this yields the ordinary propagator (13.6b) of a massive vector boson in the unitary limit $\xi\to\infty$, and the associated ultraviolet behaviour (13.9). However for all finite values of ξ the behaviour as the (Euclidean) momentum $\bar{p}\to\infty$ is

$$\tilde{\Delta}_{\mathrm{F}\rho\sigma}(p)\sim|\bar{p}|^{-2} \quad (13.80)$$

just as in the massless case (13.7). It is for this reason that the R_ξ gauge is 'manifestly renormalisable'. To demonstrate that the theory really is sensible it is necessary to show that the poles at $p^2=\xi m_A^2$ cancel from S-matrix elements. This was done by 't Hooft[5].

13.6 The Higgs mechanism in non-Abelian theories

We have seen that when the local U(1) invariance of scalar electrodynamics is spontaneously broken the gauge boson acquires a mass by 'eating' the Goldstone boson associated with the global symmetry, while the renormalisability of the theory is preserved. A similar phenomenon occurs when a local non-Abelian gauge invariance is spontaneously broken. Since electroweak theory is believed to be just such a theory, it is worthwhile to explore the general case in a little detail.

As in (13.42) we consider a general non-Abelian gauge symmetry G and n real scalar fields $\varphi(x)$. We are now concerned with a local gauge invariance so we must replace the derivative ∂_μ in (13.47) by the covariant derivative (matrix) (9.15)

$$\mathrm{D}_\mu \equiv \mathbf{I}\partial_\mu + \mathrm{i}g\mathbf{T}^a A_\mu^a \tag{13.81}$$

where $\mathbf{I}$ is the unit $n \times n$ matrix, g is the coupling constant, $\mathbf{T}^a$ $(a=1,\ldots,N)$ are the $n \times n$ matrices satisfying the Lie algebra (9.21) of the group G, and A_μ^a are the gauge vector fields. Adding the (locally gauge invariant) Yang–Mills Lagrangian (9.33) for these gauge fields leads us to the non-Abelian analogue of (13.59a):

$$\mathscr{L} = \tfrac{1}{2}(\mathrm{D}_\mu\varphi)^{\mathrm{T}}(\mathrm{D}^\mu\varphi) - V(\varphi) - \tfrac{1}{4}F_{\mu\nu}^a F^{a\mu\nu} \tag{13.82a}$$

where

$$F_{\mu\nu}^a = \partial_\mu A_\nu^a - \partial_\nu A_\mu^a - gf^{abc}A_\mu^b A_\nu^c \tag{13.82b}$$

and $V(\varphi)$ satisfies (13.49) as a consequence of the symmetry. When the symmetry is spontaneously broken some or all of the fields $\hat{\varphi}$ acquire VEVs, as in (13.50). As before, (13.53), we define new fields

$$\tilde{\varphi} = \hat{\varphi} - v \tag{13.83}$$

all of which have zero VEVs. Expressed in terms of these fields

$$\begin{aligned}(\mathrm{D}_\mu\varphi)^{\mathrm{T}}(\mathrm{D}^\mu\varphi) &= (\partial^\mu\tilde{\varphi})^{\mathrm{T}}(\partial^\mu\tilde{\varphi}) + 2g(\partial_\mu\tilde{\varphi}^{\mathrm{T}})\mathrm{i}\mathbf{T}^a v A^{a\mu} \\ &\quad + g^2 A_\mu^a A^{b\mu} v^{\mathrm{T}}\mathbf{T}^a\mathbf{T}^b v + \ldots\end{aligned} \tag{13.84}$$

where ... denotes cubic and quartic interaction terms. (In deriving (13.84) we have used the antisymmetry of the matrices $\mathbf{T}^a$ deduced in §13.4.) Suppose we choose the generators $\mathbf{T}^a$, as in §13.4, so that the first M generate the maximal subgroup S of G which is left invariant after the spontaneous symmetry breaking. Then (13.58) is satisfied and we see from (13.84) that the $N-M$ Goldstone modes $\tilde{\varphi}^{\mathrm{T}}\mathrm{i}\mathbf{T}^a v$ $(a=M+1,\ldots,N)$ are mixed with the corresponding gauge bosons. We therefore anticipate that the vector fields A_μ^a $(a=1,\ldots,M)$ remain massless, as they are not mixed with Goldstone bosons, while the remaining $N-M$ vector fields acquire masses. This is clear if we presume the existence of a unitary gauge, as before, in which the Goldstone modes are transformed to zero. In other words, we assume that by an appropriate non-Abelian gauge transformation we can arrange that

$$\varphi(x) \rightarrow \varphi'(x) \tag{13.85a}$$

where

$$\tilde{\varphi}'^{\mathrm{T}} \mathrm{i} \mathbf{T}^a v = 0 \qquad (a = M+1, \ldots, N). \tag{13.85b}$$

In this gauge it follows from (13.84) that the surviving $n-N+M$ Higgs scalar fields are unmixed with gauge bosons. The third term of (13.84) is the vector bosons' mass term. The actual masses are found by diagonalising the mass matrix

$$(M_A^2)^{ab} \equiv g^2 v^{\mathrm{T}} \mathbf{T}^a \mathbf{T}^b v \qquad (a, b = 1, \ldots, N). \tag{13.86}$$

Since the matrices $\mathbf{T}^a$ satisfy (13.58a), it is clear that

$$(M_A^2)^{ab} = 0 \qquad \text{for} \quad (a, b = 1, \ldots, M)$$

so that the gauge bosons A_μ^a $(a = 1, \ldots, M)$ are indeed massless as anticipated. Further, if we restirct a, b to values larger than M, the resulting $(N-M) \times (N-M)$ sub-matrix $(\tilde{M}_A^2)^{ab}$ is positive definite. To see this we note first that $(\tilde{M}_A^2)^{ab}$ is symmetric (exercise). It follows that it may be diagonalised using a real orthogonal transformation O. Thus

$$[O(\tilde{M}_A^2)O^{\mathrm{T}}]^{ab} = \lambda^a \delta^{ab} \qquad \text{(no summation)}. \tag{13.87}$$

Using (13.86) this gives

$$\lambda^a = \sum_{p,q=M+1}^{N} (gO^{ap} \mathrm{i} \mathbf{T}^p v)^{\mathrm{T}} (gO^{aq} \mathrm{i} \mathbf{T}^q v). \tag{13.88}$$

Since any linear combination $gO^{aq}\mathrm{i}\mathbf{T}^q v$ of the linearly independent (real) vectors $\mathrm{i}\mathbf{T}^q v$ must be non-zero, it follows that λ^a is positive and we have $N-M$ massive vector particles, as anticipated. It remains only to justify the assumed existence of a unitary gauge. The proof is elegant[6] but something of an aside, so we omit it.

The unitary gauge is unsuitable for calculations because the presence of massive gauge bosons means that the theory is not manifestly renormalisable. It is desirable to perform calculations in a guage in which the renormalisability is manifest, so that the divergences encountered are no worse than QED, for example. As before, this is the case in 't Hooft's R_ξ gauge, analogous to (13.76). In the general case the gauge-fixing Lagrangian is taken to be

$$\mathscr{L}_{\mathrm{GF}} = -\frac{1}{2\xi} (\partial_\mu A^{a\mu} - \xi g \tilde{\varphi}^{\mathrm{T}} \mathrm{i} \mathbf{T}^a v)^2 \tag{13.89}$$

so that the cross term $\partial_\mu A^{a\mu} g \tilde{\varphi}^{\mathrm{T}} \mathrm{i} \mathbf{T}^a v$ cancels the corresponding term in (13.84), after dropping a total divergence. Then in the quadratic terms the gauge fields

A^a_μ are decoupled from the scalars. The choice (13.89) corresponds to the gauge conditions

$$F_a \equiv \partial_\mu A^{a\mu} - \xi g \tilde{\boldsymbol{\varphi}}^{\mathrm{T}} \mathrm{i} \mathbf{T}^a \boldsymbol{v} - f^a(x) = 0 \tag{13.90}$$

instead of (10.31). This means that the functional derivative $\delta F_a(x')/\delta\Lambda_b(x)$ is different from that derived in (10.55), and consequently that the Fadeev–Popov Lagrangian will differ from (10.58).

Before discussing these modifications we derive the Feynman rules (in the gauge boson and scalar sectors) which follow from the addition of the gauge fixing Lagrangian $\mathscr{L}_{\mathrm{GF}}$ to $\mathscr{L}$ given in (13.82). Identifying the terms which are quadratic in the gauge fields we have

$$\begin{aligned}\mathscr{L}(\text{quadratic}, A^a_\mu) = &-\frac{1}{2}(\partial_\mu A^a_\nu)(\partial^\mu A^{a\nu} - \partial^\nu A^{a\mu}) \\ &-\frac{1}{2\xi}(\partial_\mu A^{a\mu})^2 + A^a_\mu (M^2_A)^{ab} A^{b\mu}\end{aligned} \tag{13.91}$$

where $(M^2_A)^{ab}$ is given in (13.86). Thus the massless gauge bosons A^a_μ $(a = 1, \ldots, M)$ each have the propagator $\tilde{D}_{\mathrm{F}\rho\sigma}(p)$ given in (10.68) and the corresponding Feynman rule (10.69). As shown in (13.88), the remaining modes A^a_μ $(a = M+1, \ldots, N)$ yield $N-M$ massive vector fields. We denote the mass eigenstates by

$$B^a_\mu \equiv O^{ab} A^b_\mu \tag{13.92}$$

where O is the orthogonal matrix diagonalising this sector $\tilde{M}^2_A$ of the mass matrix M^2_A, as in (13.87). The eigenvalues λ^a may be written

$$\lambda^a \equiv (M^a)^2 \tag{13.93}$$

since we have verified their positivity. Then the propagator for each mode B^a_μ is

$$\tilde{\Delta}_{\mathrm{F}\rho\sigma}(p, M^a) = \frac{-g_{\rho\sigma} + (1-\xi)p_\rho p_\sigma[p^2 - \xi(M^a)^2]^{-1}}{p^2 - (M^a)^2 + \mathrm{i}\varepsilon} \tag{13.94a}$$

as in (13.79), and the corresponding Feynman rule

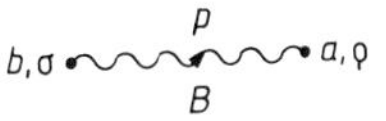

$: \quad \mathrm{i}\delta^{ab}\tilde{\Delta}_{\mathrm{F}\rho\sigma}(p, M^a)$ (no summation). (13.94b)

The terms in the augmented Lagrangian which are quadratic in the (shifted) scalar fields $\tilde{\boldsymbol{\varphi}}$ are

$$\mathscr{L}(\text{quadratic}, \tilde{\boldsymbol{\varphi}}) = \tfrac{1}{2}(\partial_\mu \tilde{\boldsymbol{\varphi}})^{\mathrm{T}}(\partial^\mu \tilde{\boldsymbol{\varphi}}) - \tfrac{1}{2}\tilde{\boldsymbol{\varphi}}^{\mathrm{T}}(\mu^2)\tilde{\boldsymbol{\varphi}} - \tfrac{1}{2}\xi g^2(\boldsymbol{\varphi}^{\mathrm{T}}\mathrm{i}\mathbf{T}^a\boldsymbol{v})^2 \tag{13.95}$$

where (μ^2) is the scalar mass matrix defined in (13.56). The scalar fields $\tilde{\boldsymbol{\varphi}}$ may be decomposed into the (unphysical) Goldstone modes $\boldsymbol{G}$ and the surviving Higgs scalars $\boldsymbol{H}$

$$\tilde{\boldsymbol{\varphi}} = \boldsymbol{G} + \boldsymbol{H} \tag{13.96}$$

as follows. Using (13.87) and (13.93) we see that the real vectors

$$\boldsymbol{e}^a \equiv \frac{g}{M^a} \sum_b O^{ab} \mathrm{i} \mathbf{T}^b \boldsymbol{v} \qquad \begin{array}{l}(a, b = M+1, \ldots, N) \\ (\text{no summation})\end{array} \tag{13.97}$$

are orthonormal:

$$\boldsymbol{e}^{a\mathrm{T}} \boldsymbol{e}^b = \delta^{ab}. \tag{13.98}$$

Since

$$(\mu^2)\boldsymbol{e}^a = \mathbf{0} \tag{13.99}$$

(from (13.57)), it is clear that the set $\{\boldsymbol{e}^a, a = M+1, \ldots, N\}$ provides a basis for the (zero-mass) Goldstone modes. The remaining scalar mass eigenstates all have positive eigenvalues of (μ^2), since $\boldsymbol{\varphi} = \boldsymbol{v}$ is a minimum of V. Further, we may choose the eigenvectors $\{\boldsymbol{f}^b: b = 1, \ldots, n - N + M\}$ to be orthonormal and orthogonal to the Goldstone basis $\{\boldsymbol{e}^a\}$. The decomposition (13.96) is then defined by

$$\boldsymbol{G} = \sum_a \boldsymbol{e}^a(\boldsymbol{e}^{a\mathrm{T}} \tilde{\boldsymbol{\varphi}}) \equiv \sum_a \boldsymbol{e}^a G^a \tag{13.100a}$$

$$\boldsymbol{H} = \sum_b \boldsymbol{f}^b(\boldsymbol{f}^{b\mathrm{T}} \tilde{\boldsymbol{\varphi}}) \equiv \sum_b \boldsymbol{f}^b H^b. \tag{13.100b}$$

Using the orthogonality of O, it follows from (13.97) that

$$\tilde{\boldsymbol{\varphi}}^{\mathrm{T}} \mathrm{i} g \mathbf{T}^a \boldsymbol{v} = \sum_b M^b O^{ba} G^b \tag{13.101}$$

and we see that the last term of (13.95) involves only the Goldstone modes. We may rewrite (13.95) as

$$\mathscr{L}(\text{quadratic}, \tilde{\boldsymbol{\varphi}}) = \tfrac{1}{2}(\partial_\mu H^b)(\partial^\mu H^b) - \tfrac{1}{2} \sum_b (\mu^b)^2 H^b H^b$$

$$+ \tfrac{1}{2}(\partial_\mu G^a)(\partial^\mu G^a) - \tfrac{1}{2} \sum_a \xi (M^a)^2 G^a G^a \tag{13.102a}$$

where

$$(\mu^2)\boldsymbol{f}^b = (\mu^b)^2 \boldsymbol{f}^b. \tag{13.102b}$$

The Goldstone mode G^a thus has a propagator

$$\tilde{\Delta}_{\mathrm{F}}(p, \sqrt{\xi} M^a) = [p^2 - \xi (M^a)^2 + \mathrm{i}\varepsilon]^{-1} \tag{13.103}$$

with the corresponding Feynman rule

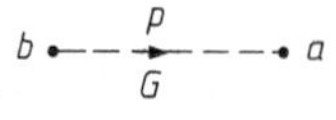

$$: \quad \mathrm{i}\delta^{ab} \tilde{\Delta}_{\mathrm{F}}(p, \sqrt{\xi} M^a) \qquad (\text{no summation}). \tag{13.104}$$

Similarly the Higgs scalar H^b has a propagator

$$\tilde{\Delta}_{\mathrm{F}}(p, \mu^b) = [p^2 - (\mu^b)^2 + \mathrm{i}\varepsilon]^{-1}$$

with the corresponding Feynman rule

$$: \quad \mathrm{i}\delta^{bc}\tilde{\Delta}_{\mathrm{F}}(p, \mu^b). \tag{13.105}$$

The remaining terms of the augmented Lagrangian characterise the interactions of the scalar particles and vector particles:

$$\begin{aligned}\mathscr{L}(\text{interaction}, \tilde{\varphi}, A) = {} & (\partial_\mu \tilde{\boldsymbol{\varphi}}^{\mathrm{T}})\mathrm{i}g\mathbf{T}^a\tilde{\boldsymbol{\varphi}}A^{a\mu} + g^2\tilde{\boldsymbol{\varphi}}^{\mathrm{T}}\mathbf{T}^a\mathbf{T}^b\boldsymbol{\upsilon}A^a_\mu A^{b\mu} \\ & + \tfrac{1}{2}g^2\tilde{\boldsymbol{\varphi}}^{\mathrm{T}}\mathbf{T}^a\mathbf{T}^b\tilde{\boldsymbol{\varphi}}A^a_\mu A^{b\mu} \\ & + \tfrac{1}{2}gf^{abc}(\partial_\mu A^a_\nu - \partial_\nu A^a_\mu)A^{b\mu}A^{c\nu} \\ & - \tfrac{1}{4}g^2 f^{abc}f^{ade}A^b_\mu A^c_\nu A^{d\mu}A^{e\nu} \\ & - [\tilde{V}(\tilde{\boldsymbol{\varphi}}) - \tfrac{1}{2}\tilde{\boldsymbol{\varphi}}^{\mathrm{T}}(\mu^2)\tilde{\boldsymbol{\varphi}}]\end{aligned} \tag{13.106}$$

where $\tilde{V}(\boldsymbol{\varphi})$ is obtained from $V(\boldsymbol{\varphi})$ by the substitution of $\boldsymbol{\upsilon} + \tilde{\boldsymbol{\varphi}}$ for $\boldsymbol{\varphi}$. This expression can be made even more complicated by casting it in terms of the mass eigenstates A^0_μ $(a = 1, \ldots, M)$, B^a_μ $(a = M+1, \ldots, N)$ defined in (13.91), G^a, H^b defined in (13.100). We shall forego this pleasure until we consider the specific spontaneous symmetry breaking required for the standard model (see §14.2).

There remains the question of the Faddeev–Popov ghost Lagrangian which is required by our gauge choice (13.90). This is controlled by the matrix **B**, defined in (10.52). With the present gauge condition

$$\begin{aligned}F_a(x') &\equiv F_a(A^{\mathrm{U}}_{b\mu}(x'), \tilde{\boldsymbol{\varphi}}^{\mathrm{U}}(x')) \\ &= \partial_{\mu x'}A^{a\mu\mathrm{U}}(x') - \xi g\tilde{\boldsymbol{\varphi}}^{\mathrm{TU}}(x')\mathrm{i}\mathbf{T}^a\boldsymbol{\upsilon} - f^a(x')\end{aligned} \tag{13.107}$$

where $A^{a\mathrm{U}}_\mu$, $\tilde{\boldsymbol{\varphi}}^{\mathrm{U}}$ are the (infinitesimally) gauge-transformed fields

$$A^{a\mu\mathrm{U}}(x') = A^{a\mu}(x') + \partial^\mu_{x'}\Lambda^a(x') + gf^{abc}\Lambda^b(x')A^{c\mu}(x') \tag{13.108a}$$

$$\begin{aligned}\tilde{\boldsymbol{\varphi}}^{\mathrm{U}}(x') &= \boldsymbol{\varphi}^{\mathrm{U}}(x') - \boldsymbol{\upsilon} \\ &= \boldsymbol{\varphi}(x') - \mathrm{i}g\mathbf{T}^b\Lambda^b(x')\boldsymbol{\varphi}(x') - \boldsymbol{\upsilon} \\ &= \tilde{\boldsymbol{\varphi}}(x') - \mathrm{i}g\mathbf{T}^b\Lambda^b(x')[\boldsymbol{\upsilon} + \tilde{\boldsymbol{\varphi}}(x')].\end{aligned} \tag{13.108b}$$

Thus

$$\begin{aligned}\frac{\delta F_a(x')}{\delta\Lambda^b(x)} = {} & \partial_{\mu x'}\{[\partial^\mu_{x'}\delta^{ab} + gf^{abc}A^{c\mu}(x')]\delta(x'-x)\} \\ & + \xi g^2(\boldsymbol{\upsilon}^{\mathrm{T}} + \tilde{\boldsymbol{\varphi}}^{\mathrm{T}}(x'))\mathbf{T}^b\mathbf{T}^a\boldsymbol{\upsilon}\delta(x'-x).\end{aligned} \tag{13.109}$$

Proceeding as before we find that the Faddeev–Popov ghost Lagrangian is now given by

$$\mathscr{L}_{\mathrm{FP}} = (\partial_\mu \eta^{a*})(\partial^\mu \eta^a) + g f^{abc} (\partial_\mu \eta^{a*}) \eta^b A^{c\mu}$$
$$- \xi \eta^{a*} (M_A^2)^{ab} \eta^b - \xi g^2 \eta^{a*} \eta^b \tilde{\boldsymbol{\varphi}}^{\mathrm{T}} \mathbf{T}^b \mathbf{T}^a \boldsymbol{v}. \qquad (13.110)$$

Evidently the ghost fields η^a $(a = 1, \ldots, M)$ remain massless and have the propagator (10.70) with the associated Feynman rule (10.71). The remaining ghost fields have a mass matrix $\xi(\tilde{M}_A^2)$ proportional to that of the corresponding gauge fields. Thus the mass eigenstates χ^a are given by the same orthogonal matrix O as appears in (13.87)

$$\chi^a \equiv O^{ab} \eta^b \qquad (a, b = M+1, \ldots, N) \qquad (13.111)$$

and the corresponding eigenvalues are $\sqrt{\xi}\, M^a$, where M^a is defined in (13.92). The propagator for this mode is therefore

$$\tilde{\Delta}_{\mathrm{F}}^{\mathrm{ghost}}(p, \sqrt{\xi}\, M^a) = [p^2 - \xi (M^a)^2 + \mathrm{i}\varepsilon]^{-1} \qquad (13.112)$$

and the corresponding Feynman rule is

$$: \quad \mathrm{i}\delta^{ab} \tilde{\Delta}_{\mathrm{F}}^{\mathrm{ghost}}(p, \sqrt{\xi}\, M^a). \qquad (13.113)$$

The remaining (interaction) terms of (13.110) may be recast in terms of η $(a = 1, \ldots, M)$, χ^a $(a = M+1, \ldots, N)$, A_μ^a $(a = 1, \ldots, M)$, B_μ^a $(a = M+1, \ldots, N)$, G^a, H^b. Again, we shall not write down the precise Feynman rules until we turn to a specific example.

13.7 Fermion masses from spontaneous symmetry breaking

The gauge theories we have encountered so far, namely QED and QCD, have all been invariant under the operation of parity, in which

$$\boldsymbol{x} \overset{P}{\rightarrow} -\boldsymbol{x}. \qquad (13.114)$$

In consequence the left and right chiral components of the fermion fields must transform in the same way under gauge transformations. To see this we group all the fermion fields into a column vector ψ and then decompose ψ into its left and right chiral components L and R as follows:

$$\psi = L + R \qquad (13.115\mathrm{a})$$

where

$$L \equiv a_{\mathrm{L}} \psi \qquad (13.115\mathrm{b})$$

$$R \equiv a_{\mathrm{R}} \psi \qquad (13.115\mathrm{c})$$

with

$$a_{\substack{L\\R}} = \tfrac{1}{2}(1 \mp \gamma_5). \tag{13.115d}$$

Then L and R are eigenstates of a chirality transformation $e^{i\alpha\gamma_5}$ since

$$R \to e^{i\alpha\gamma_5} R = e^{i\alpha} R \tag{13.116a}$$

$$L \to e^{i\alpha\gamma_5} L = e^{-i\alpha} L. \tag{13.116b}$$

Now suppose that under a gauge transformation U

$$L \to \exp(-ig\mathbf{T}_L^a \Lambda^a)L \tag{13.117a}$$

$$R \to \exp(-ig\mathbf{T}_R^a \Lambda^a)R \tag{13.117b}$$

with $\mathbf{T}_L^a$, $\mathbf{T}_R^a$ characterising the (possibly different) transformation properties of L and R, and consider the kinetic part of the Lagrangian

$$\mathscr{L}_K \equiv \bar{L}\gamma^\mu(i\partial_\mu - g\mathbf{T}_L^a A_\mu^a)L + \bar{R}\gamma^\mu(i\partial_\mu - g\mathbf{T}_R^a A_\mu^a)R. \tag{13.118}$$

By construction $\mathscr{L}_K$ is gauge invariant, and it follows from (13.116) that it is also invariant under the chiral transformation. However $\mathscr{L}_K$ is in general not invariant under the parity transformation, basically because parity interchanges the left and right chiral components. We can see this as follows. Under the parity transformation

$$\psi \xrightarrow{P} \varepsilon\gamma_0\psi \tag{13.119a}$$

with

$$|\varepsilon| = 1 \tag{13.119b}$$

and

$$A_\mu^a \to A^{a\mu}. \tag{13.120}$$

Then

$$L \xrightarrow{P} \varepsilon\gamma_0 R \tag{13.121a}$$

$$R \xrightarrow{P} \varepsilon\gamma_0 L \tag{13.121b}$$

and $\mathscr{L}_K$ is parity invariant if and only if

$$\mathbf{T}_L^a = \mathbf{T}_R^a \equiv \mathbf{T}^a. \tag{13.122}$$

In other words L and R transform identically under a gauge transformation.

Now consider the mass term

$$\mathscr{L}_M = -\bar{\psi} M_F \psi. \tag{13.123}$$

It is never invariant under the chiral transformation (13.116), but it is gauge invariant if (13.122) is satisfied and if

$$[M_{\mathrm{F}}, \mathbf{T}^a]=0. \tag{13.124}$$

In the case of QCD, for example, parity invariance ensures that (13.122) is satisfied and (13.123) requires all colours of a given flavour to have the same mass, as in (9.35).

It is well known that the weak interactions are not invariant under parity. Thus when we construct the electroweak gauge theory in Chapter 14 it should come as no surprise to find that the left and right chiral components of the fermion fields behave differently under gauge transformations, as envisaged in (13.117). Then $\mathscr{L}_M$ given in (13.123) is *not* gauge invariant. If we demand gauge invariance, the fermions are massless and the fermionic Lagrangian is chirally invariant. Clearly we must break these invariances, since with the exception of the neutrinos all fermions are known to have masses.

We have seen that it is possible to break the gauge invariance (spontaneously) by introducing scalar fields which acquire non-zero VEVS. However this does not of itself induce the chiral symmetry breaking which is essential if the fermions are to acquire masses. What is needed is an interaction of the fermion which is chirally non-invariant but gauge invariant. Then, when the gauge invariance is spontaneously broken, we shall expect fermion masses to emerge, since both invariances will have been broken. This may be achieved by including a Yukawa coupling of the fermions to the n real scalar fields $\varphi_1, \ldots, \varphi_n$ as follows. We take

$$\mathscr{L}_{\mathrm{Y}}=\bar{L}Y_pR\varphi_p+\bar{R}Y_p^{\dagger}L\varphi_p \tag{13.125}$$

where Y_p $(p=1, \ldots, n)$ are matrices chosen so that $\mathscr{L}_{\mathrm{Y}}$ is gauge invariant. The chiral non-invariance follows immediately from (13.116), and the gauge invariance requires that

$$\mathbf{T}_{\mathrm{L}}^aY_p-Y_p\mathbf{T}_{\mathrm{R}}^a=(\mathbf{T}_{\varphi}^a)_{pq}Y_q \tag{13.126}$$

where $\mathbf{T}_{\mathrm{L}}^a$, $\mathbf{T}_{\mathrm{R}}^a$ are defined in (13.117) and under the same gauge transformation

$$\boldsymbol{\varphi} \rightarrow \boldsymbol{\varphi}'=\exp(-\mathrm{i}g\mathbf{T}_{\varphi}^a\Lambda^a)\boldsymbol{\varphi}. \tag{13.127}$$

When the symmetry is spontaneously broken we must re-express $\mathscr{L}_{\mathrm{Y}}$ in terms of the field $\tilde{\boldsymbol{\varphi}}$, as in (13.83). Then

$$\mathscr{L}_{\mathrm{Y}}=\bar{L}Y_pR(v_p+\tilde{\varphi}_p)+\bar{R}Y_p^{\dagger}L(v_p+\tilde{\varphi}_p) \tag{13.128}$$

and we see that the symmetry breaking has generated mass terms for the fermions, as required and anticipated. The mass matrix M_{F} in (13.123) is given by

$$\boldsymbol{Y}^{\dagger}\cdot\boldsymbol{v}=\boldsymbol{Y}\cdot\boldsymbol{v}=-M_{\mathrm{F}}. \tag{13.129}$$

13.8 Magnetic monopoles

We have now derived the techniques necessary to apply spontaneous symmetry breaking to the case of actual interest. However, before doing so it is irresistible to note some further consequences which arise at the classical level when a non-Abelian gauge invariance is spontaneously broken.

As we have already said, spontaneous symmetry breaking occurs when some scalar fields $\varphi(x)$ acquire non-zero VEVS:

$$\langle 0|\tilde{\varphi}(x)|0\rangle = \boldsymbol{v} \tag{13.130}$$

and we argued that $\boldsymbol{v}$ is independent of x because the vacuum is observed to be spatially homogeneous. We might therefore wonder whether other states of finite energy might exist in which the scalar fields have spatially varying expectation values. Let us call such a state M. Then we want

$$\langle M|\hat{\varphi}(x)|M\rangle = \boldsymbol{\varphi}_M(x) \tag{13.131}$$

and $\boldsymbol{\varphi}_M(x)$ to describe a state of finite energy. We consider first a continuous global symmetry, such as that discussed in §13.4. The energy density T^0_0 for a single scalar field $\varphi(x)$ was derived in (3.45). In the present case we have

$$T^0_0 = \tfrac{1}{2}(\partial_0\boldsymbol{\varphi}^{\mathrm{T}})(\partial_0\boldsymbol{\varphi}) + \tfrac{1}{2}(\partial_i\boldsymbol{\varphi}^{\mathrm{T}})(\partial_i\boldsymbol{\varphi}) + V(\boldsymbol{\varphi}) \tag{13.132}$$

where to be definite we take

$$V(\boldsymbol{\varphi}) = \tfrac{1}{4}\alpha(\boldsymbol{\varphi}^{\mathrm{T}}\boldsymbol{\varphi} - v^2)^2. \tag{13.133}$$

Since we require the state $|M\rangle$ to have finite energy, it is clear that the density T^0_0 must approach zero as $r \equiv |\boldsymbol{x}| \to \infty$. In particular this requires that $V \to 0$, so

$$\boldsymbol{\varphi}^{\mathrm{T}}_M\boldsymbol{\varphi}_M \to v^2 \qquad \text{as} \quad r \to \infty. \tag{13.134}$$

This does not necessarily require that $\boldsymbol{\varphi}_M$ approaches a constant, since (13.134) requires only that the magnitude of $\boldsymbol{\varphi}_M$ approaches a constant; its orientation is arbitrary. We might therefore hope that it would be possible to arrange that $\boldsymbol{\varphi}_M$ is tied to a non-trivial topological structure as $r \to \infty$. For example in an SO(3) symmetric theory in which the fields $\boldsymbol{\varphi}$ constitute a three-dimensional representation we might hope to find a field configuration $\boldsymbol{\varphi}_M$ in which

$$\boldsymbol{\varphi}_M \to v\hat{\boldsymbol{r}} \qquad \text{as} \quad r \to \infty. \tag{13.135}$$

Unfortunately any such field configurations have infinite energy[7]. We can see this as follows. For large values of r we may approximate $\boldsymbol{\varphi}_M$ by the form

$$\boldsymbol{\varphi}_M(r, \theta, \varphi, t) \sim v\boldsymbol{\psi}(\theta, \varphi, t). \tag{13.136}$$

Then

$$(\partial_i\boldsymbol{\varphi}^{\mathrm{T}}_M)(\partial_i\boldsymbol{\varphi}_M) \sim \frac{1}{r^2}\left(\frac{\partial\boldsymbol{\psi}^{\mathrm{T}}}{\partial\theta}\frac{\partial\boldsymbol{\psi}}{\partial\theta} + \frac{1}{\sin^2\theta}\frac{\partial\boldsymbol{\psi}^{\mathrm{T}}}{\partial\varphi}\frac{\partial\boldsymbol{\psi}}{\partial\varphi}\right)$$
$$= \mathrm{O}(r^{-2}). \tag{13.137}$$

It follows from (13.132) that the energy is given by

$$E=\int d^3x T^0_0 > \int d^3x \tfrac{1}{2}(\partial_i \boldsymbol{\varphi}_M^{\mathrm{T}})(\partial_i \boldsymbol{\varphi}) = \infty \tag{13.138}$$

since $d^3x = r^2\, dr\, d\Omega$ and the radial integration is divergent.

In any case it is easy to see that any such *static* solution having finite energy is unstable. The energy of a static solution is given from (13.132) by

$$E = T[\boldsymbol{\varphi}_M] + V[\boldsymbol{\varphi}_M] \tag{13.139a}$$

where

$$T[\boldsymbol{\varphi}_M] \equiv \int d^Dx \tfrac{1}{2}(\partial_i \boldsymbol{\varphi}_M^{\mathrm{T}})(\partial_i \boldsymbol{\varphi}_M) \tag{13.139b}$$

and

$$V[\boldsymbol{\varphi}_M] = \int d^Dx\, V(\boldsymbol{\varphi}_M) \tag{13.139c}$$

with $D(=3)$ the number of spatial dimensions. We denote by $\boldsymbol{\varphi}_{M\lambda}(\boldsymbol{x})$ the scale-transformed solution

$$\boldsymbol{\varphi}_{M\lambda}(\boldsymbol{x}) = \boldsymbol{\varphi}_M(\lambda \boldsymbol{x}). \tag{13.140}$$

Then

$$\partial_i \boldsymbol{\varphi}_{M\lambda}(\boldsymbol{x}) = \lambda \partial_i \boldsymbol{\varphi}_M(\lambda \boldsymbol{x}) \tag{13.141}$$

and it follows, changing integration variables from $\boldsymbol{x}$ to $\lambda \boldsymbol{x}$, that

$$T[\boldsymbol{\varphi}_{M\lambda}] = \lambda^{2-D} T[\boldsymbol{\varphi}_M] \tag{13.142}$$

and

$$V[\boldsymbol{\varphi}_{M\lambda}] = \lambda^{-D} V[\boldsymbol{\varphi}_M]. \tag{13.143}$$

Since E must be stationary with respect to variation of λ we have

$$\left.\frac{dE}{d\lambda}\right|_{\lambda=1} = 0 = (2-D)T - DV. \tag{13.144}$$

For the solution to be stable we also require that E is a local minimum with respect to variation of λ. So

$$\begin{aligned} 0 < \left.\frac{d^2E}{d\lambda^2}\right|_{\lambda=1} &= (2-D)(1-D)T + D(D+1)V \\ &= (2-D)2T \end{aligned} \tag{13.145}$$

using (13.144). Since T is always positive, we see that in $D=3$ dimensions (13.145) is not satisfied, and any such static solution is consequently unstable. This is 'Derrick's theorem'[8].

However, if we consider a local, rather than a global, symmetry, we must amend the above argument to include the effects of the gauge fields. We leave this as an exercise, and merely quote the results. Requiring that E is a minimum with respect to variation of the scale parameter λ now gives

$$(2-D)T_\varphi+(4-D)T_A-DV=0 \tag{13.146a}$$

$$(2-D)(1-D)T_\varphi+(4-D)(3-D)T_A+D(D+1)V>0 \tag{13.146b}$$

where now

$$E=T_\varphi+T_A+V \tag{13.146c}$$

with

$$T_\varphi\equiv\int \mathrm{d}^D x\tfrac{1}{2}(D_i\boldsymbol{\varphi}_M^{\mathrm{T}})(D_i\boldsymbol{\varphi}_M) \tag{13.146d}$$

$$T_A\equiv\int \mathrm{d}^D x\tfrac{1}{4}F^a_{ijM}F^a_{ijM} \tag{13.146e}$$

and V is given in (13.139c). Eliminating V and putting $D=3$ gives

$$T_A>\tfrac{1}{2}T^\varphi>0 \tag{13.147}$$

which *can* be satisfied.

All that is happening is that the decrease in T_φ and V as λ is increased is being (more than) compensated for by the increase in T_A. The inclusion of the gauge fields also enables us to avoid the infinite energy of the purely scalar configuration. This is because we can arrange a cancellation so that although $|\partial_i\boldsymbol{\varphi}_M|=\mathrm{O}(r^{-1})$ and $|A_{iM}|=\mathrm{O}(r^{-1})$ as $r\to\infty$, the covariant derivative $|\mathrm{D}_i\boldsymbol{\varphi}_M|=\mathrm{O}(r^{-2})$, which is sufficient to ensure the convergence of T_φ.

To see how, we consider the particular case[9] already mentioned when the symmetry is SO(3) and the scalar fields $\boldsymbol{\varphi}$ transform as a three-dimensional representation. We further assume that a spherically symmetric configuration exists and that $\boldsymbol{\varphi}_M$ is radial everywhere, not just as $r\to\infty$ as in (13.138). Thus we assume $\boldsymbol{\varphi}_M$ has the form

$$\boldsymbol{\varphi}_M(\boldsymbol{r})=v\hat{\boldsymbol{r}}\Phi(r). \tag{13.148}$$

Since v is the only dimensionful scale in the system it must scale the r dependence. Further, since the r dependence is controlled by the covariant derivative D_i, it is clear that $\Phi(r)$ is a function of gr. Thus without loss of generality we may write instead of (13.148)

$$\boldsymbol{\varphi}_M(\boldsymbol{r})=v\hat{\boldsymbol{r}}\frac{H(vgr)}{vgr}=\frac{1}{gr^2}H(vgr)\boldsymbol{r} \tag{13.149}$$

and we insist that

$$\xi^{-1}H(\xi)\to 1 \qquad \text{as} \quad \xi\to\infty \tag{13.150}$$

so that (13.135) is satisfied. As before, $\partial_i \boldsymbol{\varphi}_M = O(r^{-1})$ for large r, so this potentially divergent behaviour must be cancelled by the gauge field contribution to

$$D_i \boldsymbol{\varphi}_M = \partial_i \boldsymbol{\varphi}_M - g \boldsymbol{A}_{iM} \wedge \boldsymbol{\varphi}_M \tag{13.151}$$

where we have used (13.81) and the fact that

$$(\mathbf{T}^a)_{ij} = i\varepsilon_{iaj} \tag{13.152}$$

in the regular representation of SO(3). To achieve the cancellation $\boldsymbol{A}_{iM}$ must have some component perpendicular to $\boldsymbol{\varphi}_M$ and we assume that

$$A^a_{iM}(\boldsymbol{r}) = -\varepsilon_{ail} \frac{r_l}{gr^2} [1 - K(vgr)]. \tag{13.153}$$

Then it follows that

$$D_i \varphi_{jM}(\boldsymbol{r}) = H(vgr)K(vgr) \frac{1}{gr^4} (\delta_{ij} r^2 - r_i r_j) + [vgrH'(vgr) - H(vgr)] \frac{r_i r_j}{gr^4} \tag{13.154}$$

and provided

$$\xi^2 \frac{d}{d\xi} [\xi^{-1} H(\xi)] \to 0 \qquad \text{as} \quad \xi \to \infty \tag{13.155a}$$

and

$$\xi K(\xi) \to 0 \qquad \text{as} \quad \xi \to \infty \tag{13.155b}$$

we see that $D_i \varphi_{jM} = O(r^{-2})$ for large r, as required. The gauge field tensor F^a_{ijM}, may be calculated, directly using (13.153), or from the commutator $[D_i, D_j]$ using (9.28). We find

$$F^a_{ijM} = g^{-1} \left[\frac{K^2 - 1}{r^2} \varepsilon_{iaj} + \left(\frac{K'}{r^2} - \frac{K^2 - 1}{r^4} \right) (\varepsilon_{iap} r_p r_j - \varepsilon_{jap} r_p r_i) \right]. \tag{13.156}$$

Then substituting (13.149) and (13.156) into (13.146) gives the energy of the configuration as a functional of H and K:

$$E[H, K] = \frac{4\pi v}{g} \int_0^\infty d\xi \, \xi^{-2} \left(\frac{1}{2} (\xi H' - H)^2 + H^2 K^2 \right.$$
$$\left. + (\xi K')^2 + \frac{1}{2} (K^2 - 1)^2 + \frac{\alpha}{4g} (H^2 - \xi^2)^2 \right). \tag{13.157}$$

Minimising with respect to variation of H and K, or directly from the Euler–Lagrange equations, gives

$$\xi^2 K'' = KH^2 + K(K^2 - 1) \tag{13.158a}$$

$$\xi^2 H'' = 2K^2 H + \frac{\alpha}{g^2} H(H^2 - \xi^2) \tag{13.158b}$$

which must be solved subject to the boundary conditions (13.155). Further, in order that the integral in (13.157) is convergent at $\xi=0$ we require the boundary conditions

$$H(\xi)\leqslant O(\xi) \qquad \text{as} \quad \xi\to 0 \tag{13.159a}$$

and

$$K(\xi)-1\leqslant O(\xi) \qquad \text{as} \quad \xi\to 0. \tag{13.159b}$$

In general the functions H and K satisfying (13.158) must be found numerically, and the energy of the field configuration (interpreted as its mass) then has the form:

$$M=\frac{4\pi v}{g}f\left(\frac{\alpha}{g^2}\right) \tag{13.160}$$

where f is found to be a slowly varying function satisfying (for example)

$$f(0)=1 \tag{13.161a}$$

$$f(0.5)=1.42 \tag{13.161b}$$

$$f(10)=1.44. \tag{13.161c}$$

The first of these values may be verified directly, since in the limit that $\alpha\to 0$ (and $\mu^2\to 0$, so that $v\neq 0$) (13.158) has the analytic solution

$$H(\xi)=\xi\coth\xi-1 \qquad K(\xi)=\xi\operatorname{cosech}\xi \tag{13.162}$$

discovered by Prasad and Sommerfield[10].

We should emphasise that the field configurations (13.149) and (13.153) we have discovered are purely classical. The 'mass' calculated in (13.160) is just the energy of this classical field configuration and is not (except in conjecture[11]) the mass of any field quantum. We can however think of the field configuration as describing a localised system with the energy concentrated in a region around the origin. In fact for large values of ξ, $H\sim\xi$ from (13.150), so (13.158a) gives

$$K''\sim K \tag{13.163}$$

since $K\to 0$, from (13.155b). Thus the asymptotic form of K is

$$K\sim e^{-\xi}=e^{-vgr} \tag{13.164}$$

and we see that the field energy is essentially concentrated in the sphere of radius $(gv)^{-1}$ centred at the origin. It is easy to see (from (13.160), for example) that

$$m_A\equiv gv\sim\frac{g^2}{4\pi}M \tag{13.165}$$

is just the mass of the (transverse) vector field which acquires mass by the spontaneous symmetry breaking. Thus the size of the field configuration we have been discussing is just the Compton wavelength of this massive vector field, and *not* the Compton wavelength associated with its 'mass'.

At distances which are large compared with m_A^{-1} the SO(3) local symmetry is broken, according to (13.135), and the residual symmetry is just the U(1) group of rotations about the radial direction $\hat{\boldsymbol{r}}$. If SO(3) had been the electroweak symmetry, which it is *not* (see Chapter 14), this surviving U(1) gauge symmetry would have been the electromagnetic gauge invariance, and presumably $g=e$. In this limit the surviving gauge field tensor $\hat{r}^a F^a_{ijM}$ is

$$\hat{r}^a F^a_{ijM} \sim \frac{-1}{er^3} \varepsilon_{iaj} r^a \qquad (evr \gg 1) \tag{13.166}$$

which corresponds to a magnetic field

$$\boldsymbol{B} \sim \frac{-1}{er^2} \hat{\boldsymbol{r}}. \tag{13.167}$$

Thus the field configuration we have been discussing is that of a magnetic monopole, since at large distances the field in everywhere radial. In fact the total magnetic charge of the monopole configuration is

$$\int_\Sigma \boldsymbol{B} \cdot \boldsymbol{n} \, \mathrm{d}\Sigma = -4\pi/e \tag{13.168}$$

where $\boldsymbol{n}$ is normal to the surface Σ enclosing the origin, all points of which have $evr \gg 1$.

The monopole configuration(s) are of no great importance in the context of electroweak theory, because they do not arise in what is now the generally accepted electroweak symmetry. However, they do occur in the grand unified theories (GUTS) which will be discussed in Chapter 16. To see why this is so we must first appreciate that the field configuration (texture) which has been developed is important not just because it has finite energy, but also because of its topological stability. We noted at the outset that $\boldsymbol{\varphi}_M$ was tied to a non-trivial topological structure as $r \to \infty$. This means that the asymptotic form (13.135) of $\boldsymbol{\varphi}_M$ cannot be continuously deformed to the trivial texture

$$\boldsymbol{\varphi}_M = v\hat{\boldsymbol{z}} \tag{13.169}$$

for example. Another reflection of this is that the magnetic charge (13.168) is 'topologically conserved'—in fact its value must be an integral multiple of $4\pi/e$, for purely geometric reasons. We can see this as follows. For large $r(evr \gg 1)\boldsymbol{\varphi}(\boldsymbol{r})$ defines a mapping

$$\boldsymbol{\varphi}: \Sigma \to \mathscr{M}^0 \tag{13.170}$$

where Σ is any closed surface enclosing the origin, and on all points of which r is large, and $\mathscr{M}^0$ is the manifold of all values of $\boldsymbol{\varphi}$ which minimise $V(\boldsymbol{\varphi})$. In the present case $V(\boldsymbol{\varphi})$ is given by (13.133) and $\mathcal{M}^0$ is the surface of the three-dimensional sphere

$$\boldsymbol{\varphi}\cdot\boldsymbol{\varphi}=v^2. \tag{13.171}$$

Thus in this case $\mathcal{M}^0=\mathrm{S}^2$. Now suppose that the points on Σ are parametrised by u and w

$$\Sigma=\{\boldsymbol{r}(u, w)\colon u\in\mathrm{U},\ w\in\mathrm{W}\}. \tag{13.172}$$

This gives the mapping $\boldsymbol{\varphi}$ in terms of the parameters u, w. The normal $\boldsymbol{n}_2$ to S^2 is given in terms of the partial derivatives $\boldsymbol{\varphi}_u$ and $\boldsymbol{\varphi}_w$ by

$$\boldsymbol{\varphi}_u\wedge\boldsymbol{\varphi}_w\,\mathrm{d}u\,\mathrm{d}w=\boldsymbol{n}_2\,\mathrm{d}S^2 \tag{13.173}$$

where $\mathrm{d}S^2$ is the element of area on S^2. It follows that

$$\boldsymbol{\varphi}\cdot\boldsymbol{\varphi}_u\wedge\boldsymbol{\varphi}_w\,\mathrm{d}u\,\mathrm{d}w=v\,\mathrm{d}S^2. \tag{13.174}$$

As (u, w) covers $\mathrm{U}\times\mathrm{W}$, $\boldsymbol{r}(u, w)$ covers the closed surface Σ, and $\boldsymbol{\varphi}(\boldsymbol{r}(u, w))$ must also cover a closed surface which must be contained in S^2. Evidently this can only be S^2 itself, or an integral multiple of S^2. Thus

$$4\pi v^3 N=\int_{\mathrm{U}\times\mathrm{W}}\boldsymbol{\varphi}\cdot\boldsymbol{\varphi}_u\wedge\boldsymbol{\varphi}_w\,\mathrm{d}u\,\mathrm{d}w \tag{13.175}$$

where N is an integer. Precisely because N is an integer it cannot be varied continuously. Thus continuous deformation of $\boldsymbol{\varphi}$ leaves N invariant, which is why we say that N is a topologically conserved quantum number.

Finally we must demonstrate the connection between the topological charge and the magnetic charge. This is done by transforming (13.175) to an integral over Σ. If the normal to Σ is $\boldsymbol{n}$, then as in (13.173)

$$\boldsymbol{n}\,\mathrm{d}\Sigma=\boldsymbol{r}_u\wedge\boldsymbol{r}_w\,\mathrm{d}u\,\mathrm{d}w \tag{13.176}$$

which implies

$$\varepsilon_{ijk}n_i\,\mathrm{d}\Sigma=\frac{\partial(r_j, r_k)}{\partial(u, w)}\,\mathrm{d}u\,\mathrm{d}w. \tag{13.177}$$

Also

$$\boldsymbol{\varphi}_u\wedge\boldsymbol{\varphi}_w=\frac{1}{2}\,\partial_j\boldsymbol{\varphi}\wedge\partial_k\boldsymbol{\varphi}\,\frac{\partial(r_j, r_k)}{\partial(u, w)}. \tag{13.178}$$

Substituting these into (13.175) gives

$$4\pi v^3 N=\int_\Sigma \tfrac{1}{2}\varepsilon_{ijk}n_i\boldsymbol{\varphi}\cdot\partial_j\boldsymbol{\varphi}\wedge\partial_k\boldsymbol{\varphi}\,\mathrm{d}\Sigma. \tag{13.179}$$

On Σ we have already noted (13.171) that $\boldsymbol{\varphi}\cdot\boldsymbol{\varphi}=v^2$, and that $\mathrm{D}_i\boldsymbol{\varphi}=\mathrm{O}(r^{-2})$. In fact because of the exponential fall off we can set

$$\mathrm{D}_i\boldsymbol{\varphi}=0. \tag{13.180}$$

Then solving (13.151) we can find the component of $\boldsymbol{A}_i$ perpendicular to $\boldsymbol{\varphi}$:

$$\boldsymbol{A}_i=\frac{1}{gv^2}\,\boldsymbol{\varphi}\wedge\partial_i\boldsymbol{\varphi}+\frac{1}{v}\,a_i\boldsymbol{\varphi} \tag{13.181}$$

where a_i is arbitrary. Then, as before, the field tensor $\boldsymbol{F}_{jk}$ is parallel to the axis $\boldsymbol{\varphi}$ of the residual U(1) symmetry and we find

$$\frac{1}{v}\,\boldsymbol{\varphi}\cdot\boldsymbol{F}_{jk}=\frac{1}{gv^3}\,\boldsymbol{\varphi}\cdot\partial_j\boldsymbol{\varphi}\wedge\partial_k\boldsymbol{\varphi}+\partial_j a_k\partial_k a_j \tag{13.182}$$

and the magnetic field is

$$B_i=-\frac{1}{2v}\,\varepsilon_{ijk}\boldsymbol{\varphi}\cdot\boldsymbol{F}_{jk}. \tag{13.183}$$

Then substituting into (13.179) gives

$$4\pi v^3 N=-\int_\Sigma \mathrm{d}\Sigma\, n_i[B_i+(\mathrm{curl}\ \boldsymbol{a})_i]gv^3. \tag{13.184}$$

The unknown curl $\boldsymbol{a}$ does not contribute and we see that the magnetic charge

$$\int_\Sigma \boldsymbol{B}\cdot\boldsymbol{n}\,\mathrm{d}\Sigma=-N\left(\frac{4\pi}{e}\right) \tag{13.185}$$

remembering that $g=e$. Thus the topological charge measures the magnetic charge in units of $(-4\pi/e)$, and the magnetic charge is conserved because of topological reasons.

13.9 The effective potential in one-loop order

We have seen in §13.3 and the following sections that spontaneous symmetry breaking requires that some scalar field develops a vacuum expectation value (VEV). This VEV is determined by the minimisation of the effective potential as was shown in (4.79). (In the case that the classical field $\varphi_c(x)$ varies in space, as was the case in §13.8, it is obtained by minimising the effective action $\Gamma[\varphi_c]$.) So far in this chapter we have ignored quantum effects, so that the effective potential is given entirely by *the* potential V which appears in the Lagrangian. (It was for this reason that we minimised the expressions (13.21b) and (13.32), for example.) We could of course include some of the neglected terms by working to some (finite) order in perturbation theory. However, spontaneous symmetry breaking is a non-perturbative effect, as we have already observed. Thus some other expansion parameter is needed if we are to improve upon the calculations we have done so far. One of the few available alternatives is the loop expansion.

We start with the generating functional $X[J]$ of the connected Green functions in a scalar field theory, defined in (4.58) and (4.25), and insert a factor $\hbar^{-1}$, so that

$$\begin{aligned} W[J] &= \exp \mathrm{i}\hbar^{-1} X[J] \\ &= N' \int \mathscr{D}\varphi \exp \mathrm{i}\hbar^{-1} \int \mathrm{d}^4x(\mathscr{L} + J\varphi) \end{aligned} \tag{13.186}$$

where N' is chosen so that

$$W[0] = 1 \qquad (X[0] = 0). \tag{13.187}$$

(We have hitherto set $\hbar = 1$, so this insertion makes no differences to the calculations we have already performed.) Since the $\hbar^{-1}$ on the second line of (13.185) multiplies the whole Lagrangian, not just the interaction part, each of the V vertices in a diagram will carry a factor $\hbar^{-1}$, while each of the I internal lines will carry a factor $\hbar$. In calculating the connected Green function $\tilde{G}^{(E)}$, with E externa lines, as in §6.2, each of the external lines has a propagator. Thus overall we have a factor

$$(\hbar)^{-V+I+E} = (\hbar)^{L+1-E} \tag{13.188}$$

using (7.7), and any diagram in the expansion of $\hbar^{-1}X[J]$ has a factor $(\hbar)^{L-E}$. The one-particle-irreducible (OPI) Green functions $\Gamma^{(n)}$ are generated by the effective action $\Gamma[\varphi_c]$ via (4.73). As we found in Chapter 6 the OPI Green functions $\tilde{\Gamma}^{(E)}$ have *no* propagators associated with the E external legs, so these Green functions are multiplied by a factor $(\hbar)^L$. In other words the power of $\hbar$ in $\tilde{\Gamma}^{(E)}$ counts the number of loops. With no loops ($L = 0$) the only non-zero OPI Green functions are

$$\tilde{\Gamma}^{(2)}(p, -p) = p^2 - \mu^2 \tag{13.189}$$

and

$$\tilde{\Gamma}^{(4)}(p_1, p_2, p_3, p_4) = -\lambda. \tag{13.190}$$

Then from (4.81) we can write down the corresponding approximation to $V(\varphi_c)$:

$$V_0(\varphi_c) = \frac{1}{2}\mu^2\varphi_c^2 + \frac{1}{4!}\lambda\varphi_c^4 \tag{13.191}$$

which is precisely the 'classical' potential (13.21b) which was expected. In this connection it is perhaps worth emphasising that the insertion of $\hbar$ is purely conventional. We do not have to assume that $\hbar$ is 'small', and it is introduced merely as an expansion parameter. Indeed, we can write down $\Gamma[\varphi_c]$ in the same approximation by using (6.36). The terms involving $\Delta_F(0)$ derive from (divergent) loop integrations, so they do not contribute in zeroth order. Thus

$$\Gamma_0[\varphi_c] = -\frac{1}{2}\int dx\,\varphi_c(x)(\partial_\mu\partial^\mu + \mu^2)\varphi_c(x) - \frac{1}{4!}\lambda\int dx[\varphi_c(x)]^4 \tag{13.192}$$

since $N=1$ so that $\Gamma[0]=0$.

We now wish to proceed beyond this approximation and to calculate the effective potential V and the effective action Γ accurate to order $\hbar$, i.e. at the one-loop order. First we shift the functional integration variable φ in (13.186) by writing

$$\varphi(x) = \varphi_0(x) + \tilde{\varphi}(x) \tag{13.193}$$

where φ_0 is the zeroth order approximation to φ_c. It therefore satisfies

$$(\partial_\mu\partial^\mu + \mu^2)\varphi_0(x) + \frac{\lambda}{6}\varphi_0^3(x) = J(x). \tag{13.194}$$

Using the Lagrangian density given in (3.34)

$$\mathscr{L}(\varphi) \equiv \frac{1}{2}(\partial_\mu\varphi)(\partial^\mu\varphi) - \frac{1}{2}\mu^2\varphi^2 - \frac{1}{4!}\lambda\varphi^4 \tag{3.34}$$

and the change of variables (13.193), it follows that

$$\int d^4x(\mathscr{L} + J\varphi) = \int d^4x(\mathscr{L}(\varphi_0) + J\varphi_0) + \int d^4x\left[(\partial_\mu\tilde{\varphi})(\partial^\mu\varphi_0) - \mu^2\tilde{\varphi}\varphi_0 - \frac{1}{6}\lambda\tilde{\varphi}\varphi_0^3 + J\tilde{\varphi}\right] + \int d^4x\left[\mathscr{L}_{\text{quad}}(\tilde{\varphi},\varphi_0) - \frac{1}{6}\lambda\varphi^3\varphi_0 - \frac{1}{4}\lambda\tilde{\varphi}^4\right] \tag{13.195a}$$

where

$$\mathscr{L}_{\text{quad}}(\tilde{\varphi},\varphi_0) \equiv \tfrac{1}{2}(\partial_\mu\tilde{\varphi})(\partial^\mu\tilde{\varphi}) - \tfrac{1}{2}\mu^2\tilde{\varphi}^2 - \tfrac{1}{4}\lambda\varphi_0^2\tilde{\varphi}^2 \tag{13.195b}$$

contains all the terms quadratic in $\tilde{\varphi}$ after the shift. The term which is linear in $\tilde{\varphi}$ in (13.195a) vanishes by virtue of (13.194) (this merely reflects the fact that φ_0 minimises the classical action). Next we rescale the new functional integration variable by

$$\tilde{\varphi} = \hbar^{1/2}\varphi \tag{13.196}$$

and substituting back into (13.186) gives

$$W[J]=N' \exp i\hbar^{-1} \int d^4x[\mathscr{L}(\varphi_0)+J\varphi_0]$$
$$\times \int \mathscr{D}\varphi \exp i \int d^4x \left(\mathscr{L}_{\text{quad}}(\varphi,\varphi_0) - \frac{\lambda}{6}\hbar^{1/2}\varphi^3\varphi^0 - \frac{1}{24}\lambda\hbar\varphi^4 \right). \tag{13.197}$$

The terms proportional to $\hbar^{1/2}$ and $\hbar$ in (13.197) may be neglected, if we only want to retain the first-order corrections, and this leaves a functional integration which may be continued to the Gaussian integral already encountered in Chapters 1 and 4. Writing

$$\int d^4x \mathscr{L}_{\text{quad}}(\varphi, \varphi_0) = -\tfrac{1}{2} \int d^4x\, d^4x' \varphi(x') A(x', x, \varphi_0) \varphi(x) \tag{13.198}$$

where

$$A(x', x, \varphi_0) = [-\partial_{x'\mu}\partial_x^\mu + \mu^2 + \tfrac{1}{2}\lambda\varphi_0^2]\delta(x'-x) \tag{13.199}$$

the result is that

$$W[J] \simeq N' \exp i\hbar^{-1} \int d^4x[\mathscr{L}(\varphi_0)+J\varphi_0]$$
$$\times \exp[-\tfrac{1}{2} \operatorname{Tr} \ln A(x', x, \varphi_0)]. \tag{13.200}$$

Using (13.187) we have that

$$W[0] = 1 \simeq N' \exp\left(-\frac{i}{2} \operatorname{Tr} \ln \mathrm{A}(x', x, 0) \right) \tag{13.201}$$

since

$$\varphi_0[0] = 0. \tag{13.202}$$

Then comparing with (13.186) we find

$$X_0[J] = \int d^4x[\mathscr{L}(\varphi_0)+J\varphi_0] \tag{13.203}$$

$$X_1[J] = \frac{i}{2} \operatorname{Tr} \ln[A(x', x, \varphi_0)/A(x', x, 0)] \tag{13.204}$$

where A is defined in (13.199). Of course, to go beyond this first-order approximation we should have to retain the $O(\hbar^{1/2})$ and $O(\hbar)$ terms in (13.202). Finally we can compute the effective action

$$\Gamma[\varphi_c] = \Gamma_0[\varphi_c] + \hbar\Gamma_1[\varphi_c] \tag{13.205}$$

in the same approximation of neglecting $O(\hbar^2)$. From (4.64) we have that

$$\varphi_c(x) \equiv \frac{\delta X}{\delta J(x)} \simeq \varphi_0(x) + O(\hbar). \tag{13.206}$$

Thus using (4.68) we have

$$\begin{aligned}\Gamma_0[\varphi_c] &= X_0[J] - \int d^4x J\varphi_0 \\ &= \int d^4x \mathscr{L}(\varphi_0) \\ &= \int d^4x \left(\frac{1}{2}(\partial_\mu \varphi_c)(\partial^\mu \varphi^c) - \frac{1}{2}\mu^2\varphi_c^2 - \frac{1}{4!}\lambda\varphi_c^4\right)\end{aligned} \tag{13.207}$$

in agreement with (13.192). Next we have

$$\begin{aligned}\hbar\Gamma_1[\varphi_c] &= X_0[J] - \Gamma_0[\varphi_c] - \int d^4x J\varphi_c + \hbar X_1[J] \\ &= \int d^4x[\mathscr{L}(\varphi_0) + J\varphi_0] - \int d^4x[\mathscr{L}(\varphi_c) + J\varphi_c] \\ &\quad + \frac{i\hbar}{2}\operatorname{Tr}\ln[A(x', x, \varphi_0)/A(x', x, 0)].\end{aligned} \tag{13.208}$$

The difference on the second line is of order $(\varphi_0 - \varphi_c)^2 = O(\hbar^2)$ since φ_0 satisfies (13.194), and in the last term we may replace φ_0 by φ_c at the required accuracy. Hence

$$\Gamma_1[\varphi_c] = \frac{i}{2}\operatorname{Tr}\ln[A(x', x, \varphi_c)/A(x', x, 0)]. \tag{13.209}$$

The effective potential $V(\varphi_c)$ is obtained from $\Gamma[\varphi_c]$ by taking φ_c to be constant. Then

$$\Gamma[\varphi_c] = -\int d^4x V(\varphi_c). \tag{13.210}$$

Now with φ_c constant we can evaluate Γ_1, and thereby V_1. To define the logarithm in (13.209) we must first diagonalise $A(x', x, \varphi_c)$:

$$\begin{aligned}A(x', x, \varphi_c) &\equiv (-\partial_{x'\mu}\partial_x^\mu + \mu^2 + \tfrac{1}{2}\lambda\varphi_c^2)\delta(x' - x) \\ &= \int \frac{d^4k}{(2\pi)^4}(-\partial_{x'\mu}\partial_x^\mu + \mu^2 + \tfrac{1}{2}\lambda\varphi_c^2)\, e^{ik(x'-x)} \\ &= \int \frac{d^4k}{(2\pi)^4}(-k^2 + \mu^2 + \tfrac{1}{2}\lambda\varphi_c^2)\, e^{ik(x'-x)}\end{aligned}$$

$$= \int \mathrm{d}^4k\, \mathrm{d}^4k' [(2\pi)^{-2}\, \mathrm{e}^{\mathrm{i}x'k'}][(-k^2+\mu^2+\tfrac{1}{2}\lambda\varphi_c^2)\delta(k'-k)]$$
$$\times [(2\pi)^{-2}\, \mathrm{e}^{-\mathrm{i}kx}]. \tag{13.211}$$

Thus

$$\ln A(x', x, \varphi_c) = \int \mathrm{d}^4k\, \mathrm{d}^4k' [(2\pi)^{-2}\, \mathrm{e}^{\mathrm{i}x'k'}] \ln(-k^2+\mu^2+\tfrac{1}{2}\lambda\varphi_c^2)\delta(k'-k)$$
$$\times [(2\pi)^{-2}\, \mathrm{e}^{-\mathrm{i}kx}] \tag{13.212}$$

and

$$\mathrm{Tr} \ln A = \int \mathrm{d}^4x\, \mathrm{d}^4x' \delta(x'-x) \ln A(x', x, \varphi_c)$$
$$= \int \mathrm{d}^4x \int \frac{\mathrm{d}^4k}{(2\pi)^4} \ln(-k^2+\mu^2+\tfrac{1}{2}\lambda\varphi_c^2). \tag{13.213}$$

Thus from (13.209), (13.210) the one-loop order contribution to the effective potential is given by

$$V_1(\varphi_c) = \frac{-\mathrm{i}}{2} \int \frac{\mathrm{d}^4k}{(2\pi)^4} \ln\left(\frac{-k^2+\mu^2+\frac{1}{2}\lambda\varphi_c^2}{-k^2+\mu^2}\right) \tag{13.214}$$

and to this order the effective potential is

$$V(\varphi_c) \simeq V_0(\varphi_c) + V_1(\varphi_c)$$
$$= \frac{1}{2}\mu^2\varphi_c^2 + \frac{1}{4!}\lambda\varphi_c^4 - \frac{\mathrm{i}}{2}\hbar \int \frac{\mathrm{d}^4k}{(2\pi)^4} \ln\left(1 - \frac{\frac{1}{2}\lambda\varphi_c^2}{k^2-\mu^2}\right). \tag{13.215}$$

The last term is ultraviolet divergent, so we regularise the integral by evaluating it in 2ω-dimensional space–time, as in §7.2. Also, the parameters μ^2, λ, φ_c which appear in the expression are those of the original *bare* Lagrangian (3.34). They should more carefully have been written as μ_B^2, λ_B, φ_{cB}, as we did in (7.3) and subsequently. When we express V in terms of the renormalised quantities defined in (7.40), (7.42) and (7.43) we obtain

$$V(\varphi_c) = \frac{1}{2}\mu^2\varphi_c^2 + \frac{1}{4!}\lambda\varphi_c^4 + \frac{1}{2}\delta\mu^2\varphi_c^2 + \frac{1}{4!}\delta\lambda\varphi_c^4$$
$$-\frac{\mathrm{i}}{2}\hbar \int \frac{\mathrm{d}^{2\omega}k}{(2\pi)^{2\omega}} \ln\left(1 - \frac{\frac{1}{2}\lambda\varphi_c^2}{k^2-\mu^2}\right) \tag{13.216}$$

where now μ^2, λ, φ_c are the *renormalised* parameters, defined as in §7.5 by imposing boundary conditions upon V. In the single loop order to which we are working the k integration yields only simple poles in $2-\omega$. Thus in (7.70) we have

$$a_v = 0 = b_v \qquad (v > 1). \tag{13.217}$$

The poles in $2-\omega$ are cancelled by taking

$$a_1 = \frac{3}{32\pi^2}\hat{\lambda}^2 \qquad b_1 = \frac{1}{32\pi^2}\hat{\lambda} \tag{13.218a}$$

where

$$\hat{\lambda} \equiv \lambda M^{2\omega-4} \tag{13.218b}$$

as in §7.5. The remaining expression is finite as $\omega \to 2$ giving

$$\begin{aligned} V(\varphi_c) = {} & \frac{1}{2}\mu^2\varphi_c^2\left[1 + b_0 - \frac{\lambda}{32\pi^2}\left(\frac{3}{2} + \Gamma'(1) + \ln 4\pi\right)\right] \\ & + \frac{1}{4!}\varphi_c^4\left[\lambda + a_0 - \frac{3\lambda^2}{32\pi^2}\left(\frac{3}{2} + \Gamma'(1) + \ln 4\pi\right)\right] \\ & + \frac{1}{64\pi^2}\left[\left(\mu^2 + \frac{1}{2}\lambda\varphi_c\right)^2 \ln\left(\frac{\mu^2 + \frac{1}{2}\lambda\varphi_c^2}{M^2}\right) - \mu^4 \ln\left(\frac{\mu^2}{M^2}\right)\right]. \end{aligned} \tag{13.219}$$

In the MS scheme the counter terms remove only the poles in $\omega - 2$, so

$$a_0^{\mathrm{MS}} = b_0^{\mathrm{MS}} = 0 \tag{13.220}$$

and the counter terms are identical to those in (7.74a). (This could have been anticipated, since although they were derived perturbatively the expressions are clearly the only one-loop contributions.) In the $\overline{\mathrm{MS}}$ scheme we also take $a_0^{\overline{\mathrm{MS}}}$ and $b_0^{\overline{\mathrm{MS}}}$ as given in (7.76) and (7.77), so as to remove the $\Gamma'(1) + \ln 4\pi$ factors.

We may also renormalise $V(\varphi_c)$ by expressing it in terms of the 'physical' mass and coupling constant. This is done by choosing a_0^{phys} and b_0^{phys} such that

$$\left.\frac{\mathrm{d}^2 V}{\mathrm{d}\varphi_c^2}\right|_{\varphi_c = 0} = \mu^2 \tag{13.221a}$$

$$\left.\frac{\mathrm{d}^4 V}{\mathrm{d}\varphi_c^4}\right|_{\varphi_c = M} = \lambda. \tag{13.221b}$$

We leave this as an exercise, and quote the result only in the case that μ^2 is 'small'. It may then happen that the radiative corrections significantly modify the tree contributions to the effective potential. For example, it is possible that the radiative corrections can generate spontaneous symmetry breaking when this is not present in V_0. Clearly this requires μ^2 to be small. Using the physical renormalisation (13.221) we find

$$V(\varphi_c) = \frac{1}{2}\mu^2\varphi_c^2 + \frac{\lambda}{4!}\varphi_c^4 + \frac{\lambda^2\varphi_c^4}{256\pi^2}\left(\ln\frac{\varphi_c^2}{M^2} - \frac{25}{6}\right) \tag{13.222a}$$

where

$$\mu^2 \ll \tfrac{1}{2}\lambda\varphi_c^2 \tag{13.222b}$$

in the radiative corrections. The parameter M may be eliminated in favour of the value v of φ_c at the minimum of $V(\varphi_c)$:

$$V'(v)=0. \tag{13.223}$$

This gives[12]

$$V(\varphi_c)=B\left(\frac{1}{2}\alpha v^2\varphi_c^2-\frac{1}{4}(\alpha+2)\varphi_c^4+\varphi_c^4\ln\frac{\varphi_c^2}{v^2}\right) \tag{13.224}$$

where

$$B\equiv\frac{\lambda^2}{256\pi^2} \qquad \alpha=\frac{\mu^2}{Bv^2}. \tag{13.225}$$

Then the mass m_H of the physical Higgs particle is given by

$$m_H^2=\left.\frac{d^2V}{d\varphi_c^2}\right|_{\varphi_c=v}=2Bv^2(4-\alpha) \tag{13.226}$$

(which is positive provided μ^2 is small enough). For $V(v)$ to be the global minimum we require $V(v)<0$ which is satisfied if $\alpha<2$.

The foregoing example is really only of academic interest, since, if the single-loop contributions (which are of order λ^2) are large enough to modify the tree contributions to the effective potential (of order λ), then presumably the two- and higher-order loop contributions, which have been neglected, are of equal importance. As an example of a model which is not open to this objection we consider next the Abelian Higgs model, which has already been explored in some detail in §13.5.

The Lagrangian is $\mathscr{L}+\mathscr{L}_{GF}$ where $\mathscr{L}$ is given in (13.59) and the gauge fixing Lagrangian $\mathscr{L}_{GF}$ is given in (13.75). We first shift the (complex) field $\varphi(x)$ by a constant amount which, without loss of generality, we take to be real. Then, as in (13.193), after the approximation (13.205), we have

$$\varphi(x)=\frac{1}{\sqrt{2}}[\varphi_c+\tilde{\varphi}_1(x)+i\tilde{\varphi}_2(x)]. \tag{13.227}$$

Next we rescale the U(1) gauge field A_μ as well as the scalar fields $\tilde{\varphi}_i$ $(i=1,2)$ by the factor $\hbar^{1/2}$ as in (13.196). In the one-loop approximation we need retain only the terms $\mathscr{L}_{quad}(\varphi_c)$ in the Lagrangian which are quadratic in the rescaled fields. Then the one-loop contribution $V_1(\varphi_c)$ to the effective potential is now given by

$$\exp\left(-i\int d^4x\,V_1(\varphi_c)\right)=\int\mathscr{D}\varphi\mathscr{D}A^\mu\exp\left(i\int d^4x\,\mathscr{L}_{quad}(\varphi_c)\right). \tag{13.228}$$

The choice of $\mathscr{L}_{\mathrm{GF}}$ ensured that there are no $\tilde{\varphi}A^{\mu}$ cross terms so we may write

$$\int \mathrm{d}^4x\, \mathscr{L}_{\mathrm{quad}}(\varphi_{\mathrm{c}}) = -\tfrac{1}{2}\int \mathrm{d}^4x'\, \mathrm{d}^4x[\varphi_i(x')A_{ij}(x', x, \varphi_{\mathrm{c}})\varphi_j(x)$$

$$+ A_{\mu}(x')B^{\mu\nu}(x', x, \varphi_{\mathrm{c}})A_{\nu}(x)] \tag{13.229}$$

where A_{ij} is diagonal and (without summation over i)

$$A_{ij}(x', x, \varphi_i) = [-\partial_{x'\mu}\partial^{\mu}_{x} + M^2_{\mathrm{S}i}(\varphi_{\mathrm{c}})]\delta_{ij}\delta(x'-x) \tag{13.230}$$

with

$$M^2_{\mathrm{S1}}(\varphi_{\mathrm{c}}) = \mu^2 + \tfrac{3}{4}\lambda\varphi_{\mathrm{c}}^2 \tag{13.231a}$$

$$M^2_{\mathrm{S2}}(\varphi_{\mathrm{c}}) = \mu^2 + \tfrac{1}{4}\lambda\varphi_{\mathrm{c}}^2 + \xi q^2\varphi_{\mathrm{c}}^2. \tag{13.231b}$$

Similarly

$$B^{\mu\nu}(x', x, \varphi_{\mathrm{c}}) = [(\xi^{-1}-1)\partial^{\mu}_{x'}\partial^{\nu}_{x} + g^{\mu\nu}\partial_{x'\mu}\partial^{\mu}_{x} - q^2\varphi_{\mathrm{c}}^2 g^{\mu\nu}]\delta(x'-x). \tag{13.232}$$

(As a check, note that when we set $\varphi_{\mathrm{c}} = v$ and use (13.34b) we retrieve the masses given in (13.78).) The functional integral may be evaluated as before and we find

$$\int \mathrm{d}^4x\, V_1(\varphi_{\mathrm{c}}) = -\frac{\mathrm{i}}{2}[\mathrm{Tr}\ln A(\varphi_{\mathrm{c}})/A(0) + \mathrm{Tr}\ln B(\varphi_{\mathrm{c}})/B(0)] \tag{13.233}$$

where the trace is now with respect to the internal symmetry labels (i, j) and Lorentz labels (μ, ν) as well as the space–time labels (x', x). This aspect is easily handled by diagonalising the mass matrices and we find

$$V_1(\varphi_{\mathrm{c}}) = -\frac{\mathrm{i}}{2}\int \frac{\mathrm{d}^4k}{(2\pi)^4}\left(\ln\frac{k^2-\mu^2-\frac{3}{4}\lambda\varphi_{\mathrm{c}}^2}{k^2-\mu^2} + \ln\frac{k^2-\mu^2-\frac{1}{4}\lambda\varphi_{\mathrm{c}}^2-\xi q^2\varphi_{\mathrm{c}}^2}{k^2-\mu^2}\right.$$

$$\left. + 3\ln\frac{k^2-q^2\varphi_{\mathrm{c}}^2}{k^2} + \ln\frac{k^2-\xi q^2\varphi_{\mathrm{c}}^2}{k^2}\right). \tag{13.234}$$

In the Landau gauge ($\xi = 0$), in which we shall now work, the coefficients of the surviving terms count the number of helicity states; one for each scalar mode, and three for each (massive) vector mode. The integrals are all of the form already encountered in (13.214), and if we again take $\mu^2 \ll \lambda\varphi_{\mathrm{c}}^2$ in the radiative corrections we have

$$V(\varphi_{\mathrm{c}}) = \frac{1}{2}\mu^2\varphi_{\mathrm{c}}^2 + \frac{1}{16}\lambda\varphi_{\mathrm{c}}^4 + B\varphi_{\mathrm{c}}^4\left(\ln\frac{\varphi_{\mathrm{c}}^2}{M^2} - \frac{25}{6}\right) \tag{13.235}$$

using the renormalisation scheme

$$\left.\frac{\mathrm{d}^2V}{\mathrm{d}\varphi_{\mathrm{c}}^2}\right|_{\varphi_{\mathrm{c}}=0} = \mu^2 \tag{13.236}$$

$$\left.\frac{\mathrm{d}^4 V}{\mathrm{d}\varphi_{\mathrm{c}}^4}\right|_{\varphi_{\mathrm{c}}=M}=\frac{3}{2}\lambda \tag{13.237}$$

analogous to (13.221), and now

$$B=\frac{1}{64\pi^2}\left(\frac{5}{8}\lambda^2+3q^4\right). \tag{13.238}$$

If we assume λ is of order q^4, and small, it is legitimate to neglect the order λ^2 contributions to B, but not, of course, the order q^4 contribution. Thus in the case of a gauge theory the quantum effects can easily modify the tree contributions to the effective potential. As before, we may eliminate M in favour of the value v of φ_{c} at the minimum of $V(\varphi_{\mathrm{c}})$. This gives the form (13.224) with B given now by (13.238). The mass m_{H} of the physical Higgs particle is given by (13.226), which implies

$$m_{\mathrm{H}}^2 \geqslant 4Bv^2 \tag{13.239}$$

since $\alpha<2$. Neglecting λ^2 compared with q^4 this gives

$$m_{\mathrm{H}} \geqslant \frac{\sqrt{3}\,q}{4\pi}\,m_A \tag{13.240a}$$

where

$$m_A = qv \tag{13.240b}$$

is the mass of the vector particle. In the particular case $\mu^2=0$, so $\alpha=0$, (13.226) yields

$$m_{\mathrm{H}} = \frac{\sqrt{6}\,q}{4\pi}\,m_A. \tag{13.241}$$

This feature, that we are able to bound, or, in the case that $\mu^2=0$, to predict the ratio of the Higgs boson to vector boson masses, is common to all gauge theories.

The generalisation of this technique to the non-Abelian gauge theories, which are our principal concern, is straightforward. The generalisation of (13.227) is now

$$\boldsymbol{\varphi}(x)=\boldsymbol{\varphi}_{\mathrm{c}}+\tilde{\boldsymbol{\varphi}}(x) \tag{13.242}$$

where $\boldsymbol{\varphi}(x)$ are the real scalar fields defined in §13.6. The only novelty is that the functional integral in (13.228) must be generalised to include integration over the Fadeev–Popov ghost fields η^a, η^{a*} which are inescapable in a non-Abelian theory, and the fermion fields ψ, $\bar{\psi}$, as well as the scalar fields $\tilde{\boldsymbol{\varphi}}$ and the gauge fields A_μ^a. The upshot is that (13.233) is generalised to

$$\int \mathrm{d}^4x\, V_1(\varphi_{\mathrm{c}}) = -\frac{\mathrm{i}}{2}\,[\mathrm{Tr}\ln \mathbf{A}(\varphi_{\mathrm{c}})/\mathbf{A}(0)+\mathrm{Tr}\,\mathbf{B}(\varphi_{\mathrm{c}})/\mathbf{B}(0)$$
$$-2\,\mathrm{Tr}\ln \mathbf{C}(\varphi_{\mathrm{c}})/\mathbf{C}(0)-2\,\mathrm{Tr}\ln \mathbf{D}(\varphi_{\mathrm{c}})/\mathbf{D}(0)] \tag{13.243}$$

where the matrices **A**, **B**, **C**, **D** specify the contributions to $\mathscr{L}_{\text{quad}}(\varphi_c)$ from the scalar, vector, ghost and fermion fields respectively. The extra factor of -2 from the ghost and fermion fields is because these fields are complex Grassmann variables, as explained in (8.17) of §8.1. It follows from (13.95) that the (ij) element of $A(x', x, \varphi)$ is

$$A_{ij}(x', x, \varphi_c) = [-\delta_{ij}\partial_{x'\mu}\partial^\mu_x + M_S^2(\boldsymbol{\varphi}_c)_{ij}]\delta(x'-x) \tag{13.244}$$

with the scalar mass matrix given by

$$M_S^2(\varphi_c)_{ij} = \left.\frac{\partial^2 V_0}{\partial\varphi_i\,\partial\varphi_j}\right|_{\boldsymbol{\varphi}=\boldsymbol{\varphi}_c} + \xi g^2(\mathbf{T}^a\boldsymbol{\varphi}_c\boldsymbol{\varphi}_c^{\mathrm{T}}\mathbf{T}^a)_{ij} \tag{13.245}$$

and $V_0(\boldsymbol{\varphi})$ the tree graph approximation to the effective potential. Similarly from (13.90)

$$B_{ab}^{\mu\nu}(x', x, \varphi_c) = [\delta_{ab}(\xi^{-1}\partial^\mu_{x'}\partial^\nu_x - \partial^\nu_{x'}\partial^\mu_x + g^{\mu\nu}\partial_{x'\lambda}\partial^\lambda_x) - g^{\mu\nu}M_A^2(\varphi_c)_{ab}]\delta(x'-x) \tag{13.246}$$

with

$$M_A^2(\boldsymbol{\varphi}_c)_{ab} = g^2\boldsymbol{\varphi}_c^{\mathrm{T}}\mathbf{T}^a\mathbf{T}^b\boldsymbol{\varphi}_c. \tag{13.247}$$

The ghost contribution follows from (13.110):

$$C_{ab}(x', x, \boldsymbol{\varphi}_c) = [-\delta_{ab}\partial_{x'\mu}\partial^\mu_x + \xi M_A^2(\boldsymbol{\varphi}_c)_{ab}]\delta(x'-x). \tag{13.248}$$

Finally the fermion contribution, from (13.118) and (13.139), is

$$\mathbf{D}(x', x, \boldsymbol{\varphi}_c) = [\mathrm{i}\mathbf{I}\gamma^\mu\partial_{x\mu} + \mathbf{M}_{\mathrm{F}}(\boldsymbol{\varphi}_c)]\delta(x'-x) \tag{13.249}$$

where **I** is the unit matrix and the fermion mass matrix

$$\mathbf{M}_{\mathrm{F}}(\boldsymbol{\varphi}_c) = \boldsymbol{Y}^{\mathrm{T}}\boldsymbol{\varphi}_c. \tag{13.250}$$

Evaluating the required traces gives the following generalisation of (13.234):

$$\begin{aligned} V_1(\boldsymbol{\varphi}_c) = \frac{-\mathrm{i}}{2}\int\frac{\mathrm{d}^4k}{(2\pi)^4}\Bigg[&\sum_S \ln\frac{k^2 - m_S^2(\boldsymbol{\varphi}_c)}{k^2 - m_S^2(\mathbf{0})} \\ &+\sum_A\left(3\ln\frac{k^2 - m_A^2(\boldsymbol{\varphi}_c)}{k^2 - m_A^2(\mathbf{0})} + \ln\frac{k^2 - \xi m_A^2(\boldsymbol{\varphi}_c)}{k^2 - \xi m_A^2(\mathbf{0})}\right) \\ &-2\sum_A \ln\frac{k^2 - \xi m_A^2(\boldsymbol{\varphi}_c)}{k^2 - \xi m_A^2(\mathbf{0})} - 4\sum_F \ln\frac{k^2 - m_F^2(\boldsymbol{\varphi}_c)}{k^2 - m_F^2(\mathbf{0})}\Bigg] \end{aligned} \tag{13.251}$$

where $m_S^2(\boldsymbol{\varphi}_c), m_A^2(\boldsymbol{\varphi}_c), m_F(\boldsymbol{\varphi}_c)$ are respectively the eigenvalues of $M_S^2(\boldsymbol{\varphi}_c)$, $M_A^2(\boldsymbol{\varphi}_c)$, $M_{\mathrm{F}}(\boldsymbol{\varphi}_c)$, and the sums are over all eigenvalues. As before, in the Landau gauge the coefficients of the surviving terms count the number of helicity states; the four for each spin 1/2 fermion mode is because there are two helicity states for both particle and antiparticle. Also as before, the effective potential may be cast in the form (13.222) or (13.235) by writing

$$\boldsymbol{\varphi}_c = \varphi_c v^{-1} \boldsymbol{v} \tag{13.252}$$

where $\boldsymbol{v}$ is the value of $\boldsymbol{\varphi}_c$ which minimises $V(\boldsymbol{\varphi}_c)$, and neglecting the contributions to $m_S^2(\boldsymbol{\varphi}_c)$ which are independent of $\boldsymbol{\varphi}_c$. Then

$$V(\varphi_c) = V_0(\varphi_c) + B\varphi_c^4\left(\ln\frac{\varphi_c^2}{M^2} - \frac{25}{6}\right) \tag{13.253}$$

with

$$B = \frac{1}{64\pi^2 v^4}\left(\sum g_B m_B^4 - \sum g_F m_F^4\right) \tag{13.254}$$

where $g_B = 1, 3$ for the scalar, vector boson modes, and $g_F = 4$ for the spin 1/2 fermion modes; $m_B^2 = m_S^2(\boldsymbol{v}), m_A^2(\boldsymbol{v})$ for the scalar, vector fields, and $m_F = m_F(\boldsymbol{v})$. If we again assume that the masses of the vector fields dominate B then the bound (13.240) on the mass of the Higgs scalar is generalised to

$$m_H \geqslant \frac{\sqrt{3}}{4\pi v}\left(\sum_A m_A^4\right)^{1/2}. \tag{13.255}$$

In the special case $\mu^2 = 0$ the generalisation of (13.241) is

$$m_H = \frac{\sqrt{6}}{4\pi v}\left(\sum_A m_A^4\right)^{1/2}. \tag{13.256}$$

In this special case there is another interesting effect which we note. Since $\mu^2 = 0$ we may write

$$V_0(\varphi_c) = \frac{1}{4!}\lambda\varphi_c^4. \tag{13.257}$$

Then $V(\varphi_c)$ can also be written in the form (13.224) (with $\alpha = 0$) so

$$V_{\text{eff}}(\varphi_c) = B\varphi_c^4\left(\ln\frac{\varphi_c^2}{v^2} - \frac{1}{2}\right) \tag{13.258}$$

with B now given by (13.254). Comparing (13.253) with (13.258) gives

$$\frac{1}{4!}\lambda - \frac{25}{6}B = B\left(-\frac{1}{2} + \ln\frac{M^2}{v^2}\right) \tag{13.259}$$

(which is just the condition (13.223)). Now if we *choose* the renormalisation scale

$$M = v \tag{13.260}$$

this gives

$$\frac{\lambda}{4!} = \frac{11}{3}B = \frac{11}{64\pi^2 v^4}\sum_A m_A^4 \tag{13.261}$$

where the last equality follows if we again assume that the vector field contributions dominate B. Thus we are able to eliminate the dimensionless coupling constant λ in favour of the dimensionful quantity v, with the result that the effective potential (13.258) involves only g and v.

This phenomenom has been called 'dimensional transmutation' by Coleman and Weinberg[12], and it is an inevitable consequence of symmetry breaking in a massless theory. In its original formulation the theory was described by two parameters (g and λ), ignoring the fermion couplings. Since the spontaneous symmetry breakdown necessarily generates a mass scale v, we must be able to trade in one of the original parameters in favour of v, as we have done in (13.258).

13.10 Instantons

We have seen in §13.8 that spontaneously broken non-Abelian gauge theories may possess topologically non-trivial magnetic monopole solutions, although these do not arise in the standard electroweak theory. However a rather similar (topologically non-trivial) classical gauge field configuration *does* occur in QCD, which is *not* a spontaneously broken gauge symmetry. The existence of these solutions seriously affects our view of the QCD vacuum, and this in turn is stringently constrained by phenomenology, as we shall see. In the attempt to circumvent these constraints in a natural way we are led to the existence of pseudoscalar 'axion' particles.

The gauge field configurations with which we are concerned are called 'instantons' and were discovered first as solutions of the classical pure gauge field equations in four-dimensional Euclidean space[13]:

$$\mathrm{D}_\mu \mathrm{F}_a^{\mu\nu} = 0 \tag{13.262}$$

which follows from (9.42) in the absence of the fermion fields ψ. In four-dimensional Euclidean space the action of the gauge field configuration is given, as in (4.29), by

$$S_{\mathrm{E}} = \int \mathrm{d}^4x \tfrac{1}{4} F_a^{\mu\nu} F_a^{\mu\nu} \tag{13.263a}$$

$$= \int \mathrm{d}^4x \tfrac{1}{4} \tilde{F}_a^{\mu\nu} \tilde{F}_a^{\mu\nu} \tag{13.263b}$$

where the dual field strength tensor $\tilde{F}_a^{\mu\nu}$ is given by

$$\tilde{F}_a^{\mu\nu} = \tfrac{1}{2} \varepsilon^{\mu\nu\rho\sigma} F_a^{\rho\sigma} \tag{13.264}$$

with $\varepsilon^{\mu\nu\rho\sigma}$ the totally anti-symmetric rank 4 tensor. It follows that

$$S_E = \tfrac{1}{8}\int d^4x(F_a^{\mu\nu} \pm \tilde{F}_a^{\mu\nu})^2 \mp \tfrac{1}{4}\int d^4x F_a^{\mu\nu}\tilde{F}_a^{\mu\nu} \geqslant \tfrac{1}{4}\left|\int d^4x F_a^{\mu\nu}\tilde{F}_a^{\mu\nu}\right| \tag{13.265}$$

and that the bound is saturated by self-dual or anti-self-dual gauge field strength configurations

$$F_a^{\mu\nu} = \pm\tilde{F}_a^{\mu\nu}. \tag{13.266}$$

As before in order to get a finite action we require that $F_a^{\mu\nu}$ vanishes fast enough as $r = |x| \to \infty$. It follows that the vector potential must approach a pure (vacuum) gauge transformation, i.e.

$$\begin{aligned}\mathbf{A}^\mu \equiv A_a^\mu(x)\mathbf{T}_a &\sim -ig^{-1}\mathbf{U}(x)\partial^\mu\mathbf{U}^{-1}(x)\\ &= ig^{-1}\partial^\mu\mathbf{U}(x)\mathbf{U}^{-1}(x)\end{aligned} \tag{13.267}$$

as $r \to \infty$, using the notation of (9.27). Although such a vector potential can always be transformed locally to zero, it may not be possible to do so globally if it has a non-trivial topological structure. The instanton solutions have finite and non-zero action just because they are supported topologically. Indeed the right-hand side of (13.265) is proportional to the 'Pontryagin index' of the configuration. To see this we use the notation of (9.28) to write

$$\tfrac{1}{2}F_a^{\mu\nu}\tilde{F}_a^{\mu\nu} = \mathrm{tr}(F^{\mu\nu}\tilde{F}^{\mu\nu}) \tag{13.268}$$

where

$$F^{\mu\nu} \equiv F_a^{\mu\nu}\mathbf{T}_a = \partial^\mu A^\nu - \partial^\nu A^\mu + ig[A^\mu, A^\nu] \tag{13.269a}$$

and

$$\tilde{F}^{\mu\nu} = \tfrac{1}{2}\varepsilon^{\mu\nu\rho\sigma}F^{\rho\sigma}. \tag{13.269b}$$

Then

$$\begin{aligned}\mathrm{tr}(F^{\mu\nu}\tilde{F}^{\mu\nu}) &= 2\varepsilon^{\mu\nu\rho\sigma}\,\mathrm{tr}[(\partial^\mu A^\nu + igA^\mu A^\nu)(\partial^\rho A^\sigma + igA^\rho A^\sigma)]\\ &= 2\varepsilon^{\mu\nu\rho\sigma}\,\mathrm{tr}[(\partial^\mu A^\nu)(\partial^\rho A^\sigma) + 2ig(\partial^\mu A^\nu)(A^\rho A^\sigma)]\\ &= 2\partial^\mu\varepsilon^{\mu\nu\rho\sigma}\,\mathrm{tr}\left[A^\nu\left(\partial^\rho A^\sigma + \frac{2ig}{3}A^\rho A^\sigma\right)\right]\\ &= \partial^\mu K^\mu\end{aligned} \tag{13.270a}$$

where

$$K^\mu = \varepsilon^{\mu\nu\rho\sigma} \operatorname{tr}[A^\nu(F^{\rho\sigma} - \tfrac{2}{3}igA^\rho A^\sigma)]. \tag{13.270b}$$

In deriving this we have made extensive use of the cyclic property of the trace, as well as the total antisymmetry of $\varepsilon^{\mu\nu\rho\sigma}$. Using Gauss' theorem we may then express the integral on the right-hand side of (13.265) in terms of the integral of K^μ over the surface sphere S^3 of the four-dimensional volume V

$$\begin{aligned}\int_V d^4x \operatorname{tr}(F^{\mu\nu}\tilde{F}^{\mu\nu}) &= \int_{S^3} d^3S n^\mu K^\mu \\ &= -\tfrac{2}{3}ig\varepsilon^{\mu\nu\rho\sigma}\int_{S^3} d^3Sn^\mu \operatorname{tr}(A^\nu A^\rho A^\sigma)\end{aligned} \tag{13.271}$$

provided we choose the sphere S^3 with r large enough that the contribution from $F^{\rho\sigma}S^3$ is negligible. Then, using the asymptotic form (13.267) for the vector potential, we get

$$\int_V d^4x \operatorname{tr}(F_a^{\mu\nu}\tilde{F}_a^{\mu\nu}) = -16\pi^2 q/g^2 \tag{13.272a}$$

where

$$\begin{aligned}q &= -\frac{1}{24\pi^2}\varepsilon^{\mu\nu\rho\sigma}\int_{S^3} d^3Sn^\mu \operatorname{tr}[\mathbf{U}(\partial^\nu\mathbf{U}^{-1})\mathbf{U}(\partial^\rho\mathbf{U}^{-1})\mathbf{U}(\partial^\sigma\mathbf{U}^{-1})] \\ &= -\frac{1}{24\pi^2}\int d^3S\varepsilon^{ijk} \operatorname{tr}[\mathbf{U}(\partial^i\mathbf{U}^{-1})\mathbf{U}(\partial^j\mathbf{U}^{-1})\mathbf{U}(\partial^k\mathbf{U}^{-1})]\end{aligned} \tag{13.272b}$$

is the Pontryagin index (winding number) for the map

$$\mathbf{U}: S^3 \to G \tag{13.272c}$$

and in the last expression the derivatives are with respect to the three coordinates (e.g. Euler angles) used to parametrise S^3. It is the four-dimensional generalisation of the three-dimensional result (13.175) used to demonstrate (topological) conservation of the magnetic charge. To see that q also has this property it may be shown that it is invariant under infinitesimal (and therefore continuous, finite) deformations (problem 13.11). We may also check that the prefactor is correct by verifying that q gives the winding number in the case when G is the group SU(2).

The general element $g \in \mathrm{SU}(2)$ can be written uniquely in the form

$$g = a + ib_i\tau_i \tag{13.273a}$$

with τ_i ($i = 1, 2, 3$) the usual Pauli matrices, a, b_i real and

$$a^2 + b_i b_i = 1 \tag{13.273b}$$

ensures that g is unitary and has unit determinant. Thus the parameter space of SU(2) is the unit three-sphere. If we choose

$$\mathbf{U}(x) = (x_4 + \mathrm{i}x_i\tau_i)/r \tag{13.274a}$$

where

$$r = (x_1^2 + x_2^2 + x_3^2 + x_4^2)^{1/2} \tag{13.274b}$$

then we have

$$\mathbf{U}: S^3 \to S^3 \tag{13.275}$$

maps the Euclidean space–time three-sphere S^3 in (1, 1) correspondence with the elements of the group SU(2). In this case we expect the winding number to be unity

$$q = 1. \tag{13.276}$$

To verify that our formula (13.272b) does give this value we note first that the mapping (13.275) is spherically symmetric so gives a constant integrand in (13.272b). To find this constant value it suffices to evaluate it at any point of the Euclidean space–time surface S^3, for example at

$$x^4 = r \qquad x^i = 0. \tag{13.277}$$

At this point we may as well choose $x_{1,2,3}$ to be three coordinates, so

$$\mathbf{U}(\partial^i \mathbf{U}^{-1}) = -\mathrm{i}\tau_i/r \tag{13.278}$$

and the integrand has the value $-12r^3$. Then using the result (7.15) that the surface area of S^3 is $2\pi^2 r^3$ we obtain the anticipated value (13.276) for the winding number.

The consequence of all this is that the self-dual or anti-self-dual solutions which saturate (13.265) have actions

$$S_{\mathrm{E}} = 8\pi^2|q|/g^2. \tag{13.279}$$

It follows that the instantons (with $q \neq 0$) cannot have $F^{\mu\nu}$ vanishingly small inside the surface sphere S^3. However at large distances they are supported by vacuum gauge vector potential configurations having Pontryagin index q. In this chapter we are not especially concerned with the precise instanton gauge field configurations which solve the (classical) field equations. The important point for our purposes is that there are a countable infinity of topologically distinct vacua labelled by the index q, not merely the trivial vacuum which is globally equivalent to $A_\mu = 0$.

The consequence of this observation is that we need to reconsider our path integral treatment of QCD: our previous treatment, in chapter 9, tacitly assumed that the ground state, the particle physics vacuum, is unique. Instead we now find that there are infinitely many classical vacua, which we denote by $|q\rangle$, with q an integer. The fact that the instanton configurations have finite action means that there is a finite non-zero (quantum mechanical) transition amplitude of order e^{-S_E} connecting the different vacua. It follows that the true vacuum is a linear combination of the $|q\rangle$ vacua. Now consider a gauge transformation U having unit winding number. Applying this to the $|q\rangle$ vacuum gives

$$U|q\rangle = |q+1\rangle. \tag{13.280}$$

Further, gauge invariance means that the Hamiltonian is invariant so

$$UHU^\dagger = H \tag{13.281}$$

or

$$[U, H] = 0. \tag{13.282}$$

It follows that the true (quantum mechanical) vacuum $|\theta\rangle$ is an eigenstate[14] of U with an eigenvalue which we can write as $e^{i\theta}$:

$$U|\theta\rangle = e^{i\theta}|\theta\rangle. \tag{13.283}$$

It is easy to check that the required linear combination is given by

$$|\theta\rangle = \sum_q e^{-iq\theta}|q\rangle. \tag{13.284}$$

Just such a situation arises in condensed state physics when, as in a crystal, there is a periodic potential. The true ground state (13.284) is the so-called Bloch wave. Different values of θ, corresponding to different U, label different, inaccessible sectors of the theory. The $|\theta\rangle$ to $|\theta'\rangle$ transition amplitude (in the presence of a source J) must therefore have the form

$$\begin{aligned}\langle\theta'|\theta\rangle_J &= \delta(\theta-\theta')I_J(\theta)\\ &= \sum_{q,q'} \mathrm{e}^{\mathrm{i}(q'\theta'-q\theta)}\langle q'|q\rangle_J\\ &= \sum_{q,q'} \mathrm{e}^{-\mathrm{i}(q-q')\theta}\,\mathrm{e}^{\mathrm{i}q'(\theta'-\theta)}\int(\mathscr{D}A^\mu)_{q-q'}\exp\left[\mathrm{i}\int \mathrm{d}^4x(\mathscr{L}+J_\mu A^\mu)\right].\end{aligned} \quad (13.285)$$

Thus

$$\begin{aligned}I_J(\theta) &= \sum_n \mathrm{e}^{-\mathrm{i}n\theta}\int(\mathscr{D}A^\mu)_n \exp\mathrm{i}\int \mathrm{d}^4x(\mathscr{L}+J_\mu A^\mu)\\ &= \sum_n \int(\mathscr{D}A^\mu)_n \exp\mathrm{i}\int \mathrm{d}^4x(\mathscr{L}_{\mathrm{eff}}+J_\mu A^\mu)\end{aligned} \quad (13.286a)$$

where we have written

$$\mathrm{e}^{-\mathrm{i}n\theta} = \exp\mathrm{i}\theta\frac{g^2}{16\pi^2}\int \mathrm{d}^4x\,\mathrm{tr}(F^{\mu\nu}\tilde{F}^{\mu\nu}) \quad (13.286b)$$

using (13.272a), so the effective QCD Lagrangian is

$$\mathscr{L}_{\mathrm{eff}} = \mathscr{L} + \frac{\theta g^2}{32\pi^2}F_a^{\mu\nu}\tilde{F}_a^{\mu\nu}. \quad (13.287)$$

The extra θ term is charge conjugation (C) and time-reversal (T) non-invariant. Although we have already shown in (13.270) that it may be expressed as the divergence of a current, it contributes to the action because of the topologically non-trivial instanton gauge field configurations, which provide a non-vanishing contribution at infinity.

The actual value of θ in our vacuum is *not* determined by the theory. Furthermore, as it stands it is not even observable since it can be altered by a global (axial) U(1) transformation of all quark fields. It is this connection with the 'axial U(1) problem' to which we now turn[15].

13.11 Axions

Suppose then that we have N flavours of massless quarks. Then the QCD Lagrangian posesses a $\mathrm{U}(N)_\mathrm{L} \times \mathrm{U}(N)_\mathrm{R}$ global symmetry, besides the local colour gauge invariance. Explicitly $\mathscr{L}_{\text{QCD}}$ is invariant under the transformations

$$q_{\text{L}i} \to q'_{\text{L}i} = (\mathbf{U}_{\text{L}})_{ij} q_{\text{L}j} \tag{13.288a}$$

$$q_{\text{R}i} \to q'_{\text{R}i} = (\mathbf{U}_{\text{R}})_{ij} q_{\text{R}j} \tag{13.288b}$$

where $\mathbf{U}_{\text{L,R}}$ are arbitrary $N \times N$ unitary matrices, and i, j label the flavours. Besides the $\text{SU}(N)_{\text{L}} \times \text{SU}(N)_{\text{R}}$ symmetry generated by the matrices $\mathbf{U}_{\text{L,R}}$, having unit determinant, is the $\text{U}(1)_{\text{L}} \times \text{U}(1)_{\text{R}}$ symmetry generated by a simultaneous phase change of all (chiral) fields

$$\text{U}(1)_{\text{L}}: q_{\text{L}i} \to \text{e}^{-\text{i}\theta_{\text{L}}} q_{\text{L}i} \tag{13.289a}$$

$$\text{U}(1)_{\text{R}}: q_{\text{R}i} \to \text{e}^{-\text{i}\theta_{\text{R}}} q_{\text{R}i} \tag{13.289b}$$

where

$$\theta_{\text{L}} = \Lambda - \alpha \tag{13.289c}$$

$$\theta_{\text{R}} = \Lambda + \alpha \tag{13.289d}$$

give the chiral phases in terms of the $\text{U}(1)_{\text{V}}$ and $\text{U}(1)_{\text{A}}$ (vector and axial U(1)s):

$$\text{U}(1)_{\text{V}}: q_i \to \text{e}^{-\text{i}\Lambda} q_i \tag{13.290a}$$

$$\text{U}(1)_{\text{A}}: q_i \to \text{e}^{-\text{i}\alpha\gamma_5} q_i . \tag{13.290b}$$

All of these symmetries are symmetries of the classical (massless) QCD Lagrangian. Of course, since the quarks are *not* massless we do not expect to see an exact version of the symmetry in the hadron spectrum. What we do see is an approximate $\text{SU}(N)_{\text{L}} \times \text{SU}(N)_{\text{R}}$ symmetry, and an exact $\text{U}(1)_{\text{V}}$ symmetry, corresponding to conservation of baryon number. However there is no version of the $\text{U}(1)_{\text{A}}$ symmetry; for example, the (light) pions are regarded as the Goldstone bosons associated with the spontaneous breaking of the $\text{SU}(2)_{\text{A}}$ generated by the light u, d quarks, but there is no *light* η meson associated with the spontaneous breakdown of the $\text{U}(1)_{\text{A}}$. The observed η is just too heavy.

From a theoretical perspective, what distinguishes the $\text{U}(1)_{\text{A}}$ symmetry is that, even in the massless case, it is broken in the full *quantum* theory. The way in which this occurs is analysed in §15.5, but it is this non-conservation which relates it to the θ vacua. We can express the non-conservation as a divergence of the axial current

$$j^5_\mu \equiv \sum_{i=1}^{N} \bar{q}_i \gamma_\mu \gamma_5 q_i \tag{13.291}$$

with the i labelling quark flavours, as above. Then for massless quarks

$$\partial_\mu j^{\mu 5} = \frac{Ng^2}{16\pi^2} F^a_{\mu\nu}\tilde{F}^{a\mu\nu} \tag{13.292}$$

is the QCD analogue of the U(1) result given in (15.175). We see that the non-invariance of the (quantum) QCD Lagrangian under a $\mathrm{U}(1)_\mathrm{A}$ transformation of the quark fields is controlled by a term proportional to that which generates the θ vacua. The change in the Lagrangian caused by the transformation (13.290*b*) is given, from (15.174), by

$$\delta\mathscr{L} = \alpha \frac{Ng^2}{16\pi^2} F^a_{\mu\nu}\tilde{F}^{a\mu\nu}. \tag{13.293}$$

Comparing this with the change (13.287) caused by the θ vacua, we see that by choosing

$$\alpha = -\theta/2\mathrm{N} \tag{13.294}$$

the effect of the θ vacua can be removed.

However, in reality, the quarks are not massless so the $\mathrm{U}(1)_\mathrm{A}$ transformation (15.290*b*) induces more than just the change (13.293) in the Lagrangian. Consider first the case of a single quark flavour ($N=1$). Then the chiral transformation changes the mass term.

$$m\bar{q}q \to m\bar{q}\,\mathrm{e}^{-2\mathrm{i}\alpha\gamma_5}q = m\cos 2\alpha\bar{q}q - m\sin 2\alpha\bar{q}\mathrm{i}\gamma_5 q \tag{13.295}$$

and the second term violates time-reversal invariance, just as the original θ term did. Thus for massive quarks the T non-invariance is a real effect. A general mass term for the quarks can be written as

$$\mathscr{L}_\mathrm{m} = -\bar{q}_{\mathrm{L}i} M_{ij} q_{\mathrm{R}j} + \mathrm{HC} \tag{13.296}$$

where M is an $N \times N$ (mass) matrix. The effect of a $\mathrm{U}(1)_\mathrm{A}$ transformation is that

$$M \to \mathrm{e}^{-2\mathrm{i}\alpha} M \tag{13.297}$$

so

$$\arg\det M \to \arg\det M - 2\alpha N. \tag{13.298}$$

We see that the combination

$$\bar{\theta} \equiv \theta + \arg\det M \tag{13.299}$$

is invariant under $U(1)_A$, and can in principle generate observable T-violating effects. The most severe constraint on $\bar{\theta}$ is obtained by comparing the theoretically predicted value of the electric dipole moment of the neutron with that measured experimentally.

As it stands in (13.287) it is not so obvious how the θ term contributes to the neutron's electric dipole moment. The result above, however, shows that, by a chiral transformation, we can move the whole effect into the quark mass matrix M. So suppose that the flavour q undergoes a chiral transformation

$$q \to e^{-i\alpha_q \gamma_5} q \tag{13.300}$$

with the effect that, as in (13.295), the mass Lagrangian $\mathscr{L}_m$ is transformed as

$$-\mathscr{L}_m = \sum_q m_q \bar{q} q \to \sum_q m_q \cos 2\alpha_q \bar{q} q - i \sum_q m_q \sin 2\alpha_q \bar{q} \gamma_5 q. \tag{13.301}$$

We need to specify the phases α_q so that the θ term is removed. To do this we note first that, since the θ term is a flavour group singlet, then so too must be the CP-violating part of (13.301). This requires that for all flavours

$$m_q \sin 2\alpha_q = x \tag{13.302}$$

with x a constant independent of q. Also the required $U(1)_A$ transformation has

$$\sum_q 2\alpha_q = -\theta. \tag{13.303}$$

It is simple to solve these in the case that α_q and θ are infinitesimal:

$$x = -\theta \left(\sum_q m_q^{-1} \right)^{-1} \tag{13.304}$$

so that the CP-violating part of $\mathscr{L}_m$ is

$$\mathscr{L}_\theta \equiv -i\theta \left(\sum_q m_q^{-1} \right)^{-1} \sum_q \bar{q} \gamma_5 q. \tag{13.305}$$

In the case that there are just two light flavours u, d we get

$$\mathscr{L}_\theta = -i\theta \frac{m_u m_d}{m_u + m_d} (\bar{u} \gamma_5 u + \bar{d} \gamma_5 d). \tag{13.306}$$

We leave it as an exercise (exercise 13.13) to solve for x in the general case that α_q, θ are finite.

It is clear now that $\mathscr{L}_\theta$ will generate a CP-violating pion–nucleon interaction described by the effective Lagrangian

$$\mathscr{L}_{\pi NN} = g^{\theta}_{\pi NN} \bar{N} \tau^a N \pi^a \tag{13.307}$$

where τ^a $(a = 1, 2, 3)$ are the SU(2) isospin flavour group Pauli matrices and the coupling constant is estimated as[16]

$$g^{\theta}_{\pi NN} = -\theta \frac{m_u m_d}{m_u + m_d} \frac{1}{f_\pi} \frac{m_\Xi - m_N}{2m_s - m_u - m_d}. \tag{13.308}$$

This CP-violating vertex contributes to the neutron's electric dipole moment via the diagrams shown in figure 13.3.

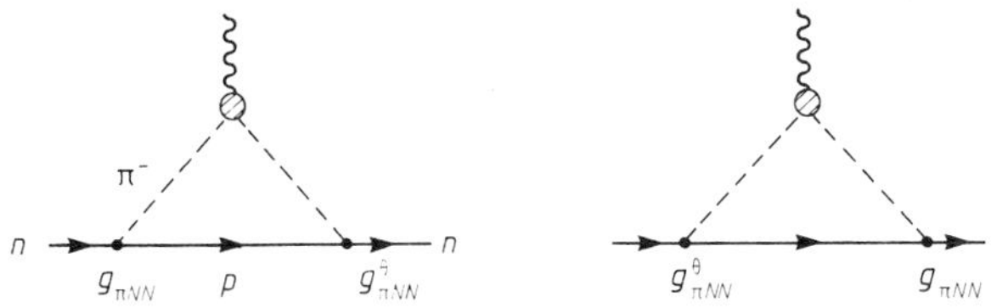

Figure 13.3 Contributions to neutron's dipole moment from CP-violating pion–nucleon interaction.

The estimate for the dipole moment is then

$$d_n = 5.2 \times 10^{-16} \theta \text{ e cm} \tag{13.309}$$

whereas the experimental data[17] give

$$d_n < 12 \times 10^{-26} \text{ e cm} \tag{13.310}$$

which shows that

$$|\bar{\theta}| < 2 \times 10^{-10}. \tag{13.311}$$

It is clear from our derivation that θ is the vacuum angle in the basis where all quark masses are real, positive and γ_5-free, which is why we have replaced θ by $\bar{\theta}$, the $\mathrm{U(1)_A}$ invariant quantity.

We need dwell no longer on the technicalities of this estimate, nor on whether other estimates are superior. The important point is that $\bar{\theta}$ is extremely small. We have already noted that it is not predicted by the theory, so it is only aesthetic considerations which demand that we explain why $\bar{\theta}$ is so small.

The clue to understanding why, is to note that if the lightest of the quarks (the u quark) happened to be massless then the effect would disappear, as is apparent in (13.306). This was already clear in (13.299) since when det M is zero its argument is indeterminate, and can be chosen to cancel arbitrary θ. However, the masses in (13.306) are the current quark masses, and the success of chiral perturbation theory in explaining so much low energy phenomenology points irrefutably to the fact that m_u is non-zero[18], in fact about 5 MeV. The solution of the 'strong CP problem' proposed by Peccei and Quinn [19] is to extend the standard model so that there is a new anomalous axial symmetry even with non-zero masses; then the new symmetry can be used to transform $\bar{\theta}$ to zero, just as described above. We shall see in the next chapter that the quark mass terms in the standard model arise from the coupling of the quark fields to scalar fields in a gauge invariant way; the spontaneous breaking of the symmetry breaks it to $SU(3)_c \times U(1)_{em}$ and generates non-zero masses for the gauge bosons and matter fermions. For our purposes all we need is the U(1) symmetry already discussed.

The left chiral components of the quark fields belong to SU(2) doublets $Q_{f\mathrm{L}} = (u_{f\mathrm{L}}, d_{f\mathrm{L}})$ where $f = 1, 2, 3$ labels the three generations. The right chiral components are SU(2) singlets denoted $U_{f\mathrm{R}}$, $D_{f\mathrm{R}}$. The scalar fields also form a doublet denoted φ. The mass terms then arise from Yukawa coupling

$$-\mathscr{L}_{\mathrm{Y}} = \bar{Q}_{f\mathrm{L}} X_{fg} \varphi D_{g\mathrm{R}} + \bar{Q}_{f\mathrm{L}} Y_{fg} \psi U_{g\mathrm{R}} + \mathrm{HC} \tag{13.312}$$

(as in (14.145)) where X, Y are matrices, and

$$\psi = \mathrm{i}\tau^2 \varphi^*. \tag{13.313}$$

The first term generates masses for the down-like quarks when the scalar doublet develops a non-zero VEV

$$\langle 0|\varphi|0\rangle = \frac{1}{\sqrt{2}}\begin{pmatrix} 0 \\ v_1\,\mathrm{e}^{\mathrm{i}\delta_1} \end{pmatrix}. \tag{13.314}$$

Similarly the second term generates masses for the up-like quarks because

$$\langle 0|\psi|0\rangle = \frac{1}{\sqrt{2}}\begin{pmatrix} v_2\,\mathrm{e}^{\mathrm{i}\delta_2} \\ 0 \end{pmatrix} \tag{13.315}$$

with

$$v_1 = v_2 \qquad \delta_1 = -\delta_2 \tag{13.316}$$

by virtue of (13.313). It follows that the determinant of the mass matrix M, appearing in (13.296), is given by

$$\det M = \det\left(\frac{1}{\sqrt{2}} v_1\, e^{i\delta_1} X\right)\det\left(\frac{1}{\sqrt{2}} v_2\, e^{i\delta_2} Y\right)$$
$$= e^{3i(\delta_1+\delta_2)}\, \tfrac{1}{8}(v_1 v_2)^3 \det(XY) \qquad (13.317)$$

so

$$\arg\det M = 3(\delta_1+\delta_2) + \arg\det(XY). \qquad (13.318)$$

In the standard model (13.316) holds, so arg det M is given entirely by the matrices X, Y.

The solution of the strong CP problem which is proposed by Peccei and Quinn is to drop the requirement (13.313); in other words we make a (minimal) extension of the standard electroweak theory and introduce *independent* scalar doublets φ, ψ. Then the VEVs are no longer related by (13.316). *Provided* we are free to adjust the phases δ_1 and δ_2 so that

$$3(\delta_1+\delta_2) = -\arg\det(XY) - \theta \qquad (13.319)$$

then

$$\bar{\theta} \equiv \theta + \arg\det M = 0 \qquad (13.320)$$

and the strong CP problem is solved. Of course we have to ensure that the required rephasing *is* a $U(1)_{PQ}$ symmetry of the theory. Since φ, ψ are independent, suppose that

$$\varphi \to e^{i\alpha\Gamma_1}\varphi \qquad \psi \to e^{i\alpha\Gamma_2}\psi \qquad (13.321)$$

where Γ_1, Γ_2 are the Peccei–Quinn (PQ) charges of φ, ψ. Under the same transformation, suppose

$$Q_L \to e^{i\alpha\Gamma_Q} Q_L \qquad U_R \to e^{i\alpha\Gamma_u} U_R \qquad D_R \to e^{i\alpha\Gamma_d} D_R. \qquad (13.322)$$

Then (13.312) is invariant under the $U(1)_{PQ}$, provided

$$\Gamma_1 + \Gamma_d = \Gamma_Q \qquad \Gamma_2 + \Gamma_u = \Gamma_Q. \qquad (13.323)$$

Thus provided

$$\Gamma_1 + \Gamma_2 = 2\Gamma_Q - \Gamma_u - \Gamma_d \neq 0 \qquad (13.324)$$

is non-zero, then by a suitable choice of the angle α we can adjust $\delta_1 + \delta_2$ so as to satisfy (13.319). For example, we may choose

$$\Gamma_1 = \Gamma_2 = -\Gamma_u = -\Gamma_d = 1. \tag{13.325}$$

(We may not choose the Peccei–Quinn charge to be the weak hypercharge Y, since necessarily φ, ψ have opposite values of Y.) We still need to ensure that the rest of the Lagrangian is $U(1)_{PQ}$ invariant. It is easy to choose the PQ charges of the leptons so that this is so. For the scalar potential, the invariance limits the terms allowed in $V(\varphi, \psi)$. For example, a term proportional to $\varphi i\tau^2\psi$ conserves Y, but not the Peccei–Quinn charge.

The (extra) $U(1)_{PQ}$ invariance is spontaneously broken when φ, ψ acquire their VEVs (13.314), (13.315). Since $U(1)_{PQ}$ is not a gauge symmetry, the spontaneous symmetry breaking yields a massless Goldstone boson, called the axion and denoted a. As in §13.4 the Goldstone boson is associated with the phase angle of the field acquiring the VEV. We write the neutral components φ^0, ψ^0 of φ and ψ as

$$\varphi^0 = \frac{1}{\sqrt{2}}(v_1 + \rho_1(x))\, e^{i\theta_1(x)/v_1} \tag{13.326a}$$

$$\psi^0 = \frac{1}{\sqrt{2}}(v_2 + \rho_2(x))\, e^{i\theta_2(x)/v_2}. \tag{13.326b}$$

Then the axion field is associated with the phase of $\varphi^0\psi^0$, so

$$\begin{aligned} a(x) &= k[\theta_1(x)/v_1 + \theta_2(x)/v_2] \\ &= [v_2\theta_1(x) + v_1\theta_2(x)]/v \end{aligned} \tag{13.327a}$$

where

$$v \equiv (v_1^2 + v_2^2)^{1/2}. \tag{13.327b}$$

The orthogonal combination

$$\chi(x) = [-v_1\theta_1(x) + v_2\theta_2(x)]/v \tag{13.328}$$

is 'eaten' during the spontaneous electroweak symmetry breakdown and generates a non-zero Z-boson mass, as we shall see in chapter 14. It is now straightforward to determine the interactions of the axion with the matter fields. We expand the neutral fields φ^0, ψ^0 as

$$\sqrt{2}\varphi^0 = v_1 + \rho_1(x) + iv^{-1}[v_2 a(x) - v_1\chi(x)] + \cdots \tag{13.329a}$$

$$\sqrt{2}\psi^0 = v_2 + \rho_2(x) + iv^{-1}[v_1 a(x) + v_2\chi(x)] + \cdots \tag{13.329b}$$

and then substituting into (13.312) gives the quark–axion Yukawa coupling

$$-\mathscr{L}_{\mathrm{Y}}^{a-q} = \frac{ia(x)}{v}\left[\frac{v_2}{v_1} m_d \bar{d}\gamma_5 d + \frac{v_1}{v_2} m_u \bar{u}\gamma_5 u\right]. \tag{13.330}$$

We shall see in the next chapter that the quantity v is determined by the weak interaction data to have the value

$$v \simeq 246 \text{ GeV}. \tag{13.331}$$

In fact the axion is *not* massless; it acquires a mass by virtue of (non-perturbative) instanton effects. This can be understood qualitatively as follows. The introduction of the axion field $a(x)$ amounts to making the $\bar{\theta}$ parameter a dynamical field. Then the consequent $\bar{\theta} = 0$ corresponds to a minimum of the axion's scalar potential $V(a)$ at a value $a = 0$. However this potential is non-trivial, in fact[20]

$$V(a) \propto 1 - \cos a \tag{13.332}$$

which reflects the required periodicity property of (13.284) that the physics is invariant under

$$\bar{\theta} \to \bar{\theta} + 2\pi. \tag{13.333}$$

The mass of the axion has been calculated[21] as

$$m_a = \left(\frac{v_1}{v_2} + \frac{v_2}{v_1}\right) 74 \text{ keV}. \tag{13.334}$$

However, it seems that such a particle does *not* exist. Despite extensive searches in kaon and ψ decays, in reactor and beam damp experiments, as well as in astrophysics, no axion has been found[22].

Nevertheless, this is not the end of the story. The scale v which controls the strength of the axion coupling in (13.330) is the electroweak scale (13.331). In a grand unified theory, such as will be discussed in chapter 16, the properties of the axion may be significantly altered[23]. Suppose that in such a theory there is a complex scalar field Σ which is a singlet under the $\mathrm{SU(3)} \times \mathrm{SU(2)} \times \mathrm{U(1)}$ gauge group. Then the scalar potential may include a term

$$V^{\Sigma} = \lambda\varphi i\tau^2\psi\Sigma + \text{HC}. \tag{13.335}$$

This is also $U(1)_{PQ}$ invariant provided Σ has PQ charge

$$\Gamma_\Sigma = -\Gamma_1 - \Gamma_2. \tag{13.336}$$

Now suppose that Σ acquires a VEV

$$\langle 0|\Sigma|0\rangle = \frac{1}{\sqrt{2}} V\,\mathrm{e}^{\mathrm{i}\delta}. \tag{13.337}$$

Then writing

$$\Sigma(x) = \frac{1}{\sqrt{2}}(V + R(x))\,\mathrm{e}^{\mathrm{i}\theta(x)/V} \tag{13.338}$$

we can see that V^Σ generates a mass term for a linear combination of the fields $\theta_1, \theta_2, \theta$

$$V^\Sigma \rightarrow \frac{\lambda}{\sqrt{8}} v_1 v_2 V\left(\frac{\theta_1}{v_1} + \frac{\theta_2}{v_2} + \frac{\theta}{V}\right)^2. \tag{13.339}$$

In *this* case the axion field $\hat{a}(x)$ must be orthogonal to this massive combination, as well as to the combination $\chi(x)$ given in (13.328), which is eaten by the Z. Thus

$$\hat{a}(x) \propto \theta(x) - \frac{v_1 v_2}{vV} a(x) \tag{13.340}$$

where $a(x)$ is the (original axion) field (13.327). The relevance to grand unified theories is that in such theories the VEV V of Σ is of order 10^{15} GeV, so that

$$V \gg v_1, v_2. \tag{13.341}$$

Then the axion field $\hat{a}$ is essentially aligned with $\theta(x)$, and its coupling to the quark fields is reduced by a factor $v_1 v_2/vV$. It is therefore essentially decoupled from the light states, and has consequently been called the 'invisible axion'. Nevertheless it is essential that the axion can decay rapidly enough to avoid dominating the energy density of the universe. This astrophysical bound requires[24]

$$V < 0(10^{11}\ \text{GeV}). \tag{13.342}$$

Although such a VEV for Σ can be arranged it does require fine tuning of the parameters of the effective potential. However this is a story which must await another book.

Problems

13.1 Prove (13.10).

13.2 Show that the $N-M$ vectors $\mathbf{T}^a v$ $(a=M+1,\ldots,N)$ (defined in (13.58), are linearly independent.

13.3 Verify (13.79).

13.4 Verify (13.98).

13.5 Show that the mass term (13.123) is not invariant under the chiral transformation (13.116).

13.6 Verify that gauge invariance of the Yukawa interaction (13.125) leads to (13.126).

13.7 Verify (13.146).

13.8 Solve (13.158) in the Prasad–Sommerfield limit that $\lambda \to 0$, and show that the energy of the resulting field configuration is $4\pi v/g$.

13.9 Show that the right-hand side of (13.175) is indeed invariant under an infinitesimal change of $\boldsymbol{\varphi}$.

13.10 Find the effective potential (to one-loop order) when the renormalisation conditions (13.221) are imposed.

13.11 Show that q, defined in (13.272), is invariant under infinitesimal deformations $\delta\mathbf{U}$ which may (always) be written as

$$\delta\mathbf{U}(x)=\mathbf{U}(x)\delta\Lambda^a(x)\mathbf{T}^a$$

where $\mathbf{T}^a$ are the matrix generators.

13.12 Verify that the configuration (13.274) has $q=1$, as claimed.

13.13 Show that for two flavours and finite θ the prefactor in equation (13.306) becomes

$$\frac{m_u m_d \sin\theta}{(2m_u m_d \cos\theta + m_u^2 + m_d^2)^{1/2}}.$$

References

The following books and review articles have been most useful to us in preparing this chapter:

Aitchison I J R and Hey A J G 1982 *Gauge Theories in Particle Physics* (Bristol: Adam Hilger) Ch. 9

Coleman S 1977 *The Uses of Instantons* (reprinted 1985 *Aspects of Symmetry*) (Cambridge: Cambridge University Press)
Cheng H-Y 1988 *Phys. Rep.* **158** 1
Goddard P and Olive D I 1978 *Rep. Prog. Phys.* **41** 1357
Iliopoulos J, Itzykson C and Martin A 1975 *Rev. Mod. Phys.* **47** 165 Section III
Kim J E 1987 *Phys. Rep.* **150** 1

References in the text

1 Coleman S 1975 in *Laws of Hadronic Matter* ed A Zichichi (London: Academic Press) p. 141
Aitchison I J R and Hey A J G 1982 *Gauge Theories in Particle Physics* (Bristol: Hilger) p. 192
2 Nambu Y 1960 *Phys. Rev. Lett.* **4** 380
Nambu Y and Jona-Lasinio G 1961 *Phys. Rev.* **122** 345
3 Goldstone J 1961 *Nuov. Cim.* **19** 154
Goldstone J, Salam A and Weinberg S 1962 *Phys. Rev.* **127** 965
4 Higgs P W 1964a *Phys. Lett.* **12** 132
—— 1964b *Phys. Rev. Lett.* **13** 508
—— 1966 *Phys. Rev.* **145** 1156
Kibble T W B 1967 *Phys. Rev.* **155** 1554
Englert F and Brout R 1964 *Phys. Rev. Lett.* **13** 321
Guralnik G S, Hagen C R and Kibble T W B 1964 *Phys. Rev. Lett.* **13** 585
5 't Hooft G 1971a *Nucl. Phys.* B **33** 173
—— 1971b *Nucl. Phys.* B **35** 167
6 Abers E S and Lee B W 1973 *Phys. Rep.* **9C** 1
7 See e.g. Coleman S 1975 in *New Phenomena in Subnuclear Physics* ed A Zichichi (New York: Plenum)
8 Derrick G H 1964 *J. Math. Phys.* **5** 1252
9 't Hooft G 1974 *Nucl. Phys.* B **79** 276
Polyakov A M 1974 *JETP Lett.* **20** 194
—— 1976 *Sov. Phys. JETP* **41** 988
10 Prasad M and Sommerfield C M 1975 *Phys. Rev. Lett.* **35** 760
11 Montonen C and Olive D I 1977 *Phys. Lett.* **72B** 117
12 Coleman S and Weinberg E 1973 *Phys. Rev.* D **7** 1888
13 Belavin A A, Polyakov A M, Schwartz A S and Tyupkin Yu S 1975 *Phys. Lett.* **59B** 85
14 Jackiw R and Rebbi C 1976 *Phys. Rev. Lett.* **37** 172
Callan C G, Dashen R and Gross D J 1976 *Phys. Lett.* **63B** 334
15 't Hooft G 1976 *Phys. Rev. Lett.* **37** 8
—— 1976 *Phys. Rev.* D **14** 3432
16 Crewther R J, Di Vecchia P, Veneziano G and Witten E 1979 *Phys. Lett.* **88B** 123
—— 1980 *Phys. Lett.* **91B** 487(E)
17 Smith K F *et al* 1990 *Phys. Lett.* **234B** 191
18 Weinberg S 1977 *Festschrift for I.I. Rabi* ed L Motz (New York: Academy of Science)
Gasser J and Leutwyler H 1982 *Phys. Rep.* **87** 77

19 Peccei R D and Quinn H R 1977 *Phys. Rev. Lett.* **38** 1440
—— 1977 *Phys. Rev.* D **16** 1791
20 Gross D J, Pisarski R and Yaffe L 1981 *Rev. Mod. Phys.* **53** 43
21 Bardeen W A and Tye S H H 1978 *Phys. Lett.* **74B** 229
22 See e.g. Zehnder A 1985 *Fundamental Interactions in Low Energy Systems* ed P Dalpiaz *et al* (New York: Plenum) p 337
Particle Data Group 1990 *Phys. Lett.* **239B** 1
23 Kim J E 1979 *Phys. Rev. Lett.* **43** 103
Zhinitsky A R 1980 *Sov. J. Nucl. Phys.* **31** 260
Dine M, Fischler W and Srednicki M 1981 *Phys. Lett.* **104B** 199
Wise M B, Georgi H and Glashow S L 1981 *Phys. Rev. Lett.* **47** 402
24 For a review of the astrophysical and cosmological bounds see Cheng H-Y 1988 *Phys. Rep.* **158** 1

14

FEYNMAN RULES FOR ELECTROWEAK THEORY

14.1 SU(2) × U(1) invariance and electroweak interactions

The spectacular success of quantum electrodynamics (QED) in calculating the Lamb shift and the anomalous magnetic moment of the electron and muon, for example, stimulated many attempts to develop a quantum field theory of weak interactions. After all, low energy weak processes are characterised by the Fermi weak coupling constant G_F, which satisfies

$$G_F m_p^2 \simeq 10^{-5} \tag{14.1}$$

so we might reasonably hope that perturbation theory with this parameter would be at least as good as perturbative QED, in which the expansion parameter is

$$\alpha \equiv \frac{e^2}{4\pi} \simeq \frac{1}{137}. \tag{14.2}$$

Indeed, there are similarities between weak and electromagnetic interactions which suggest how one might proceed to construct a field theory of the weak interactions similar to QED. The principal similarity is that between the weak currents and the electromagnetic current. For the present we consider only the leptons (ν_e, e, ν_μ, μ, ν_τ, τ). It has been known for many years that the leptons enter the (effective) weak interaction in the combination known as the leptonic current L_μ,

$$L_\mu(x) = \sum_{l=e,\mu,\tau} \bar{l}(x)\gamma_\mu(1-\gamma_5)\nu_l(x) \tag{14.3a}$$

and its hermitian conjugate

$$L_\mu^\dagger(x) = \sum_{l=e,\mu,\tau} \bar{\nu}_l(x)\gamma_\mu(1-\gamma_5)l(x) \tag{14.3b}$$

where $l(x)$, $\nu_l(x)$ are the fields of l and ν_l. It is beyond the scope of this book to give any details of how this was derived[1] from the phenomenology of weak processes. (The interested reader is referred to Bailin[2], and references therein, for a detailed account.) The currents (14.3) differ from the electromagnetic current (3.112) in that they are 'charged' currents; the neutrinos ν_l and antineutrinos $\bar{\nu}_l$ are, of course, neutral, while the leptons l have charge -1 (in units of the proton charge) and their antiparticles $\bar{l}$ have charge $+1$. Thus L_μ has charge 1 and $L_\mu^\dagger$ charge -1, whereas j_μ in (3.112) has zero charge. It follows that any field theory of the weak interactions similar to QED will couple these charged currents to *charged* vector fields $W_\mu^\pm$, so as to conserve charge. In principle it is possible to proceed to construct a renormalisable theory of the weak interactions *alone*. However this is not sensible. Although we could calculate weak radiative corrections to arbitrary accuracy with such a theory, they would be small, at least at low energies, compared with the electromagnetic radiative corrections. So long as the only charged particles are leptons we know that these latter corrections may be calculated using QED. The trouble is that we now have in addition the charged vector fields $W_\mu^\pm$ which must also interact with the electromagnetic field. It is known that these fields are *not* massless, and the result is that there are extra divergences, generated by their electromagnetic interactions, which make the theory unrenormalisable. Thus we would be unable to calculate the electromagnetic corrections to our renormalisable weak theory. For this reason the only sensible course is to construct a *unified* renormalisable theory of weak *and* electromagnetic corrections, as we shall now proceed to do.

The weak currents involve *only* the left chiral components of the fields, whereas the electromagnetic current involves both components (since both spin states of the electron, say, have equal charge). We can make this explicit using the left and right chiral projection operators a_L and a_R

$$a_\mathrm{L} \equiv \tfrac{1}{2}(1-\gamma_5) \tag{14.4a}$$

$$a_\mathrm{R} \equiv \tfrac{1}{2}(1+\gamma_5). \tag{14.4b}$$

For *any* field ψ we define

$$\psi_\mathrm{L} \equiv a_\mathrm{L}\psi \qquad \psi_\mathrm{R} \equiv a_\mathrm{R}\psi. \tag{14.5}$$

Then

$$\bar{\psi}_\mathrm{L} = \psi_\mathrm{L}^\dagger\gamma_0 = \psi^\dagger a_\mathrm{L}\gamma_0 = \bar{\psi}a_\mathrm{R} \tag{14.6}$$

using (3.75), (3.80) and (3.92). Similarly

$$\bar{\psi}_\mathrm{R} = \bar{\psi}a_\mathrm{L}. \tag{14.7}$$

It follows that

$$\tfrac{1}{2}L_\mu = \sum_l \bar{l}_\mathrm{L}\gamma_\mu \nu_{l\mathrm{L}} \tag{14.8a}$$

$$\tfrac{1}{2}L_\mu^+ = \sum \bar{\nu}_{l\mathrm{L}}\gamma_\mu l_\mathrm{L}. \tag{14.8b}$$

The resemblance of these currents to the isospin raising and lowering operators suggests that we define a weak isospin, i.e. SU(2), group with the field components $\nu_{l\mathrm{L}}$ and l_{L} constituting a two-dimensional representation. Thus we define

$$E_{l\mathrm{L}} \equiv \begin{pmatrix} \nu_l \\ l \end{pmatrix}_{\mathrm{L}} \qquad (l = \mathrm{e}, \mu, \tau) \tag{14.9}$$

and

$$T^i_\mu \equiv \sum_l \bar{E}_{l\mathrm{L}} \gamma_\mu \tfrac{1}{2} \tau_i E_{l\mathrm{L}} \qquad (i = 1, 2, 3) \tag{14.10}$$

where τ_i $(i = 1, 2, 3)$ are the Pauli matrices, so that

$$\tfrac{1}{2} L_\mu = T^1_\mu - \mathrm{i} T^2_\mu \tag{14.11a}$$

and

$$\tfrac{1}{2} L^\dagger_\mu = T^1_\mu + \mathrm{i} T^2_\mu. \tag{14.11b}$$

Thus the SU(2) group generates the weak currents (14.3) but not, of course, the electromagnetic current

$$j_\mu = -\sum_l \bar{l} \gamma_\mu l \tag{14.12}$$

since j_μ involves both left and right chiral components. Thus any group which is to include the weak *and* electromagnetic currents must contain at least one generator in addition to those given in (14.10). The simplest possibility[3] (which, amazingly, turns out to be correct) is therefore to enlarge SU(2) to SU(2) × U(1) where the new U(1) is similar, but *not* identical, to the U(1) already encountered in QED. The current Y_μ associated with it must be invariant under SU(2); that is what is meant by SU(2) × U(1). Thus *a priori* it can be any linear combination

$$Y_\mu = \sum_l (x_l \bar{E}_{l\mathrm{L}} \gamma_\mu \mathbf{I}_2 E_{l\mathrm{L}} + y_l \bar{l}_{\mathrm{R}} \gamma_\mu l_{\mathrm{R}}) \tag{14.13}$$

of SU(2) invariants. The first terms are SU(2) invariant because of the unit 2 × 2 matrix $\mathbf{I}_2$, and the second terms involve only the right chiral components which, by assumption, are SU(2) invariant. (We adopt the view that the components $\nu_{l\mathrm{R}}$ $(l = \mathrm{e}, \mu, \tau)$ do not exist—otherwise they too could contribute to Y_μ.) Using the notation of (14.6) and (14.7), we now decompose the electromagnetic current (14.12) into isovector and isoscalar pieces

$$j_\mu = \sum_l [\tfrac{1}{2}(\bar{\nu}_{l\mathrm{L}} \gamma_\mu \nu_{l\mathrm{L}} - \bar{l}_{\mathrm{L}} \gamma_\mu l_{\mathrm{L}})$$

$$- \tfrac{1}{2}(\bar{\nu}_{l\mathrm{L}} \gamma_\mu \nu_{l\mathrm{L}} + \bar{l}_{\mathrm{L}} \gamma_\mu l_{\mathrm{L}}) - \bar{l}_{\mathrm{R}} \gamma_\mu \gamma_{\mathrm{R}}] \tag{14.14}$$

$$= T^3_\mu + Y_\mu \tag{14.15}$$

where T^3_μ is given in (14.10) and Y_μ in (14.13) with the values

$$x_l = -\tfrac{1}{2} \qquad y_l = -1. \tag{14.16}$$

(This is where any possible contribution from $\nu_{l\mathrm{R}}$ would have disappeared had we included it.) The notation (14.15) is reminiscent of the Gell–Mann–Nishijima formula, although we have dropped the conventional '$\tfrac{1}{2}$' in front of Y_μ. Thus we say that $\nu_{l\mathrm{L}}$, l_L have 'weak isospin' $\tfrac{1}{2}$ with T_3 respectively $+\tfrac{1}{2}$, $-\tfrac{1}{2}$, and 'weak hypercharge' $-\tfrac{1}{2}$, while l_R are weak isoscalars with weak hypercharge -1.

The next step is straightforward. We have to write down a Lagrangian which is locally SU(2) × U(1) gauge invariant, just as QED is locally U(1) gauge invariant. We have already seen how do to this for a general non-Abelian gauge group in §9.2. For the U(1) gauge group we use §9.1 with the 'charges' of the fermions now being the weak hypercharge. The group SU(2) has three generators and therefore there are three gauge bosons W^a_μ $(a = 1, 2, 3)$ associated with it. The gauge boson of the U(1) group is denoted B_μ to avoid confusion with the electromagnetic field A_μ, which will be identified later. Thus the SU(2) × U(1) gauge invariant Lagrangian containing the lepton fields with the weak isospin and hypercharges already assigned is

$$\begin{aligned}\mathscr{L} = \sum_l \,&(\bar{E}_{l\mathrm{L}} \mathrm{i}\gamma^\mu \mathrm{D}_\mu E_{l\mathrm{L}} + \bar{l}_\mathrm{R} \mathrm{i}\gamma^\mu \mathrm{D}_\mu l_\mathrm{R}) \\ &-\tfrac{1}{4} W^a_{\mu\nu} W^{a\mu\nu} - \tfrac{1}{4} B_{\mu\nu} B^{\mu\nu}\end{aligned} \tag{14.17a}$$

where we have used (9.11), (9.36) and

$$\mathrm{D}_\mu E_{l\mathrm{L}} = (\partial_\mu + \mathrm{i}g\tfrac{1}{2}\tau^a W^a_\mu - \mathrm{i}g'\tfrac{1}{2}B_\mu) E_{l\mathrm{L}} \tag{14.17b}$$

$$\mathrm{D}_\mu l_\mathrm{R} = (\partial_\mu - \mathrm{i}g' B_\mu) l_\mathrm{R} \tag{14.17c}$$

$$W^a_{\mu\nu} = \partial_\mu W^a_\nu - \partial_\nu W^a_\mu - g\varepsilon^{abc} W^b_\mu W^c_\nu \tag{14.17d}$$

$$B_{\mu\nu} = \partial_\mu B_\nu - \partial_\nu B_\mu. \tag{14.17e}$$

The quantities g and g' are the coupling constants associated with the SU(2) and U(1) gauge groups respectively, and in (14.17d) we have used the fact that the structure constants of SU(2) are ε^{abc}, the totally antisymmetric rank 3 tensor in three dimensions with

$$\varepsilon^{123} = +1. \tag{14.18}$$

The most obvious difference between (14.17) and the examples given in Chapter 9 is the absence of mass terms for the fermion fields. It is easy to see that the addition of any such mass terms would violate the gauge invariance. For example

$$m_l \bar{l} l = m_l (\bar{l}_\mathrm{L} l_\mathrm{R} + \bar{l}_\mathrm{R} l_\mathrm{L}) \tag{14.19}$$

violates the invariance because l_R is an isoscalar while l_L has weak isospin $\tfrac{1}{2}$.

(The reason for the appearance of such mass terms in Chapter 9 was the tacit assumption, in (9.16), that both chiral components behave in the same way under an infinitesimal gauge transformation.) This absence of mass terms for the neutrino fields ν_l is quite acceptable, since they are thought to be massless. However the charged leptons l are known not to be massless, which means that the $SU(2) \times U(1)$ symmetry must be broken. The absence of mass terms for the gauge bosons also follows from the gauge invariance, as explained in Chapter 9. This is of course entirely acceptable for the electromagnetic field. However *we* have a total for four massless gauge fields, and there is ample evidence that no such massless weak bosons exist. Indeed the very short range of weak interactions (not to mention the recent discovery of massive vector particles) indicates that in this respect also the electroweak symmetry is known to be broken.

It is easy to verify that the interaction of the leptons with the gauge fields contained in (14.17) does indeed couple the observed currents as anticipated, because, using (14.11)

$$g T^a_\mu W^{a\mu} + g' Y_\mu B^\mu = \frac{g}{2\sqrt{2}} (L_\mu W^{-\mu} + L^\dagger_\mu W^{+\mu})$$
$$+ (g \sin\theta_W T^3_\mu + g' \cos\theta_W Y_\mu) A^\mu$$
$$+ (g \cos\theta_W T^3_\mu - g' \sin\theta_W Y_\mu) Z^\mu \qquad (14.20a)$$

where

$$W^\pm_\mu \equiv \frac{1}{\sqrt{2}} (W^1_\mu \; iW^2_\mu) \qquad (14.20b)$$

and we choose θ_W so that the electromagnetic field

$$A_\mu \equiv \cos\theta_W B_\mu + \sin\theta_W W^3_\mu \qquad (14.20c)$$

and the orthogonal combination is

$$Z_\mu \equiv -\sin\theta_W B_\mu + \cos\theta_W W^3_\mu. \qquad (14.20d)$$

The requirement that A^μ is coupled to the electromagnetic current j_μ given in (14.12) with strength e as in (9.1) gives

$$g \sin\theta_W = g' \cos\theta_W = e \qquad (14.21)$$

and then we may rewrite (14.20) as

$$g T^a_\mu W^{a\mu} + g' Y_\mu B^\mu = \frac{g}{2\sqrt{2}} (L_\mu W^{-\mu} + L^\dagger_\mu W^{+\mu})$$
$$+ e j_\mu A^\mu + \frac{g}{\cos\theta_W} (T^3_\mu - \sin^2\theta_W j_\mu) Z^\mu. \qquad (14.22)$$

It is now straightforward to write down the Feynman rules for the lepton–

gauge boson vertices just as we did in Chapter 10. The rules thus obtained are those we shall eventually obtain. However at this stage it is not clear that this is worthwhile, since the SU(2)×U(1) gauge invariant theory to which they apply is manifestly deficient in the respects previously described. We therefore postpone this exercise until we have a phenomenologically acceptable theory.

14.2 Spontaneous breaking of SU(2)×U(1) local gauge invariance

For the reasons discussed in Chapter 13 the only known method of breaking the SU(2)×U(1) local gauge invariance while maintaining renormalisability is to do so 'spontaneously'. This requires the introduction of a scalar field which has a non-zero vacuum expectation value (VEV) and which transforms non-trivially under the action of the group. In the present case the simplest choice, which, again amazingly, turns out to be correct[4], is that the scalar field has weak isospin $\frac{1}{2}$. The group S which must be left invariant is of course the electromagnetic U(1) gauge invariance, which in our present notation is generated by T_3+Y. Thus from (13.58) we deduce that if the field acquiring a VEV has $T_3=-\frac{1}{2}$, (by convention), then the isospin $\frac{1}{2}$ multiplet to which it belongs must have weak hypercharge $+\frac{1}{2}$. We denote the scalar multiplet by

$$\boldsymbol{\varphi}\equiv\begin{pmatrix} G^+ \\ \varphi^0 \end{pmatrix}=\frac{1}{\sqrt{2}}(\boldsymbol{\varphi}_1+\mathrm{i}\boldsymbol{\varphi}_2) \tag{14.23}$$

where $\boldsymbol{\varphi}_1$ and $\boldsymbol{\varphi}_2$ are real two-component column vectors. Under an SU(2)×U(1) transformation parameterised by $\Lambda^a(x)$ $(a=1,2,3)$ and $\Lambda(x)$

$$\boldsymbol{\varphi}(x)\rightarrow \exp[-\mathrm{i}g\tfrac{1}{2}\tau^a\Lambda^a(x)-\mathrm{i}g'\tfrac{1}{2}\mathbf{I}_2\Lambda(x)]\boldsymbol{\varphi}(x). \tag{14.24}$$

We now use the general treatment presented in §13.6. We define a four-component column vector $\boldsymbol{\Phi}$ of *real* scalar fields

$$\boldsymbol{\Phi}\equiv\begin{pmatrix} \boldsymbol{\varphi}_1 \\ \boldsymbol{\varphi}_2 \end{pmatrix}. \tag{14.25}$$

Then under the SU(2)×U(1) transformation (14.24)

$$\boldsymbol{\Phi}(x)\rightarrow \exp[-\mathrm{i}g\mathbf{t}^a\Lambda^a(x)-\mathrm{i}g'\mathbf{Y}\Lambda(x)]\boldsymbol{\Phi}(x)$$

where $\mathbf{t}^a$, $\mathbf{Y}$ are 4×4 matrices generating the SU(2), U(1) groups in the representation (14.25). It is easy to see (problem 14.1) that

$$\mathbf{t}^a=\tfrac{1}{2}\mathrm{i}\begin{pmatrix} \mathrm{Im}\,\tau_a & \mathrm{Re}\,\tau_a \\ -\mathrm{Re}\,\tau_a & \mathrm{Im}\,\tau_a \end{pmatrix} \tag{14.26}$$

$$\mathbf{Y}=\tfrac{1}{2}\mathrm{i}\begin{pmatrix} \mathbf{0} & \mathbf{I}_2 \\ -\mathbf{I}_2 & \mathbf{0} \end{pmatrix}. \tag{14.27}$$

Thus in this case the SU(2) × U(1) covariant derivative is given by

$$D_\mu \mathbf{\Phi} = (\partial_\mu + \mathrm{i}g\mathbf{t}^a W_\mu^a + \mathrm{i}g'\mathbf{Y}B_\mu)\mathbf{\Phi} \tag{14.28}$$

and the SU(2) × U(1) gauge invariant Lagrangian analogous to (13.82) is

$$\mathscr{L}_\varphi = \tfrac{1}{2}(D_\mu \mathbf{\Phi})^{\mathrm{T}}(D^\mu \mathbf{\Phi}) - \tfrac{1}{2}\mu^2 \mathbf{\Phi}^{\mathrm{T}}\mathbf{\Phi} - \tfrac{1}{16}\lambda(\mathbf{\Phi}^{\mathrm{T}}\mathbf{\Phi})^2. \tag{14.29}$$

(We have omitted the pure gauge field contributions, as they have already been written down in (14.17).) As before, spontaneous symmetry breaking occurs when $\mu^2 < 0$. With our choice of basis we require that the field φ^0 in (14.23) acquires a VEV

$$\langle 0|\hat{\varphi}^0(x)|0\rangle = \frac{1}{\sqrt{2}}\, v. \tag{14.30}$$

That is, we assume that the ground state of the theory (when $\mu^2 < 0$) is characterised by a non-zero value v of φ_{12}, so that from (14.25)

$$\langle 0|\mathbf{\Phi}(x)|0\rangle \equiv \boldsymbol{v} = \begin{pmatrix} 0 \\ v \\ 0 \\ 0 \end{pmatrix} \tag{14.31}$$

where, as before,

$$v = (-4\mu^2/\lambda)^{1/2}. \tag{14.32}$$

The invariant subgroup $S = \mathrm{U}(1)_{\mathrm{em}}$ is generated by

$$\mathbf{T}^1 \equiv \sin\theta_{\mathrm{W}}\mathbf{Q} = (\mathbf{t}_3 + \mathbf{Y})\sin\theta_{\mathrm{W}} \tag{14.33}$$

using the notation of §13.6, and we can check from (14.26), (14.27) that

$$\mathbf{T}^1 \boldsymbol{v} = \mathbf{0} \tag{14.34}$$

as required (by (13.58), for example). The constant of proportionality ($\sin\theta_{\mathrm{W}}$) is so that $\mathbf{T}^1$ is coupled with strength $g\sin\theta_{\mathrm{W}} = e$ (from (14.21)). The remaining three generators are, from (14.22),

$$\mathbf{T}^2 = \mathbf{t}^1 \qquad \mathbf{T}^3 = \mathbf{t}^2 \qquad \mathbf{T}^4 = \frac{1}{\cos\theta_{\mathrm{W}}}(\mathbf{t}^3\cos^2\theta_{\mathrm{W}} - \mathbf{Y}\sin^2\theta_{\mathrm{W}}). \tag{14.35}$$

We define new (shifted) fields

$$\tilde{\mathbf{\Phi}} \equiv \mathbf{\Phi} - \boldsymbol{v} \tag{14.36}$$

as in (13.83), which means that the three Goldstone modes $\tilde{\mathbf{\Phi}}^{\mathrm{T}}\mathrm{i}\mathbf{T}^a\boldsymbol{v}$ ($a = 2,3,4$) are just the fields $\varphi_{11}, \varphi_{21}, \varphi_{22}$ which do not acquire VEVs. Thus in (14.23) G^+ and the imaginary piece of φ^0 are the would-be Goldstone modes. They may be removed by a gauge transformation (to the unitary gauge) as in (13.85) and are therefore unphysical fields. The fourth field is associated with the physical Higgs scalar. It is easy to see that in this basis the mass matrix (13.86) is already diagonal.

$$(M_A^2)^{ab} \equiv g^2 \mathbf{v}^{\mathrm{T}} \mathbf{T}^a \mathbf{T}^b \mathbf{v} \qquad (a, b = 1, 2, 3, 4). \tag{13.86}$$

The zero eigenvalue, for the photon mass, is associated with $a = b = 1$, while

$$(M_A^2)^{22} = (M_A^2)^{33} = \tfrac{1}{4} g^2 v^2 \equiv m_{\mathrm{W}}^2 \tag{14.37a}$$

$$(M_A^2)^{44} = \tfrac{1}{4} \sec^2 \theta_{\mathrm{W}} g^2 v^2 \equiv m_{\mathrm{Z}}^2. \tag{14.37b}$$

Thus, as anticipated, the three remaining gauge fields have acquired masses, showing that the SU(2)×U(1) gauge symmetry is indeed broken. The gauge-fixing Lagrangian is chosen precisely as in (13.89), so the Feynman rules for the gauge particle propagators may be read off from (13.94). They are

$$: \quad \mathrm{i}(p^2 + \mathrm{i}\varepsilon)^{-1}[-g_{\mu\nu} + (1 - \xi) p_\mu p_\nu / p^2] \tag{14.38}$$

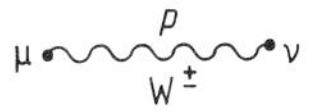

$$: \quad \mathrm{i}\tilde{\Delta}_{\mathrm{F}\mu\nu}(p, m_{\mathrm{W}}) \tag{14.39}$$

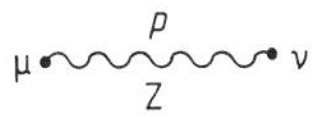

$$: \quad \mathrm{i}\tilde{\Delta}_{\mathrm{F}\mu\nu}(p, m_{\mathrm{Z}}) \tag{14.40}$$

with

$$\tilde{\Delta}_{\mathrm{F}\mu\nu}(p, m) \equiv (p^2 - m^2 + \mathrm{i}\varepsilon)^{-1}[-g_{\mu\nu} + (1 - \xi)(p^2 - \xi m^2)^{-1} p_\mu p_\nu] \tag{14.41}$$

as in (13.93). It is easy to verify from (13.97), remembering that $O^{ab} = \delta^{ab}$ in this case, that

$$e^2 = \begin{pmatrix} 0 \\ 0 \\ 1 \\ 0 \end{pmatrix} \qquad e^3 = \begin{pmatrix} 1 \\ 0 \\ 0 \\ 0 \end{pmatrix} \qquad e^4 = \begin{pmatrix} 0 \\ 0 \\ 0 \\ 1 \end{pmatrix}. \tag{14.42}$$

We therefore choose the remaining basis vector

$$f^1 = \begin{pmatrix} 0 \\ 1 \\ 0 \\ 0 \end{pmatrix} \tag{14.43}$$

and note that it is indeed an eigenvector of the matrix (μ^2), defined in (13.56), as demanded by (13.102b):

$$(\mu^2) = \mathrm{diag}(0, -2\mu^2, 0, 0) \tag{14.44}$$

where we have used (14.32). Thus from (13.104) we obtain the Feynman rule for the Goldstone boson propagators

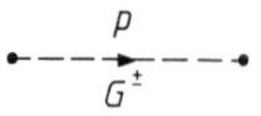

$$: \quad \mathrm{i}(p^2 - \xi m_W^2 + \mathrm{i}\varepsilon)^{-1} \tag{14.45}$$

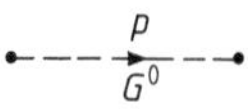

$$: \quad \mathrm{i}(p^2 - \xi m_Z^2 + \mathrm{i}\varepsilon)^{-1} \tag{14.46}$$

and from (13.105) the Feynman rule for the Higgs particle propagator is

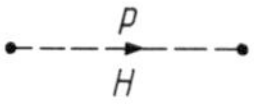

$$: \quad \mathrm{i}(p^2 - m_H^2 + \mathrm{i}\varepsilon)^{-1} \tag{14.47}$$

with

$$m_H^2 \equiv -2\mu^2 = \tfrac{1}{2}\lambda v^2. \tag{14.48}$$

The Faddeev–Popov ghost propagators follow immediately from (13.112). Evidently there is a massless ghost associated with the photon mode ($a = 1$) and massive ghosts associated with $W^{\pm}$ and Z. Thus the Feynman rules for the ghost propagators are

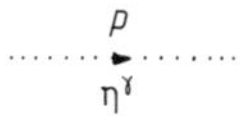

$$: \quad \mathrm{i}(p^2 + \mathrm{i}\varepsilon)^{-1} \tag{14.49}$$

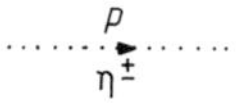

$$: \quad \mathrm{i}(p^2 - \xi m_W^2 + \mathrm{i}\varepsilon)^{-1} \tag{14.50}$$

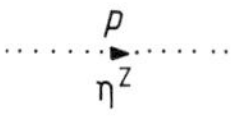

$$: \quad \mathrm{i}(p^2 - \xi m_Z^2 + \mathrm{i}\varepsilon)^{-1}. \tag{14.51}$$

(We remind the reader that (from Chapter 10), because the ghost fields are Grassmann variables, the closed loops in Feynman diagrams involving them will each have a minus sign associated with it.)

Finally we have to use the spontaneous symmetry breaking to generate masses for the charged leptons l, using the technique developed in §13.7. From (13.128) we can write down a Yukawa interaction of the lepton multiplets E_{lL} and l_R with the real scalar fields $\mathbf{\Phi}$. Thus

$$\mathscr{L}_Y = \sum_{l,p} \bar{E}_{lL} Y_{lp} l_R \Phi_p + \mathrm{HC} \tag{14.52}$$

where the matrices Y_{lp} are chosen so that $\mathscr{L}_Y$ is SU(2) × U(1) invariant. This may be accomplished by imposing the condition (13.129). However in the present context it is more convenient to use the known transformation properties of the two-component (complex) column vector $\boldsymbol{\varphi}$ defined in (14.23).

Since φ is a doublet with weak hypercharge $+\frac{1}{2}$, and l_R is a singlet with weak hypercharge -1, it follows that φl_R transforms as a doublet with hypercharge $-\frac{1}{2}$ just like E_{lL}. Thus $\bar{E}_{lL}\varphi l_R$ *is* SU(2) × U(1) invariant and we take

$$\mathscr{L}_Y = -\sum_l G_l(\bar{E}_{lL}\varphi l_R + \bar{l}_R\varphi^\dagger E_{lL}). \tag{14.53}$$

When the symmetry is spontaneously broken, we write

$$\varphi(x) = \begin{pmatrix} G^+ \\ \frac{1}{\sqrt{2}}(v + H + iG^0) \end{pmatrix} \tag{14.54a}$$

$$\equiv \frac{1}{\sqrt{2}}\mathbf{v} + \tilde{\varphi}(x) \tag{14.54b}$$

where

$$\mathbf{v} \equiv \begin{pmatrix} 0 \\ v \end{pmatrix} \qquad v = (-4\mu^2/\lambda)^{1/2} \tag{14.55}$$

and

$$\langle 0|\tilde{\varphi}(x)|0\rangle = 0. \tag{14.56}$$

The fermion masses arise from the parts of (14.53) involving v. In fact using (14.9) and (14.55) we find the mass Lagrangian to be

$$\mathscr{L}^M = \sum_l \frac{-G_l}{\sqrt{2}}(\bar{l}_L v l_R + \bar{l}_R v l_L). \tag{14.57}$$

Thus only the charged leptons acquire a mass

$$m_l = \frac{G_l}{\sqrt{2}} v \tag{14.58}$$

and the lepton propagators are

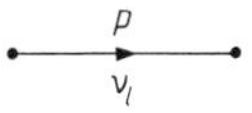

$$: \quad i(\not{p} + i\varepsilon)^{-1} a_L \tag{14.59}$$

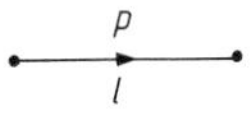

$$: \quad i(\not{p} - m_l + i\varepsilon)^{-1} \tag{14.60}$$

with m_l given in (14.58).

14.3 Feynman rules for the vertices

It remains only to extract the Feynman rules for the various vertices included in (14.17), (14.29), (13.110) and (14.52). We start with lepton–gauge boson vertices contained in (14.17). We have already expressed these interactions in terms of the gauge boson fields ($W^{\pm}_{\mu}$, Z_{μ}, A_{μ}) which remain mass eigenstates after the symmetry breaking. Thus using (14.22) and (14.8) we find the following vertices

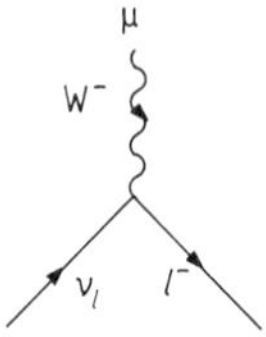

$$: \quad \frac{-ig}{2\sqrt{2}}\gamma_{\mu}(1-\gamma_5) \tag{14.61}$$

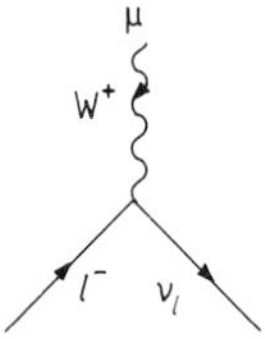

$$: \quad \frac{-ig}{2\sqrt{2}}\gamma_{\mu}(1-\gamma_5) \tag{14.62}$$

$$: \quad ie\gamma_{\mu} \tag{14.63}$$

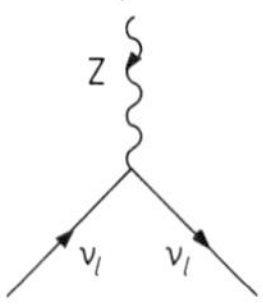

$$: \quad \frac{-ig}{4\cos\theta_{\mathrm{W}}}\gamma_{\mu}(1-\gamma_5) \tag{14.64}$$

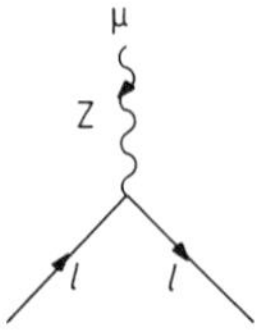

$$: \quad \frac{ig}{2\cos\theta_{\mathrm{W}}}\gamma_{\mu}(g_V-g_A\gamma_5) \tag{14.65a}$$

where

$$g_V=\tfrac{1}{2}-2\sin^2\theta_{\mathrm{W}} \tag{14.65b}$$

$$g_A=\tfrac{1}{2}. \tag{14.65c}$$

The remaining pieces of the Lagrangian (14.17) describe the self-interactions of the gauge vector bosons. The Feynman rules may be read off directly from (10.74) and (10.77). Evidently the only non-zero trilinear vertex involves the fields W_μ^1, W_μ^2 and W_μ^3. It is straightforward to cast the rules (10.74) into a more practical form involving the charge eigenstates W^+, W^- and Z, A. We find (using 14.21)) that the only non-zero vertices are

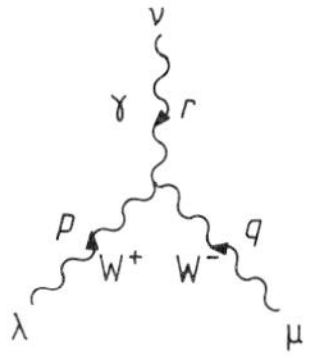

$$: \quad ie[(r-q)_\lambda g_{\mu\nu} + (q-p)_\nu g_{\lambda\mu} + (p-r)_\mu g_{\nu\lambda}] \qquad (14.66)$$

and

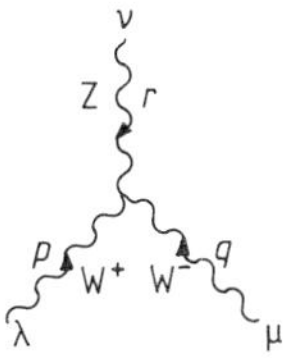

$$: \quad ig\cos\theta_W[(r-q)_\lambda g_{\mu\nu} + (q-p)_\nu g_{\lambda\mu} + (p-r)_\mu g_{\nu\lambda}]. \qquad (14.67)$$

It is also clear from (10.77) that the only non-zero quadrilinear vertices involve either four charged vector bosons or two charged and two neutrals. We find

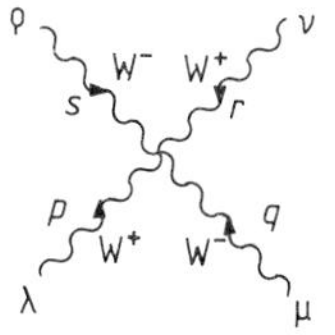

$$: \quad ig^2[2g_{\lambda\nu}g_{\mu\rho} - g_{\mu\nu}g_{\lambda\rho} - g_{\mu\lambda}g_{\nu\rho}] \qquad (14.68)$$

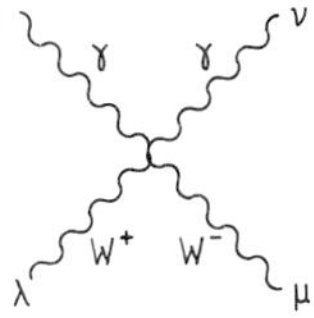

$$: \quad -ie^2[2g_{\nu\rho}g_{\mu\lambda} - g_{\mu\rho}g_{\nu\lambda} - g_{\mu\nu}g_{\lambda\rho}] \qquad (14.69)$$

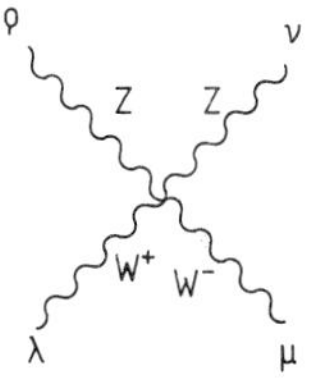

$$: \quad -ig^2\cos^2\theta_W[2g_{\nu\rho}g_{\mu\lambda} - g_{\mu\rho}g_{\nu\lambda} - g_{\mu\nu}g_{\lambda\rho}] \qquad (14.70)$$

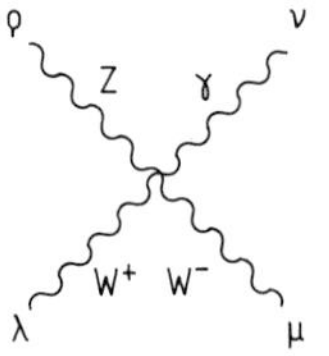

$$: \quad -ieg\cos\theta_W[2g_{\nu\rho}g_{\mu\lambda}-g_{\mu\rho}g_{\nu\lambda}-g_{\mu\nu}g_{\lambda\rho}]. \qquad (14.71)$$

Besides generating the masses (14.37) for the $W^{\pm}$, Z gauge bosons the scalar Lagrangian (14.29) also generates interactions between the scalar fields and these gauge particles, as well as the self-interactions of the scalars. We can read off the required vertices from the general formula given in (13.106), and the generators $\mathbf{T}^a$ given in (14.33) and (14.35). Alternatively we may rewrite $\mathscr{L}_\varphi$ given in (14.29) in terms of the original complex doublet $\boldsymbol{\varphi}$ given in (14.23)

$$\mathscr{L}_\varphi=(D_\mu\boldsymbol{\varphi})^\dagger(D^\mu\boldsymbol{\varphi})-\mu^2\boldsymbol{\varphi}^\dagger\boldsymbol{\varphi}-\tfrac{1}{4}\lambda(\boldsymbol{\varphi}^\dagger\boldsymbol{\varphi})^2 \qquad (14.72a)$$

where

$$D_\mu\boldsymbol{\varphi}=(\partial_\mu+ig\tfrac{1}{2}\mathbf{T}^aW^a_\mu+ig'\tfrac{1}{2}B_\mu)\boldsymbol{\varphi}. \qquad (14.72b)$$

When the symmetry is broken we make the substitution (14.54). This gives

$$\mathscr{L}_\varphi=\left|\partial_\mu G^+ + ig\frac{\cos 2\theta_W}{2\cos\theta_W}Z_\mu G^+ + ieA_\mu G^+ + \tfrac{1}{2}igW^+_\mu(v+H+iG^0)\right|^2$$
$$+\tfrac{1}{2}|\partial_\mu(H+iG^0)+igW^-_\mu G^+ - \tfrac{1}{2}ig\sec\theta_W Z_\mu(v+H+iG^0)|^2$$
$$-\tfrac{1}{2}m_H^2H^2-\tfrac{1}{2}\lambda vH[G^+G^-+\tfrac{1}{2}(G^{02}+H^2)]-\tfrac{1}{4}\lambda[G^+G^-+\tfrac{1}{2}(G^{02}+H^2)]^2. \qquad (14.73)$$

As it must, this gives the same masses for the gauge bosons (and scalars) as we found in (14.37) from the general treatment. We may also use (14.73), instead of (13.106), to obtain the interaction vertices involving scalars, if we choose. Either way we obtain the same Feynman rules. We start with vertices involving two scalars and a gauge boson.

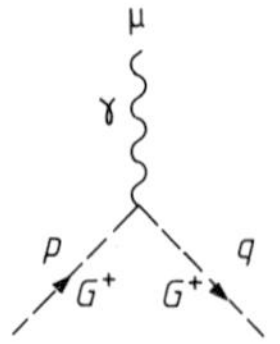

$$: \quad -ie(p+q)_\mu \qquad (14.74)$$

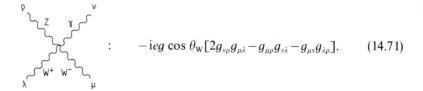

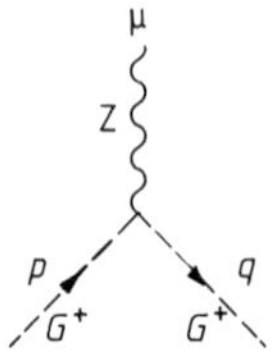

$$: \quad -ig\frac{\cos 2\theta_W}{2\cos\theta_W}(p+q)_\mu \qquad (14.75)$$

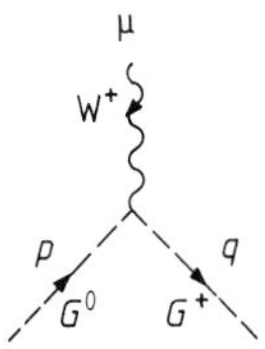

$$: \quad \tfrac{1}{2}g(p+q)_\mu \tag{14.76}$$

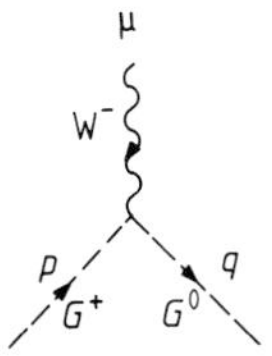

$$: \quad -\tfrac{1}{2}g(p+q)_\mu \tag{14.77}$$

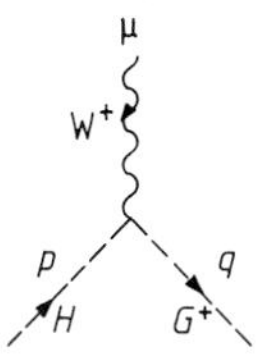

$$: \quad -\tfrac{1}{2}\mathrm{i}g(p+q)_\mu \tag{14.78}$$

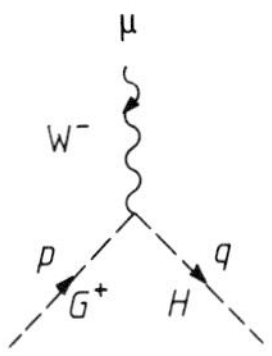

$$: \quad -\tfrac{1}{2}\mathrm{i}g(p+q)_\mu \tag{14.79}$$

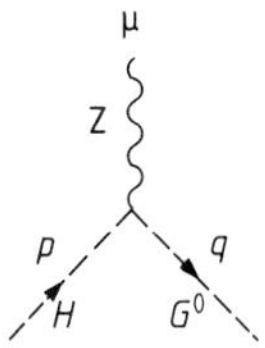

$$: \quad \frac{1}{2}\frac{g}{\cos\theta_{\mathrm{W}}}(p+q)_\mu . \tag{14.80}$$

Notice that in the unitary gauge, in which the (unphysical) Goldstone modes $G^{\pm}, G^0$ do not appear, none of these vertices occurs. Thus these vertices are only important when calculating radiative corrections (higher-order loop contributions), in which it is impossible to calculate in the unitary gauge. Next we consider vertices involving two gauge particles and a scalar. Using the mass formulae (14.37) we can write the Feynman rules in the following form:

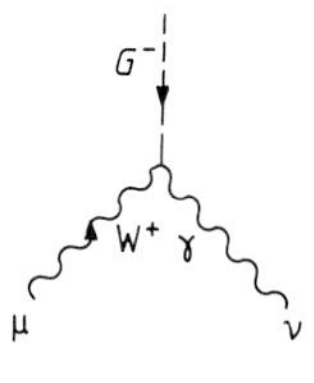

$$: \quad \mathrm{i}em_{\mathrm{W}}g_{\mu\nu} \tag{14.81}$$

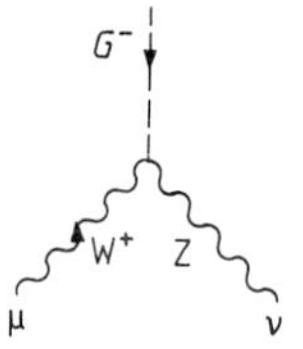

$$: \quad -igm_Z \sin^2 \theta_W g_{\mu\nu} \tag{14.82}$$

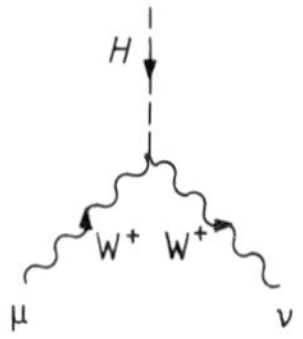

$$: \quad igm_W g_{\mu\nu} \tag{14.83}$$

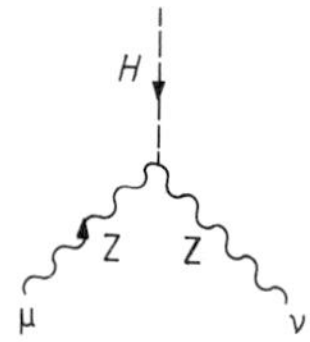

$$: \quad ig \frac{m_Z^2}{m_W} g_{\mu\nu}. \tag{14.84}$$

Note that there are no vertices involving G^0. We may also derive the Feynman rules for the vertices involving two gauge particles and two scalars. It is possibly easier to do this from (14.72) rather than (13.106). In view of the large number of such vertices we merely include them in Appendix C.

Finally there are the vertices involving the Higgs scalar H and the Goldstone modes $G^{\pm}, G^0$. Using the formula (14.48) for m_H as well as (14.37) the three-scalar vertices have the following Feynman rules

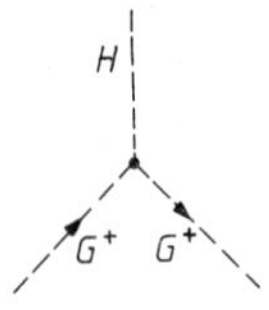

$$: \quad -\tfrac{1}{2}ig \frac{m_H^2}{m_W} \tag{14.85}$$

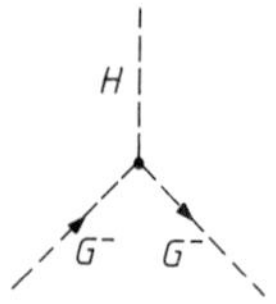

$$: \quad -\tfrac{1}{2}ig \frac{m_H^2}{m_W} \tag{14.86}$$

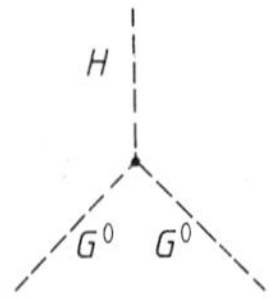

$$: \quad -\tfrac{1}{2}ig \frac{m_H^2}{m_W} \tag{14.87}$$

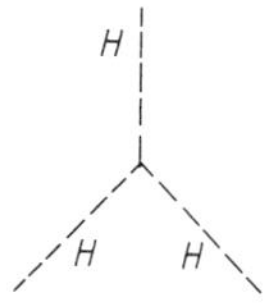

$$: \quad -\tfrac{3}{2}ig\frac{m_{\mathrm{H}}^2}{m_{\mathrm{W}}}. \tag{14.88}$$

There are also six vertices involving four scalars with a strength proportional to

$$-\tfrac{1}{4}\mathrm{i}\lambda = -\tfrac{1}{8}\mathrm{i}g^2\frac{m_{\mathrm{H}}^2}{m_{\mathrm{W}}^2}. \tag{14.89}$$

The rules for these are relegated to Appendix C.

Next we extract the Feynman rules for vertices involving the Fadeev–Popov ghost particles. Their interactions are specified by the trilinear terms of (13.110). Thus

$$\begin{aligned}\mathscr{L}_{\mathrm{FP}}(\text{interaction}, \eta^a, \eta^{a*}, \boldsymbol{\varphi}) = {} & gf^{abc}(\partial_\mu\eta^{a*})\eta^b A^{c\mu} \\ & -\xi g^2\eta^{a*}\eta^b\tilde{\boldsymbol{\varphi}}^{\mathrm{T}}\mathbf{T}^b\mathbf{T}^a\boldsymbol{v}.\end{aligned} \tag{14.90}$$

In the first term the constants f^{abc} are the structure constructs defined by the matrices $\mathbf{T}^a$:

$$[\mathbf{T}^a, \mathbf{T}^b] = \mathrm{i}f^{abc}\mathbf{T}^c \tag{14.91}$$

where $\mathbf{T}^a$ are defined in (14.33) and (14.35). We may either calculate the f^{abc} directly, or else observe that they derive from the algebra of SU(2) × U(1). The SU(2) is generated by $\mathbf{t}^1 = \mathbf{T}^2$, $\mathbf{t}^2 = \mathbf{T}^3$ and $\mathbf{t}^3 = \sin\theta_{\mathrm{W}}\mathbf{T}^1 + \cos\theta_{\mathrm{W}}\mathbf{T}^4$ and $\mathbf{Y}$ generates the U(1). Since U(1) is Abelian, there are of course no structure constants involving it. Thus the only gauge field combinations which can arise are $W^{\pm}$, from $\mathbf{t}^1$ and $\mathbf{t}^2$, and $\sin\theta_{\mathrm{W}}A + \cos\theta_{\mathrm{W}}Z$, from $\mathbf{t}^3$. Similarly the only ghost combinations which interact with the gauge bosons are $\eta^{\pm}$, $\sin\theta_{\mathrm{W}}\eta^{\gamma} + \cos\theta_{\mathrm{W}}\eta^Z$, and their conjugates. Expressing (14.90) in terms of the mass eigenstates we find

$$\begin{aligned}&\mathscr{L}_{\mathrm{FP}}(\text{interaction}, \eta^a, \eta^{a*}, \boldsymbol{\varphi}) \\ &\quad = \mathrm{i}g(\cos\theta_{\mathrm{W}}Z^\mu + \sin\theta_{\mathrm{W}}A^\mu)(\partial_\mu\eta^{*-}\eta^+ - \partial_\mu\eta^{*+}\eta^-) \\ &\qquad + \mathrm{i}g(\cos\theta_{\mathrm{W}}\partial_\mu\eta^{*Z} + \sin\theta_{\mathrm{W}}\partial_\mu\eta^{*\gamma})(\eta^- W^{+\mu} - \eta^+ W^{-\mu}) \\ &\qquad + \mathrm{i}g(W_\mu^-\partial^\mu\eta^{*+} - W_\mu^+\partial^\mu\eta^{*-})(\cos\theta_{\mathrm{W}}\eta^Z + \sin\theta_{\mathrm{W}}\eta^\gamma) \\ &\qquad - \tfrac{1}{2}g\xi m_{\mathrm{W}}H(\eta^{*+}\eta^- + \eta^{*-}\eta^+ + \sec^2\theta_{\mathrm{W}}\eta^{*Z}\eta^Z) \\ &\qquad + \tfrac{1}{2}g\xi m_{\mathrm{W}}\mathrm{i}G^0(\eta^{*+}\eta^- - \eta^{*-}\eta^+) - e\xi m_{\mathrm{W}}(G^+\eta^{*-} + G^-\eta^{*+})\eta^\gamma \\ &\qquad - \tfrac{1}{2}g\xi m_{\mathrm{W}}\frac{\cos 2\theta_{\mathrm{W}}}{\cos\theta_{\mathrm{W}}}(G^+\eta^{*-} + G^-\eta^{*+})\eta^Z + \tfrac{1}{2}g\xi m_Z\eta^{*Z}(G^+\eta^- + G^-\eta^+).\end{aligned} \tag{14.92}$$

This generates the following Feynman rules

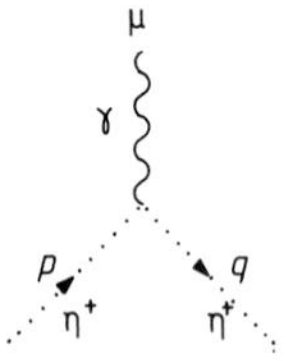

: $\quad -ieq_\mu$ (14.93)

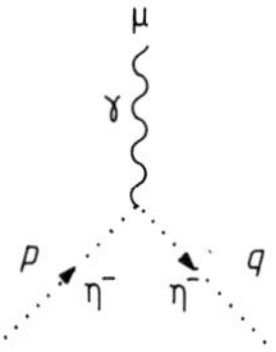

: $\quad +ieq_\mu.$ (14.94)

The couplings to Z in place of γ are obtained by replacing e by $g \cos \theta_W$

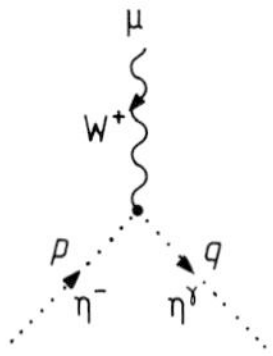

: $\quad -ieq_\mu$ (14.95)

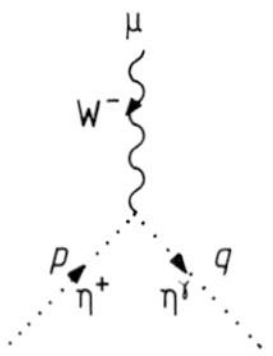

: $\quad ieq_\mu.$ (14.96)

The couplings to η^Z in place of η^γ are obtained by replacing e by $g \cos \theta_W$.

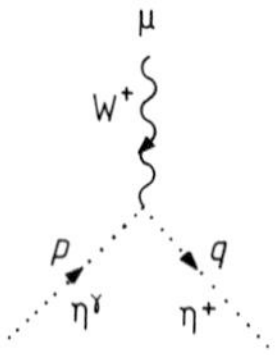

: $\quad ieq_\mu$ (14.97)

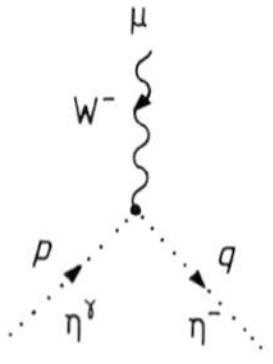

: $\quad -ieq_\mu.$ (14.98)

Again couplings to η^Z in place of η^γ are obtained by replacing e by $g \cos \theta_W$

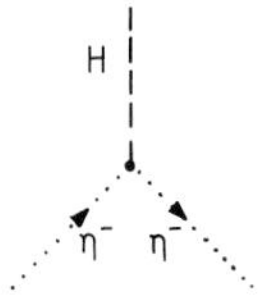

$$: \quad -\tfrac{1}{2} i g \xi m_W. \tag{14.99}$$

The H has the same coupling to η^+ instead of η^- and a coupling of $-\frac{1}{2} i \xi g m_W \sec^2 \theta_W$ if η^Z replaces $\eta^\pm$

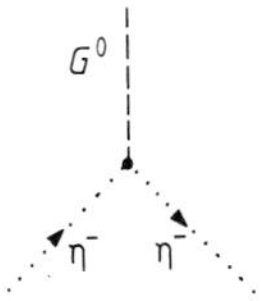

$$: \quad -\tfrac{1}{2} g \xi m_W \tag{14.100}$$

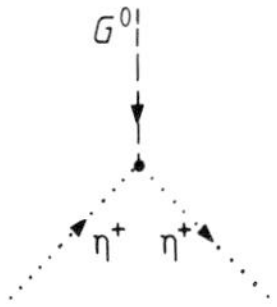

$$: \quad \tfrac{1}{2} g \xi m_W \tag{14.101}$$

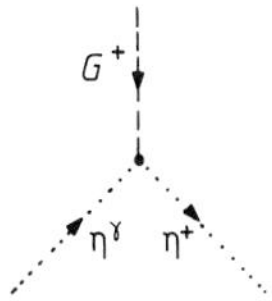

$$: \quad -i e \xi m_W \tag{14.102}$$

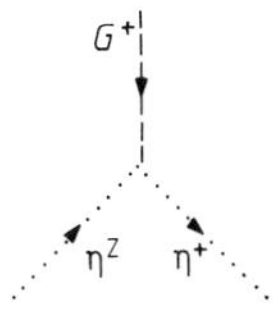

$$: \quad -\tfrac{1}{2} i g \frac{\cos 2\theta_W}{\cos \theta_W} \xi m_W \tag{14.103}$$

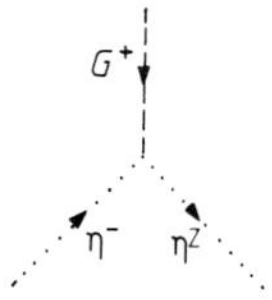

$$: \quad \tfrac{1}{2} i g \xi m_Z. \tag{14.104}$$

The couplings with G^+ replaced by G^- and $\eta^+ \leftrightarrow \eta^-$ are identical. Note that all of the scalar–ghost couplings vanish in the Landau gauge ($\xi = 0$).

Besides generating masses for the leptons, the Yukawa interaction (14.52) also generates (Yukawa) couplings of the scalars ($G^\pm, G^0, H$) to the leptons.

Using the mass formula (14.58) the Feynman rules for these can be written as follows.

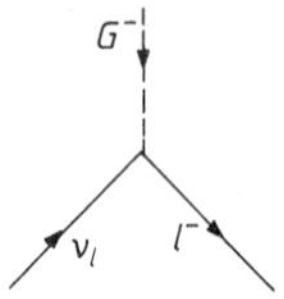

$$: \quad -\mathrm{i}\frac{g}{\sqrt{2}}\frac{m_l}{m_\mathrm{W}}\tfrac{1}{2}(1-\gamma_5) \tag{14.105}$$

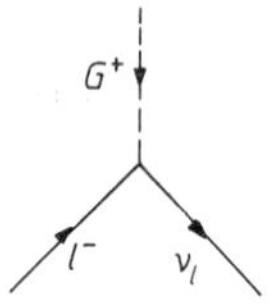

$$: \quad -\mathrm{i}\frac{g}{\sqrt{2}}\frac{m_l}{m_\mathrm{W}}\tfrac{1}{2}(1+\gamma_5) \tag{14.106}$$

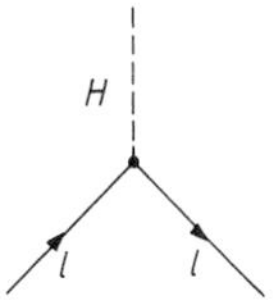

$$: \quad -\tfrac{1}{2}\mathrm{i}g\frac{m_l}{m_\mathrm{W}} \tag{14.107}$$

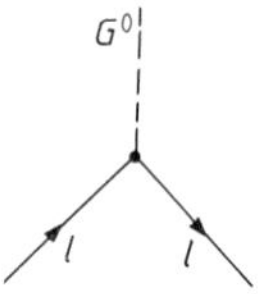

$$: \quad \tfrac{1}{2}g\frac{m_l}{m_\mathrm{W}}\gamma_5. \tag{14.108}$$

14.4 Tests of electroweak theory

The electroweak theory which we have developed so far is incomplete because we still have to include the couplings of the hadrons (quarks). Thus we are not yet able to calculate the S-matrix elements for processes such as neutron decay or pion decay. Nevertheless there are a number of weak processes involving only leptons which may be used to test the consistency of the theory so far.

At present we have introduced seven independent parameters into the theory. These are g and g', the two coupling constants associated with the SU(2) and U(1) gauge groups, the (negative) mass squared μ^2 and the φ^4 coupling constant λ, both of which appear in the scalar Lagrangian $\mathscr{L}_\varphi$ in (14.29), and the three Yukawa coupling constants G_l ($l=\mathrm{e}, \mu, \tau$), which appear in (14.57). It follows from (14.21) that g and g' are expressible in terms of e and θ_W, and of course e is very precisely determined:

$$e=\sqrt{4\pi\alpha}=0.302\,822\,1 \tag{14.109}$$

from data on the fine structure constant α. Thus only one of these two parameters (θ_W) is not immediately known. Also, using (14.32) and (14.48), we can use the parameters m_H and v, instead of μ^2 and λ. Finally, from (14.58), it is apparent that, since v is already a parameter, we can use the three masses m_l ($l=e, \mu, \tau$), rather than the Yukawa couplings G_l, and these masses also are rather well determined:

$$m_e = 0.511\,003\,4 \pm 0.000\,001\,4 \text{ MeV} \tag{14.110a}$$

$$m_\mu = 105.659\,43 \pm 0.000\,18 \text{ MeV} \tag{14.110b}$$

$$m_\tau = 1784.2 \pm 3.2 \text{ MeV}. \tag{14.110c}$$

Now it is clear from (14.37) that the parameters θ_W and v may be determined from knowledge of m_W and m_Z, since

$$\cos\theta_W = \frac{m_W}{m_Z} \tag{14.111a}$$

$$v = \frac{2m_W}{em_Z}(m_Z^2 - m_W^2)^{1/2}. \tag{14.111b}$$

The discovery[5] of $W^\pm$ and Z in 1983 and the measurement of their masses

$$m_Z = 92.9 \pm 1.6 \text{ GeV} \tag{14.112a}$$

$$m_W = 80.8 \pm 2.7 \text{ GeV} \tag{14.112b}$$

thus means that six of the seven parameters are completely determined without any direct recourse to weak interaction data:

$$\sin^2\theta_W = 0.2433 \pm 0.022 \tag{14.113a}$$

$$v = 263 \pm 59 \text{ GeV}. \tag{14.113b}$$

The remaining unknown, m_H, will have to await the discovery of the Higgs scalar H, which is the last major component of the theory needing experimental confirmation. Nevertheless, it turns out that the weak data available to date is remarkably insensitve to m_H, and it is possible to test a lot of the theory knowing nothing about the mass of the Higgs particle.

The purely leptonic process of muon decay,

$$\mu^- \to e^- \bar{\nu}_e \nu_\mu \tag{14.114}$$

is an excellent illustration of this. It has been very carefully measured over many years, and the theory which we have described makes unambiguous predictions of both the energy spectrum of the electron in the final state, as well as the overall lifetime of the decay.

The S-matrix element is

$$S(\mu^- \to e^- \bar{\nu}_e \nu_\mu) = (2\pi)^4 \delta(\mu - e - \bar{\nu}_e - \nu_\mu)\mathcal{M}$$

where without confusion we use the same letter to denote a particle and its four-momentum. The Lorentz invariant $\mathcal{M}$ is given by the following Feynman diagrams

$$\mathscr{M} =$$

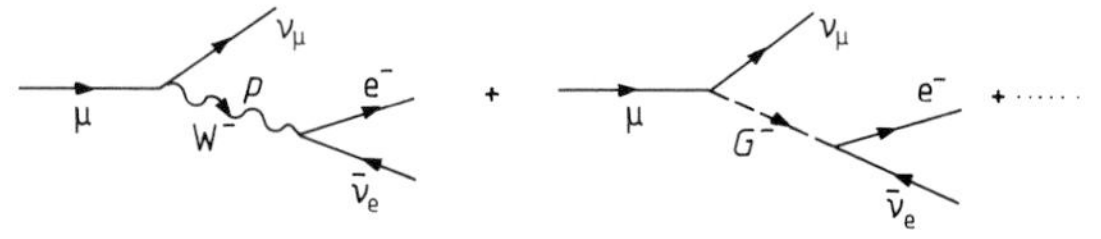

$$= \mathscr{M}_1 + \mathscr{M}_2 + \dots . \tag{14.115}$$

The ... refers to higher order diagrams, involving loops. The Feynman rules derived in the previous section show that these diagrams give contributions which are smaller than those displayed by a factor of at least O(α). Thus, as a first approximation we need to calculate only the tree diagrams displayed above. Using the Feynman rules (14.39), (14.61) and (14.62) we have

$$\mathscr{M}_1 = \left(\frac{-ig}{2\sqrt{2}}\right)\bar{u}(\nu_\mu)\gamma^\mu(1-\gamma_5)u(\mu)\mathrm{i}(p^2 - m_W^2 + \mathrm{i}\varepsilon)^{-1}$$
$$\times [-g_{\mu\nu} + (1-\xi)p_\mu p_\nu(p^2 - \xi m_W^2)^{-1}]\left(\frac{-ig}{2\sqrt{2}}\right)\bar{u}(\mathrm{e})\gamma^\nu(1-\gamma_5)v(\bar{\nu}_\mathrm{e}) \tag{14.116}$$

where, using energy–momentum conservation:

$$p \equiv \mu - \nu_\mu = \mathrm{e} + \bar{\nu}_\mathrm{e}. \tag{14.117}$$

Using (14.45), (14.105) and (14.106) we find

$$\mathscr{M}_2 = \left(\frac{-ig}{2\sqrt{2}}\right)\frac{m_\mu}{m_W}\bar{u}(\nu_\mu)(1+\gamma_5)u(\mu)\mathrm{i}(p^2 - \xi m_W^2)^{-1}$$
$$\times \left(\frac{-ig}{2\sqrt{2}}\right)\frac{m_\mathrm{e}}{m_W}\bar{u}(\mathrm{e})(1-\gamma_5)v(\bar{\nu}_\mathrm{e}). \tag{14.118}$$

Note that each term depends upon the gauge fixing parameter ξ. Physical quantities, such as S-matrix elements, must be independent of ξ. To see how this comes about consider the ξ-dependent part of the numerator of the W propagator in $\mathscr{M}_1$: $(1-\xi)p_\mu p_\nu(p^2 - \xi m_W^2)^{-1}$. The p_μ is contracted with the γ^μ matrix and

$$\begin{aligned} p_\mu \bar{u}(\nu_\mu)\gamma^\mu(1-\gamma_5)u(\mu) &= \bar{u}(\nu_\mu)\not{p}(1-\gamma_5)u(\mu) \\ &= \bar{u}(\nu_\mu)(\not{\mu} - \not{\nu}_\mu)(1-\gamma_5)u(\mu) \\ &= \bar{u}(\nu_\mu)(1+\gamma_5)\not{\mu}u(\mu) \\ &= m_\mu \bar{u}(\nu_\mu)(1+\gamma_5)u(\mu) \end{aligned} \tag{14.119}$$

where we have used (14.117) and the Dirac equation (3.88) for both $\bar{u}(\nu_\mu)$ and $u(\mu)$. Similarly

$$p_\nu \bar{u}(\mathrm{e})\gamma^\nu(1-\gamma_5)v(\bar{\nu}_\mathrm{e}) = m_\mathrm{e}\bar{u}(\mathrm{e})(1-\gamma_5)v(\bar{\nu}_\mathrm{e}). \tag{14.120}$$

Substituting (14.119), (14.120) into (14.116), and combining with (14.118) gives

$$\mathcal{M}=\mathcal{M}_1+\mathcal{M}_2=\tfrac{1}{8}\mathrm{i}g^2(p^2-m_{\mathrm{W}}^2)^{-1}\Big(\bar{u}(\nu_\mu)\gamma^\mu(1-\gamma_5)u(\mu)\bar{u}(\mathrm{e})\gamma_\mu(1-\gamma_5)v(\bar{\nu}_\mathrm{e})$$
$$-\frac{m_\mathrm{e}m_\mu}{m_\mathrm{W}^2}\bar{u}(\nu_\mu)(1+\gamma_5)u(\mu)\bar{u}(\mathrm{e})(1-\gamma_5)v(\bar{\nu}_\mathrm{e})\Big) \tag{14.121}$$

which *is* independent of ξ, as required. Since this expression is independent of m_H, all of the parameters appearing above are known, and we can proceed to calculate the decay spectrum. The experimental values (14.110) and (14.112) mean that we may certainly neglect the term proportional to $m_\mathrm{e}m_\mu/m_\mathrm{W}^2$, since it is considerably smaller than the higher order diagrams already neglected. Similarly since $p^2=\mathrm{O}(m_\mu^2)$ we may neglect p^2 compared with m_W^2. We therefore obtain

$$\mathcal{M}\simeq\frac{-\mathrm{i}}{2v^2}\bar{u}(\nu_\mu)\gamma^\mu(1-\gamma_5)u(\mu)\bar{u}(\mathrm{e})\gamma_\mu(1-\gamma_5)v(\bar{\nu}_\mathrm{e}) \tag{14.122}$$

using $g^2/8m_\mathrm{W}^2=(2v^2)^{-1}$, which follows from (14.37). Proceeding as in §6.5 we find the transition rate for a single muon is

$$\mathrm{d}\Gamma(\mu\to\mathrm{e}\bar{\nu}_\mathrm{e}\nu_\mu)=(2\pi)^{-5}\frac{1}{2\mu_0}\delta(\mu-\mathrm{e}-\bar{\nu}_\mathrm{e}-\nu_\mu)|\mathcal{M}|^2\frac{\mathrm{d}^3\mathrm{e}}{2\mathrm{e}_0}\frac{\mathrm{d}^3\nu_\mu}{2\nu_{\mu0}}\frac{\mathrm{d}^3\bar{\nu}_\mathrm{e}}{2\bar{\nu}_{\mathrm{e}0}}. \tag{14.123}$$

In practice neither $\bar{\nu}_\mathrm{e}$ nor ν_μ is observed, so we may integrate $\mathrm{d}^3\nu_\mu\,\mathrm{d}^3\bar{\nu}_\mathrm{e}$ over all of the available phase space. Also, the spin of the decaying muon and that of the final state electron are usually not measured. This means that we have to sum over the possible final spin states and average over the initial spin states. In these circumstances we may replace $|\mathcal{M}|^2$ by $\frac{1}{2}\sum_{s_\mathrm{e},s_\mu}|\mathcal{M}|^2$, where the sum is over the two spin states of both muon and electron. The technicalities of performing the phase space integration, as well as such sums are adequately dealt with in many text books[2] so we shall not dwell upon them. Instead we merely quote the results. The differential spectrum is given by

$$\frac{\mathrm{d}\Gamma}{\mathrm{d}E}=\frac{2v^{-4}}{3(2\pi)^3}m_\mu E(E^2-m_\mathrm{e}^2)^{1/2}(3W-2E-m_\mathrm{e}^2E^{-1}) \tag{14.124a}$$

where E is the energy (e^0) of the electron and W is its maximum allowed value

$$W=\frac{1}{2m_\mu}(m_\mu^2+m_\mathrm{e}^2). \tag{14.124b}$$

The total decay rate is

$$\Gamma(\mu)=\frac{m_\mu^5}{384\pi^3v^4}(1-8y+8y^3-y^4-12y^2\ln y) \tag{14.125a}$$

where

$$y\equiv m_\mathrm{e}^2/m_\mu^2. \tag{14.125b}$$

The differential spectrum is entirely consistent with the measured spectrum shape, and using the values (14.110), (14.113) we predict the mean lifetime

$$\tau_{\mu\text{th}} \equiv \Gamma(\mu)^{-1} = (2.90 \pm 2.61) \times 10^{-6}\ \text{s}. \tag{14.126}$$

The value actually measured is

$$\tau_{\mu\text{exp}} = 2.197\,138 \pm 0.000\,065 \times 10^{-6}\ \text{s}. \tag{14.127}$$

Thus, while our prediction is consistent with the data, the large errors on $\tau_{\mu\text{th}}$ means that this is hardly a rigorous test of the theory. These large errors stem from those in v in (14.113), since v enters $\tau_{\mu\text{th}}$ via v^4. It is only the recent determinations of m_W and m_Z which allow such a test. Previously muon decay data was used as input leading to the determination of v. In fact using the data (14.127) as input gives

$$v = 246\ \text{GeV}. \tag{14.128}$$

14.5 Inclusion of hadrons

For simplicity, the only fermion fields we have so far included are those of the leptons. It is now time to rectify this deficiency. We now know that there are six quark flavours (u, d, c, s, t, b) and that their participation in electroweak interactions is similar, but not identical, to the way in which the six leptons participate. There are two principal differences. First, the electric charges of the quarks are not the same as those of the leptons. In fact, in units of e (the charge of the positron)

$$Q_u = Q_c = Q_t = \tfrac{2}{3} \tag{14.129a}$$

$$Q_d = Q_s = Q_b = -\tfrac{1}{3}. \tag{14.129b}$$

Second, all six of the quarks are massive, whereas three of the leptons (ν_e, ν_μ, ν_τ) are believed to be massless. We notice that the two distinct charge eigenvalues differ by *one* unit. So

$$Q_u - Q_d = Q_{\nu_e} - Q_e = 1. \tag{14.130}$$

The difference between Q_{ν_e} and Q_e, say, reflects their different weak isospins; ν_{eL} and e_L have $T_3 = \frac{1}{2}$ and $-\frac{1}{2}$ respectively. It is therefore tempting to group the left chiral components of the quark fields into doublets, just as we did the lepton fields in (14.9). In fact this is correct, since it has been known for many years that the charged hadronic weak currents are 'left-handed', just as the charged leptonic currents (14.3) are. However, when we attempt to construct doublets, some of the complications due to the second difference (the fact that all six quarks are massive) enter. The problem arises when we ask which (left chiral) quark field is to be the $T_3 = -\frac{1}{2}$ partner of (say) u_L. It would have been nice, from the viewpoint of model building, if the answer involved a single quark flavour, d_L, for example. However the weak decays of hadrons show unambiguously that the partner of u_L involves at least two flavours (d_L and s_L). Thus, to be general, we shall suppose that the partner is an arbitrary super position of d_L, s_L and b_L. We therefore take the three left-handed quark doublets to be

$$Q_{fL} \equiv \begin{pmatrix} U_f \\ D'_f \end{pmatrix}_L \qquad (f=1, 2, 3) \tag{14.131a}$$

where U_f labels the charge $\frac{2}{3}$ quarks

$$U_1 = u \qquad U_2 = c \qquad U_3 = t \tag{14.131b}$$

and

$$D'_f = V_{fg} D_g \tag{14.131c}$$

with D_f labelling the charge $-\frac{1}{3}$ quarks

$$D_1 = d \qquad D_2 = s \qquad D_3 = b \tag{14.131d}$$

and **V** is an arbitrary 3×3 unitary matrix[6]. (The unitarity is so that the field combinations are orthogonal to one another.) As before, the suffix 'L' is to denote the left chiral component of the field. The reason it is not necessary to have such a complication in the leptonic sector is because the neutrinos are, by assumption, massless. Therefore any superposition of them is also massless, and ν_{lL} is *defined* as the partner of l_L. Should the neutrinos be found to have non-zero (and therefore presumably non-degenerate) masses, a precisely analogous mixing will have to be involved. Since the quarks all have different masses, the particular combinations D'_f of mass eigenstates which enter the weak currents is something which can in principle be ascertained from experiment. The other difference between the quarks and the leptons is that the quarks are coloured, while the leptons are not. However this presents only a trivial complication. Since the electroweak interactions are 'colour-blind' the matrix **V** does not mix colours, and the colour label of all of the three flavours involved in D'_f is the same, and the same as that of its $T_3 = +\frac{1}{2}$ partner. We therefore leave the colour label undisplayed; where necessary a sum over colour labels will be understood.

The extra weak isodoublets of course mean extra contributions to the weak isospin currents. We denote these by

$$\mathscr{T}^i_\mu \equiv \bar{Q}_{fL} \gamma_\mu \tfrac{1}{2} \tau_i Q_{fL} \qquad (i=1,2,3) \tag{14.132}$$

(summing over the repeated index f, and the suppressed colour label). Then the charged hadronic currents are $\mathscr{J}_\mu$ and $\mathscr{J}^\dagger_\mu$, where

$$\mathscr{J}^\dagger_\mu = 2(\mathscr{T}^1_\mu + i\mathscr{T}^2_\mu) = \bar{U}_f \gamma_\mu (1-\gamma_5) V_{fg} D_g \tag{14.133a}$$

$$\mathscr{J}_\mu = 2(\mathscr{T}^1_\mu - i\mathscr{T}^2_\mu) = \bar{D}_f \gamma_\mu (1-\gamma_5) V^\dagger_{fg} V_g. \tag{14.133b}$$

The third weak isospin current is given by

$$\begin{aligned}\mathscr{T}_\mu^3 &= \tfrac{1}{2}(\bar{U}_{fL}\gamma_\mu U_{fL} - \bar{D}'_{fL}\gamma_\mu D'_{fL}) \\ &= \tfrac{1}{2}(\bar{U}_{fL}\gamma_\mu U_{fL} - \bar{D}_{fL}\gamma_\mu D_{fL})\end{aligned} \tag{14.134}$$

since $\mathbf{V}$ is unitary[7]. Also, from (14.129) the hadronic electromagnetic current

$$\begin{aligned}j_\mu(q) &= \tfrac{2}{3}\bar{U}_f\gamma_\mu U_f - \tfrac{1}{3}\bar{D}_f\gamma_\mu D_f \\ &= \mathscr{T}_\mu^3 + \mathscr{Y}_\mu\end{aligned} \tag{14.135}$$

where the hadronic contribution $\mathscr{Y}_\mu$ to the weak hypercharge current is therefore given by

$$\mathscr{Y}_\mu = \tfrac{1}{6}\bar{Q}_{fL}\gamma_\mu \mathbf{I}_2 Q_{fL} + \tfrac{2}{3}\bar{U}_{fR}\gamma_\mu U_{fR} - \tfrac{1}{3}\bar{D}_{fR}\gamma_\mu D_{fR}. \tag{14.136}$$

Thus the first difference between the leptonic and hadronic sectors, namely the different charges (14.129) possessed by the quarks, manifests itself in a different structure for the weak hypercharge current; in the hadronic sector the doublets all have weak hypercharge $Y = \frac{1}{6}$, whereas the lepton doublets have $Y = -\frac{1}{2}$. The hypercharges of the right chiral quark fields correspond to their actual electric charges, since they are all singlets with respect to the $SU(2)_L$ group. It follows that the $SU(2) \times U(1)$ gauge invariant coupling of the quark field to the gauge fields (the analogue of (14.17)) is given by

$$\mathscr{L}(\text{quark}) = \bar{Q}_{fL}\mathrm{i}\gamma^\mu D_\mu Q_{fL} + \bar{U}_{fR}\mathrm{i}\gamma^\mu D_\mu U_{fR} + \bar{D}_{fR}\mathrm{i}\gamma^\mu \mathrm{D}_\mu D_{fR} \tag{14.137a}$$

where

$$\mathrm{D}_\mu Q_{fL} \equiv (\partial_\mu + \mathrm{i}g\tfrac{1}{2}\tau^a W_\mu^a + \mathrm{i}g'\tfrac{1}{6}B_\mu)Q_{fL} \tag{14.137b}$$

$$\mathrm{D}_\mu U_{fR} \equiv (\partial_\mu + \mathrm{i}g'\tfrac{2}{3}B_\mu)U_{fR} \tag{14.137c}$$

$$\mathrm{D}_\mu D_{fR} \equiv (\partial_\mu + \mathrm{i}g'(-\tfrac{1}{3})B_\mu)D_{fR}. \tag{14.137d}$$

As in (14.22), it is more useful to express the interactions in terms of the mass eigenstates

$$\begin{aligned}g\mathscr{T}_\mu^a W^{a\mu} + g'\mathscr{Y}_\mu B^\mu &= \frac{g}{2\sqrt{2}}(\mathscr{J}_\mu W^{-\mu} + \mathscr{J}_\mu^\dagger W^{+\mu}) \\ &\quad + ej_\mu(q)A^\mu + \frac{g}{\cos\theta_\mathrm{W}}(\mathscr{T}_\mu^3 - \sin^2\theta_\mathrm{W} j_\mu(q))Z^\mu.\end{aligned} \tag{14.138}$$

Then without further ado we may derive the Feynman rules for the quark–gauge boson vertices:

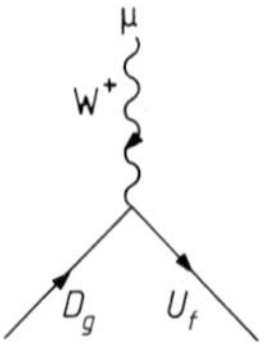

$$-\mathrm{i}\frac{g}{2\sqrt{2}}V_{fg}\gamma_\mu(1-\gamma_5) \tag{14.139}$$

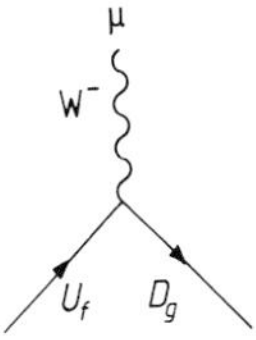

$$: \quad -\mathrm{i}\frac{g}{2\sqrt{2}} V^*_{fg}\gamma_\mu(1-\gamma_5) \qquad (14.140)$$

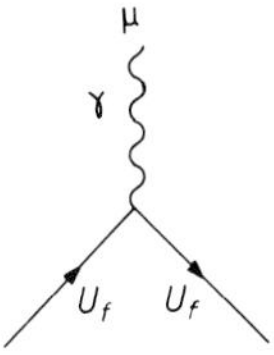

$$: \quad -\mathrm{i}\tfrac{2}{3}e\gamma_\mu \qquad (14.141\mathrm{a})$$

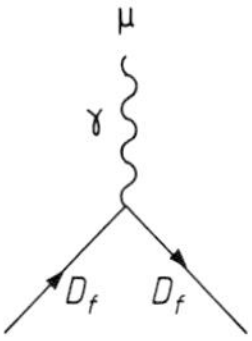

$$: \quad \mathrm{i}\tfrac{1}{3}e\gamma_\mu \qquad (14.141\mathrm{b})$$

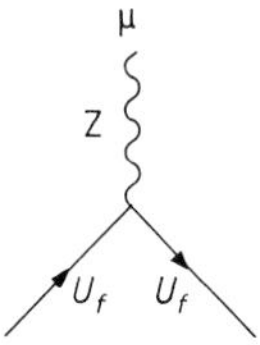

$$: \quad -\mathrm{i}\frac{g}{2\cos\theta_\mathrm{W}}\gamma_\mu(g_V^U - g_A^U\gamma_5) \qquad (14.142\mathrm{a})$$

where

$$g_V^U = \tfrac{1}{2} - \tfrac{4}{3}\sin^2\theta_\mathrm{W} \qquad g_A^U = \tfrac{1}{2}. \qquad (14.142\mathrm{b})$$

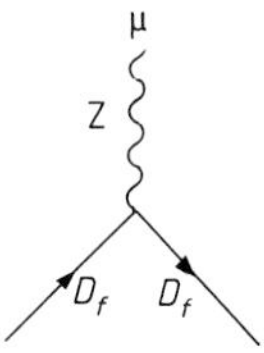

$$: \quad -\mathrm{i}\frac{g}{2\cos\theta_\mathrm{W}}\gamma_\mu(g_V^D - g_A^D\gamma_5) \qquad (14.143\mathrm{a})$$

where

$$g_V^D = -\tfrac{1}{2} + \tfrac{2}{3}\sin^2\theta_\mathrm{W} \qquad g_A^D = -\tfrac{1}{2}. \qquad (14.143\mathrm{b})$$

As before, we must also arrange that the spontaneous symmetry breaking generates masses for the fermions. The complication is that all six quarks are massive, which means that we must couple the doublets $Q_{f\mathrm{L}}$ to the singlets $U_{g\mathrm{R}}$ *and* $D_{g\mathrm{R}}$. To achieve this we need to use not only the scalar doublet $\boldsymbol{\varphi}$ having weak hypercharge $+\frac{1}{2}$, but also its charge conjugate ψ

$$\psi \equiv i\tau_2 \varphi^*. \tag{14.144}$$

We leave it as an exercise (problem 14.5) to verify that ψ *does* transform as a doublet having weak hypercharge $-\frac{1}{2}$ under an $SU(2) \times U(1)$ transformation. It follows that the most general $SU(2) \times U(1)$ invariant Yukawa coupling can be written

$$\mathscr{L}_Y = -(\bar{Q}_{fL} X_{fg} \varphi D_{gR} + \bar{D}_{gR} \varphi^\dagger X^*_{fg} Q_{fL}$$
$$+ \bar{Q}_{fL} Y_{fg} \psi U_{gR} + \bar{U}_{gR} \psi^\dagger Y^*_{fg} Q_{fL}) \tag{14.145}$$

where X_{fg} and Y_{fg} are, for the moment, arbitrary matrices. When the symmetry is broken the fermion mass terms arise from $\langle 0|\varphi|0\rangle = v$ and $\langle 0|\psi|0\rangle = i\tau_2 v$. We find that the mass Lagrangian is

$$\mathscr{L}^M = -\frac{v}{\sqrt{2}} (\bar{D}'_{fL} X_{fg} D_{gR} + \bar{D}_{gR} X^*_{fg} D'_{fL}$$
$$+ \bar{U}_{fL} Y_{fg} U_{gR} + \bar{U}_{gR} Y^*_{fg} U_{fL}). \tag{14.146}$$

Thus to ensure that the states U_f, D_f are mass eigenstates, as defined, the matrices **X**, **Y** and **V** must satisfy

$$\frac{v}{\sqrt{2}} \mathbf{V}^\dagger \mathbf{X} = \frac{v}{\sqrt{2}} \mathbf{X}^\dagger \mathbf{V} = \mathbf{m}(D) \tag{14.147a}$$

$$\frac{v}{\sqrt{2}} \mathbf{Y} = \mathbf{m}(U) \tag{14.147b}$$

where $\mathbf{m}(U)$ and $\mathbf{m}(D)$ are the diagonal mass matrices

$$\mathbf{m}(U) = \mathrm{diag}(m_u, m_c, m_t) \tag{14.148a}$$

$$\mathbf{m}(D) = \mathrm{diag}(m_d, m_s, m_b). \tag{14.148b}$$

The Feynman rules for the quark propagators are now trivial; for the quark flavour q, where $q = $ u, c, t, d, s, b, it is

$$\xrightarrow[q]{p} \quad : \quad i(\not{p} - m_q + i\varepsilon)^{-1}. \tag{14.149}$$

It follows from (14.147) that

$$\mathbf{X} = \frac{\sqrt{2}}{v} \mathbf{V}\mathbf{m}(D) \qquad \mathbf{X}^\dagger = \frac{\sqrt{2}}{v} \mathbf{m}(D)\mathbf{V}^\dagger \tag{14.150a}$$

$$\mathbf{Y} = \mathbf{Y}^\dagger = \frac{\sqrt{2}}{v} \mathbf{m}(U) \tag{14.150b}$$

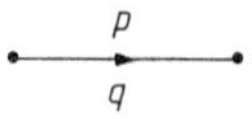

and therefore that the interaction part of $\mathscr{L}_Y$ is

$$\mathscr{L}_Y(\text{interaction}) = \frac{-g}{\sqrt{2} m_W} \Big(\bar{\mathbf{U}}[\mathbf{V}\mathbf{m}(D)a_R - \mathbf{m}(U)\mathbf{V}a_L]\mathbf{D}G^+ + \bar{\mathbf{D}}[\mathbf{m}(D)\mathbf{V}^\dagger a_L - \mathbf{V}^\dagger \mathbf{m}(U) a_R]\mathbf{U}G^- + \frac{1}{\sqrt{2}}\bar{\mathbf{D}}\mathbf{m}(D)\mathbf{D}H + \frac{1}{\sqrt{2}}\bar{\mathbf{U}}\mathbf{m}(U)\mathbf{U}H + \frac{1}{\sqrt{2}}\mathrm{i}\bar{\mathbf{D}}\mathbf{m}(D)\gamma_5\mathbf{D}\mathrm{i}G^0 - \frac{\mathrm{i}}{\sqrt{2}}\bar{\mathbf{U}}\mathbf{m}(U)\gamma_5\mathbf{U}G^0 \Big). \quad (14.151)$$

Thus the Feynman rules are

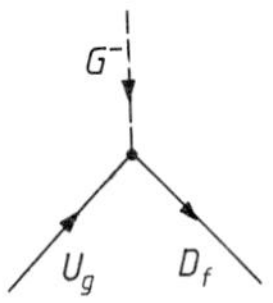

:

$$\frac{-ig}{2\sqrt{2}\, m_W}\{[\mathbf{m}(D)\mathbf{V}^\dagger - \mathbf{V}^\dagger\mathbf{m}(U)]_{fg} - \gamma_5[\mathbf{m}(D)\mathbf{V}^\dagger + \mathbf{V}^\dagger\mathbf{m}(U)]_{fg}\} \quad (14.152)$$

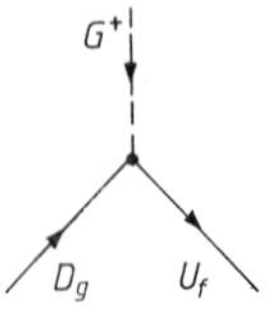

:

$$\frac{-ig}{2\sqrt{2}\, m_W}\{[\mathbf{V}\mathbf{m}(D) - \mathbf{m}(U)\mathbf{V}]_{fg} + \gamma_5[\mathbf{V}\mathbf{m}(D) + \mathbf{m}(U)\mathbf{V}]_{fg}\} \quad (14.153)$$

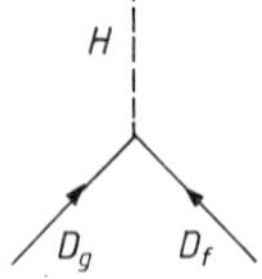

: $\frac{-ig}{2m_W}[\mathbf{m}(D)]_{fg}$ (14.154

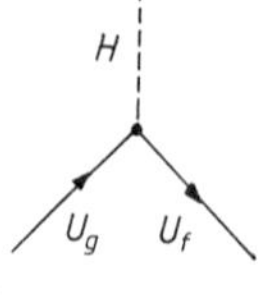

: $\frac{-ig}{2m_W}[\mathbf{m}(U)]_{fg}$ (14.1

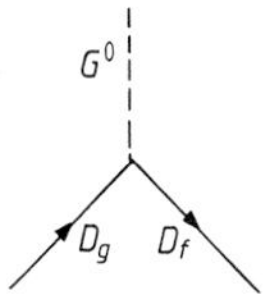

: $$\frac{g}{2m_W}[\mathbf{m}(D)]_{fg}\gamma_5 \tag{14.156}$$

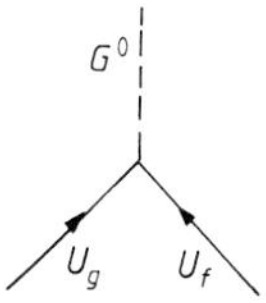

: $$\frac{-g}{2m_W}[\mathbf{m}(U)]_{fg}\gamma_5. \tag{14.157}$$

All of the above Feynman rules, as well as those for the interactions of the quarks with the gauge bosons, include an implicit unit colour matrix, reflecting the fact that the electroweak interactions are independent of the colour label of the quarks, as already noted.

Problems

14.1 Show that the matrices representing $SU(2)\times U(1)$ in the real representation (14.25) of the scalar fields are those given in (14.26), (14.27). Verify that they satisfy the Lie algebra of $SU(2)\times U(1)$.

14.2 Verify (14.34) and (14.35).

14.3 Check that both (13.106) and (14.73) yield the same Feynman rule for the G^0–H–Z vertex, for example.

14.4 Suppose that, more generally than (14.23), the scalar field φ^0 which develops a non-zero VEV belongs to a representation with weak isospin I and third component I_3. Calculate m_W and m_Z in this case and determine vlaues of I and I_3 such that $m_Z \cos\theta_W = m_W$.

14.5 Show that

$$\psi \equiv i\tau_2\varphi^*$$

with φ defined in (14.23), transforms under an $SU(2)\times U(1)$ transformation as a doublet with weak hypercharge $-\frac{1}{2}$.

14.6 Calculate $\Gamma(Z\to e^+e^-)$.

References

The classic papers upon which this chapter is founded are:

Glashow S L 1961 *Nucl. Phys.* **22** 579

Weinberg S 1967 *Phys. Rev. Lett.* **19** 1264

Salam A 1968 in *Elementary Particle Theory* ed N Svartholm (Stockholm: Almqvist and Wiksell)

We have also found most useful the following books and review articles:

Taylor J C 1976 *Gauge Theories of Weak Interactions* (Cambridge: Cambridge University Press)

Aitchison I J R and Hey A J G 1982 *Gauge Theories in Particle Physics* (Bristol: Adam Hilger)

Aoki K *et al.* 1982 *Supp. Prog. Theor. Phys.* **73** 106

References in the text

1 Feynman R P and Gell-Mann M 1959 *Phys. Rev.* **109** 143

2 Bailin D 1982 *Weak Interactions* (Bristol: Adam Hilger) Second edition

3 Glashow S L 1961 *Nucl. Phys.* **22** 579

4 Weinberg S 1967 *Phys. Rev. Lett.* **19** 1264

5 Arnison G *et al.* 1983 *Phys. Lett.* **122B** 103; **126B** 398; **129B** 273
Banner M *et al.* 1983 *Phys. Lett.* **122B** 476; **129B** 130

6 Kobayashi M and Maskawa M 1973 *Prog. Theor. Phys.* **49** 652

7 Glashow S L, Iliopoulos J and Maiani L 1970 *Phys. Rev.* D **2** 1285

15

RENORMALISATION OF ELECTROWEAK THEORY

15.1 Electroweak theory renormalisation schemes

The electroweak theory developed in the previous chapter has had considerable success in predicting accurately the results of various experiments: neutrino–electron and proton elastic scattering, $e^+e^- \to \mu^+\mu^-$, charge and neutral current deep inelastic scattering of neutrinos, deep inealstic scattering of polarised electrons, atomic parity violation, neutrino and electroproduction of pions, to name but a few. However the experimental accuracy so far achieved means that the theory is only tested at tree level. To check whether the theory is correct at the *quantum* level requires more accurate experiments, and, of course, the calculation of the predictions to one or more loop order. In this way it is to be expected that electroweak theory will be subjected to the same rigorous testing as that to which QED has been subjected during the past 35 years.

Actually to perform the calculation of the radiative corrections requires the renormalisation of the theory. Since the theory *is* renormalisable, the infinities which occur when we evaluate Feynman diagrams may be absorbed into the various renormalised parameters of the theory, and we obtain finite predictions involving these renormalised parameters. We have already observed in §7.5 that there is considerable freedom in the precise definition of the renormalised parameters; for example, the renormalised mass μ^2 has a value which is different in the various renormalisation schemes (MS, $\overline{\text{MS}}$, mom, phys), and (by definition) only in one of the schemes is μ^2 the actual physical mass of the scalar particle in the theory. The same arbitrariness exists, of course, in electroweak theory, but there is a further arbitrariness in deciding which parameters are fundamental, and which derived. We observed in §14.4 that, in the leptonic sector, the original Lagrangian involves the parameters

$$g,\ g',\ \mu^2,\ \lambda,\ G_l\ (l = \mathrm{e},\ \mu,\ \tau). \tag{15.1}$$

Because of the tree-level formulae (14.21), (14.32), (14.37), (14.58) we could instead use the equivalent sets

$$\mathrm{e},\ \sin\theta_{\mathrm{W}},\ v,\ m_{\mathrm{H}}^2,\ m_l\ (l = \mathrm{e},\ \mu,\ \tau) \tag{15.2}$$

or

$$\mathrm{e},\ m_{\mathrm{W}}^2,\ m_{\mathrm{Z}}^2,\ m_{\mathrm{H}}^2,\ m_l\ (l = \mathrm{e},\ \mu,\ \tau). \tag{15.3}$$

So long as we only make predictions at tree-level accuracy it is immaterial which set is used since the predictions can be transformed at will using the formulae relating the parameters. However, these formulae are in general not true beyond tree level, and it is necessary to select one set, and stick to it.

In view of the foregoing observations it is hardly surprising that the literature abounds[1] with well-motivated but different approaches. If we were able to calculate to infinite order then all (correct!) calculations using different approaches must agree on the predicted value of some physically measurable quantity. However, since the perturbation series is always truncated at some finite order, different approaches in general lead to different numerical predictions, even though *formally* they are in agreement. To illustrate this consider a theory with just one parameter α (like massless QED), and suppose the physical quantity to be calculated is P: for example the magnetic moment of the 'electron'. We denote by α_1 and α_2 the renormalised parameter in two different renormalisation schemes 1 and 2. The value of P is calculated to order α^2, so neglecting $O(\alpha^3)$ the predictions from the two schemes are

$$P_1 = \alpha_1 + c_1\alpha_1^2 \tag{15.4a}$$

$$P_2 = \alpha_2 + c_2\alpha_2^2 \tag{15.4b}$$

where c_1 and c_2 are independent of α. As we observed in §7.5, the renormalised α_1 and α_2 differ by a finite amount, and so, to the same accuracy,

$$\alpha_1 = \alpha_2 + k\alpha_2^2 \tag{15.5}$$

where k is a constant. Substituting into (15.4a) gives

$$P_1 = \alpha_2 + (k + c_1)\alpha_2^2 + \alpha_2^3 c_1 k(2 + k\alpha_2^3). \tag{15.6}$$

Thus the two calculations are in *formal* agreement provided

$$c_2 = k + c_1. \tag{15.7}$$

(If this is not satisfied one or both of the calculations must be incorrect.) Nevertheless the numerical values P_1 and P_2 clearly differ in general:

$$P_1 - P_2 = \alpha_2^3 c_1 k(2 + k\alpha_2^3) \tag{15.8}$$

even though the difference is formally of higher order in α. In QED the expansion parameter α is so small as to make such differences negligible, for most purposes. And in any case there is a consensus in favour of the on-shell renormalisation scheme.

In electroweak theory there is as yet no consensus. The large mass scales m_W, m_Z mean that in principle it is possible to obtain large numbers which are formally of order unity, although to date the calculations performed display a remarkable consistency in their numerical predictions. Even so, it may be thought preferable to use a 'neutral' scheme, in which the parameters have no direct physical significance. The MS and $\overline{\text{MS}}$ schemes, described in §7.5, are examples; the renormalised parameters are determined by the mass scale M which enters via dimensional regularisation, and the chosen value(s) of the gauge-fixing parameter(s) ξ. It might be argued that, by adjusting M to the characteristic energy scale of the process under consideration, one should be able to make the numerical value of the radiative corrections small compared with those arising in a different approach, or from an injudicious choice of M. (The problem of minimising radiative corrections in every order has been addressed by Stevenson[2].) Even if the preceding argument is true, the logic of it requires that the values of the various renormalised parameters should be determined from other experiments where the same value M is also the characteristic scale, and this is not always easy to arrange. Besides, the renormalised parameters in such schemes in general depend upon the gauge-fixing parameter(s) ξ. Since S-matrix elements are independent of ξ, as we showed in Appendix B, only ξ-independent combinations of the renormalised parameters can be determined from experiment or enter into theoretical predictions of scattering amplitudes. It has therefore also been argued that it is preferable to use an on-shell approach in which each renormalised parameter is automatically ξ-independent, and preferably directly ascertainable from experiment. Another more practical consideration which may affect the specific choice of renormalisation approach is the ease and convenience with which it may be applied to actual calculations of physical interest.

Without necessarily denying the arguments which have been advanced in support of the MS and $\overline{\text{MS}}$ schemes, we have, nevertheless, decided to present the on-shell scheme which has been advocated by Sirlin[3]. The simplicity of the scheme, besides making for ease of calculation, makes feasible a text-book treatment of what may otherwise become an extremely complicated topic.

15.2 Definition of the renormalised parameters

The major simplification which characterises Sirlin's scheme[3] is achieved by working throughout with *unrenormalised* fields. In other words, all wave function renormalisation constants are chosen to be *unity*. The immediate objection to such an approach is that the renormalised Green functions are then in general divergent (in the limit that the space–time dimensionality $d=2\omega$ approaches 4.) However, S-matrix elements may be rendered finite using *only* mass and coupling constant renormalisations, as we shall shortly verify. Thus provided we only wish to calculate radiative corrections to physical scattering amplitudes, we can eliminate a large number of counter terms, at a stroke. Although this elimination of counter terms obviously leads to major simplification, there is a (small) price to be paid. In the competing on-shell schemes the wavefunction renormalisation constants are *chosen* so as to ensure that propagator poles have residue unity. Then S-matrix elements are obtained immediately from the Green functions. In Sirlin's scheme, since no such choice has been made, the poles in general do not have unit residue. Thus to determine S-matrix elements it is necessary to rescale the Green functions by an appropriate amount so that *external* lines do have unit residue. (This is why the factors $Z_{R,L}$ in (15.128) for example are not equal to unity.) The ready determination of the renormalised parameters from experiment is achieved by using the parameter set (15.3) as input. The masses are all the measured physical masses of the particles and e is the measured electric charge of the positron.

We start with the bare gauge boson mass Lagrangian which may be found using (14.29) or (14.72):

$$\mathscr{L}_{\mathrm{B}}^{\mathrm{M}}(W, Z) = m_{\mathrm{WB}}^2 W_\mu^- W^{+\mu} + \tfrac{1}{2} m_{\mathrm{ZB}}^2 Z_\mu Z^\mu \tag{15.9}$$

where the suffix 'B' indicates that all quantities are bare. (We omit the suffix from the fields, since they remain throughout unrenormalised, as explained above.) The bare masses m_{WB}^2 and m_{ZB}^2 are given by (14.37):

$$m_{\mathrm{WB}}^2 = \tfrac{1}{4} g_{\mathrm{B}}^2 v_{\mathrm{B}}^2 \tag{15.10a}$$

$$m_{\mathrm{ZB}}^2 = \tfrac{1}{4}(g_{\mathrm{B}}^2 + g_{\mathrm{B}}'^2) v_{\mathrm{B}}^2 = \sec^2 \theta_{\mathrm{WB}} m_{\mathrm{WB}}^2 \tag{15.10b}$$

and the bare weak mixing angle θ_{WB} is defined by

$$\tan \theta_{\mathrm{WB}} = g'_{\mathrm{B}} / g_{\mathrm{B}} \tag{15.10c}$$

so that the mass Lagrangian is diagonalised by the combinations defined in (14.20). Counter terms are generated by the substitutions

$$m_{\mathrm{WB}}^2 = m_{\mathrm{W}}^2 + \Delta m_{\mathrm{W}}^2 = m_{\mathrm{W}}^2(1 + K_{\mathrm{W}}) \tag{15.11a}$$

$$m_{\mathrm{ZB}}^2 = m_{\mathrm{Z}}^2 + \Delta m_{\mathrm{Z}}^2 = m_{\mathrm{Z}}^2(1 + K_{\mathrm{Z}}) \tag{15.11b}$$

but *no* field renormalisations. Then

$$\mathscr{L}_{\mathrm{B}}^{\mathrm{M}}(W, Z) = \mathscr{L}^{\mathrm{M}}(W, Z) + \Delta \mathscr{L}^{\mathrm{M}}(W, Z) \tag{15.12a}$$

where the renormalised Lagrangian

$$\mathscr{L}^{\mathrm{M}}(W, Z) = m_{\mathrm{W}}^2 W_\mu^+ W^{-\mu} + \tfrac{1}{2} m_{\mathrm{Z}}^2 Z_\mu Z^\mu \tag{1512b}$$

and the counter term Lagrangian

$$\Delta \mathscr{L}^{\mathrm{M}}(W, Z) = K_{\mathrm{W}} m_{\mathrm{W}}^2 W_\mu^+ W^{-\mu} + \tfrac{1}{2} K_{\mathrm{Z}} m_{\mathrm{Z}}^2 Z_\mu Z^\mu. \tag{15.12c}$$

The renormalised quantities m_{W} and m_{Z} are so far unspecified, but we now identify them with the *physical* masses[4] of the observed W and Z particles, given in (14.112):

$$m_{\mathrm{W}} = m_{\mathrm{W,phys}} = 80.8 \pm 2.7 \text{ GeV} \tag{15.13a}$$

$$m_{\mathrm{Z}} = m_{\mathrm{Z,phys}} = 92.9 \pm 1.6 \text{ GeV}. \tag{15.13b}$$

Thus these particular (combinations of the) renormalised parameters are guaranteed to be independent of ξ, and have already been ascertained from experiment.

The form of the renormalised mass Lagrangian (15.12b) is the same as that of the unrenormalised mass Lagrangian (15.9) with the replacement of the bare masses by the renormalised masses. Thus when combined with the (unrenormalised) kinetic energy terms in (14.72) and those in the gauge-fixing Lagrangian, they lead to the same form of propagators (14.38)–(14.40), but with m_W and m_Z now having their renormalised values (15.13). The counter term Lagrangian, $\Delta\mathscr{L}^M$ generates additional vertices having the following Feynman rules:

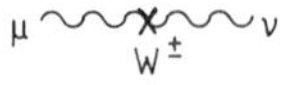

$$: \quad ig_{\mu\nu}K_W m_W^2 \tag{15.14a}$$

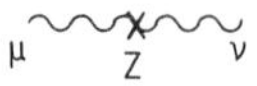

$$: \quad ig_{\mu\nu}K_Z m_Z^2. \tag{15.14b}$$

The gauge fixing Lagrangian also contains mass terms for the Goldstone modes $G^{\pm}$, G^0, with

$$m_B^2(G^+) = \xi m_{WB}^2 \tag{15.15a}$$

$$m_B^2(G^0) = \xi m_{ZB}^2. \tag{15.15b}$$

Thus the renormalisation (15.11) also renormalises the masses of these Goldstone particles. When combined with the (unrenormalised) kinetic energy terms for these particles, which are contained in (14.73), these renormalised mass terms generate propagators having the same form as (14.45), (14.46), but with m_W and m_Z now having their renormalised values (15.13). The counter term Goldstone mass Lagrangian generates additional vertices with the following Feynman rules:

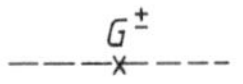

$$: \quad -iK_W \xi m_W^2 \tag{15.16a}$$

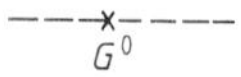

$$: \quad -iK_Z \xi m_Z^2. \tag{15.16b}$$

We turn next to the fermion–gauge boson interactions contained in (14.17). Using (14.22) and remembering that all parameters are bare, we write the interactions with the electromagnetic field A_μ as

$$\mathscr{L}_B^f(\text{em}) = -e_B j_\mu A^\mu \tag{15.17a}$$

where

$$e_B \equiv g_B g_B'(g_B^2 + g_B'^2)^{-1/2} \tag{15.17b}$$

and the electromagnetic current j_μ is given in (14.12) for the leptons and in (14.145) for the quarks. As before, the counter terms are generated by the substitution

$$e_B = e + \Delta e = e(1 + K_e) \tag{15.18}$$

(and *no* field operator renormalisations). Then

$$\mathscr{L}_B^f(\text{em}) = \mathscr{L}^f(\text{em}) + \Delta\mathscr{L}^f(\text{em}) \tag{15.19a}$$

where the renormalised Lagrangian

$$\mathscr{L}^f(\text{em}) = -ej_\mu A^\mu \tag{15.19b}$$

and the counter term Lagrangian

$$\Delta\mathscr{L}^f(\text{em}) = -K_e ej_\mu A^\mu. \tag{15.19c}$$

The renormalised parameter e is how identified with the measured physical electric charge of the positron given in (14.119):

$$e=0.302\,822\,1. \tag{15.20}$$

Thus this parameter too is guaranteed to be independent of ξ, and has already been ascertained from experiment.

The remaining fermion–gauge boson interactions given in (14.17), involving the W and Z bosons, are expressed in terms of g_{B} and θ_{WB}, or equivalently g_{B} and g'_{B}. This gives (for the leptons)

$$\begin{aligned}\mathscr{L}^{\mathrm{f}}_{\mathrm{B}}(W,Z)= &-\frac{g_{\mathrm{B}}}{2\sqrt{2}}(L_\mu W^{-\mu}+L^\dagger_\mu W^{+\mu})\\ &-(g_{\mathrm{B}}^2+g_{\mathrm{B}}'^2)^{1/2}\left(T^3_\mu-\frac{g_{\mathrm{B}}'^2}{g_{\mathrm{B}}^2+g_{\mathrm{B}}'^2}j_\mu\right)Z^\mu\end{aligned} \tag{15.21}$$

where the currents L_μ, $L^\dagger_\mu$ anf T^3_μ are defined in (14.3) and (14.13). Using (15.10) and (15.17) we can equivalently express it in terms of e_{B}, m_{WB} and m_{ZB}. Since we have now completely specified the renormalised parameters e, m_{W} and m_{Z}, it follows that the corresponding renormalised and counter term Lagrangians $\mathscr{L}^{\mathrm{f}}(W,Z)$ and $\Delta\mathscr{L}^{\mathrm{f}}(W,Z)$ are also determined. The renormalised Lagrangian is

$$\begin{aligned}\mathscr{L}^{\mathrm{f}}(W,Z)= &-\frac{g}{2\sqrt{2}}(L_\mu W^{-\mu}+L^\dagger_\mu W^{+\mu})\\ &-(g^2+g'^2)^{1/2}\left(T^3_\mu-\frac{g'^2}{g^2+g'^2}j_\mu\right)Z^\mu\end{aligned} \tag{15.22a}$$

where

$$g\equiv e m_{\mathrm{Z}}(m_{\mathrm{Z}}^2-m_{\mathrm{W}}^2)^{-1/2} \tag{15.22b}$$

$$g'\equiv e m_{\mathrm{Z}}/m_{\mathrm{W}}\equiv g\tan\theta_{\mathrm{W}} \tag{15.22c}$$

and the counter term Lagrangian is

$$\begin{aligned}\Delta\mathscr{L}^{\mathrm{f}}(W,Z)= &-\frac{Kg}{2\sqrt{2}}(L_\mu W^{-\mu}+L^\dagger_\mu W^{+\mu})\\ &-K_3(g^2+g'^2)^{1/2}T^3_\mu Z^\mu+K_4\frac{g'^2}{(g^2+g'^2)^{1/2}}j_\mu Z^\mu\end{aligned} \tag{15.23}$$

where

$$\begin{aligned}(1+K)^2&=(1+K_{\mathrm{e}})^2(1+K_{\mathrm{Z}})\sin^2\theta_{\mathrm{W}}(\sin^2\theta_{\mathrm{W}}+K_{\mathrm{Z}}-\cos^2\theta_{\mathrm{W}}K_{\mathrm{W}})^{-1}\\ (1+K_3)^2&=(1+K_{\mathrm{e}})^2(1+K_{\mathrm{Z}})^2(1+K_{\mathrm{W}})^{-1}\\ &\quad\times\sin^2\theta_{\mathrm{W}}(\sin^2\theta_{\mathrm{W}}+K_{\mathrm{Z}}-\cos^2\theta_{\mathrm{W}}K_{\mathrm{W}})^{-1}\\ (1+K_4)^2&=(1+K_{\mathrm{e}})^2(1+K_{\mathrm{W}})^{-1}\operatorname{cosec}^2\theta_{\mathrm{W}}(\sin^2\theta_{\mathrm{W}}+K_{\mathrm{Z}}-\cos^2\theta_{\mathrm{W}}K_{\mathrm{W}}).\end{aligned} \tag{15.24}$$

In (15.22) and (15.24) we have *defined* the angle θ_W by

$$\tan\theta_W \equiv g'/g. \tag{15.25}$$

Thus it follows that in Sirlin's scheme

$$\sin^2\theta_W \equiv (m_Z^2 - m_W^2)/m_Z^2 = 0.243 \pm 0.022 \tag{15.26}$$

as in (14.113). Note that θ_W is *not* the weak mixing angle; the (unrenormalised) fields are mixed with mixing angle θ_{WB}, given in (15.10c), which is not equal to θ_W.

As before, the forms of the renormalised Lagrangians $\mathscr{L}^f(\text{em})$ and $\mathscr{L}^f(W, Z)$ are the same as the forms of the bare Lagrangians with the replacement of the bare coupling constants e_B, g_B and g'_B by the renormalised values given in (15.20), (15.22). Thus the corresponding Feynman rules are also those given in (14.61) to (14.65) but with g now given by (15.22b) and θ_W by (15.26). The counter term Lagrangians generate additional vertices with the following Feynman rules:

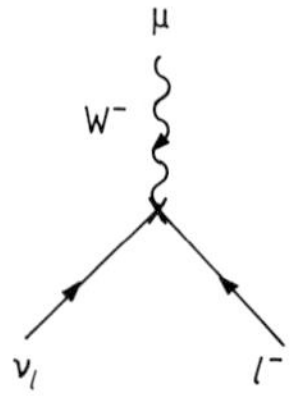

$$: \quad \frac{-iKg}{2\sqrt{2}}\gamma_\mu(1-\gamma_5) \tag{15.27}$$

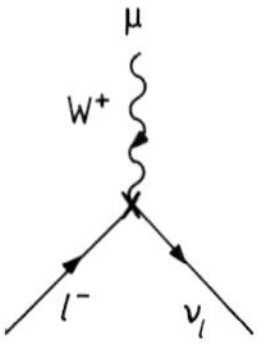

$$: \quad \frac{-iKg}{2\sqrt{2}}\gamma_\mu(1-\gamma_5) \tag{15.28}$$

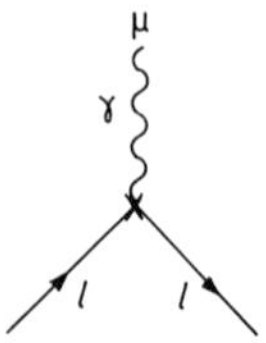

$$: \quad iK_e e\gamma_\mu \tag{15.29}$$

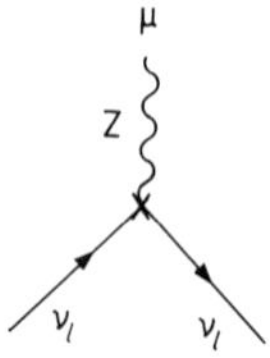

$$: \quad \frac{-iK_3 g}{4\cos\theta_W}\gamma_\mu(1-\gamma_5) \tag{15.30}$$

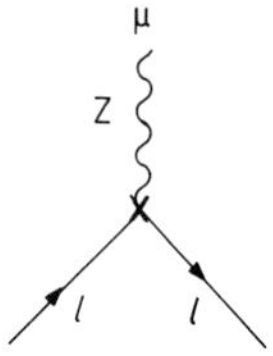

$$: \quad \frac{\mathrm{i}g}{2\cos\theta_{\mathrm{W}}}\gamma_\mu\left(\frac{K_3}{2}-2K_4\sin^2\theta_{\mathrm{W}}-\frac{K_3}{2}\gamma_5\right). \tag{15.31}$$

As before, g and θ_{W} are given in (15.22b) and (15.26). The constants K, K_3 and K_4 are given in terms of K_{e}, K_{Z}, K_{W} in (15.24).

The treatment of the quark–gauge boson interactions proceeds similarly. Again the Feynman rules for the vertices deriving from the renormalised Lagrangian have the same form as those in (14.149) to (14.153), but with g given by (15.22b) and θ_{W} by (15.26). The counter term Lagrangian generates vertices with the following Feynman rules

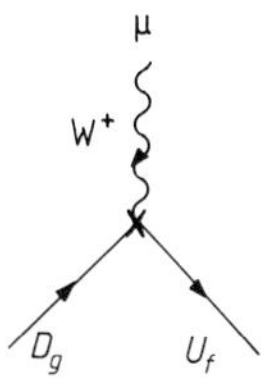

$$: \quad \frac{-\mathrm{i}Kg}{2\sqrt{2}}V_{fg}\gamma_\mu(1-\gamma_5) \tag{15.32}$$

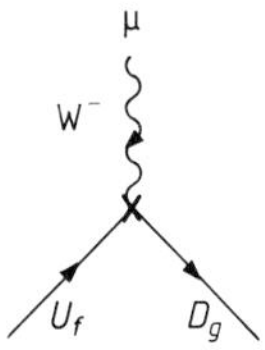

$$: \quad \frac{-\mathrm{i}Kg}{2\sqrt{2}}V_{fg}^{*}\gamma_\mu(1-\gamma_5) \tag{15.33}$$

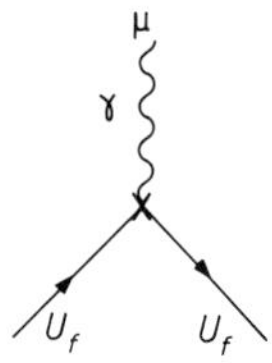

$$: \quad -\mathrm{i}\tfrac{2}{3}K_{\mathrm{e}}e\gamma_\mu \tag{15.34}$$

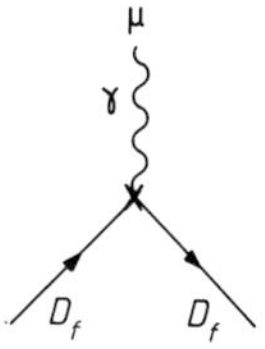

$$: \quad \mathrm{i}\tfrac{1}{3}K_{\mathrm{e}}e\gamma_\mu \tag{15.35}$$

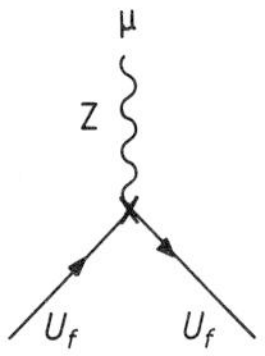

$$: \quad \frac{-ig}{2\cos\theta_W}\gamma_\mu\left(\frac{K_3}{2}-\frac{4}{3}K_4\sin^2\theta_W-\frac{K_3}{2}\gamma_5\right) \tag{15.36}$$

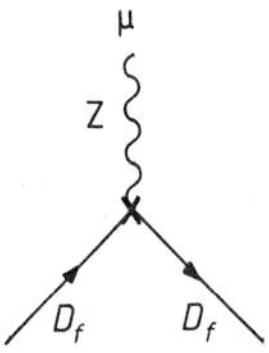

$$: \quad \frac{-ig}{2\cos\theta_W}\gamma_\mu\left(-\frac{K_3}{2}+\frac{2}{3}K_4\sin^2\theta_W+\frac{K_3}{2}\gamma_5\right). \tag{15.37}$$

The constants K, K_3, K_4 are given in terms of K_e, K_W and K_Z in (15.24).

Other pieces of the bare Lagrangian are also expressible in terms of g_B, g'_B and v_B, or equivalently m_{WB}, m_{ZB} and e_B. So, just like $\mathscr{L}^f_B(W, Z)$ the corresponding renormalised and counter term Lagrangians are also completely specified. For example, the self-interaction of the gauge bosons is expressible in terms of g_B, g'_B, when the fields corresponding to the mass eigenstates $W^\pm$, Z, A are used. Thus, as before, the Feynman rules for the renormalised Lagrangian are those given in (14.66) to (14.71); again with g and θ_W given in (15.22b) and (15.26). Proceeding as before we find that the counter term Lagrangian generates vertices with the following Feynman rules:

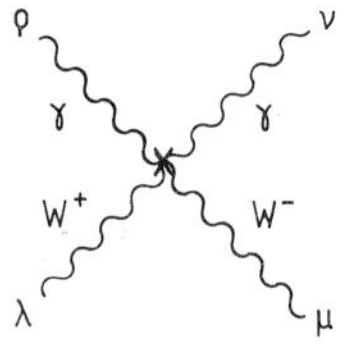

$$: \quad -i(K_e^2+2K_e)e^2[2g_{\nu\rho}g_{\lambda\mu}-g_{\mu\rho}g_{\lambda\nu}-g_{\mu\nu}g_{\lambda\rho}] \tag{15.38}$$

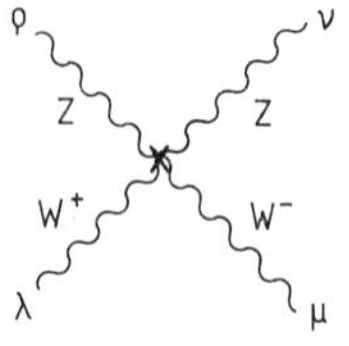

$$: \quad -i(K_5^2+2K_5)g^2\cos^2\theta_W[2g_{\nu\rho}g_{\lambda\mu}-g_{\mu\rho}g_{\lambda\nu}-g_{\mu\nu}g_{\lambda\rho}] \tag{15.39}$$

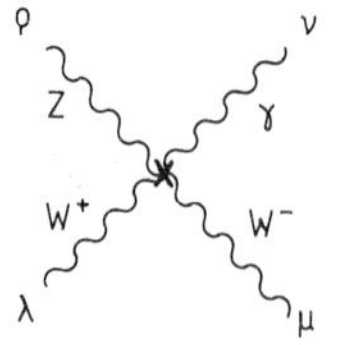

$$: \quad -i(K_5K_e+K_5+K_e)eg\cos\theta_W[2g_{\nu\rho}g_{\lambda\mu}-g_{\mu\rho}g_{\lambda\nu}-g_{\mu\nu}g_{\lambda\rho}] \tag{15.40}$$

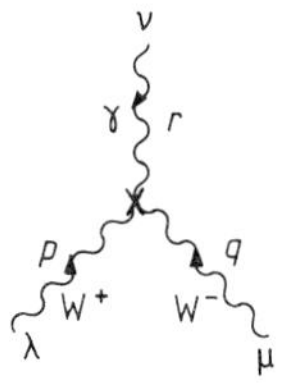

$$: \quad iK_e e[(r-q)_\lambda g_{\mu\nu}+(q-p)_\nu g_{\lambda\mu}+(p-r)_\mu g_{\nu\lambda}] \qquad (15.41a)$$

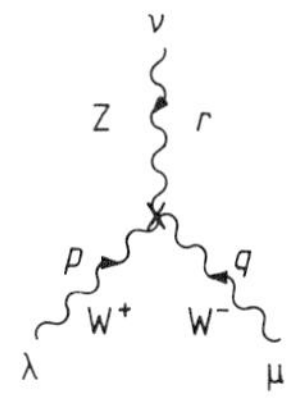

$$: \quad iK_5 g \cos\theta_W[(r-q)_\lambda g_{\mu\nu}+(q-p)_\nu g_{\lambda\mu}+(p-r)_\mu g_{\nu\lambda}] \qquad (15.41b)$$

where K_5 is given by

$$(1+K_5)^2 \equiv (1+K_e)^2(1+K_W)\sin^2\theta_W[\sin^2\theta_W+K_Z-\cos^2\theta_W K_W]^{-1} \qquad (15.42)$$

and g and θ_W are given in (15.22b) and (15.26).

We turn next to the renormalisation of the scalar Lagrangian (14.73). The first two terms, arising from the covariant derivative of the scalar doublet, are again expressible in terms of e_B, m_{WB} and m_{ZB}, and so their renormalised Lagrangian leads to the vertices (14.74) to (14.94) with g and θ_W given in (15.22b) and (15.26), and there are corresponding vertices generated by the counter term Lagrangian. Proceeding as before we find that the Feynman rules for these vertices are as follows:

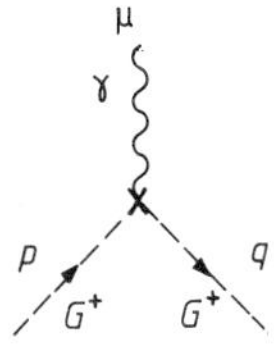

$$: \quad -ieK_e(p+q)_\mu \qquad (15.43)$$

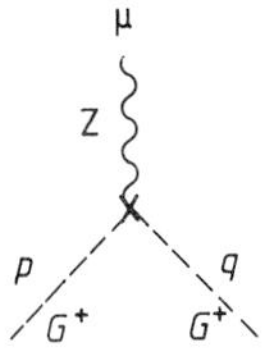

$$: \quad -iK_6 g\frac{\cos 2\theta_W}{\cos\theta_W}(p+q)_\mu \qquad (15.44)$$

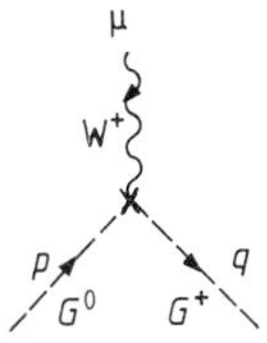

$$: \quad \tfrac{1}{2}Kg(p+q)_\mu \qquad (15.45)$$

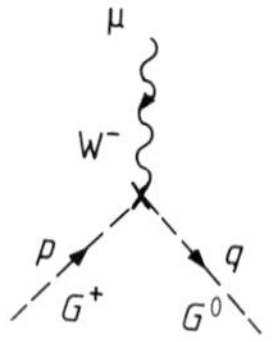

$$: \quad -\tfrac{1}{2}Kg(p+q)_{\mu} \tag{15.46}$$

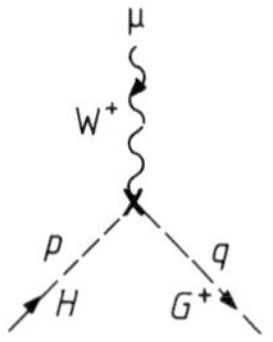

$$: \quad -\mathrm{i}\tfrac{1}{2}Kg(p+q)_{\mu} \tag{15.47}$$

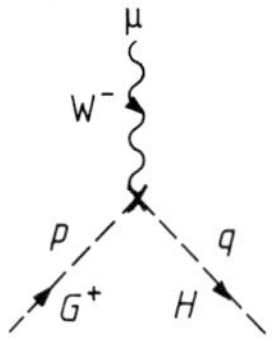

$$: \quad -\mathrm{i}\tfrac{1}{2}Kg(p+q)_{\mu} \tag{15.48}$$

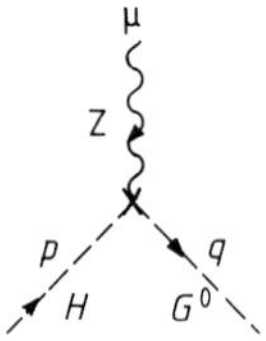

$$: \quad \tfrac{1}{2}K_3(p+q)_{\mu} \tag{15.49}$$

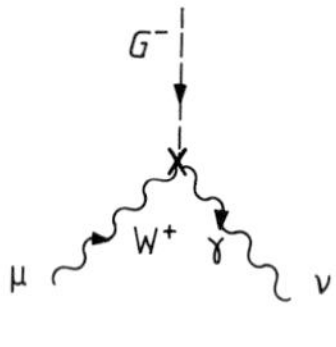

$$: \quad \mathrm{i}K_7 e m_{\mathrm{W}} g_{\mu\nu} \tag{15.50}$$

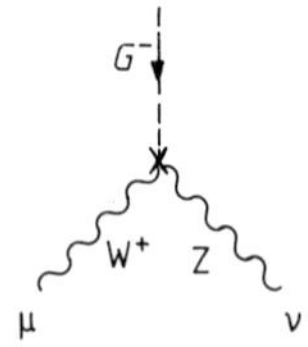

$$: \quad -\mathrm{i}K_8 g m_{\mathrm{Z}} \sin^2\theta_{\mathrm{W}} g_{\mu\nu} \tag{15.51}$$

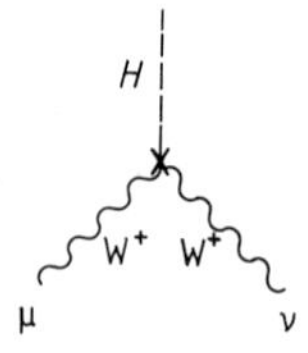

$$: \quad \mathrm{i}K_9 g m_{\mathrm{W}} g_{\mu\nu} \tag{15.52}$$

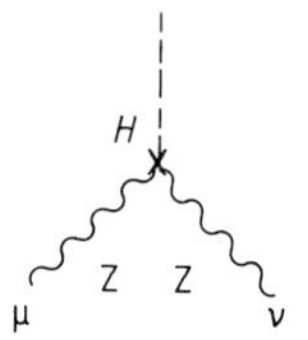

: $\quad iK_{10}gm_Z g_{\mu\nu}$ (15.53)

where K is defined in (15.24), and

$$K_6=(1+K_Z)^{-1}\sec 2\theta_W[(K_3-K_Z)\cos 2\theta_W+2\cos^2\theta_W(1+K_3)(K_W-K_Z)] \tag{15.54a}$$

$$1+K_7=(1+K_e)(1+K_W)^{1/2} \tag{15.54b}$$

$$(1+K_8)^2=(1+K_e)^2\operatorname{cosec}^2\theta_W[\sin^2\theta_W+K_Z-K_W\cos^2\theta_W] \tag{15.54c}$$

$$1+K_9=(1+K)(1+K_W)^{1/2} \tag{15.54d}$$

$$1+K_{10}=(1+K)(1+K_Z)^{1/2} \tag{15.54e}$$

with K_3 as defined in (15.24).

The remaining terms in (14.73) all involve the (bare) scalar self-interaction coupling constant λ_B, or equivalently the (bare) Higgs particle mass. We start with the Higgs mass Lagrangian

$$\mathscr{L}_B^M(H)=-\tfrac{1}{2}m_{HB}^2H^2 \tag{15.55}$$

and renormalise in a manner directly analogous to (15.11):

$$m_{HB}^2=m_H^2(1+K_H) \tag{15.56}$$

with *no* field renormalisation. Then

$$\mathscr{L}_B^M(H)=\mathscr{L}^M(H)+\Delta\mathscr{L}^M(H) \tag{15.57}$$

where the renormalised Lagrangian

$$\mathscr{L}^M(H)=-\tfrac{1}{2}m_H^2H^2 \tag{15.58}$$

and the counter term Lagrangian

$$\Delta\mathscr{L}^M(H)=-\tfrac{1}{2}K_Hm_H^2\,H^2. \tag{15.59}$$

As before, the renormalised mass m_H is now identified with the *physical* mass of the (so far unobserved) Higgs particle:

$$m_H=m_{H,\text{phys}}. \tag{15.60}$$

Assuming that the Higgs particle is eventually discovered, the renormalised parameter m_H will be directly ascertainable from experiment, and is guaranteed to be gauge-parameter independent. As before, this means that the Higgs particle propagator has the same form as (14.47), but now m_H has its renormalised value. The counter term Lagrangian generates an additional

vertex having the Feynman rule

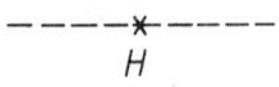

$$: \quad -\mathrm{i}K_{\mathrm{H}}m_{\mathrm{H}}^2. \tag{15.61}$$

Since

$$m_{\mathrm{HB}}^2 = \tfrac{1}{2}\lambda_{\mathrm{B}}v_{\mathrm{B}}^2 \tag{15.62}$$

we may express λ_{B} in terms of m_{HB} and the parameters e_{B}, m_{WB}, m_{ZB}, whose renormalisation has already been defined. Thus the renormalisation of the remaining (interaction) terms in (14.73) is completely specified. The renormalised Lagrangian is

$$\begin{aligned}\mathscr{L}_\varphi(G,H) = &-\tfrac{1}{2}\lambda v[G^+G^- + \tfrac{1}{2}(G^{02}+H^2)]H \\ &-\tfrac{1}{4}\lambda[G^+G^- + \tfrac{1}{2}(G^{02}+H^2)]^2\end{aligned} \tag{15.63}$$

where

$$\lambda \equiv \frac{e^2 m_{\mathrm{H}}^2 m_{\mathrm{Z}}^2}{2m_{\mathrm{W}}^2(m_{\mathrm{Z}}^2 - m_{\mathrm{W}}^2)} \tag{15.64a}$$

$$\lambda v \equiv \frac{e m_{\mathrm{H}}^2 m_{\mathrm{Z}}}{m_{\mathrm{W}}(m_{\mathrm{Z}}^2 - m_{\mathrm{W}}^2)^{1/2}} \tag{15.64b}$$

and the counter term Lagrangian is

$$\begin{aligned}\Delta\mathscr{L}_\varphi(G,H) = &-\tfrac{1}{2}K_{11}\lambda v[G^+G^- + \tfrac{1}{2}(G^{02}+H^2)]H \\ &-\tfrac{1}{4}K_\lambda\lambda[G^+G^- + \tfrac{1}{2}(G^{02}+H^2)]^2\end{aligned} \tag{15.65}$$

where

$$\begin{aligned}1+K_\lambda = &(1+K_{\mathrm{e}})^2(1+K_{\mathrm{H}})(1+K_{\mathrm{Z}})(1+K_{\mathrm{W}})^{-1}\sin^2\theta_{\mathrm{W}} \\ &\times(\sin^2\theta_{\mathrm{W}} + K_{\mathrm{Z}} - K_{\mathrm{W}}\cos^2\theta_{\mathrm{W}})^{-1}\end{aligned} \tag{15.66a}$$

$$\begin{aligned}1+K_{11} = &(1+K_{\mathrm{e}})(1+K_{\mathrm{H}})(1+K_{\mathrm{Z}})^{1/2}(1+K_{\mathrm{W}})^{-1/2}\sin\theta_{\mathrm{W}} \\ &\times(\sin^2\theta_{\mathrm{W}} + K_{\mathrm{Z}} - K_{\mathrm{W}}\cos^2\theta_{\mathrm{W}})^{-1/2}.\end{aligned} \tag{15.66b}$$

Clearly the renormalised Lagrangian gives vertices having the Feynman rules (14.95) to (14.98), but now with renormalised parameters. The counter term Lagrangian generates analogous vertices for the (three-scalar) vertices with the substitution

$$g m_{\mathrm{H}}^2/m_{\mathrm{W}}^2 \to K_{11} g m_{\mathrm{H}}^2/m_{\mathrm{W}}^2. \tag{15.67}$$

(Note that, using (15.22b), $g m_{\mathrm{H}}^2/m_{\mathrm{W}} = \lambda v$.) The counter term four-scalar vertices are obtained by the substitution

$$\lambda = \tfrac{1}{2}g^2 m_{\mathrm{H}}^2/m_{\mathrm{W}}^2 \to K_\lambda \tfrac{1}{2} g^2 m_{\mathrm{H}}^2 m_{\mathrm{W}}^2. \tag{15.68}$$

We have seen in (14.49) and (14.50) that the ghost particles $\eta^\pm, \eta^{\mathrm{Z}}$, and their conjugates, have squared masses ξm_{W}^2 and ξm_{Z}^2 respectively. Thus the renormalisation (15.11) of the gauge particle masses also renormalises the masses of the ghosts. As before the effect is to give ghost propagators of the same form, but with renormalised masses, and to generate counter term interactions with Feynman rules

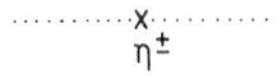

$$: \quad -\mathrm{i}K_{\mathrm{W}}\xi m_{\mathrm{W}}^2 \tag{15.69}$$

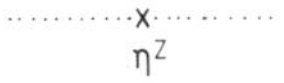

$$: \quad -\mathrm{i}K_{\mathrm{Z}}\xi m_{\mathrm{Z}}^2. \tag{15.70}$$

The previously introduced renormalisations of e_{B}, m_{WB}, m_{ZB} also renormalise the interaction part (14.102) of the Faddeev–Popov Langrangian. The renormalised Lagrangian generates vertices having Feynman rules of the same form as (14.103) to (14.114), but with renormalised parameters $e, g, \theta_{\mathrm{W}}, m_{\mathrm{W}}, m_{\mathrm{Z}}$ (the renormalised values of g and θ_{W} are defined in (15.22b) and (15.26)). The counter term Lagrangian generates analogous vertices for which the Feynman rules are obtained by the following substitutions:

$$e \rightarrow K_{\mathrm{e}}e \tag{15.71a}$$

$$\xi g m_{\mathrm{W}} \rightarrow K_9 \xi g m_{\mathrm{W}} \tag{15.71b}$$

$$\xi g m_{\mathrm{W}} \frac{\cos 2\theta_{\mathrm{W}}}{\cos \theta_{\mathrm{W}}} \rightarrow [(1+K_6)(1+K_{\mathrm{W}})^{1/2}-1]\xi g m_{\mathrm{W}} \frac{\cos 2\theta_{\mathrm{W}}}{\cos \theta_{\mathrm{W}}} \tag{15.71c}$$

$$\xi g m_{\mathrm{Z}} \rightarrow K_{10}\xi g m_{\mathrm{Z}} \tag{15.71d}$$

where K, K_6, K_9, K_{10} are defined in (15.24) and (15.54). (Since the ghost particles only arise in closed loops, these ghost counter terms contribute first only in two-loop order.)

Finally we renormalise the Yukawa sector of the theory. The (bare) lepton mass Lagrangian (14.57) may be written in the form

$$\mathscr{L}_{\mathrm{B}}^{\mathrm{M}}(l) = -\sum_l m_{l\mathrm{B}}\bar{l}l \tag{15.72}$$

using the relation (14.58). We generate counter terms by making the substitutions

$$m_{l\mathrm{B}} = m_l(1+k_l) \qquad (l=\mathrm{e}, \mu, \tau) \tag{15.73}$$

where the renormalised masses are identified with the *physical* masses (14.120) of the observed leptons. Evidently these renormalised parameters are guaranteed to be gauge-parameter independent, and have already been ascertained from experiment. As before these renormalisations ensure that the renormalised lepton propagators have the same form as (14.59) and (14.60), but with m_l now the renormalised mass. The counter terms lead to the following Feynman rules

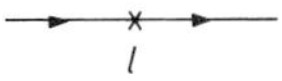

$$: \quad -\mathrm{i}k_l m_l. \tag{15.74}$$

The renormalisation of the remaining Yukawa interactions in (14.57) is now completely specified by the above renormalisation of $m_{l\mathrm{B}}$ and the previously introduced renormalisations of e_{B}, m_{WB} and m_{ZB}. The renormalised vertices are given by (14.115) to (14.118), but with g, m_l and m_{W} now having their renormalised values. The corresponding counter terms are obtained by the substitution

$$g\frac{m_l}{m_{\mathrm{W}}} \rightarrow [(1+K)(1+k_l)(1+K_{\mathrm{W}})^{-1/2}-1]g\frac{m_l}{m_{\mathrm{W}}} \qquad (15.75)$$

in all four vertices.

The inclusion of the quarks is only slightly more complicated. The (bare) quark mass Lagrangian (14.146) may be written in the form

$$\mathscr{L}_{\mathrm{B}}^{\mathrm{M}}(q) = -[\bar{D}_f m_{\mathrm{B}}(D)_{fg} D_g + \bar{U}_f m_{g\mathrm{B}}(U)_{fg} U_g] \qquad (15.76)$$

using (14.150). The matrices $\mathbf{m}_{\mathrm{B}}(U)$, $\mathbf{m}_{\mathrm{B}}(D)$ are the diagonal quark mass matrices given in (14.148). The renormalisation is achieved by the substitutions

$$\mathbf{m}_{\mathrm{B}}(D) = \mathbf{m}(D)[1+\mathbf{k}(D)] \qquad (15.77a)$$

$$\mathbf{m}_{\mathrm{B}}(U) = \mathbf{m}(U)[1+\mathbf{k}(U)] \qquad (15.77b)$$

where $\mathbf{m}(D)$, $\mathbf{m}(U)$, $\mathbf{k}(D)$, $\mathbf{k}(U)$ are all diagonal. Then the quark propagators have the same form as (14.149), but with m_q now the renormalised mass. As before, this is chosen to be the 'physical' mass of the 'observed' quark. The counter term vertices are

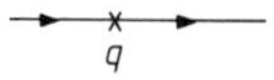

$$: \quad -ik_q m_q \qquad (15.78)$$

where k_q is the appropriate element of the matrix $\mathbf{k}(U)$ or $\mathbf{k}(D)$. Then the renormalised Yukawa vertices are given by (14.152) to (14.157), but with g, $\mathbf{m}(U)$, $\mathbf{m}(D)$ and m_{W} now having their renormalised values. The corresponding counter terms are obtained by the substitutions

$$\frac{g}{m_{\mathrm{W}}}\mathbf{m}(D) \rightarrow \{(1+K)[1+\mathbf{k}(D)](1+K_{\mathrm{W}})^{-1/2}-1\}\frac{g}{m_{\mathrm{W}}}\mathbf{m}(D) \qquad (15.79a)$$

$$\frac{g}{m_{\mathrm{W}}}\mathbf{m}(U) \rightarrow \{(1+K)[1+\mathbf{k}(U)](1+K_{\mathrm{W}})^{-1/2}-1\}\frac{g}{m_{\mathrm{W}}}\mathbf{m}(U) \qquad (15.79b)$$

in all six vertices.

15.3 Evaluation of the renormalisation constants

In the previous section we gave precise definitions of the fundamental renormalised parameters (e, m_{W}, m_{Z}, m_{H}, m_l, m_q) which characterise Sirlin's electroweak renormalisation scheme. Each fundamental renormalised parameter has an associated counter term vertex, given in (15.16), (15.29), (15.61), (15.74), (15.78), and all other counter term vertices generated by the renormalisation are expressed in terms of the constants $K_e, K_W, K_Z, K_H, k_l, k_q$. Thus in order to calculate S-matrix elements it is necessary to evaluate the renormalisation constants. In this section we shall indicate how to determine them to one-loop order.

We start with K_Z, defined in (15.11), and consider the S-matrix element for the process in which a single (physical) incoming Z boson of momentum p_1 goes to a single outgoing Z of momentum p_2; momentum conservation of course ensures that $p_1 = p_2$. To one-loop order

$S_{fi} =$

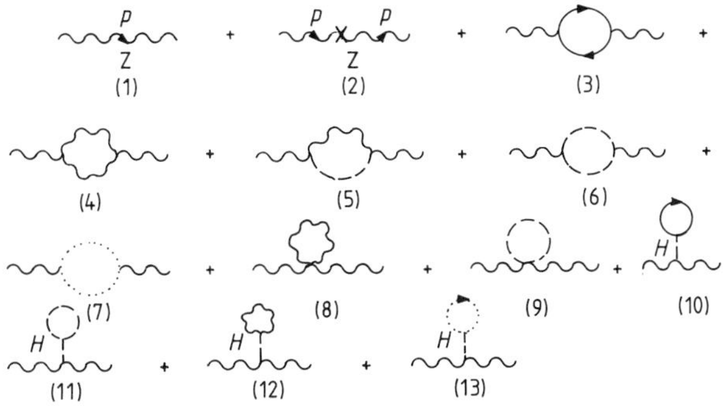

(15.80)

The large number of diagrams which contribute even at one-loop order has necessitated some economy of notation. In the above equation the external lines are all (physical on-shell) Z boson lines, but the internal lines are to be understood to include all allowed particles with the convention that continuous lines ——▸—— represent fermions, wavy lines ∿∿∿ represent gauge vector boson, dashed lines – – – – represent scalars (Goldstone and Higgs), and dotted lines ····▸···· represent ghosts. Thus in the above example the first fermion loop diagram represents the contribution from all three colours of all six quark flavours, plus all six leptons. It follows from the Feynman rules given in (14.115) to (14.118) that in the last four ('tadpole') diagrams the exchanged scalar must be a Higgs particle (so in the fermion tadpole there is no contribution from the neutrinos assuming they have zero mass). The first diagram represents the no-scattering contribution which is present in all S-matrix elements. Since there *is* no scattering for the process $Z \to Z$ this must give the *entire* contribution. So

$$S_{fi} = S_{fi}(1) \tag{15.81}$$

by definition. It follows that the remaining contributions to S_{fi} must sum to zero:

$$\sum_{n=2}^{13} S_{fi}(n) = 0. \tag{15.82}$$

Thus the counter term contribution $S_{\mathrm{fi}}(2)$, proportional to K_Z, must be adjusted so as to cancel the remaining contributions which, as explained in the previous section, are completely determined by the (physical) parameters e, m_W, m_Z, m_H, m_l, m_q. We denote the sum of the contributions from diagrams (3), (4), (5), (6), (7), (8) and (9) *without the external polarisation vectors and for a general momentum* p by

$$\Pi_{Z\mu\nu}(p) \equiv A_Z(p^2)\left(g_{\mu\nu} - \frac{p_\mu p_\nu}{p^2}\right) + B_Z(p^2)\frac{p_\mu p_\nu}{p^2}. \tag{15.83}$$

Thus $\Pi_{Z\mu\nu}(p)$ is essentially the OPI two-point function $\Gamma^{(2)}_{\mu\nu}(p, -p)$. Similarly let us denote the sum of the tadpole contributions (10), (11), (12), (13) *without the external polarisation vectors and for general* p by

$$T_{Z\mu\nu}(p) = T_Z g_{\mu\nu}. \tag{15.84}$$

The form (15.83) is the most general allowed by Lorentz covariance and the form (15.84) follows from the form (14.84) of the ZZH vertex. The requirement (15.82) that S-matrix element contributions (2) to (13) cancel means that

$$\varepsilon^{(1)\mu}[\mathrm{i}g_{\mu\nu}K_Z m_Z^2 + \Pi_{Z\mu\nu}(p) + T_{Z\mu\nu}(p)]\varepsilon^{(2)\nu} = 0 \tag{15.86}$$

when $p^2 = m_Z^2$ and $\varepsilon^{(1)}$ and $\varepsilon^{(2)}$ are the physical polarisation vectors of the initial and final Z boson. It follows from (13.3) that the polarisation vectors of a massive vector boson satisfy

$$\varepsilon^{(1)} \cdot p = 0 = \varepsilon^{(2)} \cdot p. \tag{15.87}$$

Then substituting (15)84) and (15.85) into (15.86) we find

$$[\mathrm{i}K_Z m_Z^2 + A_Z(m_Z^2) + T_Z]\varepsilon^{(1)} \cdot \varepsilon^{(2)} = 0. \tag{15.88}$$

Since this must be satisfied for all choices of $\varepsilon^{(1)}$ and $\varepsilon^{(2)}$ (including $\varepsilon^{(1)} = \varepsilon^{(2)}$), we see that K_Z is determined by the (calculable) quantities $(A_Z m_Z^2)$ and T_Z:

$$K_Z = \mathrm{i}m_Z^{-2}[A_Z(m_Z^2) + T_Z]. \tag{15.89}$$

Thus 'all' that is required to evaluate K_Z is the calculation of the loop diagrams in (15.80). Before discussing the details of these calculations we note that a precisely analogous argument can be presented for the S-matrix element for the process in which a single incoming W boson goes to a single outgoing W boson. Then the renormalisation constant K_W is given by

$$K_W = \mathrm{i}m_W^{-2}[A_W(m_W^2) + T_W] \tag{15.90}$$

where $A_W(p^2)$ and T_W are defined in exact analogy to $A_Z(p^2)$ and T_Z in (15.83), (15.84).

The loop diagrams displayed in (15.80) all have divergent momentum integrals, in four dimensions, and it is necessary to regulate them. As in previous chapters we use the dimensional regularisation introduced in §7.2. This involves performing the integrations in 2ω-dimensional space–time.

Then the divergence of the four-dimensional integrals is reflected by the appearance of poles at $\varepsilon \equiv 2-\omega = 0$ in the corresponding 2ω-dimensional integrals. Also, the diagrams in general depend upon the gauge fixing parameter(s) ξ which enters via the Feynman rules for the propagators and vertices given in Chapter 14. We are only concerned with S-matrix elements, which we have demonstrated in Appendix B are independent of ξ. Thus we may choose any convenient value of the gauge parameter. We always use the 'Feynman gauge' in which

$$\xi = 1. \tag{15.91}$$

Using the techniques of Chapters 7 and 11, the evaluation of the loop diagrams displayed in (15.80) is now straightforward in principle. However in practice it is tedious and very protracted. General formulae for the various types of Feynman integral which are encountered have been given by Aoki *et al.*[5], but even so the sheer complexity of the results makes a complete and detailed treatment impossible within the limited space of a single chapter. For example, Aoki *et al.*[5] find that the quantity $A_Z(p^2)$, defined in (15.83), is given by

$$\begin{aligned} A_Z(p^2) = \frac{-i\hat{e}^2}{32\pi^2 m_W^2 (m_Z^2 - m_W^2)} \Bigg[& \bigg(\sum_i m_Z^4 [-\tfrac{1}{6} p^2 (\eta_i^2 + 1) + m_i^2] \\ & - 2m_Z^2 (m_Z^4 + 2m_Z^2 m_W^2 - 4m_W^4) - \tfrac{1}{3} p^2 (m_Z^4 - 2m_Z^2 m_W^2 - 18 m_W^4) \bigg) (\varepsilon^{-1} + \Gamma'(1) + \ln 4\pi) \\ & + \sum_i m_Z^4 [(\eta_i^2 + 1) p^2 F(m_i, m_i, p^2) - m_i^2 F_0(m_i, m_i, p^2)] - \tfrac{1}{3} p^2 (m_Z^4 - 2m_W^2 m_Z^2 + 4m_W^4) \\ & + m_W^2 (m_Z^4 - 4m_Z^2 m_W^2 + 16 m_W^4) \ln m_W^2/M^2 + \tfrac{1}{2} m_Z^4 (m_Z^2 \ln m_Z^2/M^2 \\ & + m_H^2 \ln m_H^2/M^2) - 10 m_W^4 p^2 F_0(m_W, m_W, p^2) + p^2 (m_Z^4 - 4m_Z^2 m_W^2 \\ & + 24 m_W^4) F(m_W, m_W, p^2) + m_W^2 (3m_Z^4 - 4m_Z^2 m_W^2 - 16 m_W^4) F_0(m_W, m_W, p^2) \\ & + m_Z^4 [2m_Z^2 F_0(m_H, m_Z, p^2) - m_Z^2 F_1(m_H, m_Z, p^2) + p^2 F(m_H, m_Z, p^2)] \\ & - m_H^2 m_Z^4 F_1(m_Z, m_H, p^2) \Bigg] \end{aligned} \tag{15.92}$$

where $\sum_i$ is a sum over fermion flavours (leptonic and hadronic) and, in the case of quark flavours, a colour sum must also be included. The constants η_i have the values

$$\eta_{\nu_l} = 1 \qquad (l = e, \mu, \tau) \tag{15.93a}$$

$$\eta_l = (4m_W^2 - 3m_Z^2)/m_Z^2 \qquad (l = e, \mu, \tau) \tag{15.93b}$$

$$\eta_{u,c,t} = (8m_W^2 - 5m_Z^2)/3m_Z^2 \tag{15.93c}$$

$$\eta_{d,s,b} = (4m_W^2 - m_Z^2)/3m_Z^2. \tag{15.93d}$$

The functions F_n ($n=0, 1, 2$) are defined by

$$F_n(m_1, m_2, p^2) = \int_0^1 \mathrm{d}x\, x^n \ln[m_1^2(1-x) + m_2^2 x - p^2 x(1-x)]/M \quad (15.94)$$

and

$$F \equiv F_1 - F_2. \quad (15.95)$$

For this reason in the remainder of this chapter we shall merely outline the calculations which have to performed to carry out the renormalisation of the theory. The reader who requires more precise details of these calculations is referred to Aoki *et al.*[5], or one of several similar treatments which appear in the literature[6]. The momentum integrations occurring in the tadpole contributions represented by diagrams (10), (11), (12), (13) of (15.80) are essentially the same as that arising in diagram (8), and we find that the combined tadpole contribution is

$$\begin{aligned} T_Z = \frac{\mathrm{i}\hat{e}^2 m_Z^4}{32\pi^2 (m_Z^2 - m_W^2) m_W^2 m_H^2} \Bigg[\Bigg(& -4 \sum_i m_i^4 + 3(2m_W^4 + m_Z^4) \\ & + \tfrac{1}{2} m_H^2 (2m_W^2 + m_Z^2 + 3m_H^2) \Bigg) (\varepsilon^{-1} + \Gamma'(1) + \ln 4\pi + 1) \\ & + 4 \sum_i m_i^4 \ln m_i^2/M^2 - 2(2m_W^4 + m_Z^4) \\ & - m_W^2 (6m_W^2 + m_H^2) \ln m_W^2/M^2 - \tfrac{1}{2} m_Z^2 (6m_Z^2 + m_H^2) \ln m_Z^2/M^2 \\ & - \tfrac{3}{2} m_H^4 \ln m_H^2/M^2 \Bigg]. \end{aligned} \quad (15.96)$$

Then putting $p^2 = m_Z^2$ in (15.92) and substituting both (15.92) and (15.96) back into (15.89) determines the renormalisation constant K_Z.

Similar calculations may be performed to determine $A_W(p^2)$, T_W and hence the renormalisation constant K_W. Because of its complexity we shall not even quote the result for $A_W(p^2)$. However the tadpole contribution is easily found with no extra work; the coupling of the Higgs scalar H to the W is given in (14.93), and the momentum integrations occurring in the tadpole contributions is precisely the same as those occurring in the Z tadpoles. Thus

$$T_W = (m_W^2/m_Z^2) T_Z. \quad (15.97)$$

Having determined $A_W(p^2)$, and hence $A_W(m_W^2)$, and T_W, the renormalisation constant K_W is now given by (15.90).

The renormalisation constant K_H is found in a very similar manner. We consider the process $H \to H$ and denote the sum of the OPI contributions, analogous to diagrams (2) to (9) of (15.80), for a *general momentum* p by $A_H(p^2)$. The tadpole contributions, analogous to diagrams (10) to (13), are denoted by

T_H. Then, as before, the on-shell renormalisation scheme requires that the counter term contribution cancels the sum of the loop contributions, and

$$K_H = -im_H^{-2}[A_H(m_H^2) + T_H] \tag{15.98}$$

analogously to (15.89).

The determination of the fermion mass renormalisation constants is only slightly more complicated. As before, we consider the S-matrix element for a single incoming fermion f going to an identical outgoing fermion. Then

$S_{fi} =$

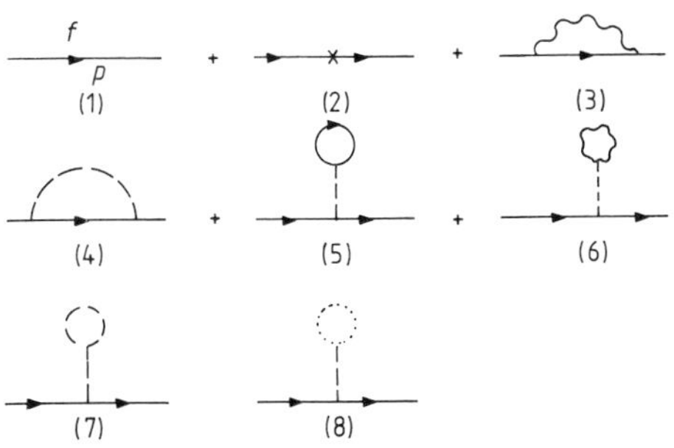

(15.99)

where we are using the same convention as in (15.80) with regard to the lines appearing in the above diagrams. The contribution from diagrams (3), (4), *without the external spinors and for a general momentum* p, is

$$\Sigma^f(p) \equiv K_1^f(p^2)\mathbf{I} + K_5^f(p^2)\gamma_5 + K_\gamma^f(p^2)\not{p} + K_{5\gamma}^f(p^2)\not{p}\gamma_5 \tag{15.100}$$

and the tadpole contribution, from diagrams (5), (6), (7), (8), *without the external spinors and for a general momentum* p is

$$\tau^f \equiv T_1^f \mathbf{I} + T_5^f \gamma_5. \tag{15.101}$$

As before, the diagram (1) gives the entire contribution to the S-matrix element, so the counter term contribution (2) must cancel the single loop contributions. Thus

$$\bar{u}_2^f(p)[-ik_f m_f + \Sigma^f(p) + \tau^f]u_1^f(p) = 0 \tag{15.102}$$

where $p^2 = m_f^2$ and $u_1^f(p)$, $u_2^f(p)$ are the spinors associated with the initial and final fermion. These spinors satisfy

$$\not{p}u^f(p) = m_f u^f(p) \tag{15.103a}$$

$$\bar{u}_2^f(p)\gamma_5 u_1^f(p) = 0 \tag{15.103b}$$

so (15.102) gives

$$k_f = -im_f^{-1}[K_1^f(m_f^2) + m_f K_\gamma^f(m_f^2) + T_1^f]. \tag{15.104}$$

Thus the fermion mass renormalisation constants k_f ($f = l, q$) are completely determined, once the calculations necessary to find $\Sigma^f(p)$, τ^f (and hence

$K_1^f(m_f^2)$, $K_\gamma^f(m_f^2)$, T_1^f) have been performed. Again we shall not quote the results.

It only remains for us to calculate the charge renormalisation constant K_e, defined in (15.18). We consider the S-matrix element for the process in which a single incoming electron is scattered by an infinitely heavy external source to produce a final outgoing electron. To one-loop order

$S_{\text{fi}} =$

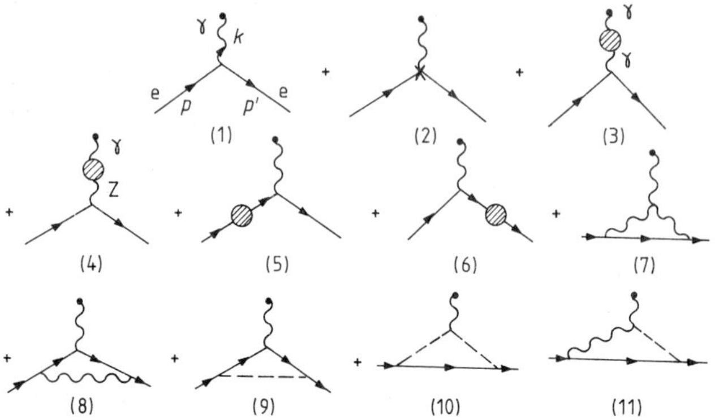

(15.105)

The infinitely heavy external source is represented by the dot at the end of the external γ line. Diagram (2) is the counter term contribution. Diagrams (3), (4), (5) and (6) represent the diagrams in which there is a counter term vertex or a loop inserted in one of the external lines; thus diagram (3), for example, represents the contribution when diagrams analogous to (2)–(13) of (15.80) are inserted in the external gauge field line. Diagram (4) represents the contribution from similar diagrams connecting γ and Z lines. Similarly diagrams (5) and (6) represent the contributions when diagrams (2) to (8) of (15.99) are inserted in the external electron lines.

The contribution from diagram (1) is

$$S_{\text{fi}} = \bar{u}^{\text{e}}(p')\text{i}e\gamma_\mu u^{\text{e}}(p)\frac{-\text{i}}{(p'-p)^2}j^\mu(2\pi)^4\delta(p-p'-k). \tag{15.106}$$

The definition of $-e$ (that it is the physical electric charge of the electron) means that, in the limit $p-p'\to 0$, $S_{\text{fi}}(1)$ gives the *entire* contribution to S_{fi}. Thus

$$\lim_{p-p'\to 0}\,[S_{\text{fi}} - S_{\text{fi}}(1)] = 0 \tag{15.107}$$

and the counter term contribution (2) must cancel the remaining contributions (3)–(11) (in the limit $p-p'\to 0$).

We may write the contribution from diagrams (7)–(11) in the form

$$\text{i}e\bar{u}^{\text{e}}(p')\Gamma_\mu(p,p')\frac{-\text{i}}{(p-p')^2}j^\mu(2\pi)^4\delta(p-p'-k) \tag{15.108}$$

where, without loss of generality,

$$\Gamma_\mu(p, p') = F_1(k^2)\gamma_\mu + F_2(k^2)\mathrm{i}\sigma_{\mu\nu}k^\nu + F_3(k^2)k_\mu$$
$$+ G_1(k^2)\gamma_\mu\gamma_5 + G_2(k^2)k_\mu\gamma_5 + G_3(k^2)\mathrm{i}\sigma_{\mu\nu}k^\nu\gamma_5. \qquad (15.109)$$

In the limit $p - p' \to 0, k \to 0$, so only the terms proportional to $F_1(0)$ and $G_1(0)$ survive. The point is that $F_1(0)$ and $G_1(0)$ are well defined and calculable, once we have decided how the divergences are to be regulated, and so, therefore, is the contribution from diagrams (7)–(11).

The contributions from diagrams (3) and (4) are also well defined, and calculable. Using the notation of (15.83), but now for a γ line, we find

$$S_{\mathrm{fi}}(3) = \bar{u}^{\mathrm{e}}(p')\mathrm{i}e\gamma_\mu u^{\mathrm{e}}(p)\frac{-\mathrm{i}}{k^2}\left[A_\gamma(k^2)\left(g^{\mu\nu} - \frac{k^\mu k^\nu}{k^2}\right) + B_\gamma(k^2)\frac{k^\mu k^\nu}{k^2}\right]\frac{-\mathrm{i}}{k^2}j_\nu(2\pi)^4\delta(p - p' - k)$$

$$= \bar{u}^{\mathrm{e}}(p')\mathrm{i}e\gamma_\mu u^{\mathrm{e}}(p)\frac{-\mathrm{i}}{k^2}A_\gamma(k^2)\frac{-\mathrm{i}}{k^2}j^\mu(2\pi)^4\delta(p - p' - k) \qquad (15.110)$$

using the Dirac equation.

In fact, $A_\gamma(0) = 0$ (since the photon has zero rest mass) so $A(k^2)/k^2 \to A'_\gamma(0)$ as $k^2 \to 0$, and

$$S_{\mathrm{fi}}(3) = -\mathrm{i}A'_\gamma(0)S_{\mathrm{fi}}(1). \qquad (15.111)$$

Clearly the effects of the insertion in the photon line will be shared between the electron vertex and the interaction with the external source. Thus for the purposes of determining the charge renormalisation we should halve the contribution from this diagram (since the other half renormalises the other vertex), at least in lowest order. Then

$$S_{\mathrm{fi}}(3) \to -\tfrac{1}{2}\mathrm{i}A'_\gamma(0)S_{\mathrm{fi}}(1). \qquad (15.112)$$

(The reader may convince himself of this by replacing the source by a heavy charged particle, and considering the appropriate diagrams.) Using the Feynman rule (14.65), the contribution from diagram (4) may be found straightforwardly.

$$S_{\mathrm{fi}}(4) = \bar{u}^{\mathrm{e}}(p')\frac{\mathrm{i}g}{2\cos\theta_{\mathrm{W}}}\gamma_\mu(g_V - g_A\gamma_5)u^{\mathrm{e}}(p)\frac{-\mathrm{i}}{k^2 - m_{\mathrm{Z}}^2}$$
$$\times\left[A_{\gamma\mathrm{Z}}(k^2)\left(g^{\mu\nu} - \frac{k^\mu k^\nu}{k^2}\right) + B_{\gamma\mathrm{Z}}(k^2)\frac{k^\mu k^\nu}{k^2}\right]\frac{-\mathrm{i}}{k^2}j_\nu(2\pi)^4\delta(p - p' - k) \qquad (15.113\mathrm{a})$$

where

$$g_V = \tfrac{1}{2} - 2\sin^2\theta_{\mathrm{W}} = \frac{4m_{\mathrm{W}}^2 - 3m_{\mathrm{Z}}^2}{2m_{\mathrm{Z}}^2} \qquad (15.113\mathrm{b})$$

$$g_A = -\tfrac{1}{2}. \qquad (15.113\mathrm{c})$$

In the limit $k \to 0$,

$$S_{\text{fi}}(4) \to \bar{u}^{\text{e}}(p')\text{i}e\gamma_\mu \frac{m_{\text{Z}}^2}{2m_{\text{W}}(m_{\text{Z}}^2 - m_{\text{W}}^2)^{1/2}} (g_V - g_A\gamma_5)u^{\text{e}}(p)$$

$$\times \text{i}A_{\gamma\text{Z}}(0)m_{\text{Z}}^{-2} \frac{-\text{i}}{k^2} j^\mu(2\pi)^4\delta(p - p' - k). \qquad (15.114)$$

However the contribution from diagrams (5) and (6) is *a priori* ill-defined. Consider, for example, diagram (5). The Feynman rules give

$$S_{\text{fi}}(5) = \bar{u}^{\text{e}}(p')\text{i}e\gamma_\mu \frac{\text{i}}{\not{p} - m_{\text{e}}} [-\text{i}k_{\text{e}}m_{\text{e}} + \Sigma^{\text{e}}(p) + \tau^{\text{e}}]u^{\text{e}}(p)$$

$$\times \frac{-\text{i}}{k^2} j^\mu(2\pi)^4\delta(p - p' - k) \qquad (15.115)$$

where Σ^{e} and τ^{e} are defined in (15.100) and (15.101) (for a general fermion f). We insert the identity

$$\mathbf{I} \equiv (\not{p} + m_{\text{e}})^{-1} \sum_s u^{\text{e}}(p, s)\bar{u}^{\text{e}}(p, s) \qquad (15.116)$$

where the sum is over the two possible polarisation states $s = \uparrow, \downarrow$ of the electron. Then

$$S_{\text{fi}}(5) = \sum_s \bar{u}^{\text{e}}(p')\text{i}e\gamma_\mu \frac{\text{i}}{\not{p} - m_{\text{e}}} (\not{p} + m_{\text{e}})^{-1}u^{\text{e}}(p, s)$$

$$\times \bar{u}^{\text{e}}(p, s)[-\text{i}k_{\text{e}}m_{\text{e}} + \Sigma^{\text{e}}(p) + \tau^{\text{e}}]u^{\text{e}}(p) \frac{-\text{i}}{k^2} j^\mu(2\pi)^4\delta(p - p' - k)$$

(15.117a)

$$= \sum_s \bar{u}^{\text{e}}(p')\text{i}e\gamma_\mu \frac{\text{i}}{p^2 - m_{\text{e}}^2} u^{\text{e}}(p, s)$$

$$\times \bar{u}^{\text{e}}(p, s)[-\text{i}k_{\text{e}}m_{\text{e}} + \Sigma^{\text{e}}(p) + \tau^{\text{e}}]u^{\text{e}}(p) \frac{-\text{i}}{k^2} j^\mu(2\pi)^4\delta(p - p' - k).$$

(15.117b)

The indeterminacy arises arises because p is the (physical) momentum of the external electron, so $p^2 - m_{\text{e}}^2 = 0$ and the first line is infinite; however, by definition the mass renormalisation constant k_{e} is such that (15.102) is satisfied, so the second line is zero. Thus the value of $S_{\text{fi}}(5)$, and similarly $S_{\text{fi}}(6)$, is ill-defined. The problem derives from the fact that we are concerned with S-matrix elements, which are defined by a limiting procedure applied to (off-shell) Green functions. Had we first renormalised the Green functions, using non-trivial wave function/field operator renormalisation constants, we should have avoided this difficulty. Let us investigate the ambiguous part of (15.115) in more detail.

Using the definitions (15.100), (15.101) and (15.104), we may write

$$-\mathrm{i}k_{\mathrm{e}}m_{\mathrm{e}}+\Sigma^{\mathrm{e}}(p)+\tau^{\mathrm{e}}=A_5^{\mathrm{e}}(p^2)\gamma_5+K_{5\gamma}^{\mathrm{e}}(p^2)\not{p}\gamma_5 + A^{\mathrm{e}}(p^2)(p^2-m_{\mathrm{e}}^2)+K_\gamma^{\mathrm{e}}(p^2)(\not{p}-m_{\mathrm{e}}) \tag{15.118a}$$

where

$$A_5^{\mathrm{e}}(p^2)\equiv T_5^{\mathrm{e}}+K_5^{\mathrm{e}}(p^2) \tag{15.118b}$$

and

$$(p^2-m_{\mathrm{e}}^2)A^{\mathrm{e}}(p^2)\equiv K_1^{\mathrm{e}}(p^2)-K_1^{\mathrm{e}}(m_{\mathrm{e}}^2)+m_{\mathrm{e}}[K_\gamma^{\mathrm{e}}(p^2)-K_\gamma^{\mathrm{e}}(m_{\mathrm{e}}^2)]. \tag{15.118c}$$

Now suppose we ignore the fact that $(\not{p}-m_{\mathrm{e}})$ is singular, since $p^2=m_{\mathrm{e}}^2$; then the contribution from the first two terms in (15.118a) to (15.115) is (fairly) well defined since

$$\frac{\mathrm{i}}{\not{p}-m_{\mathrm{e}}}[A_5^{\mathrm{e}}(p^2)\gamma_5+K_{5\gamma}^{\mathrm{e}}(p^2)\not{p}\gamma_5]u^{\mathrm{e}}(p)=\frac{\mathrm{i}}{\not{p}-m_{\mathrm{e}}}[A_5^{\mathrm{e}}(m_{\mathrm{e}}^2)-K_{5\gamma}^{\mathrm{e}}(m_{\mathrm{e}}^2)m_{\mathrm{e}}]\gamma_5u^{\mathrm{e}}(p)$$

$$=\frac{-\mathrm{i}}{2m_{\mathrm{e}}}[A_5^{\mathrm{e}}(m_{\mathrm{e}}^2)-K_{5\gamma}^{\mathrm{e}}(m_{\mathrm{e}}^2)m_{\mathrm{e}}]\gamma_5u^{\mathrm{e}}(p) \tag{15.119}$$

and we have assumed

$$f(\not{p})u^{\mathrm{e}}(p)=f(m_{\mathrm{e}})u^{\mathrm{e}}(p). \tag{15.120}$$

If we were to cancel the $(\not{p}-m_{\mathrm{e}})^{-1}$ propagator against the two remaining terms in (15.118a) we would have

$$\frac{\mathrm{i}}{\not{p}-m_{\mathrm{e}}}[(p^2-m_{\mathrm{e}}^2)A^{\mathrm{e}}(p^2)+K_\gamma^{\mathrm{e}}(p^2)(\not{p}-m_{\mathrm{e}})]$$

$$=\mathrm{i}(\not{p}+m_{\mathrm{e}})A^{\mathrm{e}}(p^2)+\mathrm{i}K_\gamma^{\mathrm{e}}(p^2) \tag{15.121}$$

and, combining with (15.119), we would find

$$\frac{\mathrm{i}}{\not{p}-m_{\mathrm{e}}}[-\mathrm{i}k_{\mathrm{e}}m_{\mathrm{e}}+\Sigma^{\mathrm{e}}(p)+\tau^{\mathrm{e}}]u^{\mathrm{e}}(p)$$

$$=\mathrm{i}\left(-\frac{1}{2m_{\mathrm{e}}}A_5^{\mathrm{e}}(m_{\mathrm{e}}^2)\gamma_5+\tfrac{1}{2}K_{5\gamma}^{\mathrm{e}}(m_{\mathrm{e}}^2)\gamma_5+2m_{\mathrm{e}}A^{\mathrm{e}}(m_{\mathrm{e}}^2)+K_\gamma^{\mathrm{e}}(m_{\mathrm{e}}^2)\right)u^{\mathrm{e}}(p). \tag{15.122}$$

In the same way we would find

$$\bar{u}^{\mathrm{e}}(p')[-\mathrm{i}k_{\mathrm{e}}m_{\mathrm{e}}+\Sigma^{\mathrm{e}}(p')+\tau^{\mathrm{e}}]\frac{\mathrm{i}}{\not{p}'-m_{\mathrm{e}}}$$

$$=\bar{u}^{\mathrm{e}}(p')\left(-\frac{1}{2m_{\mathrm{e}}}A_5^{\mathrm{e}}(m_{\mathrm{e}}^2)\gamma_5-\tfrac{1}{2}K_{5\gamma}^{\mathrm{e}}(m_{\mathrm{e}}^2)\gamma_5+2m_{\mathrm{e}}A^{\mathrm{e}}(m_{\mathrm{e}}^2)+K_\gamma^{\mathrm{e}}(m_{\mathrm{e}}^2)\right)\mathrm{i}. \tag{15.123}$$

Thus the combined contributions of diagrams (5) and (6) would be

$$S_{\mathrm{fi}}(5+6)=\bar{u}^{\mathrm{e}}(p')\mathrm{i}e\gamma_{\mu}\mathrm{i}[K^{\mathrm{e}}_{5\gamma}(m_{\mathrm{e}}^2)\gamma_5+4m_{\mathrm{e}}A^{\mathrm{e}}(m_{\mathrm{e}}^2)+2K^{\mathrm{e}}_{\gamma}(m_{\mathrm{e}}^2)]\,u^{\mathrm{e}}(p)$$

$$\times\left(\frac{-\mathrm{i}}{k^2}\right)j^{\mu}(2\pi)^4\delta(p-p'-k)\qquad \textbf{(FALSE)}. \qquad (15.124)$$

Actually this last expression is not quite correct. The coefficients of $K^{\mathrm{e}}_{\gamma}(m_{\mathrm{e}}^2)$ and of $A^{\mathrm{e}}(m_{\mathrm{e}}^2)$ must be *halved*. We can see this most easily by considering the (hypothetical) case that the electron has zero mass, and the theory has chiral symmetry. Then in the notation of (15.100)

$$\Sigma(p)=K_{\gamma}(p^2)\not p+K_{5\gamma}(p^2)\not p\gamma_5$$

$$=\not p[K_{\gamma}(p^2)+K_{5\gamma}(p^2)]a_{\mathrm{R}}+\not p[K_{\gamma}(p^2)-K_{5\gamma}(p^2)]a_{\mathrm{L}}. \qquad (15.125)$$

The full propagator is then given by

$$S(p)=\frac{\mathrm{i}}{\not p}+\frac{\mathrm{i}}{\not p}\Sigma(p)\frac{\mathrm{i}}{\not p}+\frac{\mathrm{i}}{\not p}\Sigma(p)\frac{\mathrm{i}}{\not p}\Sigma(p)\frac{\mathrm{i}}{\not p}+\dots \qquad (15.126)$$

$$=[1-\mathrm{i}(K_{\gamma}(0)+K_{5\gamma}(0))]^{-1}a_{\mathrm{R}}\frac{\mathrm{i}}{\not p}$$

$$+[1-\mathrm{i}(K_{\gamma}(0)-K_{5\gamma}(0))]^{-1}a_{\mathrm{L}}\frac{\mathrm{i}}{\not p}+\dots \qquad (15.127)$$

where we have retained only the pole contributions at $p^2=0$. Evidently the effect of the radiative corrections is to 'renormalise' the right (left) component of the propagator with a factor $Z_{\mathrm{R}}(Z_{\mathrm{L}})$ where

$$Z_{\substack{\mathrm{R}\\ \mathrm{L}}}=\{1-\mathrm{i}[K_{\gamma}(0)\pm K_{5\gamma}(0)]\}^{-1}. \qquad (15.128)$$

Thus the effect of the radiative corrections on external (electron) lines is to renormalise their contribution by a factor $Z^{1/2}_{\mathrm{R,L}}$. So if we scatter a right chiral electron, for example, from an infinitely heavy source, the radiative corrections contained in diagrams (5) and (6) have the effect of multiplying the uncorrected contribution by a factor Z_{R}. It follows that

$$S_{\mathrm{fi}}(5+6)_{\mathrm{R}}=(Z_{\mathrm{R}}-1)S_{fi}(1)_R \qquad (15.129\mathrm{a})$$

and to lowest order in α

$$Z_{\mathrm{R}}-1\simeq\mathrm{i}[K_{\gamma}(0)+K_{5\gamma}(0)]. \qquad (15.129\mathrm{b})$$

Since $\gamma_5u_{\mathrm{R}}=+u_{\mathrm{R}}$, the coefficient of $K_{5\gamma}$ in (15.124) agrees with the correct result (15.129) (when $m_{\mathrm{e}}=0$), whereas the coefficient of K^{e}_{γ} in (15.124) is a factor 2 too large. A similar argument applies to $A^{\mathrm{e}}(m_{\mathrm{e}}^2)$, and the correct result is

$$S_{\mathrm{fi}}(5+6)=\bar{u}^{\mathrm{e}}(p')\mathrm{i}e\gamma_{\mu}\mathrm{i}[K^{\mathrm{e}}_{5\gamma}(m_{\mathrm{e}}^2)\gamma_5+2m_{\mathrm{e}}A^{\mathrm{e}}(m_{\mathrm{e}}^2)$$

$$+K^{\mathrm{e}}_{\gamma}(m_{\mathrm{e}}^2)]u^{\mathrm{e}}(p)\frac{-\mathrm{i}}{k^2}j^{\mu}(2\pi)^4\delta(p-p'-k). \qquad (15.130)$$

We note in passing that a similar argment could have been presented to justify the modification (by a factor $\frac{1}{2}$) of the contribution from the photon line insertions. These have the effect of renormalising the uncorrected amplitude by a factor $[1+iA'_{\gamma}(0)]^{-1}$. Thus the contribution to each vertex is $[1+iA'_{\gamma}(0)]^{-1/2} \simeq 1-\frac{1}{2}iA'_{\gamma}(0)$, as claimed in (15.112). The requirement that the counter term contribution $S_{fi}(2)$ cancels the contributions from diagrams (3) to (11), in the limit $p-p' \to 0$, then gives

$$K_e + F_1(0) + G_1(0)\gamma_5 - i\tfrac{1}{2}A'_{\gamma}(0) + i\tfrac{1}{2}\frac{A_{\gamma Z}(0)}{m_W(m_Z^2-m_W^2)^{1/2}}(g_V - g_A\gamma_5)$$
$$+ i[K^e_{5\gamma}(m_e^2)\gamma_5 + 2m_e A^e(m_e^2) + K^e_{\gamma}(m_e^2)] = 0. \quad (15.131)$$

The terms proportional to γ_5 arise because of different contributions to the scattering of the right and left chiral components of the electron. This derives from the fact that electroweak theory is a parity non-conserving theory. However, we know that the right and left chiral components have the same electric charge. Thus the parity-violating contributions to (15.131) must cancel separately. Hence

$$G_1(0) + iK^e_{5\gamma}(m_e^2) - i\frac{A_{\gamma Z}(0)}{2m_W(m_Z^2-m_W^2)^{1/2}}g_A = 0 \quad (15.132)$$

and substituting from (15.118c) we find

$$K_e = -F_1(0) + i\tfrac{1}{2}A'_{\gamma}(0) - iK^e_{\gamma}(m_e^2)$$
$$-i\frac{A_{\gamma Z}(0)}{2m_W(m_Z^2-m_W^2)^{1/2}}g_V - i2m_e[K^{e\prime}_1(m_e^2) + m_e K^{e\prime}_{\gamma}(m_e^2)]. \quad (15.133)$$

So the electric charge renormalisation constnat K_e is completely determined once the calculations necessary to find $F_1(0)$, $A_{\gamma}(0)$, as well as $K^e_{\gamma}(m_e^2)$, $K^{e\prime}_{\gamma}(m_e^2)$ and $K^{e\prime}_1(m_e^2)$, have been performed. Again we shall not quote the result. We leave it as an exercise (problem 15.5) to verify that (15.132) *is* satisfied.

15.4 Radiative corrections to muon decay

We have indicated in the previous section how all of the renormalisation constants (K_e, K_W, K_Z, K_H, k_l, k_q) which fix the counter term vertices are determined. In principle, therefore, we may evaluate the S-matrix element for any electroweak process to arbitrary accuracy, including the effects of all electroweak radiative corrections. We shall illustrate this by considering the order α radiative corrections to muon decay

$$\mu^- \to e^- \bar{\nu}_e \nu_\mu$$

which was computed in lowest order in §14.4. As before, we shall not actually evaluate the corrections, which are complicated and not in themselves especially illuminating. Rather, we shall discuss the various contributions, and in particular how the loop corrections and the counter term contributions combine to give a finite, calculable result.

The lowest order contribution to the process is given by the diagrams

$S_{fi}^{(0)} =$

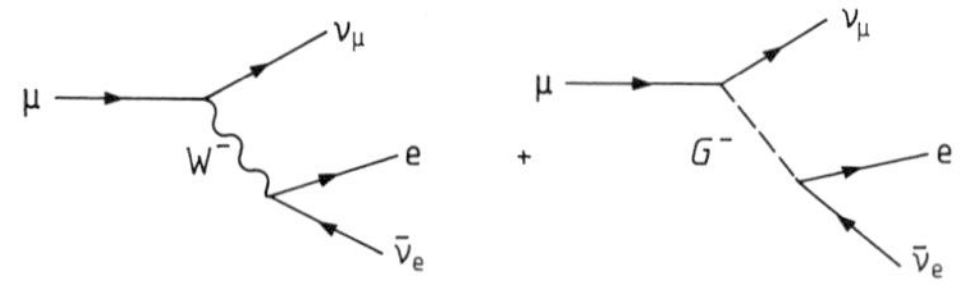

(15.134)

and, as we showed in (14.118), the contribution from the second diagram is smaller than the first by a factor of order $m_e m_\mu / m_W^2$. We shall therefore neglect this contribution, since it is smaller than the O(α) terms we are about to consider. The first set of radiative corrections are generated by self-energy, tadpole, and counter term insertions into the various lines

$\Delta S_{fi}^{(1)} =$

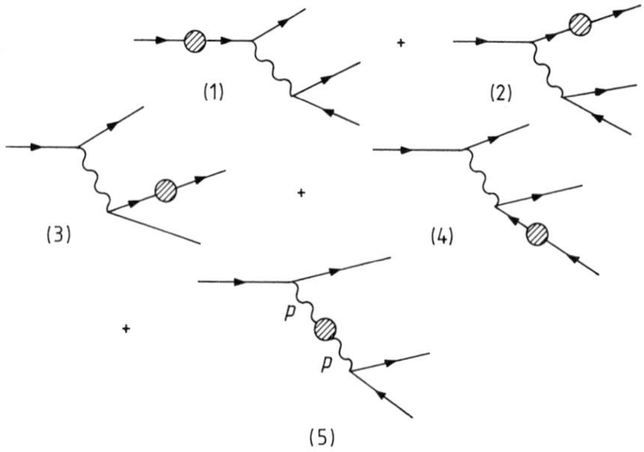

(15.135)

We are using the same shorthand as in (15.105), where the shaded blob signifies the sum of the self-energy loop, tadpole and counter term insertion. As explained in the previous section, the *external* line insertions are *a priori* ambiguous in our approach. However, the analysis given there shows how it is resolved. Then the contribution from diagram (3) is

$$\Delta S_{fi}^{(1)}(3) = i\Bigg(\tfrac{1}{2}K_\gamma^{e}(m_e^2) + m_e K_1^{e\prime}(m_e^2) + m_e^2 K_\gamma^{e\prime}(m_e^2) - \tfrac{1}{2}K_{5\gamma}^{e}(m_e^2) - \frac{1}{2m_e}K_5^{e}(m_e^2) - \frac{1}{2m_e}T_5^{e}\Bigg)S_{fi}^{(0)} \tag{15.136}$$

and the contributions from diagrams (1), (2) and (4) may be obtained in a similar manner (problem 15.6). The contribution from diagram (5) is determined by the W self-energy, tadpole and mass counter terms. In the Feynman gauge, the propagator of the W is

$$\tilde{\Delta}_{\mu\nu}(p, m_W) = -ig_{\mu\nu}(p^2 - m_W^2)^{-1} \tag{15.137}$$

and the above insertions amount to the replacement

$$\tilde{\Delta}_{\mu\nu}(p, m_W) \to \tilde{\Delta}_{\mu\rho}(p, m_W)[ig^{\rho\sigma}K_W m_W^2 + \Pi_W^{\rho\sigma}(p) + T_W g^{\rho\sigma}]\tilde{\Delta}_{\sigma\nu}(p, m_W) \tag{15.138}$$

where Π_W is defined analoguously to Π_Z in (15.83). Using (15.90) it follows that

$$ig^{\rho\sigma}K_W m_W^2 + \Pi_W^{\rho\sigma}(p) + T_W g^{\rho\sigma}$$
$$= [A_W(p^2) - A_W(m_W^2)]g^{\rho\sigma} + [B_W(p^2) - A_W(p^2)]\frac{p^\rho p^\sigma}{p^2}. \tag{15.139}$$

The terms proportional to $p^\rho p^\sigma$ lead to lepton mass terms via the Dirac equation, as shown in (14.120). Neglecting them we find

$$\Delta S_{fi}^{(1)}(5) = -i\frac{A_W(p^2) - A_W(m_W^2)}{p^2 - m_W^2} S_{fi}^{(0)}. \tag{15.140}$$

The next set of radiative corrections arises from vertex insertions and counter terms

$\Delta S_{fi}^{(2)} =$ (15.141)

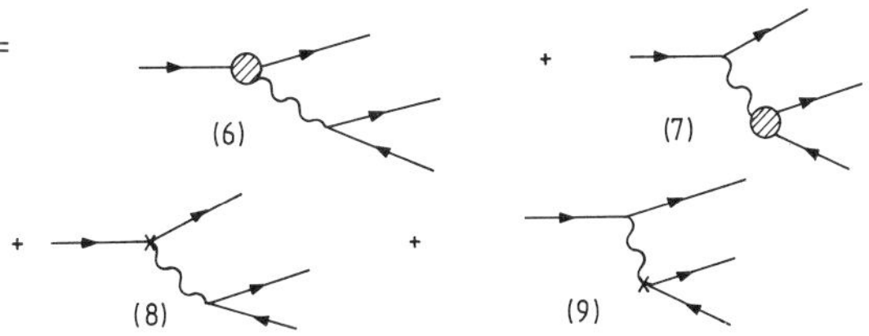

Diagrams (6) and (7) include all three-point vertex corections, analogous to those displayed in diagrams (7)–(11) of (15.105). The counter term vertex which appears in diagrams (8) and (9) is that given in (15.27) with K defined in (15.24). Thus (8) and (9) give

$$\Delta S_{fi}^{(2)}(8+9) = 2KS_{fi}^{(0)}. \tag{15.142}$$

Since all of the counter terms are of order e^2 we may expand (15.24) accurate to this order and obtain

$$K = K_e + \tfrac{1}{2}\cot^2\theta_W(K_W - K_Z)$$
$$= K_e + \tfrac{1}{2}m_W^2(m_Z^2 - m_W^2)^{-1}(K_W - K_Z). \tag{15.143}$$

If we like, $K_W - K_Z$ can be re-expressed using (15.89), (15.90) and (15.97) to give

$$K_W - K_Z = -i[m_Z^{-2}A_Z(m_Z^2) - m_W^{-2}A_W(m_W^2)]. \tag{15.144}$$

The presence of a neutrino in both of the three-point vertex corrections means that the analogue of (15.109) may be written in the form

$$\Gamma_\mu^{(e\bar{\nu}_e)}(p_e, p_{\bar{\nu}}) = [f_1^{(e\bar{\nu})}(p^2)\gamma_\mu + f_2^{(e\bar{\nu})}(p^2)\mathrm{i}\sigma_{\mu\nu}p^\nu + f_3^{(e\bar{\nu})}(p^2)p_\mu](1-\gamma_5) \tag{15.145}$$

and we can neglect the f_3 term, since p_μ leads to lepton mass terms which we shall drop. Thus, for example, the invariant amplitude for diagram (7) is

$$\mathscr{M}(7) = \frac{\mathrm{i}g^2}{8(p^2 - m_W^2)}\,\bar{u}(\nu_\mu)\gamma^\mu(1-\gamma_5)u(\mu)$$

$$\times\,\bar{u}(e)[f_1^{(e\nu)}(p^2)\gamma_\mu + f_2^{(e\nu)}(p^2)\mathrm{i}\sigma_{\mu\nu}p^2](1-\gamma_5)v(\bar{\nu}_e) \tag{15.146}$$

and a similar expression may be obtained for $\mathscr{M}(6)$.

The final set of diagrams are the four-point functions. These are too numerous to display in full and we merely illustrate the gauge boson contributions

$\Delta S_{\mathrm{fi}}^{(3)} =$ 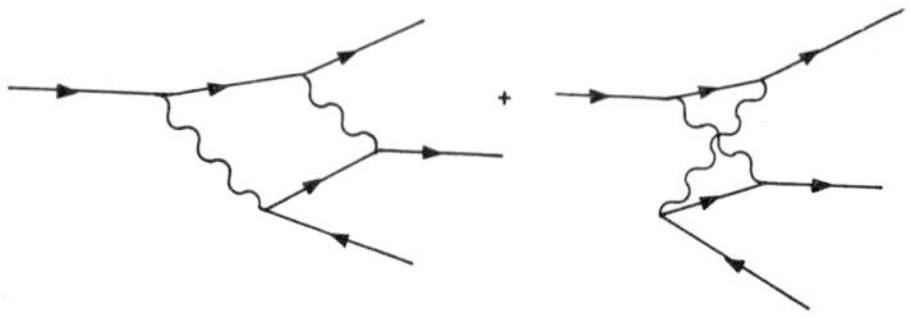 $+\ldots$

(15.147)

where the exchanged lines must be W, Z or W, γ. The undisplayed diagrams include all contributions from scalar exchanges. We shall not need any details of these amplitudes. The interested reader is referred to Sirlin[3] and references therein. Suffice it to say that all of these diagrams are (ultraviolet) finite. In other words they contain no poles in $2-\omega$ which derive from the large momentum behaviour of the integrands. (The infrared behaviour is another matter, which we shall not address.)

The ultraviolet divergent contributions arise from the terms already displayed. Neglecting the masses of the leptons, the only divergent terms in (15.136) are K_γ^e and $K_{5\gamma}^e$, with similar divergences from the other self-energy contributions. We leave it as an exercise to verify that when these contributions are combined with the other divergent contributions, contained in (15.140), (15.142), (15.146), they lead to a *finite* correction to the muon decay amplitude. (This is just a verification that electroweak theory is a renormalisable theory, as we have claimed but not proved.) Numerical details of these calculations are contained in Sirlin, who concludes that the order α corrections have the effect of increasing the decay rate by about 7% compared with the tree graph prediction (14.125).

15.5 Anomalies

We have shown in Appendix B that S-matrix elements are independent of the gauge-fixing parameter ξ. In consequence the ghost particles and would-be

Goldstone scalars, whose masses depend upon ξ, decouple from the physical states. This means that the physical states are complete and that the S-matrix is unitary. The proof of this ξ-independence utilised the gauge (and BRS) invariance of the functional integration measure, and this invariance may be verified for the (non-chiral) theories envisaged in Chapters 9 and 10.

However, the electroweak theory with which we are now concerned is a *chiral* theory; the left and right chiral components of fermion fields transform differently under gauge transformations. For these theories the fermionic integration measure is *not* in general gauge invariant[7]. In consequence the Slavnov–Taylor identity (B.9) is violated, and (supposed) S-matrix elements will be ξ-dependent, because of the appearance of unphysical particles (e.g. Goldstone scalars) in physical processes. Clearly such a theory is nonsense; the gauge fixing and ghost terms in the Lagrangian were merely technical devices to permit the formulation of the *quantum* field theory, and nothing physical should depend upon ξ, if the theory is to make sense. Thus the gauge non-invariance of the fermionic measure is a disaster for the (chiral) gauge theories in which it appears. It means that these theories are non-quantisable[8]. Only theories in which this problem can be evaded can be considered as candidates to describe reality.

We illustrate the problem by considering first the case of 'axial electrodynamics', which is described by

$$\mathscr{L}=\bar{\psi}\mathrm{i}\gamma^{\mu}(\partial_{\mu}+\mathrm{i}qV_{\mu}+\mathrm{i}gA_{\mu}\gamma_{5})\psi-\tfrac{1}{4}F_{\mu\nu}F^{\mu\nu}-\tfrac{1}{4}G_{\mu\nu}G^{\mu\nu} \tag{15.148a}$$

where

$$F_{\mu\nu}\equiv\partial_{\mu}V_{\nu}-\partial_{\nu}V_{\mu} \tag{15.148b}$$

$$G_{\mu\nu}\equiv\partial_{\mu}A_{\nu}-\partial_{\nu}A_{\mu}. \tag{15.148c}$$

Clearly $\mathscr{L}$ is invariant with respect to local $\mathrm{U}(1)_V\times\mathrm{U}(1)_A$ gauge transformations in which

$$\psi(x)\rightarrow\psi'(x)=\exp[-\mathrm{i}q\Lambda(x)-\mathrm{i}g\alpha(x)\gamma_{5}]\psi(x) \tag{15.149a}$$

$$V_{\mu}(x)\rightarrow V'_{\mu}(x)=V_{\mu}(x)+\partial_{\mu}\Lambda(x) \tag{15.149b}$$

$$A_{\mu}(x)\rightarrow A'_{\mu}(x)=A_{\mu}(x)+\partial_{\mu}\alpha(x). \tag{15.149c}$$

Then the left and right chiral components of ψ transform according to

$$\psi_{\mathrm{L}}(x)\equiv a_{\mathrm{L}}\psi(x)\rightarrow\exp[-\mathrm{i}\theta_{\mathrm{L}}(x)]\psi_{\mathrm{L}}(x) \tag{15.150a}$$

$$\psi_{\mathrm{R}}(x)\equiv a_{\mathrm{R}}\psi(x)\rightarrow\exp[-\mathrm{i}\theta_{\mathrm{R}}(x)]\psi_{\mathrm{R}}(x) \tag{15.150b}$$

with

$$\theta_{\substack{\mathrm{L}\\ \mathrm{R}}}=q\Lambda\mp g\alpha. \tag{15.150c}$$

The transformation law (15.149a) means that the fermionic measure $\mathscr{D}\psi$

transforms according to

$$\mathscr{D}\psi \rightarrow \{\mathrm{Det}\exp[-iq\Lambda(x)-ig\alpha(x)\gamma_5]\}^{-1}\mathscr{D}\psi \tag{15.151}$$

where the *inverse* of the determinant appears because integration over the Grassmann variable $\psi(x)$ is defined (in Chapter 3) as the left-differentiation with respect to $\psi(x)$. The determinant Det, with a capital D, is taken over both Dirac spinor indices and the space–time labels x. In the same way, since (15.149a) shows that

$$\bar{\psi}(x) \rightarrow \bar{\psi}'(x)=\bar{\psi}(x)\exp[iq\Lambda(x)-ig\alpha(x)\gamma_5] \tag{15.152}$$

the fermionic measure

$$\mathscr{D}\bar{\psi} \rightarrow \{\mathrm{Det}\exp[iq\Lambda(x)-ig\alpha(x)\gamma_5]\}^{-1}\mathscr{D}\bar{\psi}. \tag{15.153}$$

Hence

$$\mathscr{D}\psi\mathscr{D}\bar{\psi} \rightarrow \mathrm{Det}\exp[2ig\alpha(x)\gamma_5]\mathscr{D}\psi\mathscr{D}\bar{\psi}. \tag{15.154}$$

Thus, as anticipated, the measure is *not* invariant, because the left and right chiral components transform differently:

$$2g\alpha=\theta_{\mathrm{R}}-\theta_{\mathrm{L}}\neq 0. \tag{15.155}$$

Formally, we may write

$$\mathrm{Det}\exp[2ig\alpha(x)\gamma_5]=\exp\mathrm{Tr}[2ig\alpha(x)\gamma_5] \tag{15.156}$$

where, as before, the trace Tr, with a capital T, is taken over both Dirac and space–time labels. (The trace tr will be used when the trace is only over Dirac and any internal labels.) We may rewrite the trace by introducing a complete orthonormal set of energy eigenfunctions φ_n satisfying

$$i\gamma^\mu \mathrm{D}_\mu \varphi_n=\lambda_n\varphi_n \tag{15.157a}$$

$$\sum_n \varphi_n(x)\varphi_n^\dagger(y)=\delta(x-y)\mathbf{I} \tag{15.157b}$$

where D_μ is the covariant derivative

$$\mathrm{D}_\mu=\partial_\mu+iqV_\mu+igA_\mu\gamma_5. \tag{15.157c}$$

Then

$$\begin{aligned}\mathrm{Tr}[2ig\alpha(x)\gamma_5]&=\mathrm{Tr}[2ig\alpha(x)\delta(x-y)\gamma_5]\\ &=\mathrm{Tr}\left(2ig\alpha(x)\gamma_5\sum_n\varphi_n(x)\varphi_n^\dagger(y)\right)\\ &=\int \mathrm{d}x\,\mathrm{d}y\,\delta(x-y)\,\mathrm{tr}\left(2ig\alpha(x)\gamma_5\sum_n\varphi_n(x)\varphi_n^\dagger(y)\right)\\ &=2ig\int\mathrm{d}x\,\alpha(x)A(x)\end{aligned} \tag{15.158a}$$

where

$$A(x) \equiv \sum_n \varphi_n^\dagger(x)\gamma_5\varphi_n(x) \tag{15.158b}$$

measures the 'anomaly'. To evaluate $A(x)$ we need to regulate the large eigenvalues, since as it stands the expression is only conditionally convergent. Thus we define

$$A(x) = \lim_{M\to\infty} \sum_n \varphi_n^\dagger(x)\gamma_5 f\left(\frac{\lambda_n^2}{M^2}\right)\varphi_n(x) \tag{15.159}$$

where $f(z)$ is any smooth function which rapidly approaches zero as $z \to -\infty$:

$$f(-\infty) = f'(-\infty) = f''(-\infty) = \ldots = 0 \tag{15.160a}$$

and

$$f(0) = 1 \tag{15.160b}$$

so that the contribution from any fixed eigenvalue λ_n approaches unity as $M \to \infty$. For example we could choose

$$f(z) = \mathrm{e}^z. \tag{15.161}$$

Using (15.157a), we can write

$$\begin{aligned} A(x) &= \lim_{M\to\infty} \sum_n \varphi_n^\dagger(x)\gamma_5 f\left(-\frac{\not{D}^2}{M^2}\right)\varphi_n(x) \\ &= \lim_{M\to\infty} \operatorname{tr} \int \frac{\mathrm{d}^4k}{(2\pi)^4}\, \mathrm{e}^{-ikx}\gamma_5 f\left(-\frac{\not{D}^2}{M^2}\right)\mathrm{e}^{ikx} \end{aligned} \tag{15.162a}$$

where we define

$$\not{D} = \gamma^\mu \mathrm{D}_\mu \tag{15.162b}$$

and we have changed from the energy eigenfunction basis $\varphi_n(x)$ to a plane-wave basis e^{ikx}. The definition (15.157c) of the covariant derivative D_μ gives the identity

$$\not{D}^2 = \gamma^\mu D_\mu \gamma^\nu \mathrm{D}_\nu = \gamma^\mu\gamma^\nu \tilde{\mathrm{D}}_\mu \mathrm{D}_\nu \tag{15.163a}$$

where

$$\begin{aligned} \tilde{\mathrm{D}}_\mu &\equiv \partial_\mu + \mathrm{i}qV_\mu - \mathrm{i}gA_\mu\gamma_5 \\ &= \mathrm{D}_\mu - 2\mathrm{i}gA_\mu\gamma_5. \end{aligned} \tag{15.163b}$$

Then

$$\begin{aligned} \not{D}^2 &= [\tfrac{1}{2}\{\gamma_\mu, \gamma_\nu\} + \tfrac{1}{2}[\gamma_\mu, \gamma_\nu]](\mathrm{D}^\mu - 2\mathrm{i}gA^\mu\gamma_5)\mathrm{D}^\nu \\ &= \mathrm{D}_\mu\mathrm{D}^\mu + \tfrac{1}{4}[\gamma_\mu, \gamma_\nu][\mathrm{D}^\mu, \mathrm{D}^\nu] - 2\mathrm{i}g\gamma_5 A_\mu \mathrm{D}^\mu - \mathrm{i}g[\gamma_\mu, \gamma_\nu]A^\mu\gamma_5\mathrm{D}^\nu \\ &= \mathrm{D}^2 + (\mathrm{i}/4)[\gamma_\mu, \gamma_\nu](qF^{\mu\nu} + gG^{\mu\nu}\gamma_5) + 2\mathrm{i}g\not{A}\gamma_5\not{D} \end{aligned} \tag{15.164}$$

and for any function $\chi(x)$

$$D_\mu[\chi(x)\,e^{ikx}]=[(ik_\mu+D_\mu)\chi]\,e^{ikx}. \tag{15.165}$$

Substituting back and rescaling k gives

$$A(x)=\lim_{M\to\infty}\operatorname{tr}\int\frac{d^4k}{(2\pi)^4}\,M^4\gamma_5 f\left[\left(k-\frac{iD}{M}\right)^2-\frac{i}{4M}[\gamma_\mu,\gamma_\nu](qF^{\mu\nu}+gG^{\mu\nu}\gamma_5)-\frac{2ig}{M}\not{A}\gamma_5\left(i\not{k}+\frac{\not{D}}{M}\right)\right]. \tag{15.166}$$

We now expand f in a Taylor series about k^2, and as $M\to\infty$ the only surviving terms will be up to and including the fourth derivative of f. The resulting expression may be simplified by noting that

$$\operatorname{tr}\gamma_5=\operatorname{tr}[\gamma_\mu,\gamma_\nu]=\operatorname{tr}\gamma_5[\gamma_\mu,\gamma_\nu]=0 \tag{15.167}$$

and by using symmetric integration to eliminate odd powers of k. Even so, the algebra is pretty horrendous, and we refer the reader to Balachandran *et al.*[9] for the complete treatment. We illustrate the result by dropping the gauge field A_μ associated with the local chiral transformation. This considerably simplifies the algebra, and there remain some effects which are interesting in their own right, as we shall see. We then find

$$\begin{aligned}A(x)&=-\frac{q^2}{32}\int\frac{d^4k}{(2\pi)^4}\operatorname{tr}(\gamma_5[\gamma_\mu,\gamma_\nu][\gamma_\rho,\gamma_\sigma])F^{\mu\nu}F^{\rho\sigma}f''(k^2)\\&=-\tfrac{1}{2}iq^2\varepsilon_{\mu\nu\rho\sigma}F^{\mu\nu}F^{\rho\sigma}\int\frac{d^4k}{(2\pi)^4}f''(k^2).\end{aligned} \tag{15.168}$$

The momentum integration is transformed to a Euclidean one. Then, as in Chapter 7,

$$\begin{aligned}k^2&=-\bar{k}^2=-|\bar{k}|^2\\d^4k&=i\,d^4\bar{k}=i\pi^2|\bar{k}|^2\,d|\bar{k}|^2.\end{aligned} \tag{15.169a}$$

Integrating by parts gives

$$i\int d^4\bar{k}\,f''(-|\bar{k}|^2)=i\pi^2\int_0^\infty dx\,f'(-x)=i\pi^2 \tag{15.169b}$$

using the properties (15.160) of the function f. Thus

$$A(x)=\frac{q^2}{32\pi^2}\varepsilon_{\mu\nu\rho\sigma}F^{\mu\nu}(x)F^{\rho\sigma}(x) \tag{15.170}$$

and we note that the anomaly is finite, and well defined. Substituting back into

(15.158) and using (15.156) and (15.154), we find that the non-invariance of the fermionic measure is given by

$$\mathscr{D}\psi\mathscr{D}\bar{\psi} \to \exp\left(\mathrm{i}\int \mathrm{d}^4x\, g\alpha(x)\,\frac{q^2}{16\pi^2}\,\varepsilon_{\mu\nu\rho\sigma}F^{\mu\nu}F^{\rho\sigma}\right)\mathscr{D}\psi\mathscr{D}\bar{\psi}. \qquad (15.171)$$

This non-invariance indicates that the anomaly is specifically a quantum effect, since functional integration is the ingredient required to proceed from classical to quantum field theory. This quantum effect leads to a violation of the classical conservation law associated with the chiral transformation.

In the absence of an associated gauge field, the Lagrangian density (and therefore the action) is no longer invariant under a local chiral transformation. In fact, for the (massless) Lagrangian (15.148), the non-invariance is given by

$$\mathscr{L} \to \mathscr{L} + g(\partial_\mu\alpha)\bar{\psi}\gamma^\mu\gamma_5\psi. \qquad (15.172)$$

However, the quantum effects we have been discussing generate an additional variation of the generating functional W. From (15.171) we see that the total effect of the chiral transformation in such a theory is given by

$$W \to \int \mathscr{D}\psi\mathscr{D}\bar{\psi}\,\exp\mathrm{i}\left[\int \mathrm{d}^4x\left(\mathscr{L} + g(\partial_\mu\alpha)\bar{\psi}\gamma^\mu\gamma_5\psi + g\alpha\,\frac{q^2}{16\pi^2}\,\varepsilon_{\mu\nu\rho\sigma}F^{\mu\nu}F^{\rho\sigma}\right)\right]. \qquad (15.173)$$

We require that W is invariant under such transformations

$$\frac{\delta W}{\delta\alpha(x)} = 0. \qquad (15.174)$$

Hence

$$\partial_\mu(\bar{\psi}\gamma^\mu\gamma_5\psi) = \frac{q^2}{16\pi^2}\,\varepsilon_{\mu\nu\rho\sigma}F^{\mu\nu}F^{\rho\sigma} \qquad (15.175)$$

and we see that the (classical) conservation of the axial vector current, associated with the invariance of a massless fermionic theory under chiral transformations, is violated by the (quantum) anomaly.

Fujikawa's derivation of the anomaly, which we have presented, makes it clear that the anomaly is a non-perturbative effect, even though the original discovery of it was noted in a perturbative context[10]. In fact (15.175) indicates that the anomaly is given *exactly* by the lowest order radiative corrections to the axial vector vertex. This is illustrated in figure 15.1. Each vertex with the gauge field V_μ is associated with a factor q, and we have seen that the effect is entirely due to fermionic quantum fluctuations. Thus only the single triangle loops illustrated contribute. All higher order radiative corrections to these are non-anomalous. In this treatment the anomaly is associated with the (linear)

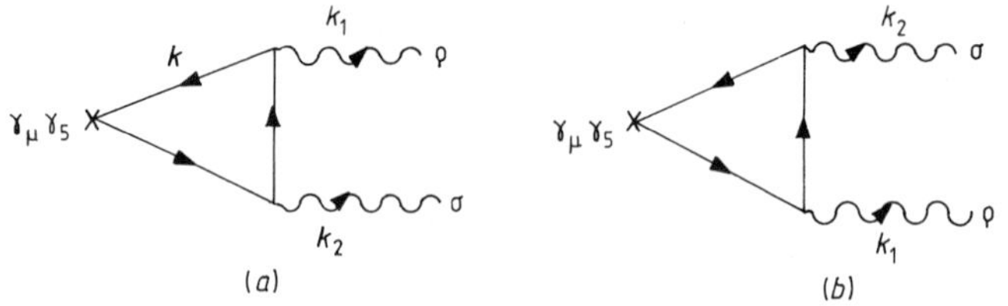

Figure 15.1 Fermion triangle diagrams responsible for the anomaly.

divergence of the fermion loop integration. We can see this as follows: the matrix element represented by figure 15.1(*a*) is

$$M^{(a)}_{\mu\rho\sigma}=\frac{q^2}{(2\pi)^4}\int \mathrm{d}^4\mathrm{k}\ \mathrm{tr}[\gamma_\mu\gamma_5 \not{k}^{-1}\gamma_\rho(\not{k}+\not{k}_1)^{-1}\gamma_\sigma(\not{k}+\not{k}_1+\not{k}_2)^{-1}]. \quad (15.176)$$

Using the trivial identity

$$\not{k}_1+\not{k}_2=(\not{k}+\not{k}_1+\not{k}_2)-\not{k} \quad (15.177)$$

it follows that

$$(k_1+k_2)^\mu M^{(a)}_{\mu\rho\sigma}=\frac{q^2}{(2\pi)^4}\int \mathrm{d}^4k\ \mathrm{tr}[\gamma_5\not{k}^{-1}\gamma_\rho(\not{k}+\not{k}_1)^{-1}\gamma_\sigma]$$
$$-\frac{q^2}{(2\pi)^4}\int \mathrm{d}^4k\ \mathrm{tr}[\gamma_5(\not{k}+\not{k}_1)^{-1}\gamma_\sigma(\not{k}+\not{k}_1+\not{k}_2)^{-1}\gamma_\rho]. \quad (15.178)$$

Both of these are divergent, but *formally* we may translate the integration in the second integral ($k\to k-k_1$) which gives

$$(k_1+k_2)^\mu M^{(a)}_{\mu\rho\sigma}=\frac{q^2}{(2\pi)^4}\int \mathrm{d}^4k\ \mathrm{tr}[\gamma_5\not{k}^{-1}\gamma_\rho(\not{k}+\not{k}_1)^{-1}\gamma_\sigma]$$
$$-\frac{q^2}{(2\pi)^4}\int \mathrm{d}^4k\ \mathrm{tr}[\gamma_5\not{k}^{-1}\gamma_\sigma(\not{k}+\not{k}_2)^{-1}\gamma_\rho]. \quad (15.179)$$

The evaluation of $(k_1+k_2)^\mu M^{(b)}_{\mu\rho\sigma}$, the contribution from figure 15.1(*b*), follows immediately by interchanging

$$k_1,\rho\leftrightarrow k_2,\sigma. \quad (15.180)$$

Since (15.179) is antisymmetric it follows that the sum $M_{\mu\rho\sigma}$ of the two diagrams *formally* satisfies

$$(k_1+k_2)^\mu M_{\mu\rho\sigma}=0 \quad (15.181)$$

in accordance with the classical conservation law (15.173).

However the above arguments are purely formal, because $M_{\mu\rho\sigma}$ is divergent. It might be thought that one could avoid this objection by using dimensional regularisation, which typically has the attractive property of preserving such conservation laws while rendering the integrals finite. In this case, however, dimensional regularisation is complicated by the (active) presence of the γ_5 matrix. γ_5 is specifically tied to four dimensions; by definition, it is the product of the four gamma matrices $\gamma^0, \gamma^1, \gamma^2, \gamma^3$, and its anticommutation properties with them then follows from the Clifford algebra. In a higher (2ω)-dimensional space–time, the actual definition of γ_5 is problematic: do we take

$$\gamma_5 = \mathrm{i}\gamma^0\gamma^1 \ldots \gamma^{(2\omega)} \qquad (2\omega \text{ even}) \tag{15.182a}$$

or

$$\gamma_5 = \mathrm{i}\gamma^0\gamma^1\gamma^2\gamma^3? \tag{15.182b}$$

In the first case, the axial vector current remains an axial 2ω-vector, but $M_{\mu\rho\sigma}$ is *zero*. In the second, the axial vector becomes an antisymmetric tensor of rank $2\omega - 3$, but its divergence is non-zero because γ_5 no longer anticommutes with all of the gamma matrices. In fact, the non-zero value is precisely that already obtained in (15.175), and the same (finite) non-zero value is obtained if we use an old-fashioned covariant cut-off and remain in four dimensions throughout.

It is therefore clear that the anomaly is a real effect, and not just some artifice of the particular regularisation scheme adopted. For the reasons given at the beginning of this section, anomalies will destroy the ξ-independence of S-matrix elements, and make the theory nonsensical. The key to evading this problem derives from noting that the anomaly is independent of the *mass* of the fermion field. (The fermion in (15.148) is massless, but this was not used in the derivation of the anomaly.) It is therefore possible to arrange that contributions from different fermions cancel against each other, by virtue of the group-theoretic structure of the model.

In a general (non-Abelian) gauge theory each of the vertices of the triangle diagrams will carry an internal symmetry matrix, and the anomaly will be multiplied by a factor

$$A^{abc} \equiv \mathrm{tr}(\mathbf{t}^a\{\mathbf{t}^b\mathbf{t}^c + \mathbf{t}^c\mathbf{t}^b\}) \tag{15.183}$$

where now the trace is over the internal symmetry space. Thus provided A^{abc} vanishes for all axial–vector–vector triangles (and all axial–axial–axial triangles) the full theory will be anomaly-free.

In electroweak theory it is easy to see that this is what happens. The SU(2) gauge bosons have a fermionic vertex $\gamma_\mu\Gamma^a$, where

$$\Gamma^a = g\tfrac{1}{2}\tau^a a_{\mathrm{L}} \qquad (a = 1, 2, 3) \tag{15.184}$$

while the U(1) gauge boson–fermion vertex is $\gamma_\mu\Gamma^4$, where

$$\begin{aligned}\Gamma^4 &= g'(Ya_{\mathrm{L}} + Qa_{\mathrm{R}}) \\ &= g'(Y + \tfrac{1}{2}\tau^3 a_{\mathrm{R}})\end{aligned} \tag{15.185}$$

since $Q = \frac{1}{2}\tau^3 + Y$. Note that on all four vertices γ_5 appears only in conjunction with one of the τ matrices. Thus the A–A–A anomaly is proportional to

$$\begin{aligned}&\mathrm{tr}(\tau^a\{\tau^b\tau^c+\tau^c\tau^b\}) \qquad (\mathrm{a,b,c}=1,2,3)\\&=2\delta^{bc}\,\mathrm{tr}\,\tau^a\\&=0.\end{aligned} \tag{15.186}$$

The A–V–V anomaly cancellation is not automatic, because although A is associated with a τ matrix the V may have a τ matrix *or* the unit matrix. The potentially dangerous cases are when two τ matrices are involved (444, 4*bc*). In this case the anomaly is proportional to

$$\begin{aligned}\mathrm{tr}(\tau^a\{\tau^bY+Y\tau^b\})&\propto\delta^{ab}\,\mathrm{tr}\,Y\\&=\delta^{ab}\,\mathrm{tr}\,Q.\end{aligned} \tag{15.187}$$

Thus to make the theory anomaly-free we need that the sum of the electric charges of all the fermions should vanish

$$\sum_f Q_f=0. \tag{15.188}$$

Remembering that each quark flavour has three colours, we see that the above requirement is satisfied in each family, since

$$Q_{\nu_l}+Q_l+3Q_U+3Q_D=0. \tag{15.189}$$

We have emphasised the form of the anomaly when the axial currents are not associated with gauge fields, and also when the (vector) gauge group is U(1). Relaxing these constraints leads to more anomalous diagrams. This is fairly obvious from (15.175), for example. In a non-Abelian theory we should expect that $F_{\mu\nu}$ is replaced by its non-Abelian analogue $F^a_{\mu\nu}$, which has terms linear *and* quadratic in the gauge fields. Thus we anticipate anomalies in the square ($AVVV$, $AAAV$) and pentagon ($AVVVV$, $AAAVV$, $AAAAA$) diagrams, besides the triangle (AVV, AAA) diagram already discussed. Also, it is apparent from (15.166) that the axial gauge fields generate extra contributions with a different structure from those we have kept. In fact, both of these expectations are fulfilled; the interested reader is referred to Balachandran *et al.*[9] for details, and also to Einhorn and Jones[11] for a very clear analysis of the opposite case considered in the text, namely when the vector gauge field V_μ (rather than A_μ) is dropped.

Problems

15.1 Verify the form of $(1+K)^2$ in (15.24).

15.2 Verify the Feynman rules (15.31), (15.41) and (15.44).

15.3 Verify (15.102) and (15.104).

15.4 Justify (15.112) by considering the scattering of an electron by a heavy charged particle.

15.5 By calculating the quantities involved, or by using the results of reference 5, check that (15.132) *is* satisfied.

15.6 Determine the contributions of diagrams (1), (2) and (4) of (15.135).

15.7 Show that the fermionic functional integration measure *is* gauge invariant in QCD.

15.8 After dropping A_μ, show that (15.168) follows from (15.166).

References

The books and articles we have used most extensively in preparing this chapter are

Lynn B W and Wheater J F (eds) 1984 *Radiative Corrections in* $SU(2)_2 \times U(1)$ (Singapore: World Scientific)

Sirlin A 1980 *Phys. Rev.* D **22** 971

Aoki K, Hioki Z, Kawabe R, Konuma M and Muta T 1982 *Supp. Prog. Theor. Phys.* **73** 106

Fujikawa K 1980 *Phys. Rev.* D **21** 2848

References in the text

1 See, for example,
Ross D A and Taylor J C 1973 *Nucl. Phys.* B **51** 116
Sirlin A 1974 *Phys. Rev. Lett.* **32** 966
Angerson W 1974 *Nucl. Phys.* B **69** 493
Salmonson P and Ueda Y 1975 *Phys. Rev.* D **11** 2606
Green M and Veltman M 1980 *Nucl. Phys.* B **169** 137
Consoli M 1979 *Nucl. Phys.* B **160** 208
Sakakibara S 1981 *Phys. Rev.* D **24** 1149
Cole J P 1985 *Progress in Particle and Nuclear Physics* **12**
Wheater J F and Llewellyn Smith C H 1982 *Nucl. Phys.* B **208** 27

2 Stevenson P M 1981 *Phys. Rev.* D **23** 2916

3 Sirlin A 1980 *Phys. Rev.* D **22** 971

4 Arnison G *et al.* 1983 *Phys. Lett.* **122B** 103; **126B** 398
Banner M *et al.* 1983 *Phys. Lett.* **122B** 476

5 Aoki K *et al.* 1982 *Supp. Prog. Theor. Phys.* **73** 106

6 See, for example, Sakakitara S 1981 *Z. Phys.* C **11** 43
or Cole J P *op cit* (reference 1)

7 Fujikawa K 1979 *Phys. Rev. Lett.* **42** 1195
—— 1980 *Phys. Rev.* D **21** 2848; D **22** 1499 (E)

8 We are grateful to Paul Frampton for impressing this point upon us.

9 Balachandran A P, Marmo G, Nair V P and Trachern C G 1982 *Phys. Rev.* D **25** 2713

10 Adler S L 1969 *Phys. Rev.* **177** 2426
Bell J and Jackiw R 1969 *Nuov. Cim.* **60A** 47
Bardeen W 1969 *Phys. Rev.* **184** 1848

11 Einhorn M B and Jones D R T 1984 *Phys. Rev.* D **29** 331

16

GRAND UNIFIED THEORY

16.1 Philosophy

We have seen in Chapter 14 how the electromagnetic and weak interactions may be unified into electroweak theory. This unification leaves something to be desired. First, there are two independent coupling constants in the theory, g and g', and, second, no unification with the strong interactions (QCD) has occurred. The aim of grand unified theory[1,2] is to rectify this by unifying strong, weak and electromagnetic interactions in a grand unified gauge theory with a single coupling constant. Once such a unification has been achieved, the coupling constants g, g' and g_s (the QCD coupling constant) will be related by group theory factors to a single grand unified coupling constant. Though such a statement flows easily from the pen, it needs some sharpening up. We have seen in Chapter 7 that the values of the renormalised coupling constants depend on the renormalisation scale, M, and we have discussed this dependence in detail in §12.7, for QCD and QED. Thus, we must decide at what renormalisation scale M_G the coupling constants g, g' and g_s satisfy the group theoretical relationships associated with the embedding of electroweak theory and QCD in the grand unified gauge group. Once we have fixed the renormalisation scale M_G (the grand unification scale) it is clear that these relationships between g, g' and g_s will not hold at lower renormalisation scales M. This is because these coupling constants vary with M in different ways when the extra gauge fields associated with the grand unification may be ignored. On the other hand, these relationships will hold at higher renormalisation scales, because for mass scales large compared with the masses of the new gauge fields all gauge fields are on the same footing. Then there must be a single coupling constant g_G developing according to the renormalisation group equation of the grand unified theory.

A priori, the grand unification scale M_G might be an ordinary mass ($\lesssim 100$ GeV). After all, (12.128) shows that for renormalisation scales of 4 or 5 GeV, the ratio of $g_s(M)$ to $e(M)$ is 5 or 6. This is the kind of number which just possibly could be attributable to group theory factors in the embedding of QCD and electroweak theory in the grand unified theory. However, at least for grand unified theories of reasonably low rank, it turns out that a ratio of g_s to e of 5 or 6 does not come out naturally from the group theory, but smaller numbers arise. If the grand unification is to succeed, it must therefore be that unification occurs at some larger mass scale, and the desired ratio of g_s to e arises when the renormalisation group equations are used to continue $g_s(M)$

and $e(M)$ back to ordinary mass scales. (Remember that $g_s^2(M)$ decreases and $e^2(M)$ increases as M increases.) Since coupling constants vary only logarithmically with M in gauge field theories, the grand unification scale is likely to have to be extremely large to achieve a significant deviation from the group theoretical value of g_s/e in this way. We shall discuss this in detail in §16.3. This large grand unification scale is a good thing, because typical grand unified theories put leptons and quarks in the same multiplets and contain baryon number violating exchanges. If the masses of the new gauge fields which mediate baryon number violating processes were not be very large, the proton would decay at an unacceptably fast rate. This point is discussed in §16.6.

16.2 SU(5) grand unified theory

Since $SU_C(3) \times SU_L(2) \times U(1)$ of QCD and electroweak theory has rank 4, it follows that any grand unified theory will have to be based on a semi-simple Lie group of rank at least 4. (The grand unification group has to be either simple, or semi-simple, with a discrete symmetry superimposed, to obtain a single coupling constant.) It turns out[1,2] that there is only one semi-simple rank 4 group which allows QCD and electroweak theory to be embedded in a way consistent with the quantum numbers of the quarks and leptons, namely SU(5). We shall restrict attention to SU(5) grand unified theory here, though other grand unifications are possible with higher rank Lie groups, e.g. SO(10) with rank 5.

The generators T_a of SU(5) for the fundamental five-dimensional representation may conveniently be represented by the natural generalisation of the Gell-Mann matrices (see Appendix D) which we denote by λ_a, $a = 1, \ldots, 24$, and normalise in the conventional way so that

$$\mathrm{Tr}(\lambda_a \lambda_b) = 2\delta_{ab}. \tag{16.1}$$

(A good discussion of the properties of SU(N) Gell-Mann matrices is given by Macfarlane *et al.*[3]) Thus

$$T_a = \frac{\lambda_a}{2} \qquad a = 1, \ldots, 24. \tag{16.2}$$

The colour group $SU_C(3)$ is conveniently identified with the λ matrices which have non-zero entries in only the first three rows and columns, viz. $\lambda_1, \ldots, \lambda_8$, and the $SU_L(2)$ group with the λ matrices which utilise only the last two rows and columns, viz. λ_{22}, λ_{23} and $(\sqrt{10}\,\lambda_{24} - \sqrt{6}\,\lambda_{15})/4$. Thus, the generators of $SU_C(3)$ are

$$T_a^C = \frac{\lambda_a}{2} \qquad a = 1, \ldots, 8 \tag{16.3}$$

and the generators of $SU_L(2)$ are

$$T_1^L = \frac{\lambda_{22}}{2} \qquad T_2^L = \frac{\lambda_{23}}{2} \qquad T_3^L = \frac{(\sqrt{10}\,\lambda_{24} - \sqrt{6}\,\lambda_{15})}{8}. \tag{16.4}$$

It remains to identify the weak hypercharge, related to the charge Q and T_3^L by

$$Q = T_3^L + Y. \tag{16.5}$$

We shall see later that consistency with the charges of the quarks and leptons requires the identification

$$Q = -\sqrt{\tfrac{2}{3}}\,\lambda_{15} \tag{16.6}$$

so that

$$Y = -\left(\sqrt{10}\,\lambda_{24} + \frac{5\sqrt{6}}{3}\,\lambda_{15}\right)\Big/ 8. \tag{16.7}$$

(Q must certainly be a generator of SU(5) since the photon is a gauge field.)

The gauge fields A_a^μ belong to the 24-dimensional adjoint representation of SU(5), and may be written as a 5×5 matrix

$$A^\mu \equiv A_a^\mu T_a = A_a^\mu \frac{\lambda_a}{2} \tag{16.8}$$

as in (9.25), where T_a are the generators of SU(5) for the five-dimensional fundamental representation as in (16.2). Having regard to the identifications of generators made above, we may write (see problem 16.1)

$$\sqrt{2}\,A^\mu = \left(\begin{array}{ccc|cc}
\frac{1}{\sqrt{2}}A_3^\mu + \frac{1}{\sqrt{6}}A_8^\mu - \frac{2B^\mu}{\sqrt{30}} & \frac{1}{\sqrt{2}}A_{1-i2}^\mu & \frac{1}{\sqrt{2}}A_{4-i5}^\mu & \bar{X}_1^\mu & \bar{Y}_1^\mu \\
\frac{1}{\sqrt{2}}A_{1+i2}^\mu & -\frac{A_3^\mu}{\sqrt{2}} + \frac{A_8^\mu}{\sqrt{6}} - \frac{2B^\mu}{\sqrt{30}} & \frac{A_{6-i7}^\mu}{\sqrt{2}} & \bar{X}_2^\mu & \bar{Y}_2^\mu \\
\frac{1}{\sqrt{2}}A_{4+i5}^\mu & \frac{1}{\sqrt{2}}A_{6+i7}^\mu & -\sqrt{\frac{2}{3}}A_8^\mu - \frac{2B^\mu}{\sqrt{30}} & \bar{X}_3^\mu & \bar{Y}_3^\mu \\
\hline
X_1^\mu & X_2^\mu & X_3^\mu & \frac{W_3^\mu}{\sqrt{2}} + \frac{3B^\mu}{\sqrt{30}} & W_+^\mu \\
Y_1^\mu & Y_2^\mu & Y_3^\mu & W_-^\mu & -\frac{W_3^\mu}{\sqrt{2}} + \frac{3B^\mu}{\sqrt{30}}
\end{array}\right) \tag{16.9}$$

where A_a^μ, $a = 1, \ldots, 8$, are the colour gluons of $SU_C(3)$, and W_3^μ, W_+^μ, W_-^μ and B^μ are the electroweak gauge fields of §14.1. In terms of SU(5) gauge fields,

$$W_3^\mu = (\sqrt{10}\, A_{24}^\mu - \sqrt{6}\, A_{15}^\mu)/4 \tag{16.10}$$

$$W_\pm^\mu = (A_{22}^\mu \mp \mathrm{i} A_{23}^\mu)/\sqrt{2} \tag{16.11}$$

and

$$B^\mu = -(\sqrt{\tfrac{5}{2}}\, A_{15}^\mu + \sqrt{\tfrac{3}{2}}\, A_{24}^\mu)/2. \tag{16.12}$$

The identifications (16.10)–(16.12) are made by observing that B^μ is the correctly normalised field coupled to weak hypercharge etc.

The new gauge fields, X_i^μ, $i=1,2,3$ and Y_i^μ, $i=1,2,3$ are anti-triplets under $\mathrm{SU_C}(3)$, and are defined by

$$X_1^\mu = \frac{1}{\sqrt{2}}(A_9^\mu + \mathrm{i} A_{10}^\mu) \qquad X_2^\mu = \frac{1}{\sqrt{2}}(A_{11}^\mu + \mathrm{i} A_{12}^\mu) \qquad X_3^\mu = \frac{1}{\sqrt{2}}(A_{13}^\mu + \mathrm{i} A_{14}^\mu) \tag{16.13}$$

and

$$Y_1^\mu = \frac{1}{\sqrt{2}}(A_{16}^\mu + \mathrm{i} A_{17}^\mu) \qquad Y_2^\mu = \frac{1}{\sqrt{2}}(A_{18}^\mu + \mathrm{i} A_{19}^\mu) \qquad Y_3^\mu = \frac{1}{\sqrt{2}}(A_{20}^\mu + \mathrm{i} A_{21}^\mu). \tag{16.14}$$

They are often referred to as lepto–quark (or diquark) bosons for reasons which will become clear shortly.

The first generation of quarks and leptons $(\mathrm{u}_i, \mathrm{d}_i, \mathrm{e}^-, \nu_\mathrm{e})$ where $i=1,2,3$ is the colour index of the quarks, has fifteen helicity states, because the neutrino is massless. These states fit into a $\bar{\mathbf{5}}$ of SU(5), $(\Psi_p)_\mathrm{L}$, $p=1, \ldots, 5$, and a **10** of SU(5), x_L^{pq}, $p, q=1, \ldots, 5$, as follows, with corresponding assignments for further generations, $(\mathrm{c}_i, \mathrm{s}_i, \mu^-, \nu_\mu)$, $(\mathrm{t}_i, \mathrm{b}_i, \tau^-, \nu_\tau)$ etc. (Cabibbo angles have been suppressed, but are easily reintroduced.)

$$(\Psi_p)_\mathrm{L} = \begin{pmatrix} \mathrm{d}_1^\mathrm{C} \\ \mathrm{d}_2^\mathrm{C} \\ \mathrm{d}_3^\mathrm{C} \\ \mathrm{e} \\ \nu_\mathrm{e} \end{pmatrix}_\mathrm{L} \tag{16.15}$$

and

$$x_\mathrm{L}^{pq} = \frac{1}{\sqrt{2}} \begin{pmatrix} 0 & \mathrm{u}_3^\mathrm{C} & -\mathrm{u}_2^\mathrm{C} & -\mathrm{u}_1 & -\mathrm{d}_1 \\ -\mathrm{u}_3^\mathrm{C} & 0 & \mathrm{u}_1^\mathrm{C} & -\mathrm{u}_2 & -\mathrm{d}_2 \\ \mathrm{u}_3^\mathrm{C} & -\mathrm{u}_1^\mathrm{C} & 0 & -\mathrm{u}_3 & -\mathrm{d}_3 \\ \mathrm{u}_1 & \mathrm{u}_2 & \mathrm{u}_3 & 0 & -\mathrm{e}^\mathrm{C} \\ \mathrm{d}_1 & \mathrm{d}_2 & \mathrm{d}_3 & \mathrm{e}^\mathrm{C} & 0 \end{pmatrix}_\mathrm{L} \tag{16.16}$$

where all fields are left-handed, and the right-handed components for a given particle have been introduced as the left-handed components of the charge conjugate field. For any fermion field ψ the notation, as in (14.4)–(14.7), is

$$\psi_{\mathrm{L}} \equiv \tfrac{1}{2}(1-\gamma_5)\psi \qquad \psi_{\mathrm{R}} \equiv \tfrac{1}{2}(1+\gamma_5)\psi \tag{16.17}$$

and

$$\psi_{\alpha}^{\mathrm{C}} \equiv C_{\alpha\beta}\bar{\psi}_{\beta} \tag{16.18}$$

with $C_{\alpha\beta}$ having the defining property

$$\mathbf{C}^{-1}\gamma^{\mu}\mathbf{C} = -(\gamma^{\mu})^{\mathrm{T}}. \tag{16.19}$$

The form of (16.16) is dictated by the fact that the **10** occurs as the anti-symmetric part of the decomposition of the product of two **5**'s. Noticing that the generators $\bar{T}_a$ for $\bar{\mathbf{5}}$ are

$$\bar{T}_a = -\frac{\lambda_a^*}{2} \qquad \text{for } \bar{\mathbf{5}} \tag{16.20}$$

and the generators $\tilde{T}_a$ for **10** are

$$\tilde{T}_a = \frac{\lambda_a}{2} \otimes 1 + 1 \otimes \frac{\lambda_a}{2} \qquad \text{for } \mathbf{10} \tag{16.21}$$

where in the first term λ_a acts on the index p of x_{L}^{pq}, and in the second term it acts on the index q, it is easy to check that the identification (16.6) for the charge operator in the **5** leads to the correct quark and lepton charges. Using the criteria (15.183) one sees that there is a cancellation of Adler anomalies between the $\bar{\mathbf{5}}$ and the **10**.

The assignment of quarks and leptons to multiplets of the grand unified group SU(5) explains things which were left unexplained by QCD and electroweak theory. For instance, the third-integral quark charges are a consequence of the fact that the charge Q is a generator of SU(5) (as in (16.6)), and so traceless. Thus the sum of the charges of the particles in any representation of SU(5) must be zero. For the $\bar{\mathbf{5}}$ this means that

$$-3Q_{\mathrm{d}} + Q_{\mathrm{e}} = 0 \tag{16.22}$$

and for the **10**

$$3Q_{\mathrm{d}} - Q_{\mathrm{e}} = 0. \tag{16.23}$$

Either way, we get

$$Q_{\mathrm{d}} = -Q_{\mathrm{e}}/3. \tag{16.24}$$

The 3 has arisen because quarks come in three colours.

Because quarks and leptons occur in the same irreducible representations of SU(5), there are vertices involving the lepto–quark gauge boson X and Y of (16.13) and (16.14) which change a quark into a lepton, e.g. those shown in

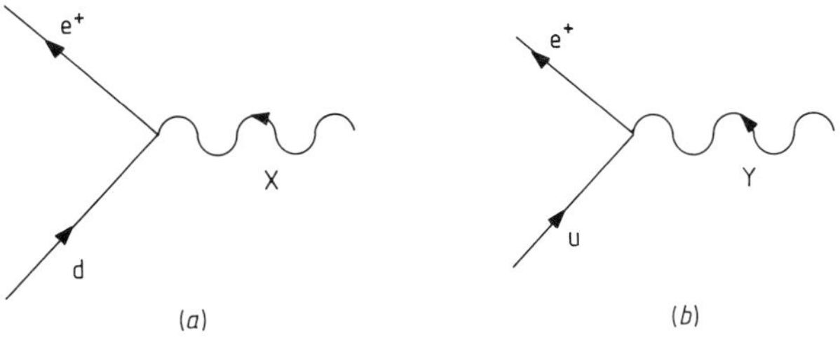

Figure 16.1 Vertices changing a quark into a lepton.

figure 16.1. Because quark fields and their conjugate fields both occur in the **10**, there are also vertices in which a pair of quarks annihilate, as, for example, in figure 16.2. These two types of vertices may be combined to produce baryon number violation, as, for example, in figure 16.3, where we have added a spectator d quark. Since baryon number is known to be conserved to a very good approximation, these processes must be suppressed by a very large mass for the X and Y lepto–quarks. We shall see in §16.6 that this is indeed the case.

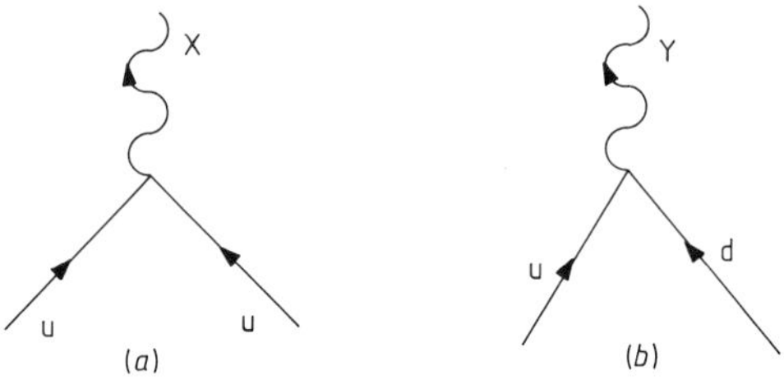

Figure 16.2 Vertices with two quarks annihilating.

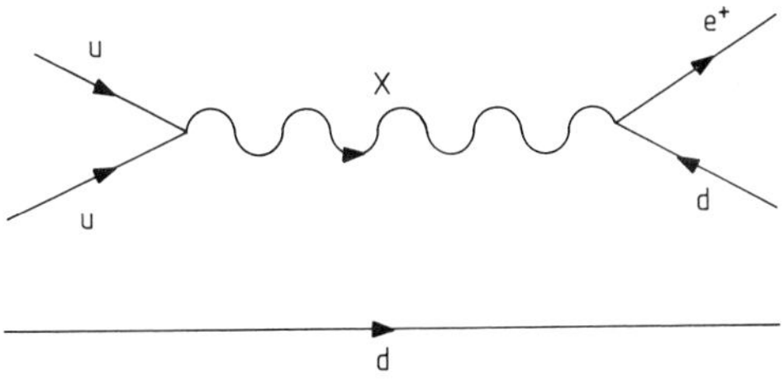

Figure 16.3 Diagram for $p \to \pi^0 e^+$.

The QCD and electroweak coupling constants, g_s, g and g' may be related to the grand unified coupling constant g_G (at the unification scale) by writing down the SU(5) gauge invariant couplings of the quarks and leptons. The appropriate terms (as in (9.36) and (9.15)) are

$$\mathscr{L}^{\text{fermion}} = \mathrm{i}(\bar{\Psi}_p)_L \gamma^\mu (D^\mu \Psi_p)_L + \mathrm{i}\bar{x}_L^{pq} \gamma^\mu D_\mu x_L^{pq} \tag{16.25}$$

where

$$\mathrm{D}_\mu(\bar{\Psi}_p)_\mathrm{L} = \partial_\mu(\bar{\Psi}_p)_\mathrm{L} - \mathrm{i}g_\mathrm{G}\frac{(\lambda_a^*)_{pq}}{2}(\Psi_q)_\mathrm{L}A_\mu^a$$

$$= \partial_\mu(\Psi_p)_\mathrm{L} - \mathrm{i}g_\mathrm{G}(A_\mu^*)_{pq}(\Psi_q)_\mathrm{L} \tag{16.26}$$

and

$$\mathrm{D}_\mu x_\mathrm{L}^{pq} = \partial_\mu x_\mathrm{L}^{pq} + \mathrm{i}g_\mathrm{G}A_\mu^a\left[\left(\frac{\lambda_a}{2}\right)_{pr} x_\mathrm{L}^{rq} + \left(\frac{\lambda_a}{2}\right)_{qs} x_\mathrm{L}^{ps}\right]$$

$$= \partial_\mu x_\mathrm{L}^{pq} + \mathrm{i}g_\mathrm{G}[(A_\mu)_{pr}x_\mathrm{L}^{rq} + (A_\mu)_{qs}x_\mathrm{L}^{ps}] \tag{16.27}$$

where A^μ is as in (16.8) and (16.9). (Fermion masses will be introduced in §16.5 after spontaneous symmetry breaking has been discussed.) Using the antisymmetry of x_L in its indices (16.27) simplifies to

$$\mathrm{D}_\mu x_\mathrm{L}^{pq} = \partial_\mu x_\mathrm{L}^{pq} + 2\mathrm{i}g_\mathrm{G}(A_\mu)_{pr}x_\mathrm{L}^{rq} \tag{16.28}$$

when it is coupled to $\bar{x}_\mathrm{L}^{pq}$. With the aid of the identity for any Dirac field ψ,

$$\bar{\psi}_\mathrm{L}^\mathrm{C}\gamma^\mu\partial_\mu\psi_\mathrm{L}^\mathrm{C} = \bar{\psi}_\mathrm{R}\gamma^\mu\partial_\mu\psi_\mathrm{R} \tag{16.29}$$

one may easily check that the g_G independent terms in (16.25) are the standard kinetic energy terms for quarks and leptons. The interaction terms in (16.25) are

$$\mathscr{L}_\mathrm{I}^\mathrm{fermion} = g_\mathrm{G}(\bar{\Psi}_p)_\mathrm{L}\gamma^\mu(A_\mu^*)_{pq}(\Psi_q)_\mathrm{L} - 2g_\mathrm{G}\bar{x}_\mathrm{L}^{pq}\gamma^\mu(A_\mu)_{pr}x_\mathrm{L}^{rq} \tag{16.30}$$

with A_μ as in (16.9), $(\Psi_p)_\mathrm{L}$ as in (16.15), and x_L^{pq} as in (16.16). It is clear from (16.9) that $\mathrm{SU_C}(3)$ and $\mathrm{SU}_2(2)$ are embedded in SU(5) in the natural way, so that

$$g_\mathrm{s} = g = g_\mathrm{G}. \tag{16.31}$$

We may identify the coupling constant g' for the U(1) of weak hypercharge by retaining only B^μ dependent terms in (16.30). Then we obtain

$$\mathscr{L}_\mathrm{I}^\mathrm{fermion} = -\sqrt{\tfrac{3}{5}}\,g_\mathrm{G}B_\mu[\tfrac{2}{3}(\bar{\mathrm{u}}_i)_\mathrm{R}\gamma^\mu(\mathrm{u}_i)_\mathrm{R} + \tfrac{1}{6}(\bar{\mathrm{u}}_i)_\mathrm{L}\gamma^\mu(\mathrm{u}_i)_\mathrm{L}$$

$$+\tfrac{1}{6}(\bar{\mathrm{d}}_i)_\mathrm{L}\gamma^\mu(\mathrm{d}_i)_\mathrm{L} - \bar{\mathrm{e}}_\mathrm{R}\gamma^\mu\mathrm{e}_\mathrm{R} - \tfrac{1}{3}(\bar{\mathrm{d}}_i)_\mathrm{R}\gamma^\mu(\mathrm{d}_i)_\mathrm{R}$$

$$-\tfrac{1}{2}\bar{\mathrm{e}}_\mathrm{L}\gamma^\mu\mathrm{e}_\mathrm{L} - \tfrac{1}{2}\bar{\nu}_\mathrm{L}\gamma^\mu\nu_\mathrm{L}] \tag{16.32}$$

where we have used the identity for any Dirac field ψ,

$$\bar{\psi}_\mathrm{L}^\mathrm{C}\gamma^\mu\psi_\mathrm{L}^\mathrm{C} = -\bar{\psi}_\mathrm{R}\gamma^\mu\psi_\mathrm{R}. \tag{16.33}$$

Comparing with §14.1, we see that

$$g' = \sqrt{\tfrac{3}{5}}\,g_\mathrm{G}. \tag{16.34}$$

16.3 The grand unification scale and θ_W

The scale at which grand unification occurs, with a single coupling constant for all gauge field interactions, may be determined using the renormalisation group equation to extrapolate[4,5] the known values of the QCD and electroweak coupling constants $g_s(M)$, $g(M)$, and $g'(M)$ at an 'ordinary' renormalisation scale. At the SU(5) grand unification scale, the coupling constants must satisfy (16.31) and (16.34). Since g_G is in the first instance unknown, there are two constraints. These may be used, for example, to determine the grand unification scale M_G from the known values of g_s and e at ordinary energies, and to predict the ratio g'/g, and so the weak mixing angle, θ_W, at ordinary energies. Equations (12.29b), (12.25) and (12.18) with $s = M/M_G$ enable us to relate coupling constants at some 'ordinary' renormalisation scale M and the grand unification scale M_G. For the U(1), $SU_L(2)$ and $SU_C(3)$ factors of electroweak theory and QCD we have

$$(g')^{-2}(M) - (g')^{-2}(M_G) = 2b_1 \ln(M/M_G) \tag{16.35}$$

$$g^{-2}(M) - g^{-2}(M_G) = 2b_2 \ln(M/M_G) \tag{16.36}$$

and

$$g_s^{-2}(M) - g_s^{-2}(M_G) = 2b_3 \ln(M/M_G) \tag{16.37}$$

where, using a slight generalisation of (12.15) (see problem 16.2),

$$16\pi^2 b_1 = -\frac{20}{9} N_G \tag{16.38}$$

$$16\pi^2 b_2 = \frac{22}{3} - \frac{4N_G}{3} \tag{16.39}$$

$$16\pi^2 b_3 = 11 - \frac{4N_G}{3}. \tag{16.40}$$

We have assumed N_G generations of fermions (u_i, d_i, e^-, ν_e), $(c_i, s_i, \bar{\mu}, \nu_\mu)$, $(t_i, b_i, \bar{\tau}, \nu_\tau)$ etc, where $i = 1, 2, 3$ is the colour index, and there is the usual assignment to multiplets of $SU_L(2) \times U(1)$ as in §14.1. Possible contributions of Higgs scalar loops to (16.39) and (16.40) have been ignored. It turns out (see problem (16.3) that such contributions make only a small difference. From (16.31) and (16.34),

$$g_s(M_G) = g(M_G) = \sqrt{\tfrac{5}{3}}\, g'(M_G) = g_G(M_G). \tag{16.41}$$

Also, from §14.21,

$$g(M) \sin\theta_W(M) = g'(M) \cos\theta_W(M) = e(M) \tag{16.42}$$

where $\theta_W(M)$ is the value of the weak mixing angle at the renormalisation scale

M. Thus, (16.35)–(16.37) imply that

$$e^{-2}(M)\sin^2\theta_W(M)-2b_2\ln(M/M_G)=g_s^{-2}(M)-2b_3\ln(M/M_G)$$
$$=\tfrac{3}{5}[e^{-2}(M)\cos^2\theta_W(M)-2b_1\ln(M/M_G)]. \tag{16.43}$$

The two results following from (16.43) are

$$\ln(M_G/M)=\frac{3(1-8\alpha(M)/3\alpha_s(M))}{8\pi\alpha(M)(8b_3-3b_1-3b_2)} \tag{16.44}$$

and

$$\sin^2\theta_W(M)=g_s^{-2}e^2+2e^2(b_2-b_3)\ln(M/M_G)$$
$$=\frac{3(b_3-b_2)}{8b_3-3(b_1+b_2)}+\frac{(5b_2-3b_1)\alpha(M)/\alpha_s(M)}{8b_3-3(b_1+b_2)} \tag{16.45}$$

where $\alpha(M)$ and $\alpha_s(M)$ are the fine structure constant and QCD fine structure constant as in (12.126) and (12.122). With b_1, b_2 and b_3 as in (16.38)–(16.40), we see that, at this order in perturbation theory, the predictions for M_G/M and θ_W are independent of N_G, the number of generations of fermions, and are

$$\ln(M_G/M)=\frac{\pi(1-8\alpha(M)/3\alpha_s(M))}{11\alpha(M)} \tag{16.46}$$

and

$$\sin^2\theta_W(M)=\frac{1}{6}+\frac{5}{9}\frac{\alpha(M)}{\alpha_s(M)}. \tag{16.47}$$

With the values of $\alpha_s(m_Z)$ and $\alpha(m_Z)$ given by (12.121) and (12.129), we obtain

$$M_G\simeq 5\times 10^{14}\ \text{GeV} \tag{16.48}$$

and

$$\sin^2\theta_W(M)\simeq 0.206 \qquad M=m_Z. \tag{16.49}$$

The extremely large value of the grand unification scale M_G has arisen from the logarithmic dependence in (16.46) with a right-hand side of order 30. The predicted value of $\sin^2\theta_W(M)$ at $M=m_Z$ differs substantially from the value $\frac{3}{8}$ at the unification scale $M=M_G$ which follows from (16.31) and (16.34). Even when many small corrections to the equations of this section are made, including higher-order perturbation theory, effects of quark thresholds, and the contributions of Higgs scalars, the predicted value of $\sin^2\theta_W$ cannot be brought into agreement with the experimental value of 0.233 at $M=m_Z$. (However, in supersymmetric grand unified theory, which is outside our scope here, agreement can be achieved.)

With the value of M_G in (16.48) and the value of $\alpha_s(m_Z)$ of (12.121) the grand unified fine structure constant is given by

$$\alpha_G(M_G) \equiv g_G^2(M_G)/4\pi = g_s^2(M_G)/4\pi \simeq 2.4 \times 10^{-2} \qquad (16.50)$$

where we have assumed three generations of fermions.

16.4 Spontaneous symmetry breaking for SU(5) grand unified theory

Since the only massless gauge fields we want are the photon and the colour gluons, we must introduce enough Higgs scalar multiplets into the theory to break the SU(5) gauge symmetry to $SU_C(3) \times U_Q(1)$, where $U_Q(1)$ is the U(1) of electromagnetism. There must therefore be Higgs scalars to give masses of order 100 GeV to $W^{\pm}$ and Z^0 and Higgs scalars to give masses of order 10^{15} GeV to the lepto–quark gauge bosons X and Y of (16.13) and (16.14) (because, as observed in §16.1, the masses of these new gauge bosons must be of the order of the grand unification scale (16.48)). Since two very different mass scales are involved, we are going to need at least two multiplets of Higgs scalars with very different vacuum expectation values. A suitable choice is a **24** of Higgs scalars (corresponding to the adjoint representation) to break SU(5) to $SU_C(3) \times SU_L(2) \times U(1)$, and a **5** of Higgs scalars (corresponding to the fundamental representation of SU(5)) to break $SU_C(3) \times SU_L(2) \times U(1)$ to $SU_C(3) \times U_Q(1)$.

$$SU(5) \underset{\mathbf{24}}{\rightarrow} SU_C(3) \times SU_L(2) \times U(1) \underset{\mathbf{5}}{\rightarrow} SU_C(3) \times U_Q(1). \qquad (16.51)$$

The expectation values are chosen so that the **24** gives masses to the lepto–quark bosons X and Y of order 10^{15} GeV, and the **5** gives masses to Z^0 and $W^{\pm}$ of order 10^2 GeV. (There is also a negligible contribution from **5** to the masses of X and Y.) The $SU_L(2)$ doublet of Higgs scalars introduced in §14.2 is contained in the **5**, and it is in these components of the **5** that the VEV must develop.

To realise this scheme of spontaneous symmetry breaking[5] we have to write down a suitable Higgs scalar Lagrangian containing the **24** of Higgses, which we write as a 5×5 matrix,

$$\Phi \equiv \sum_{a=1}^{24} \phi_a T_a \qquad (16.52)$$

where T_a are the generators of the five-dimensional fundamental representation of SU(5) as in (16.2), and the **5** of Higgses, which we write as a column vector,

$$H \equiv \begin{pmatrix} H_1 \\ H_2 \\ H_3 \\ H_4 \\ H_5 \end{pmatrix}. \tag{16.53}$$

(It is convenient to write the adjoint representation of Higgs scalars as a matrix, as in (16.52), just as we often write the adjoint representation of gauge fields as a matrix, as in (16.8) and (16.9).) For simplicity, we first consider the Higgs multiplet Φ in isolation, and then the multiplet H in isolation, returning later to the coupling between these two sectors.

For the adjoint representation of Higgs scalars Φ the most general Lagrangian (apart from coupling to fermions) is

$$\mathscr{L}_\Phi = \mathrm{Tr}(\mathrm{D}_\mu\Phi)^2 - m_1^2\,\mathrm{Tr}\,\Phi^2 - \lambda_1(\mathrm{Tr}\,\Phi^2)^2 \\ -\lambda_2\,\mathrm{Tr}\,\Phi^4 \tag{16.54}$$

where we have imposed a discrete symmetry under $\Phi \to -\Phi$ to avoid a $\mathrm{Tr}\,\Phi^3$ term.

Using (9.15) and (9.41) for the covariant derivative of scalar fields in the adjoint representation,

$$\mathrm{D}_\mu\phi_a = \partial_\mu\phi_a - gf_{abc}(A_\mu)_b\phi_c \tag{16.55}$$

so that

$$\mathrm{D}_\mu\Phi = \partial_\mu\Phi + \mathrm{i}g[A_\mu, \Phi]. \tag{16.56}$$

Corresponding to (16.54), the effective potential at tree approximation is

$$V_\Phi = m_1^2\,\mathrm{Tr}\,\Phi_\mathrm{C}^2 + \lambda_1(\mathrm{Tr}\,\Phi_\mathrm{C}^2)^2 + \lambda_2\,\mathrm{Tr}\Phi_\mathrm{C}^4. \tag{16.57}$$

Minimisation of this expression shows that the desired spontaneous symmetry breaking to $\mathrm{SU_C}(3) \times \mathrm{SU_L}(2) \times \mathrm{U}(1)$ occurs provided

$$\lambda_2 > 0 \qquad \lambda_1 > \frac{-7}{30}\lambda_2 \tag{16.58}$$

the second condition being necessary for the potential to be bounded below. The form of expectation value corresponding to this symmetry breaking is

$$\Phi_\mathrm{C} = \frac{\phi_\mathrm{C}}{\sqrt{15}}\,\mathrm{diag}\{1, 1, 1, -\tfrac{3}{2}, -\tfrac{3}{2}\} \tag{16.59}$$

and at the minimum

$$\phi_\mathrm{C}^2 = v_\phi^2 = -m_1^2/(\lambda_1 + \tfrac{7}{30}\lambda_2). \tag{16.60}$$

The covariant derivative term in (16.54) produces gauge field mass terms

$$\mathscr{L}_{\text{mass}} = -g_G^2 \operatorname{Tr}([A_\mu, \Phi_C]^2). \tag{16.61}$$

With Φ_C at the SU(3) × SU(3) × U(1) invariant minimum given by (16.59) and (16.60), and A_μ given by (16.9), one verifies (see problem 16.3) that only the lepto–quark gauge fields acquire masses and these masses are given by

$$\mathscr{L}_{\text{mass}} = \tfrac{5}{12} g_G^2 v_\phi^2 \sum_{i=1}^{3} (\bar{X}_i^\mu X_\mu^i + \bar{Y}_i^\mu Y_\mu^i). \tag{16.62}$$

Thus all three colours of X or Y lepto–quarks have the same mass m_X or m_Y, and

$$m_X^2 = m_Y^2 = \tfrac{5}{12} g_G^2 v_\phi^2. \tag{16.63}$$

As discussed in §16.1, the grand unification scale M_G is of the order of the lepto–quark masses m_X, m_Y. Thus, using (16.48) and (16.50), we find that the VEV for the adjoint of Higgses is given by

$$v_\phi = \frac{5 \times 10^{14} \text{ GeV}}{(\sqrt{5/12}) g_G} \approx 1.5 \times 10^{15} \text{ GeV}. \tag{16.64}$$

The general Lagrangian for the **5** of Higgs scalars H is

$$\mathscr{L}_H = \frac{-m_2^2}{2} H^\dagger H - \frac{\lambda_3}{4} (H^\dagger H)^2. \tag{16.65}$$

To break $SU_C(3) \times SU_L(2) \times U(1)$ to $SU_C(3) \times U_Q(1)$ we must take the VEV in the neutral, $SU_L(2)$ doublet, colour singlet component of H. With the identification of generators for the **5** of (16.3), (16.4) and (16.6), the appropriate component is H_5. The tree approximation effective potential corresponding to (16.65) is

$$V_H = \frac{m_2^2}{2} H_C^\dagger H_C + \frac{\lambda_3}{4} (H_C^\dagger H_C)^2 \tag{16.66}$$

and at the asymmetric minimum

$$(H_5)_C^2 = v_H^2 = -m_2^2/\lambda_3. \tag{16.67}$$

This VEV gives masses to $W^\pm$ and Z^0 as in §14.2. (In the notation of §14.23 and §14.30, H_5 is ϕ^0, and v_H is $v/\sqrt{2}$.)

In the absence of cross-terms coupling Φ and H, there arises the difficulty (see problem 16.5) that there remains a massless combination of the colour triplet components of H and Φ, after taking account of the Higgs mechanism. Such massless scalars would, amongst other things, lead to disastrously rapid baryon number violating processes. This difficulty is overcome when the cross-terms are included. The most general terms coupling Φ to H are

$$\mathscr{L}_{\Phi H} = -\lambda_5 H^\dagger H \operatorname{Tr} \Phi^2 - \lambda_6 H^\dagger \Phi^2 H \tag{16.68}$$

where we have again imposed a discrete symmetry, $\Phi \to -\Phi$. The corresponding tree approximation effective potential terms are

$$V_{\Phi H} = \lambda_5 H_C^\dagger H_C \operatorname{Tr} \Phi_C^2 + \lambda_6 H_C^\dagger \Phi_C^2 H_C. \tag{16.69}$$

When the complete effective potential made up from (16.57), (16.66) and (16.69) is minimised, the troublesome colour triplet of Higgses acquires a mass of order m_X, and there are small corrections to the form of the VEV (16.59). It is necessary to ensure that v_H (the VEV of the **5**) remains of order 100 GeV, while v_Φ (the VEV of the **24**) is of order 10^{15} GeV, so that the appropriate mass hierarchy for $W^\pm$, Z^0 and X, Y is retained. This turns out to require a fine tuning of the parameters in the effective potential to 24 orders of magnitude! This unnatural fine tuning is readily disturbed by radiative corrections and constitutes one of the major aesthetic objections to grand unified theories (the hierarchy problem). The resolution of this difficulty may require the use of supersymmetry, which produces miraculous cancellations of radiative corrections. However, the large subject of supersymmetric gauge field theories is outside the scope of this book.

16.5 Fermion masses in SU(5)

As in electroweak theory, it is not possible to introduce fermion mass terms directly into the SU(5) grand unified theory consistently with the gauge symmetry. The reason for this is as follows. For a general Dirac spinor field ψ, the mass term is constructed from $\bar{\psi}\psi$ and (see problem 16.6)

$$\bar{\psi}\psi = \bar{\psi}_R \psi_L + \bar{\psi}_L \psi_R = -\psi_L^T \mathbf{C} \psi_L^C + \text{h.c.} \tag{16.70}$$

where we have used (16.18), and $\mathbf{C}$ is the charge conjugation matrix with defining property (16.19). With (a generation of) fermions in a $\bar{\mathbf{5}}$ and a **10**, as (16.15) and (16.16), mass terms could *a priori* arise from $\bar{\mathbf{5}} \otimes \mathbf{10}$ and $\mathbf{10} \otimes \mathbf{10}$. However,

$$\bar{\mathbf{5}} \otimes \mathbf{10} = \mathbf{45} + \mathbf{5} \tag{16.71}$$

and

$$\mathbf{10} \otimes \mathbf{10} = \overline{\mathbf{45}} + \bar{\mathbf{5}} + \mathbf{50}. \tag{16.72}$$

In neither case can we construct a gauge singlet, and if mass terms were introduced directly into the Lagrangian they would break the gauge invariance (and so spoil the renormalisability).

On the other hand, since **5** or $\bar{\mathbf{5}}$ is contained in (16.71) or (16.72), it is possible to write down gauge invariant Yukawa couplings of the fermions to the **5** of Higgses. Then spontaneous symmetry breaking can give masses to the fermions. For the first generation of fermions (and ignoring Cabibbo angles which mix generations) the gauge invariant Yukawa interactions are

$$\mathscr{L}_{\mathrm{Y}} = -G_1^{\mathrm{e}}((\Psi_p)_{\mathrm{L}}^{\mathrm{T}} \mathbf{C} x_{\mathrm{L}}^{pq} H_q^{\dagger} + \text{h.c.}) \\ - G_2^{\mathrm{e}}(\varepsilon_{pqrst}(x_{\mathrm{L}}^{pq})^{\mathrm{T}} \mathbf{C} x_{\mathrm{L}}^{rs} H^t + \text{h.c.}) \qquad (16.73)$$

where ε_{pqrst} is the totally antisymmetric Levi–Civita symbol, and the superscript e on the Yukawa coupling constants G_1^{e} and G_2^{e} labels the electron's generation of fermions. After spontaneous symmetry breaking the first term gives mass to d quarks and electrons, and the second term gives masses to u quarks (and similarly for other generations). Explicitly, using (16.15), (16.16) and (16.67), we find the mass terms

$$\mathscr{L}_{\mathrm{m}} = -\frac{G_1^{\mathrm{e}} v_H}{\sqrt{2}} \left(\sum_{i=1}^{3} \bar{\mathrm{d}}_i \mathrm{d}_i + \bar{\mathrm{e}}\mathrm{e} \right) - 4G_2^{\mathrm{e}} v_H \sum_{i=1}^{3} \bar{\mathrm{u}}_i \mathrm{u}_i . \qquad (16.74)$$

There are thus the masses

$$m_{\mathrm{d}} = m_{\mathrm{e}} = G_1^{\mathrm{e}} v_H / \sqrt{2} \qquad (16.75)$$

and

$$m_{\mathrm{u}} = 4G_2^{\mathrm{e}} v_H . \qquad (16.76)$$

The equality of the quark and electron masses applies at renormalisation scales greater than or equal to the grand unification scale. (The SU(5) invariant Yukawa interactions of (16.73) presuppose that we are working at a scale at which the grand unified symmetry group is applicable.) There will be corresponding results for other generations of fermions so that

$$m_{\mathrm{s}} = m_{\mu} \qquad (16.77)$$

and

$$m_{\mathrm{b}} = m_{\tau} . \qquad (16.78)$$

It is possible to use the renormalisation group equation (12.21) to turn the predictions for fermion masses (16.75), (16.77) and (16.78) at a renormalisation scale greater than or equal to the grand unification scale, into predictions[5,6] at an 'ordinary' renormalisation scale at which we usually define renormalised masses for leptons and quarks. From (12.21),

$$\bar{m}(s)/m = s^{-1} \exp\left(-\int_1^s \frac{\mathrm{d}s'}{s'} \gamma_m(\bar{g}(s')) \right). \qquad (16.79)$$

Using (12.19) to change the variable to $\bar{g}$,

$$\bar{m}(s)/m = s^{-1} \exp\left(-\int_g^{\bar{g}(s)} \frac{\mathrm{d}\bar{g}}{\beta_g(\bar{g})} \gamma_m(\bar{g}) \right) \qquad (16.80)$$

and inserting (12.10) and (12.12) (for four dimensions, $\varepsilon=0$),

$$\bar{m}(s)=s^{-1}(\bar{g}^2(s)/g^2)^{b_m/2b}m. \tag{16.81}$$

In terms of the renormalised mass $m(sM)$ for renormalisation scale sM, given by (12.27),

$$m(sM)=(\bar{g}^2(s)/g^2)^{b_m/2b}m(M). \tag{16.82}$$

For

$$s=\tilde{M}/M \tag{16.83}$$

we have

$$m(\tilde{M})=[g^2(\tilde{M})/g^2(M)]^{b_m/2b}m(M) \tag{16.84}$$

where we have used (12.25) and (12.118).

In the case of SU(5) grand unified theory at renormalisation scales below the grand unification scale there are contributions to the mass renormalisation from the gauge fields of each of the factors $SU_C(3)$, $SU_L(2)$ and U(1). Thus, for any fermion mass m,

$$m(\tilde{M})/m(M)=(g_s^2(\tilde{M})/g_s^2(M))^{b_m^3/2b_3}(g^2(\tilde{M})/g^2(M))^{b_m^2/2b_2} \times (g'^2(\tilde{M})/g'^2(M))^{b_m^1/2b_1} \tag{16.85}$$

with b_3, b_2, b_1 as in (16.38)–(16.40), and b_m^3, b_m^2, b_m^1 determined from (12.12) using the assignment of the fermion to representations of $SU_C(3)$, $SU_L(2)$ and U(1), respectively. If, for instance, we wish to continue the prediction (16.78) to an 'ordinary' renormalisation scale, we require

$$16\pi^2 b_m^3=4 \qquad 16\pi^2 b_m^2=\tfrac{9}{4} \qquad 16\pi^2 b_m^1=\tfrac{1}{12} \qquad \text{for b quark} \tag{16.80}$$

and

$$16\pi^2 b_m^3=0 \qquad 16\pi^2 b_m^2=\tfrac{9}{4} \qquad 16\pi^2 b_m^1=\tfrac{3}{4} \qquad \text{for } \tau \text{ lepton.} \tag{16.81}$$

With the aid of (16.38)–(16.40) and (16.85) this leads to

$$\frac{m_b(\tilde{M})/m_b(M)}{m_\tau(\tilde{M})/m_\tau(M)}=(g_s^2(\tilde{M})/g_s^2(M))^{4/(11-\frac{4}{3}N_G)}(g'^2(\tilde{M})/g'^2(M))^{3/4N_G} \tag{16.88}$$

where N_G is the number of generations of fermions. We may now obtain the required prediction for m_b/m_τ at an 'ordinary' renormalisation scale by taking $M=M_G$ and, for instance, $\tilde{M}=m_Z$. At $M=M_G$ the relation (16.78) holds between the masses and the relations (16.31) and (16.34) hold between the coupling constants. The variation of $g_s^2(\tilde{M})$ and $g'^2(\tilde{M})$ with $\tilde{M}$ is given by (16.37) and (16.35). Using the empirical values

$$\alpha_s(m_Z)=0.113 \qquad \alpha^{-1}(m_Z)=127.9 \qquad \sin^2\theta_W(m_Z)=0.233 \tag{16.89}$$

together with (16.42), the result for three generations of fermions, $N_G = 3$, is

$$m_b(m_Z)/m_\tau(m_Z) \simeq 2.2. \tag{16.90}$$

For a τ mass of about 1.8 GeV, $m_b(m_Z)$ is about 4.0 GeV compared with a value of m_b from states containing bottom quarks of 5 GeV. Quark masses m_q as derived from current algebra are often defined at a scale $\tilde{M}$ given by

$$\tilde{M} = 2m_q(\tilde{M}) \tag{16.91}$$

which corresponds to the threshold for producing quark–antiquark pairs (if such a thing were possible.) The continuation from $\tilde{M} = m_Z$ as in (16.90) to this $\tilde{M}$ does not make much difference to the value of m_b/m_τ since it involves less than an order of magnitude in $\tilde{M}$, whereas the extrapolation to the grand unification scale is over 15 orders of magnitude. Similar, but less successful predictions may be made for m_s/m_μ and m_d/m_e.

16.6 Proton decay

It was observed in §16.2 that there are processes in the SU(5) grand unified theory (e.g. that of figure 16.3) which can produce proton decay[1]. It is easy to estimate the order of magnitude of the proton lifetime. For processes such as this one, with low energies associated with the external legs, the amplitude may be approximated by using a current–current four-fermion interaction proportional to $\alpha_G m_X^{-2}$ (or $\alpha_G m_Y^{-2}$) with the grand unified fine structure constant given by (16.50), and the lepto–quark mass m_X (or m_Y) given by

$$m_X \approx M_G \tag{16.92}$$

with M_G as in (16.48). Squaring the amplitude to get the decay rate, and getting the dimensions right using the proton mass m_p, we may estimate the proton lifetime τ_p to be

$$\begin{aligned} \tau_p &\approx \alpha_G^{-2} m_X^4 m_p^{-5} \\ &\approx 3 \times 10^{30} \text{ years} \end{aligned} \tag{16.93}$$

where the numerical values of (16.48) and (16.50) have been used. Although more sophisticated treatments can increase τ_p to as much as 10^{31} years, consistency cannot be achieved with the observed lower bound of 6×10^{32} years. However, in supersymmetric grand unified theory, which is outside our scope here, a sufficiently long proton lifetime can be obtained. The particles into which the proton may decay are limited by symmetry principles. For example, the amplitude of figure 16.3 obeys the selection rule

$$\Delta(B - L) = 0 \tag{16.94}$$

where B, L are baryon number and lepton number, respectively. This is true of all amplitudes that can mediate proton decay in the SU(5) grand unified theory. (See problem 16.7.)

Problems

16.1 Check the assignment (16.9) of the gauge fields to the adjoint representation of SU(5).

16.2 Show that when the left- and right-handed components of fermions are assigned to different representations of a gauge group, then (12.15) generalises to

$$b=(16\pi^2)^{-1}\left(\frac{11}{3}c_1-\frac{2}{3}\sum_R c_2^R(\mathrm{L})-\frac{2}{3}\sum_R c_2^R(\mathrm{R})\right)$$

where the group theory factors $c_2^R(\mathrm{L})$ and $c_2^R(\mathrm{R})$ are for left- and right-handed fermion fields assigned to representation R of the gauge group, respectively.

16.3 Calculate the contribution of Higgs scalar loops to (16.39).

16.4 Verify that the Higgs scalar expectation value given by (16.59) and (16.60) gives masses only to the lepto–quark gauge fields (as expected from the symmetry of this expectation value).

16.5 Show that there is a massless combination of the colour triplet components of H and Φ, after spontaneous symmetry breaking, in the absence of cross-terms coupling Φ and H.

16.6 Check (16.70) for a fermion mass term.

16.7 Show that the selection rule $\Delta(B-L)=0$ applies to all amplitudes mediating proton decay in the SU(5) grand unified theory.

References

1 Georgi H and Glashow S L 1974 *Phys. Rev. Lett.* **32** 438
2 For a review see Langacker P 1981 *Phys. Rep.* **72C** 185
3 MacFarlane A J, Sudbery A and Weisz P H 1968 *Commun. Math. Phys.* **11** 77
4 Georgi H, Quinn H R and Weinberg S 1974 *Phys. Rev. Lett.* **33** 451
5 Buras A, Ellis J, Gaillard M K and Nanopoulos D V 1978 *Nucl. Phys.* B **135** 66
6 Chanowitz M S, Ellis J and Gaillard M K 1977 *Nucl. Phys.* B **128** 506

17

FIELD THEORIES AT FINITE TEMPERATURE

17.1 The partition function for scalar field theory

In our discussion of field theories at finite temperature we shall not attempt to treat dynamical properties, but shall content ourselves with a study of the equilibrium thermodynamic properties of the system[1-4]. One of the fundamental objects of statistical thermodynamics at finite temperature T is the partition function Z, defined by

$$Z = \mathrm{Tr}\, \mathrm{e}^{-\beta \hat{H}} \tag{17.1}$$

where $\hat{H}$ is the Hamiltonian operator,

$$\beta = (k_{\mathrm{B}} T)^{-1} = T^{-1} \tag{17.2}$$

in units with the Boltzmann constant k_{B} set equal to 1, and the trace in (17.1) means to sum the matrix elements of $\mathrm{e}^{-\beta \hat{H}}$ between all independent states of the system. Once the partition function Z has been evaluated, the (Helmholtz) free energy F is given by

$$Z = \mathrm{e}^{-\beta F}. \tag{17.3}$$

As usual in thermodynamics, the free energy is related to the internal energy E and the entropy S through

$$F = E - TS \tag{17.4}$$

and the pressure P and the entropy S are obtained from

$$P = -\left.\frac{\partial F}{\partial V}\right|_T \tag{17.5}$$

and

$$S = -\left.\frac{\partial F}{\partial T}\right|_V. \tag{17.6}$$

For a scalar field theory, the partition function may be formulated as a path integral by the following series of steps[4]. First, we take the independent states of the system to be the eigenstates of the Schrödinger picture field operator. In §4.1, we introduced eigenstates $|\phi(\boldsymbol{x}), t\rangle$ of the Heisenberg picture field operator $\hat{\phi}(t, \boldsymbol{x})$

$$\hat{\phi}(t, \boldsymbol{x})|\phi(\boldsymbol{x}), t\rangle = \phi(\boldsymbol{x})|\phi(\boldsymbol{x}), t\rangle. \tag{17.7}$$

The Schrödinger picture field operator is $\hat{\phi}(t=0,\boldsymbol{x})$, and the corresponding eigenstates $|\phi(\boldsymbol{x}), t=0\rangle$ are given by

$$\hat{\phi}(t=0,\boldsymbol{x})|\phi(\boldsymbol{x}), t=0\rangle = \phi(\boldsymbol{x})|\phi(\boldsymbol{x}), t=0\rangle. \tag{17.8}$$

Then the partition function of (17.1) may be written explicitly as a 'summation' over the eigenstates:

$$Z = \sum_{\phi(\boldsymbol{x})} \langle \phi(\boldsymbol{x}), t=0|\mathrm{e}^{-\beta\hat{H}}|\phi(\boldsymbol{x}), t=0\rangle. \tag{17.9}$$

Second, we make an analogy with the zero temperature field theory of a scalar field. From (4.2), and the field theory analogue of (2.5),

$$\begin{aligned}&\langle \phi''(\boldsymbol{x}), t''|\phi'(\boldsymbol{x}), t'\rangle \\ &= \langle \phi''(\boldsymbol{x}), t=0|\mathrm{e}^{-\mathrm{i}\hat{H}(t''-t')}|\phi'(\boldsymbol{x}), t=0\rangle \\ &\propto \int \mathscr{D}\phi \int \mathscr{D}\pi \exp \mathrm{i} \int_{t'}^{t''} \mathrm{d}t \int \mathrm{d}^3x \left(\pi \frac{\partial \phi}{\partial t} - \mathscr{H}(\pi, \phi)\right)\end{aligned} \tag{17.10}$$

where the path integral is over all functions $\pi(t, \boldsymbol{x})$ and over functions satisfying the boundary conditions (4.3). If, heuristically we introduce a variable

$$\tau = \mathrm{i}t = \mathrm{i}x^0 \tag{17.11}$$

and take the limits of integration in (17.10) to be

$$t' = 0 \qquad t'' = -\mathrm{i}\beta \tag{17.12}$$

we obtain

$$\begin{aligned}&\langle \phi''(\boldsymbol{x}), t=0|\mathrm{e}^{-\beta\hat{H}}|\phi'(\boldsymbol{x}), t=0\rangle \\ &\qquad \propto \int \mathscr{D}\phi \int \mathscr{D}\pi \exp \int_0^{\beta} \mathrm{d}\tau \int \mathrm{d}^3x \left(\mathrm{i}\pi \frac{\partial \phi}{\partial \tau} - \mathscr{H}(\pi, \phi)\right)\end{aligned} \tag{17.13}$$

where ϕ and π are now regarded as functions of τ and $\boldsymbol{x}$, and the path integral is over all functions $\pi(\tau, \boldsymbol{x})$, and over functions $\phi(\tau, \boldsymbol{x})$ satisfying the boundary conditions

$$\phi(\beta, \boldsymbol{x}) = \phi''(\boldsymbol{x}) \qquad \phi(0, \boldsymbol{x}) = \phi'(\boldsymbol{x}). \tag{17.14}$$

It should be noted that, although the introduction of the variable τ here is formally similar to the introduction of the variable $\bar{x}^0$ in (4.10), the interpretation is quite different. In Chapter 4, we introduced $\bar{x}^0$ and continued to Euclidean space in order to make the path integral well defined, but at the end of the day we continued back to Minkowski space to obtain the physical generating functional and Green functions. Here, we introduce the variable τ to make a bridge from field theory to statistical mechanics. There will be no question of continuing back to the variable t, because our goal is to obtain the partition function, which is a thermodynamic object with no time dependence.

The final step is to take

$$|\phi''(\boldsymbol{x}), t=0\rangle = |\phi'(\boldsymbol{x}), t=0\rangle = |\phi(\boldsymbol{x}), t=0\rangle \tag{17.15}$$

in (17.13), and 'sum' over all eigenstates, as in (17.9). Then we obtain

$$Z \propto \int_{\text{periodic}} \mathscr{D}\phi \int \mathscr{D}\pi \exp \int_0^\beta \mathrm{d}\tau \int \mathrm{d}^3x \left(\mathrm{i}\pi \frac{\partial \phi}{\partial \tau} - \mathscr{H}(\pi, \phi) \right). \tag{17.16}$$

The boundary conditions (17.14), together with (17.15), mean that the path integral is now restricted to functions $\phi(\tau, \boldsymbol{x})$ which are periodic in τ with period β,

$$\phi(\tau=0, \boldsymbol{x}) = \phi(\tau=\beta, \boldsymbol{x}) \tag{17.17}$$

and the fact that we sum over all eigenstates means that all such functions are to be integrated over. The path integral over π is still over all $\pi(\tau, \boldsymbol{x})$.

When the Lagrangian and Hamiltonian densities take the form

$$\mathscr{L}(\phi, \partial_\mu \phi) = \tfrac{1}{2}(\partial_0 \phi)^2 + f(\phi, \nabla\phi) \tag{17.18}$$

and

$$\mathscr{H} = \tfrac{1}{2}\pi^2 - f(\phi, \nabla\phi) \tag{17.19}$$

the integration over π may be carried out explicitly with the aid of (1.14) (much as in §4.1) to obtain

$$\begin{aligned} Z &= \tilde{N}(\beta) \int_{\text{periodic}} \mathscr{D}\phi \exp - \int_0^\beta \mathrm{d}\tau \int \mathrm{d}^3x \left[\frac{1}{2}\left(\frac{\partial \phi}{\partial \tau}\right)^2 - f(\phi, \nabla\phi) \right] \\ &= \tilde{N}(\beta) \int_{\text{periodic}} \mathscr{D}\phi \exp \int_0^\beta \mathrm{d}\tau \int \mathrm{d}^3x \, \mathscr{L}(\phi, \bar{\partial}_\mu \phi) \end{aligned} \tag{17.20}$$

where

$$\bar{\partial}_\mu \phi \equiv \left(\mathrm{i} \frac{\partial \phi}{\partial \tau}, \nabla\phi \right) \tag{17.21}$$

and $\tilde{N}(\beta)$ is a temperature dependent normalisation, arising from the operator determinant when the path integral over π is carried out.

17.2 Partition function for free scalar field theory

For the free scalar field theory, the path integral (17.20) becomes a Gaussian integral, and may be carried out exactly. The appropriate Lagrangian is

$$\mathscr{L}(\phi, \bar{\partial}_\mu \phi) = -\frac{1}{2}\left(\frac{\partial \phi}{\partial \tau}\right)^2 - \frac{1}{2}(\nabla\phi)^2 - \frac{m^2}{2}\phi^2 \tag{17.22}$$

where we have denoted the mass by m rather than μ to avoid any possible confusion with the chemical potential, in the thermodynamic context.

Thus

$$Z=\tilde{N}(\beta)\int_{\text{periodic}}\mathscr{D}\phi\,\exp\left(-\frac{1}{2}\int_0^\beta d\tau'\int d^3x'\int_0^\beta d\tau\int d^3x\,\phi(\bar{x}')A(\bar{x}',\bar{x})\phi(\bar{x})\right) \tag{17.23}$$

where

$$A(\bar{x}',\bar{x})=(-\bar{\partial}'_\mu\bar{\partial}^\mu+m^2)\delta(\bar{x}'-\bar{x}) \tag{17.24}$$

with the shorthand notations

$$\delta(\bar{x}'-\bar{x})\equiv\delta(\tau'-\tau)\delta(\boldsymbol{x}'-\boldsymbol{x}) \tag{17.25}$$

$$\bar{x}\equiv(-i\tau,\boldsymbol{x}) \tag{17.26}$$

and $\bar{\partial}_\mu$ as in (17.21). Following (1.6) we obtain

$$Z=\tilde{N}(\beta)\exp(-\tfrac{1}{2}\operatorname{Tr}\ln\mathbf{A}) \tag{17.27}$$

where the operator trace is to be carried out in a way appropriate to functions $\phi(\tau,\boldsymbol{x})$ obeying the periodicity condition (17.17). To evaluate this trace, we first observe that the periodicity of ϕ in $0<\tau<\beta$ means that it can be expressed in the Fourier expansion

$$\phi(\bar{x})=\frac{1}{\beta}\sum_n\int\frac{d^3p}{(2\pi)^3}\,e^{-i\omega_n\tau}\,e^{i\boldsymbol{p}\cdot\boldsymbol{x}}\,\tilde{\phi}(\omega_n,\boldsymbol{p}) \tag{17.28}$$

with the Matsubara frequencies for bosons,

$$\omega_n=2\pi n/\beta \qquad \text{bosons} \tag{17.29}$$

with n an integer.

This can be made shorthand by writing

$$\phi(\bar{x})=\frac{1}{\beta}\sum_n\int\frac{d^3p}{(2\pi)^3}\,e^{-i\bar{p}\cdot\bar{x}}\,\tilde{\phi}(\bar{p}) \tag{17.30}$$

with

$$\bar{p}\equiv(i\omega_n,\boldsymbol{p}) \tag{17.31}$$

and

$$\bar{p}\cdot\bar{x}\equiv\omega_n\tau-\boldsymbol{p}\cdot\boldsymbol{x}. \tag{17.32}$$

Correspondingly we write (see problem 17.1)

$$\delta(\bar{x}'-\bar{x})=\frac{1}{\beta}\sum_n\int\frac{d^3p}{(2\pi)^3}\,e^{-i\bar{p}\cdot(\bar{x}'-\bar{x})} \tag{17.33}$$

and

$$A(\bar{x}',\bar{x})=\frac{1}{\beta}\sum_n\int\frac{d^3p}{(2\pi)^3}\,e^{-i\bar{p}\cdot(\bar{x}'-\bar{x})}(-\bar{p}^2+m^2) \tag{17.34}$$

where

$$\bar{p}^2 \equiv -(\omega_n^2 + \boldsymbol{p}^2). \tag{17.35}$$

Thus, setting $\bar{x}' = \bar{x}$ and integrating (and summing) over all values of $\bar{x}$,

$$\begin{aligned}\operatorname{Tr}\ln \mathbf{A} &= \int_0^\beta \mathrm{d}\tau \int \mathrm{d}^3x \frac{1}{\beta} \sum_n \int \frac{\mathrm{d}^3p}{(2\pi)^3} \ln(-\bar{p}^2 + m^2) \\ &= \int \mathrm{d}^3x \sum_n \int \frac{\mathrm{d}^3p}{(2\pi)^3} \ln(\omega_n^2 + \boldsymbol{p}^2 + m^2). \end{aligned} \tag{17.36}$$

Such frequency sums are easily done. (See Appendix E.)

The result is

$$\begin{aligned}\operatorname{Tr}\ln \mathbf{A} = \int \mathrm{d}^3x \int \frac{\mathrm{d}^3p}{(2\pi)^3} \{&\beta\sqrt{\boldsymbol{p}^2+m^2} + 2\ln[1-\exp(-\beta\sqrt{\boldsymbol{p}^2+m^2})] \\ &+ (\sqrt{\boldsymbol{p}^2+m^2})\text{-independent constant}\}. \end{aligned} \tag{17.37}$$

Using (17.37) in (17.27),

$$-\beta F = \ln Z = -\int \mathrm{d}^3x \int \frac{\mathrm{d}^3p}{(2\pi)^3} \left(\frac{\beta}{2}\sqrt{\boldsymbol{p}^2+m^2} + \ln[1-\exp(-\beta\sqrt{\boldsymbol{p}^2+m^2})] \right) \tag{17.38}$$

after cancellation of the constant in (17.37) against $\tilde{N}(\beta)$, as discussed in Appendix E.

When the mass of the scalar field is negligible compared with the temperature, the integral in (17.38) is particularly easy, and we find for the free energy density $\mathscr{F}$ (apart from an additive temperature independent constant corresponding to the zero-point energy of the vacuum)

$$\mathscr{F} = \frac{-\pi^2}{90\beta^4} = -\frac{\pi^2 T^4}{90} \qquad T \gg m \tag{17.39}$$

where we have written

$$F = \int \mathrm{d}^3x\, \mathscr{F}. \tag{17.40}$$

From (17.5) and (17.6), the corresponding pressure and entropy density $\mathscr{S}$ of the ideal ultrarelativistic boson gas are

$$P = \pi^2 T^4/90 \tag{17.41}$$

and

$$\mathscr{S} = 2\pi^2 T^3/45 \tag{17.42}$$

where

$$S=\int d^3x\,\mathcal{S}. \tag{17.43}$$

From (17.4), the energy density ρ is

$$\rho=\pi^2T^4/30. \tag{17.44}$$

17.3 Partition function for gauge vector bosons

A subtlety which arises for gauge fields is that there are only two independent degrees of freedom for a massless vector field, but in a typical renormalisable gauge the Lagrangian involves four degrees of freedom. The two extra degrees of freedom are not physical and cannot be in equilibrium with a heat bath. There are also the Faddeev–Popov ghosts, which do not correspond to physical particles and lead to the same difficulty. The resolution of the problem is obtained by noticing that there are gauges (for instance axial gauge) in which each gauge field has only two degrees of freedom, and in which there are no Faddeev–Popov ghosts. In such gauges there should be no difficulty, and the analogue of (17.20) should be correct (with two factors of $\tilde{N}(\beta)$, one for each gauge field degree of freedom) and should equal $\mathrm{Tr}\,e^{-\beta\hat{H}}$. In other gauges, we may continue to use this expression, with the Faddeev–Popov ansatz for $\mathcal{L}$. (These points are discussed in detail by Bernard[4].) However, in general Z, is not equal to $\mathrm{Tr}\,e^{-\beta\hat{H}}$, because the trace would involve unphysical states, which cannot be in equilibrium with a heat bath. Thus, for gauge fields

$$Z=[\tilde{N}(\beta)]^{2d_G}\int_{\text{periodic}}\mathcal{D}A^\mu\int_{\text{periodic}}\mathcal{D}\eta^*\mathcal{D}\eta$$
$$\times\exp\int_0^\beta d\tau\int d^3x\,\mathcal{L}(A_a^\mu,\eta_a) \tag{17.45}$$

with $\mathcal{L}$ as in (10.57), (10.58) and (10.59) with the fields functions of $\bar{x}\equiv(-i\tau,\boldsymbol{x})$, and derivatives ∂_μ replaced by $\bar{\partial}_\mu$ (as in (17.21)), and d_G the number of gauge fields (the dimension of the adjoint representation of the gauge group). The Faddeev–Popov ghost fields are treated as having the same periodicity in τ as the gauge fields, rather than in the way we shall treat fermion fields in the next section. The reason is that the (unphysical) ghost fields arise from a determinant defined in the space of gauge fields. (See (10.40), (10.43) and (10.56).)

In the free-field limit ($g\to0$) we may again perform the Gaussian path integrals exactly. For notational simplicity, we consider the Abelian case. In the non-Abelian case we will just have to multiply by the number of gauge fields the result for F.

For $g=0$,

$$Z=[\tilde{N}(\beta)]^2 \int_{\text{periodic}} \mathscr{D}A^\mu \exp \int d\bar{x}\left(-\frac{1}{4}\bar{F}^{\mu\nu}\bar{F}_{\mu\nu}-\frac{1}{2\xi}(\bar{\partial}_\mu A^\mu)^2\right)$$

$$\times \int_{\text{periodic}} \mathscr{D}\eta^* \mathscr{D}\eta \exp \int d\bar{x}\, \bar{\partial}_\mu \eta^* \bar{\partial}^\mu \eta \tag{17.46}$$

where

$$\bar{F}^{\mu\nu} \equiv \bar{\partial}^\mu A^\nu - \bar{\partial}^\nu A^\mu \tag{17.47}$$

and we have adopted the notation

$$\int d\bar{x} \equiv \int_0^\beta d\tau \int d^3x. \tag{17.48}$$

Thus,

$$Z=[\tilde{N}(\beta)]^2 \int_{\text{periodic}} \mathscr{D}A^\mu \exp -\tfrac{1}{2} \int d\bar{x}' \int d\bar{x}\, A_\mu(\bar{x}')B^{\mu\nu}(\bar{x}',\bar{x})A_\nu(\bar{x})$$

$$\times \int_{\text{periodic}} \mathscr{D}\eta^* \mathscr{D}\eta \exp - \int d\bar{x}' d\bar{x}\, \eta^*(\bar{x}')C(\bar{x}',\bar{x})\eta(\bar{x}) \tag{17.49}$$

where

$$B^{\mu\nu}(\bar{x}',\bar{x}) = (g^{\mu\nu}\bar{\partial}^\rho_{x'}\bar{\partial}_{\rho x} - (1-\xi^{-1})\bar{\partial}^\nu_{x'}\bar{\partial}^\mu_x)\delta(\bar{x}'-\bar{x}) \tag{17.50}$$

and

$$C(\bar{x}',\bar{x}) = \bar{\partial}^\rho_{x'}\bar{\partial}_{\rho x}\delta(\bar{x}'-\bar{x}) \tag{17.51}$$

with $\delta(\bar{x}'-\bar{x})$ as in (17.25). Performing the Gaussian path integrals we obtain

$$Z=[\tilde{N}(\beta)]^2 \exp(-\tfrac{1}{2}\operatorname{Tr}\ln \mathbf{B}) \exp(\operatorname{Tr}\ln \mathbf{C}). \tag{17.52}$$

Fourier transforming as in (17.33) and (17.34) we write

$$B^{\mu\nu}(\bar{x}',\bar{x}) = \frac{1}{\beta}\sum_n \int \frac{d^3p}{(2\pi)^3} e^{-i\bar{p}\cdot(\bar{x}'-\bar{x})}$$

$$\times [\bar{p}^2(g^{\mu\nu}-\bar{p}^{-2}\bar{p}^\mu\bar{p}^\nu)+\bar{p}^2\xi^{-1}\bar{p}^{-2}\bar{p}^\mu\bar{p}^\nu] \tag{17.53}$$

where we have separated into projection operators. The logarithm of **B** may now be taken by taking the logarithm of the coefficient of each projection operator. Taking the trace both in $\bar{x}$ space and in the space of Lorentz indices, we obtain

$$\operatorname{Tr}\ln \mathbf{B} = \frac{1}{\beta}\int d\bar{x} \sum_n \int \frac{d^3p}{(2\pi)^3} [3\ln \bar{p}^2 + \ln(\xi^{-1}p^{-2})]$$

$$= \int d^3x \sum_n \int \frac{d^3p}{(2\pi)^3} 4\ln(\omega_n^2+\boldsymbol{p}^2) \tag{17.54}$$

where we have dropped an additive temperature independent (infinite) constant. Also,

$$C(\vec{x}', \vec{x}) = \frac{1}{\beta} \sum_n \int \frac{d^3p}{(2\pi)^3} e^{-i\vec{p}\cdot(\vec{x}'-\vec{x})} \bar{p}^2 \tag{17.55}$$

leading to

$$\mathrm{Tr} \ln \mathbf{C} = \int d^3x \sum_n \int \frac{d^3p}{(2\pi)^3} \ln(\omega_n^2 + \boldsymbol{p}^2). \tag{17.56}$$

Substituting (17.54) and (17.56) in (17.52) gives

$$Z = [\tilde{N}(\beta)]^2 \exp\left(-\frac{1}{2} \int d^3x \sum_n \int \frac{d^3p}{(2\pi)^3} 2 \ln(\omega_n^2 + \boldsymbol{p}^2) \right). \tag{17.57}$$

This is just what we had in §17.2 for $m = 0$, but with the exponent doubled. Thus, we obtain for the free energy density

$$\mathscr{F} = -2\pi^2 T^4/90. \tag{17.58}$$

The Faddeev–Popov ghosts have cancelled the contribution from the two non-physical degrees of freedom of the gauge field.

17.4 Partition function for fermions

In extending the discussion of §17.1 to fermions we recall that physical observables always involve even powers of the Dirac field ψ (because ψ changes sign under a rotation through 2π). Thus, the eigenstates $|\pm\psi(\boldsymbol{x}), t=0\rangle$ of the Schrödinger picture field operator $\hat{\psi}(t=0, \boldsymbol{x})$ correspond to the same values of the physical observables and describe the same state. There is therefore some ambiguity deriving a path integral formulation of the partition function. To obtain a prescription which is consistent with Fermi statistics when thermal averages are calculated, it turns out to be necessary to start from

$$Z = \sum_{\psi(\boldsymbol{x})} \langle \psi(\boldsymbol{x}), t=0 | e^{-\beta \hat{H}} | -\psi(\boldsymbol{x}), t=0 \rangle. \tag{17.59}$$

The analgoue of (17.20) for fermions is then

$$Z = N'(\beta) \int_{\text{antiperiodic}} \mathscr{D}\bar{\psi}\mathscr{D}\psi \exp \int_0^\beta d\tau \int d^3x\, \mathscr{L}(\psi) \tag{17.60}$$

where in $\mathscr{L}$ the field ψ is understood to be a function of τ and $\boldsymbol{x}$ antiperiodic in $0 < \tau < \beta$,

$$\psi(\tau = 0, \boldsymbol{x}) = -\psi(\tau = \beta, \boldsymbol{x}) \tag{17.61}$$

and derivatives ∂_μ are understood to be replaced by $\bar{\partial}_\mu$ as in (17.21). An appropriate Fourier expansion for $\psi(\bar{x})$ is therefore

$$\psi(\bar{x})=\frac{1}{\beta}\sum_n\int\frac{d^3p}{(2\pi)^3}\,e^{-i\bar{p}\cdot\bar{x}}\tilde{\psi}(\bar{p}) \tag{17.62}$$

where the Matsubara frequencies for fermions are

$$\omega_n=\frac{(2n+1)\pi}{\beta}\qquad\text{fermions} \tag{17.63}$$

with n an integer, and $\bar{x}$, $\bar{p}^\mu$ and $\bar{p}\cdot\bar{x}$ are as in (17.26), (17.31) and (17.32).

For the free-field case, the appropriate Lagrangian is

$$\mathscr{L}(\psi)=\bar{\psi}(\bar{x})(i\gamma^\mu\bar{\partial}_\mu-m)\psi(\bar{x}). \tag{17.64}$$

Thus

$$Z=N'(\beta)\int_{\text{antiperiodic}}\mathscr{D}\bar{\psi}\mathscr{D}\psi\,\exp\left(-\int d\bar{x}\int d\bar{x}\,\bar{\psi}(\bar{x}')D(\bar{x}',\bar{x})\psi(\bar{x})\right) \tag{17.65}$$

with

$$D(\bar{x}',\bar{x})=(i\gamma^\mu\bar{\partial}_\mu+m)\delta(\bar{x}'-\bar{x}). \tag{17.66}$$

Performing the Gaussian path integral,

$$Z=N'(\beta)\exp(\mathrm{Tr}\ln\mathbf{D}). \tag{17.67}$$

The trace is evaluated by Fourier transforming, much as in §17.2, except that the Matsubara frequencies are given by (17.63) and we have to take a trace on the Dirac indices as well as on $\bar{x}$.

$$D(\bar{x}',\bar{x})=\frac{1}{\beta}\sum_n\int\frac{d^3p}{(2\pi)^3}\,e^{-i\bar{p}\cdot(\bar{x}'-\bar{x})}(-\bar{\not{p}}+m) \tag{17.68}$$

leading to (see problem 17.2)

$$\begin{aligned}\mathrm{Tr}\ln\mathbf{D}&=\int_0^\beta d\tau\int d^3x\,\frac{1}{\beta}\sum_n\int\frac{d^3p}{(2\pi)^3}\,2\ln(m^2-\bar{p}^2)\\&=2\int d^3x\sum_n\int\frac{d^3p}{(2\pi)^3}\ln(\omega_n^2+\boldsymbol{p}^2+m^2).\end{aligned} \tag{17.69}$$

Performing the (fermion) Matsubara frequency sum (see Appendix E) gives

$$\begin{aligned}\mathrm{Tr}\ln\mathbf{D}=2\int d^3x\int\frac{d^3p}{(2\pi)^3}\,&\{\beta\sqrt{\boldsymbol{p}^2+m^2}+2\ln[1+\exp(-\beta\sqrt{\boldsymbol{p}^2+m^2})]\\&+(\sqrt{\boldsymbol{p}^2+m^2})\text{-independent constant}\}.\end{aligned} \tag{17.70}$$

Using (17.70) in (17.67)

$$-\beta F=\ln Z=2\int d^3x\int\frac{d^3p}{(2\pi)^3}\,\{\beta\sqrt{\boldsymbol{p}^2+m^2}+2\ln[1+\exp(-\beta\sqrt{\boldsymbol{p}^2+m^2})]\} \tag{17.71}$$

after cancellation of the constant in (17.70) against $N'(\beta)$. For $T \gg m$, we find for the free energy density (apart from an additive temperature-independent constant)

$$\mathscr{F} = -7\pi^2 T^4/180. \tag{17.72}$$

For massless fermions (with only one helicity state) a similar calculation, using Weyl spinors, gives half the above answer. (See problem 17.3.)

We can summarise the results of the last three sections for the free energy density of an ideal ultrarelativistic gas ($T \gg m$) as

$$\mathscr{F} = -\pi^2 T^4(N_B + \tfrac{7}{8}N_F)/90 \tag{17.73}$$

where N_B and N_F are the number of bosonic and fermionic degrees of freedom, respectively. ($N_B = 1$ for a neutral scalar field, $N_B = 2$ for a neutral gauge field, $N_F = 4$ for a Dirac field where there are two helicity states for the particle and two for the antiparticle, and $N_F = 2$ for a Weyl field.) Correspondingly, using (17.5) and (17.6), the pressure P and entropy density $\mathscr{S}$ are

$$P = \pi^2 T^4(N_B + \tfrac{7}{8}N_F)/90 \tag{17.74}$$

and

$$\mathscr{S} = 2\pi^2 T^3(N_B + \tfrac{7}{8}N_F)/45 \tag{17.75}$$

and, using (17.4), the energy density is

$$\rho = \pi^2 T^4(N_B + \tfrac{7}{8}N_F)/30. \tag{17.76}$$

17.5 Temperature Green functions and generating functionals

For simplicity of presentation we shall restrict the discussion to scalar fields, but the changes necessary to include gauge fields and fermion fields will be clear from the discussion of previous sections. In Chapter 4, when we studied Green functions and generating functionals for these Green functions at zero temperature, the Green functions had time dependence and carried dynamical information. Here we shall discuss generating functionals for temperature Green functions which contain information about the equilibrium thermodynamic properties of the finite temperature system and have *no* time dependence.

In §17.1, a path integral formulation of the partition function was obtained by introducing the variable $\tau = \mathrm{i}x^0$, and integrating our classical fields ϕ which were functions of τ. Temperature Green functions are *defined* in terms of field operators $\hat{\phi}$ which are also functions of τ. By analogy with (4.7) we introduce the temperature Green functions

$$\mathscr{G}^{(N)}(\bar{x}, \ldots, \bar{x}_N) \equiv \langle T_\tau(\hat{\phi}(\bar{x}_1) \ldots \hat{\phi}(\bar{x}_N))\rangle \tag{17.77}$$

where

$$\bar{x} \equiv (-\mathrm{i}\tau, \boldsymbol{x}) \tag{17.78}$$

and T_τ means to order the fields from right to left in order of increasing τ. The expectation value $\langle \ \rangle$ now means a thermal average rather than just a vacuum expectation value.

$$\langle T_\tau(\hat{\phi}(\bar{x}_1)\ldots\hat{\phi}(\bar{x}_N))\rangle \equiv \frac{\mathrm{Tr}[\mathrm{e}^{-\beta\hat{H}}T_\tau(\hat{\phi}(\bar{x}_1)\ldots\hat{\phi}(\bar{x}_N))]}{\mathrm{Tr}[\mathrm{e}^{-\beta\hat{H}}]} \tag{17.79}$$

where the trace means to sum the matrix elements of the operator in the square bracket between all independent states of the system.

To see why such objects might be of interest, consider $\mathscr{G}^{(2)}(\bar{x}_1, \bar{x}_2)$ and suppose we want to know the expectation value (thermal average) of some observable A represented by the operator $\hat{A}$ in the Schrödinger picture with, for instance,

$$\hat{A} = \int \mathrm{d}^3x\, \hat{\phi}^2(t=0, \boldsymbol{x}). \tag{17.80}$$

Then

$$\langle \hat{A} \rangle = \frac{\mathrm{Tr}[\mathrm{e}^{-\beta\hat{H}}\int \mathrm{d}^3x \hat{\phi}^2(t=0, \boldsymbol{x})]}{\mathrm{Tr}[\mathrm{e}^{-\beta\hat{H}}]}. \tag{17.81}$$

But

$$\lim_{\tau' \to \tau^+, \boldsymbol{x}' \to \mathrm{x}} \int \mathrm{d}^3x\, \mathscr{G}^{(2)}(\bar{x}', \bar{x}) = \frac{\int \mathrm{d}^3x\, \mathrm{Tr}[\mathrm{e}^{-\beta\hat{H}}\hat{\phi}(\tau, \boldsymbol{x})\hat{\phi}(\tau, \boldsymbol{x})]}{\mathrm{Tr}[\mathrm{e}^{-\beta\hat{H}}]}$$

$$= \frac{\int \mathrm{d}^3x\, \mathrm{Tr}[\mathrm{e}^{-\beta\hat{H}}\hat{\phi}^2(0, \boldsymbol{x})]}{\mathrm{Tr}[\mathrm{e}^{-\beta\hat{H}}]} \tag{17.82}$$

where in the last step we have used the connection (2.3) between the field operator at time t and time zero with $t \to -\mathrm{i}\tau$. Thus,

$$\langle \hat{A} \rangle = \lim_{\tau' \to \tau^+, \boldsymbol{x}' \to \mathrm{x}} \int \mathrm{d}^3x\, \mathscr{G}^{(2)}(\bar{x}', \bar{x}). \tag{17.83}$$

A path integral representation for the temperature Green functions may be obtained as follows. The field theory analogue of (2.50) is

$$\langle \phi''(\boldsymbol{x}), t=0 | \mathrm{e}^{-\mathrm{i}\hat{H}(t''-t')} T(\hat{\phi}(x_1)\ldots\hat{\phi}(x_N)) | \phi'(\boldsymbol{x}), t=0\rangle$$

$$\propto \int \mathscr{D}\phi \int \mathscr{D}\pi \phi(x_1)\ldots\phi(x_N) \exp \mathrm{i} \int_{t'}^{t''} \mathrm{d}t \int \mathrm{d}^3x \left(\pi \frac{\partial\phi}{\partial t} - \mathscr{H}(\pi, \phi)\right). \tag{17.84}$$

By the same (heuristic) steps as in §17.1, we arrive at

$$\mathscr{G}^{(N)}(\bar{x}_1,\ldots,\bar{x}_N)=\frac{\displaystyle\int_{\text{periodic}}\mathscr{D}\phi\int\mathscr{D}\pi\,\phi(\bar{x}_1)\ldots\phi(\bar{x}_N)\exp\int_0^\beta \mathrm{d}\tau\int \mathrm{d}^3x\left(\mathrm{i}\pi\frac{\partial\phi}{\partial\tau}-\mathscr{H}\right)}{\displaystyle\int_{\text{periodic}}\mathscr{D}\phi\int\mathscr{D}\pi\exp\int_0^\beta \mathrm{d}\tau\int \mathrm{d}^3x\left(\mathrm{i}\pi\frac{\partial\phi}{\partial\tau}-\mathscr{H}\right)} \tag{17.85}$$

and if $\mathscr{L}$ is of the form (17.18),

$$\mathscr{G}^{(N)}(\bar{x}_1,\ldots,\bar{x}_N)=\frac{\displaystyle\int_{\text{periodic}}\mathscr{D}\phi\,\phi(\bar{x}_1)\ldots\phi(\bar{x}_N)\exp\int_0^\beta \mathrm{d}\tau\int \mathrm{d}^3x\,\mathscr{L}(\phi,\bar{\partial}_\mu\phi)}{\displaystyle\int_{\text{periodic}}\mathscr{D}\phi\exp\int_0^\beta \mathrm{d}\tau\int \mathrm{d}^3x\,\mathscr{L}(\phi,\bar{\partial}_\mu\phi)}. \tag{17.86}$$

By analogy with Chapter 4, we now introduce a generating functional for temperature Green functions.

$$\bar{W}[J]=\frac{\displaystyle\int_{\text{periodic}}\mathscr{D}\phi\exp\left(\int_0^\beta \mathrm{d}\tau\int \mathrm{d}^3x(\mathscr{L}(\phi,\bar{\partial}_\mu\phi)+J\phi)\right)}{\displaystyle\int_{\text{periodic}}\mathscr{D}\phi\exp\left(\int_0^\beta \mathrm{d}\tau\int \mathrm{d}^3x\,\mathscr{L}(\phi,\bar{\partial}_\mu\phi)\right)} \tag{17.87}$$

where the source J is a function of $\bar{x}$. The temperature Green functions are obtained from $\bar{W}[J]$ by functional differentiation.

$$\mathscr{G}^{(N)}(\bar{x}_1,\ldots,\bar{x}_N)=\left.\frac{\delta^N\bar{W}[J]}{\delta J(\bar{x}_N)\ldots\delta J(\bar{x}_1)}\right|_{J=0} \tag{17.88}$$

and $\bar{W}[J]$ may be expanded in temperature Green functions as

$$\bar{W}[J]=\sum_{N=0}^{\infty}\frac{1}{N!}\int \mathrm{d}\bar{x}_1\ldots\int \mathrm{d}\bar{x}_N\,\mathscr{G}^{(N)}(\bar{x}_1,\ldots,\bar{x}_N) \tag{17.89}$$

where

$$\mathrm{d}\bar{x}\equiv\int_0^\beta \mathrm{d}\tau\int \mathrm{d}^3x. \tag{17.90}$$

Fourier transformed temperature Green functions $\tilde{\mathscr{G}}^{(N)}$ may be introduced through

$$\tilde{\mathscr{G}}^{(N)}(\bar{p}_1,\ldots,\bar{p}_N)\delta(\bar{p}_1+\ldots+\bar{p}_N)(2\pi)^3\beta$$

$$=\int \mathrm{d}\bar{x}_1\ldots\int \mathrm{d}\bar{x}_N\exp[\mathrm{i}(\bar{p}_1\cdot\bar{x}_1+\ldots+\bar{p}_N\cdot\bar{x}_N)]\mathscr{G}^{(N)}(\bar{x}_1,\ldots,\bar{x}_N) \tag{17.91}$$

where

$$\bar{p} \equiv (\mathrm{i}\omega_n, \boldsymbol{p}) \tag{17.92}$$

with ω_n as in (17.29), $\bar{p}\cdot\bar{x}$ is as in (17.32) and we use the notation

$$\delta(\bar{p}_1 + \ldots + \bar{p}_N) \equiv \delta_{\omega_1+\ldots+\omega_N,0}\,\delta(\boldsymbol{p}_1 + \ldots + \boldsymbol{p}_N). \tag{17.93}$$

A generating functional $\bar{X}[J]$ for connected temperature Green functions $G^{(N)}$ may be defined through

$$\bar{W}[J] = \mathrm{e}^{\bar{X}[J]} \tag{17.94}$$

with the relations

$$G^{(N)}(\bar{x}_1, \ldots, \bar{x}_N) = \left.\frac{\delta^N \bar{X}[J]}{\delta J(\bar{x}_N)\ldots\delta J(\bar{x}_1)}\right|_{J=0} \tag{17.95}$$

and

$$\bar{X}[J] = \sum_{N=1}^{\infty} \frac{1}{N!} \int \mathrm{d}\bar{x}_1 \ldots \int \mathrm{d}\bar{x}_N\, G^N(\bar{x}_1, \ldots, \bar{x}_N). \tag{17.96}$$

17.6 Finite temperature generating functional for a free scalar field

The procedure is similar to §4.2. In the case of a free scalar field, the appropriate Lagrangian is (17.22). Thus,

$$\bar{W}_0[J] = \frac{\displaystyle\int_{\text{periodic}} \mathscr{D}\phi \exp\left(-\frac{1}{2}\int \mathrm{d}\bar{x}' \int \mathrm{d}\bar{x}\, \phi(\bar{x}')A(\bar{x}', \bar{x})\phi(\bar{x}) + \int \mathrm{d}\bar{x}\, J(\bar{x})\phi(\bar{x})\right)}{\displaystyle\int_{\text{periodic}} \mathscr{D}\phi \exp\left(-\frac{1}{2}\int \mathrm{d}\bar{x}' \int \mathrm{d}\bar{x}\, \phi(\bar{x}')A(\bar{x}', \bar{x})\phi(\bar{x})\right)} \tag{17.97}$$

with $A(\bar{x}', \bar{x})$ as in (17.24). Using (1.14), we obtain

$$\bar{W}_0[J] = \exp\left(-\frac{1}{2}\int \mathrm{d}\bar{x}' \int \mathrm{d}\bar{x}\, J(\bar{x}')\bar{\Delta}_{\mathrm{F}}(\bar{x}' - \bar{x})J(\bar{x})\right) \tag{17.98}$$

where we have written

$$\bar{\Delta}_{\mathrm{F}}(\bar{x}' - \bar{x}) \equiv -A^{-1}(\bar{x}', \bar{x}). \tag{17.99}$$

The inverse $\mathbf{A}^{-1}$ may be obtained by Fourier transforming as in (17.34). Thus,

$$\bar{\Delta}_{\mathrm{F}}(\bar{x}' - \bar{x}) = \frac{1}{\beta}\sum_n \int \frac{\mathrm{d}^3 p}{(2\pi)^3}\, \mathrm{e}^{-\mathrm{i}\bar{p}\cdot(\bar{x}' - \bar{x})}\bar{\Delta}_{\mathrm{F}}(\bar{p}) \tag{17.100}$$

with $\bar{p}$ as (17.31), and

$$\bar{\Delta}_{\mathrm{F}}(\bar{p}) = -(-\bar{p}^2 + m^2)^{-1} = -(\omega_n^2 + \boldsymbol{p}^2 + m^2)^{-1}. \tag{17.101}$$

17.7 Feynman rules for temperature Green functions

The approach used in Chapter 6 for ordinary Green functions is easily adapted to temperature Green functions. The only real differences arise because p^0 has been replaced by $\mathrm{i}\omega_n$, and because various factors of i no longer occur in (17.86), (17.98) and (17.89) compared with the zero temperature case. The resulting Feynman rules are as follows (see problem 17.4):

1 With each line carrying 'momentum' $\bar{p}=(\mathrm{i}\omega_n, \boldsymbol{p})$ we are to associate a factor $(\bar{p}^2-m^2)^{-1}$.

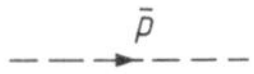

$$:\quad (\bar{p}^2-m^2)^{-1}=-(\omega_n^2+\boldsymbol{p}^2+m^2)^{-1}.$$

2 With each vertex of four lines carrying 'momenta' $\bar{p}_1, \bar{p}_2, \bar{p}_3, \bar{p}_4$ we associate a factor $-\lambda$, constraining the 'momenta' so that there is overall conservation

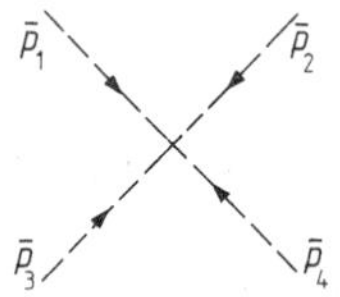

$$:\quad -\lambda \qquad (\bar{p}_1+\bar{p}_2+\bar{p}_3+\bar{p}_4=0).$$

3 Integrate and sum over each independent internal loop 'momentum' $\bar{p}=(\mathrm{i}\omega_n, \boldsymbol{p})$ with weight

$$\frac{1}{\beta}\sum_n \int \frac{\mathrm{d}^3p}{(2\pi)^3}.$$

The corresponding modifications are made to the Feynman rules for fermion fields and gauge fields. (No factors of i for vertices or propagators, $p_0 \to \mathrm{i}\omega_n$ and $\int \mathrm{d}^4p/(2\pi)^4 \to (1/\beta)\sum_n \mathrm{d}^3p/(2\pi)^3$.)

17.8 The finite temperature effective potential

By analogy with §4.4, a classical field $\phi_c(\bar{x})$ may be defined by

$$\phi_c(\bar{x})=\frac{\delta \bar{X}[J]}{\delta J(\bar{x})}. \tag{17.102}$$

From (17.87),

$$\frac{\delta \bar{W}[J]}{\delta J(\bar{x})}=\langle\hat{\phi}(\bar{x})\rangle_J \tag{17.103}$$

where $\langle\hat{\phi}(\bar{x})\rangle_J$ is the expectation value (thermal average) of $\hat{\phi}(\bar{x})$ in the presence of the source J. Using (17.94),

$$\phi_c(\bar{x}) = \langle \hat{\phi}(\bar{x}) \rangle_J / \bar{W}[J]. \qquad (17.104)$$

For zero source,

$$\phi_c(\bar{x}) = \langle \hat{\phi}(\bar{x}) \rangle \qquad J = 0 \qquad (17.105)$$

since

$$\bar{W}[0] = 1. \qquad (17.106)$$

Moreover,

$$\begin{aligned} \langle \hat{\phi}(\bar{x}) \rangle &= \mathrm{Tr}[\mathrm{e}^{-\beta \hat{H}} \hat{\phi}(\tau, \boldsymbol{x})] / \mathrm{Tr}[\mathrm{e}^{-\beta \hat{H}}] \\ &\sim \mathrm{Tr}[\mathrm{e}^{-\beta \hat{H}} \hat{\phi}(0, \boldsymbol{x})] / \mathrm{Tr}[\mathrm{e}^{-\beta \hat{H}}] \end{aligned} \qquad (17.107)$$

where we have used the connection (2.3) between the field operator at time t and time zero, with $t \to -\mathrm{i}\tau$. Combining (17.105) and (17.107),

$$\phi_c(\bar{x}) = \langle \hat{\phi}(0, \boldsymbol{x}) \rangle \qquad J = 0. \qquad (17.108)$$

Thus, for zero source, $\phi_c(\bar{x})$ is the expectation value (thermal average) of $\hat{\phi}(0, \boldsymbol{x})$, the Schrödinger picture field operator.

An effective action $\bar{\Gamma}$ is defined by analogy with (4.68),

$$\bar{\Gamma}[\phi_c] = \bar{X}[J] - \int \mathrm{d}\bar{x}\, J(\bar{x}) \phi_c(\bar{x}) \qquad (17.109)$$

and the source is given by

$$J(\bar{x}) = \frac{-\delta \bar{\Gamma}[\phi_c]}{\delta \phi_c(\bar{x})}. \qquad (17.110)$$

One-particle-irreducible temperature Green functions, $\Gamma^{(N)}$, may be defined by the expansion

$$\begin{aligned} \bar{\Gamma}[\phi_c] = \sum_{N=1}^{\infty} \frac{1}{N!} \int & \mathrm{d}\bar{x}_1 \ldots \mathrm{d}\bar{x}_N \Gamma^{(N)}(\bar{x}_1, \ldots, \bar{x}_N) \\ & \times \phi_c(\bar{x}_1) \ldots \phi_c(\bar{x}_N) \end{aligned} \qquad (17.111)$$

and momentum space OPI temperature Green functions, $\tilde{\Gamma}^{(N)}$, by

$$\begin{aligned} &\tilde{\Gamma}^{(N)}(\bar{p}_1, \ldots, \bar{p}_N) \delta(\bar{p}_1 + \ldots + \bar{p}_N)(2\pi)^3 \beta \\ &\quad = \int \mathrm{d}\bar{x}_1 \ldots \int \mathrm{d}\bar{x}_N \exp[\mathrm{i}(\bar{p}_1 \cdot \bar{x}_1 \ldots + \bar{p}_N \cdot \bar{x}_N)] \Gamma^{(N)}(\bar{x}_1, \ldots, \bar{x}_N) \end{aligned} \qquad (17.112)$$

with $\delta(\bar{p}_1 + \ldots + \bar{p}_N)$ as in (17.93).

The finite temperature effective potential $\bar{V}(\phi_c)$ may be defined by an expansion analogous to (4.77),

$$\bar{\Gamma}[\phi_c]=\int \mathrm{d}\bar{x}\left(-\bar{V}(\phi_c)+\frac{\bar{A}(\phi_c)}{2}\bar{\partial}_\mu\phi_c\bar{\partial}^\mu\phi_c+\ldots\right). \tag{17.113}$$

If the classical field has no spatial (or τ) dependence then only the $\bar{V}(\phi_c)$ term in the expansion (17.113) need to be retained, and (17.110) becomes

$$\mathrm{d}\bar{V}/\mathrm{d}\phi_c=J. \tag{17.114}$$

If we set the source term to zero, then, from (17.108), ϕ_c has the significance of the expectation value (thermal average) of the field operator, and

$$\frac{\mathrm{d}\bar{V}}{\mathrm{d}\phi_c}=0. \tag{17.115}$$

Thus, when it has no spatial variation, the expectation value of the field operator at finite temperature may be obtained by minimising the finite temperature effective potential.

Using the inverse of (17.112) in (17.111), the effective potential may be expanded in terms of Fourier transformed temperature Green functions at zero 'momenta'. (See problem 17.5.)

$$\bar{V}(\phi_c)=-\sum_{N=1}^{\infty}\tilde{\Gamma}^{(N)}(0,\ldots,0)\phi_C^N/N!. \tag{17.116}$$

17.9 Finite temperature effective potential at one-loop order

The contribution to the finite temperature effective action from one-loop diagrams in the expansion (17.111) is obtained by exact analogy with §13.9. Thus we have to shift scalar fields by their expectation values (which are now thermal averages) and isolate the terms in the (τ dependent) Lagrangian which are quadratic in all (shifted) fields including fermion fields and gauge fields. The one-loop contribution to the effective action $\bar{\Gamma}_1[\phi_c]$ is then obtained as a Gaussian path integral.

$$\exp\bar{\Gamma}_1[\phi_c]=\int_{\text{periodic}}\mathscr{D}\phi\,\mathscr{D}A^\mu\,\mathscr{D}\eta^*\,\mathscr{D}\eta\int_{\text{antiperiodic}}\mathscr{D}\bar{\psi}\,\mathscr{D}\psi$$
$$\times\exp\int_0^\beta \mathrm{d}\tau\int \mathrm{d}^3x\,\mathscr{L}_{\text{quad}}(\phi_c(\bar{x})) \tag{17.117}$$

where ϕ_i, A_a^μ, η_a and ψ_r are the scalar fields, gauge fields, Fadeev–Popov ghost fields and (Dirac) fermion fields, respectively, and $\mathscr{L}_{\text{quad}}(\phi_c(\bar{x}))$ is the quadratic term in the (τ dependent) fields, in the shifted Lagrangian. The one-loop contribution to the finite temperature effective potential $\bar{V}_1(\phi_c)$ is obtained by taking ϕ_c to be constant, and using (17.113).

$$\exp\left(-\int_0^\beta \mathrm{d}\tau \int \mathrm{d}^3x\, \bar{V}_1(\phi_c)\right) = \int_{\text{periodic}} \mathscr{D}\phi\mathscr{D}A^\mu\mathscr{D}\eta^*\mathscr{D}\eta \int_{\text{antiperiodic}} \mathscr{D}\bar{\psi}\mathscr{D}\psi$$

$$\times \exp\int_0^\beta \mathrm{d}\tau \int \mathrm{d}^3x\, \mathscr{L}_{\text{quad}}(\phi_c). \tag{17.118}$$

In general, $\mathscr{L}_{\text{quad}}(\phi_c)$ takes the form

$$\begin{aligned}\mathscr{L}_{\text{quad}}(\phi_c) = {} & \tfrac{1}{2}\bar{\partial}_\mu\phi_i\bar{\partial}^\mu\phi_i - \tfrac{1}{2}[\hat{M}_{\mathrm{S}}^2(\phi_c)]_{ij}\phi_i\phi_j \\ & -\tfrac{1}{4}(\bar{\partial}^\mu A_a^\nu - \bar{\partial}^\nu A_a^\mu)(\bar{\partial}_\mu(A_a)_\nu - \bar{\partial}_\nu(A_a)_\mu) \\ & +\tfrac{1}{2}[\hat{M}_{\mathrm{V}}^2(\phi_c)]_{ab}A_a^\mu A_\mu^b - (1/2\xi)(\bar{\partial}_\mu A_a^\mu)^2 \\ & +\bar{\partial}_\mu\eta_a^*\bar{\partial}^\mu\eta_a + \mathrm{i}\bar{\psi}_r\gamma^\mu\bar{\partial}_\mu\psi_r \\ & -[\hat{M}_{\mathrm{F}}(\phi_c)]_{rs}\bar{\psi}_r\psi_s \end{aligned} \tag{17.119}$$

where we have adopted Landau gauge, $\xi \to 0$, so as to avoid couplings of the scalar fields to the Fadeev–Popov ghosts. The mass matrices of the scalar, vector and fermion fields after spontaneous symmetry breaking are $\hat{M}_{\mathrm{S}}^2(\phi_c)$, $\hat{M}_{\mathrm{V}}^2(\phi_c)$ and $\hat{M}_{\mathrm{F}}(\phi_c)$, where ϕ_c is used as shorthand for the expectation values (thermal averages) $(\phi_i)_c$. We may write

$$\begin{aligned}\int_0^\beta \mathrm{d}\tau \int \mathrm{d}^3x\, \mathscr{L}_{\text{quad}}(\phi_c) = {} & -\frac{1}{2}\int \mathrm{d}\bar{x}' \int \mathrm{d}\bar{x}\, \phi_i(\bar{x}')A_{ij}(\bar{x}',\bar{x})\phi_j(\bar{x}) \\ & -\frac{1}{2}\int \mathrm{d}\bar{x}' \int \mathrm{d}\bar{x}\, A_\mu^a(\bar{x}')B_{ab}^{\mu\nu}(\bar{x}',\bar{x})A_\nu^b(\bar{x}) \\ & -\int \mathrm{d}\bar{x}' \int \mathrm{d}\bar{x}\, \eta^*(\bar{x}')C(\bar{x}',\bar{x})\eta(\bar{x}) \\ & -\int \mathrm{d}\bar{x}' \int \mathrm{d}\bar{x}\, \bar{\psi}_r(\bar{x}')D_{rs}(\bar{x}',\bar{x})\psi_s(\bar{x}) \end{aligned} \tag{17.120}$$

with

$$A_{ij}(\bar{x}',\bar{x}) = (-\delta_{ij}\bar{\partial}'_\mu\bar{\partial}^\mu + [\hat{M}_{\mathrm{S}}^2(\phi_c)]_{ij})\delta(\bar{x}'-\bar{x}) \tag{17.121}$$

$$B_{ab}^{\mu\nu}(\bar{x}',\bar{x}) = \{(g^{\mu\nu}\bar{\partial}_{x'}^\rho\bar{\partial}_{\rho x} - (1-\xi^{-1})\bar{\partial}_{x'}^\nu\bar{\partial}_x^\mu)\delta_{ab} - g^{\mu\nu}[\hat{M}_{\mathrm{V}}^2(\phi_c)]_{ab}\}\delta(\bar{x}'-\bar{x}) \tag{17.122}$$

$$C(\bar{x}',\bar{x}) = -\bar{\partial}_{x'}^\rho\bar{\partial}_{\rho x}\delta(\bar{x}'-\bar{x}) \tag{17.123}$$

and

$$D_{rs}(\bar{x}',\bar{x}) = (\mathrm{i}\delta_{rs}\gamma^\mu\bar{\partial}_\mu + [\hat{M}_{\mathrm{F}}(\phi_c)]_{rs})\delta(\bar{x}'-\bar{x}) \tag{17.124}$$

where $\bar{x}$ and $\delta(\bar{x}'-\bar{x})$ are as in (17.26) and (17.25). Performing the Gaussian path integrals in (17.118) leads to

$$-\int_0^\beta \mathrm{d}\tau \int \mathrm{d}^3x\, \bar{V}_1(\phi_c) = -\tfrac{1}{2}\,\mathrm{Tr}\ln\mathbf{A} - \tfrac{1}{2}\,\mathrm{Tr}\ln\mathbf{B} + \mathrm{Tr}\ln\mathbf{C} + \mathrm{Tr}\ln\mathbf{D}. \tag{17.125}$$

The traces are evaluated much as in §17.2, §17.3 and §17.4 to obtain (see problem 17.6)

$$\mathrm{Tr}\ln \mathbf{A}=\int \mathrm{d}^3x \sum_n^{\text{bosons}} \int \frac{\mathrm{d}^3p}{(2\pi)^3} \sum_i \ln[\omega_n^2+\boldsymbol{p}^2+(M_\mathrm{S}^2)_i] \tag{17.126}$$

$$\mathrm{Tr}\ln \mathbf{B}=\int \mathrm{d}^3x \sum_n^{\text{bosons}} \int \frac{\mathrm{d}^3p}{(2\pi)^3} \sum_a \{3\ln[\omega_n^2+\boldsymbol{p}^2+(M_\mathrm{V}^2)_a]+\ln[\omega_n^2+\boldsymbol{p}^2+\xi(M_\mathrm{V}^2)_a]\} \tag{17.127}$$

with $\xi\to 0$ for Landau gauge,

$$\mathrm{Tr}\ln \mathbf{C}=\int \mathrm{d}^3x \sum_n^{\text{bosons}} \int \frac{\mathrm{d}^3p}{(2\pi)^3} \sum_a \ln(\omega_n^2+\boldsymbol{p}^2) \tag{17.128}$$

$$\mathrm{Tr}\ln \mathbf{D}=2\int \mathrm{d}^3x \sum_n^{\text{fermions}} \int \frac{\mathrm{d}^3p}{(2\pi)^3} \sum_r \ln[\omega_n^2+\boldsymbol{p}^2+(M_\mathrm{F}^2)_r] \tag{17.129}$$

where $(M_\mathrm{S}^2(\phi_\mathrm{c}))_i$, $(M_\mathrm{V}^2(\phi_\mathrm{c}))_a$ and $(M_\mathrm{F}^2(\phi_\mathrm{c}))_r$ are the eigenvalues of the matrices $\hat{M}_\mathrm{S}^2(\phi_\mathrm{c})$, $\hat{M}_\mathrm{V}^2(\phi_\mathrm{c})$ and $\hat{M}_\mathrm{F}^2(\phi_\mathrm{c})$, and the Matsubara frequencies ω_n are as in (17.29) and (17.63) for boson and fermion sums respectively. Substituting (17.126)–(17.129) in (17.125), and using the Matsubara frequency sums of appendix E, gives

$$\begin{aligned}\beta\bar{V}_1(\phi_\mathrm{c})=&\frac{1}{2}\int\frac{\mathrm{d}^3p}{(2\pi)^3}\sum_i\{\beta\sqrt{\boldsymbol{p}^2+(M_\mathrm{S}^2)_i}+2\ln[1-\exp(-\beta\sqrt{\boldsymbol{p}^2+(M_\mathrm{S}^2)_i})]\}\\ &+\frac{1}{2}\int\frac{\mathrm{d}^3p}{(2\pi)^3}\sum_a\{3\beta\sqrt{\boldsymbol{p}^2+(M_\mathrm{V}^2)_a}+6\ln[1-\exp(-\beta\sqrt{\boldsymbol{p}^2+(M_\mathrm{V}^2)_a})]\\ &-\beta|\boldsymbol{p}|-2\ln(1-\mathrm{e}^{-\beta|\boldsymbol{p}|})\}\\ &-2\int\frac{\mathrm{d}^3p}{(2\pi)^3}\sum_r\{\beta\sqrt{\boldsymbol{p}^2+(M_\mathrm{F}^2)_r}+2\ln[1+\exp(-\beta\sqrt{\boldsymbol{p}^2+(M_\mathrm{F}^2)_r})]\}.\end{aligned} \tag{17.130}$$

We may separate $\bar{V}_1(\phi_\mathrm{c})$ into a part $\bar{V}_1^0(\phi_\mathrm{c})$ which survives at $T=0$, and a T dependent part $\bar{V}_1^T(\phi_\mathrm{c})$,

$$\bar{V}_1(\phi_\mathrm{c})=\bar{V}_1^0(\phi_\mathrm{c})+\bar{V}_1^T(\phi_\mathrm{c}) \tag{17.131}$$

where

$$\begin{aligned}\bar{V}_1^0(\phi_\mathrm{c})=\frac{1}{2}\int\frac{\mathrm{d}^4p}{(2\pi)^4}\Big(&\sum_i\ln(p_0^2+\boldsymbol{p}^2+(M_\mathrm{S}^2)_i)\\ &+\sum_a[3\ln(p_0^2+\boldsymbol{p}^2+(M_\mathrm{V}^2)_a)-\ln(p_0^2+\boldsymbol{p}^2)]\\ &-4\sum_r\ln(p_0^2+\boldsymbol{p}^2+(M_\mathrm{F}^2)_r)\Big)\end{aligned} \tag{17.132}$$

and

$$\bar{V}_1^T(\phi_c)=\frac{T^4}{2\pi^2}\int_0^\infty \mathrm{d}y\,y^2\Big(\sum_i \ln[1-\exp(-\sqrt{y^2+T^{-2}(M_S^2)_i})]$$
$$+\sum_a 3\ln[1-\exp(-\sqrt{y^2+T^{-2}(M_V^2)_a})-\ln(1-\mathrm{e}^{-y})]$$
$$-4\sum_r \ln[1+\exp(-\sqrt{y^2+T^{-2}(M_F^2)_r})]\Big) \qquad (17.133)$$

where

$$y\equiv\beta|\boldsymbol{p}|=T^{-1}|\boldsymbol{p}|. \qquad (17.134)$$

In arriving at (17.132), we have used the fact that

$$\int\frac{\mathrm{d}^3p}{(2\pi)^3}\sqrt{\boldsymbol{p}^2+R}=\int\frac{\mathrm{d}^4p}{(2\pi)^4}\ln(p_0^2+\boldsymbol{p}^2+R)+(\text{constant independent of } R). \qquad (17.135)$$

The expression (17.132) is just the $T=0$ radiative correction to the effective potential, at one-loop order. The evaluation and renormalisation of such terms has already been studied in §13.9.

The temperature dependent part of the one-loop effective potential, $\bar{V}_1^T(\phi_c)$, is particularly simple in two limiting cases. First, for all mass eigenvalues $(M_S^2)_i$, $(M_V^2)_a$ and $(M_F^2)_r$, very much greater than T^2, all contributions to (17.133) approach zero exponentially fast, and $\bar{V}_1^T$ becomes negligible. (This is also true of the $\ln[1-\exp(-y)]$ term since if we were to carry the gauge parameter ξ through to this stage we would have $\ln[1-\exp(-\sqrt{y^2+T^{-2}\xi(M_V^2)_a})]$. We then take the limit $\xi\to 0$ after the limit $(M_V^2)_aT^{-2}\to\infty$.) Second, in the high temperature limit, T very much greater than all the mass eigenvalues, we may use

$$\frac{T^4}{2\pi^2}\int_0^\infty \mathrm{d}y\,y^2\ln[1-\exp(-\sqrt{y^2+RT^{-2}})]\approx\frac{-\pi^2T^4}{90}+\frac{RT^2}{24} \qquad RT^{-2}\ll 1 \qquad (17.136)$$

and

$$\frac{T^4}{2\pi^2}\int_0^\infty \mathrm{d}y\,y^2\ln[1+\exp(-\sqrt{y^2+RT^{-2}})]\approx\frac{7\pi^2T^4}{720}-\frac{RT^2}{48} \qquad RT^{-2}\ll 1. \qquad (17.137)$$

Thus, for the high temperature limit,

$$\bar{V}_1^T(\phi_c)\approx-\frac{\pi^2T^4}{90}(N_B+\tfrac{7}{8}N_F)+\frac{T^2}{24}\Big(\sum_i(M_S^2)_i+3\sum_a(M_V^2)_a+2\sum_r(M_F^2)_r\Big)$$
$$=-\frac{\pi^2T^4}{90}(N_B+\tfrac{7}{8}N_F)$$
$$+\frac{T^2}{24}[\mathrm{Tr}\,\hat{M}_S^2(\phi_c)+3\,\mathrm{Tr}\,\hat{M}_V^2(\phi_c)+2\,\mathrm{Tr}\,\hat{M}_F^2(\phi_c)] \qquad (17.138)$$

where $\hat{M}_S^2(\phi_c)$, $\hat{M}_V^2(\phi_c)$ and $\hat{M}_F^2(\phi_c)$ are the scalar, vector and Dirac fermion mass matrices of (17.119). (For fermions described by Weyl spinor fields there would be no factor of 2 in front of the last term of (17.138). See problem 17.7.) The T^4 term in (17.138) is just the free energy density for an ideal ultrarelativistic gas of (17.73), with N_B and N_F the number of bosonic and fermionic degrees of freedom, respectively. If some fields are heavy and some light, on the scale of the temperature T, then N_B and N_F should be interpreted as the degrees of freedom of light fields, and the traces of mass matrices should be evaluated only for light fields, since the heavy fields do not contribute, as discussed above.

17.10 The Higgs model at finite temperatures

The simplest model, incorporating both scalar and vector fields, to study at finite temperature is the Higgs model described in §13.5. The finite temperature (effective) Lagrangian is

$$\mathscr{L} = \bar{\mathrm{D}}_\mu\phi\bar{\mathrm{D}}^\mu\phi^* - m^2\phi^*\phi - (\lambda/4)(\phi^*\phi)^2 - \tfrac{1}{4}\bar{F}_{\mu\nu}\bar{F}^{\mu\nu}$$
$$-(1/2\xi)(\bar{\partial}_\mu A^\mu)^2 + \bar{\partial}_\mu\eta^*\bar{\partial}^\mu\eta \qquad (17.139)$$

where m^2 is negative,

$$\bar{\mathrm{D}}_\mu\phi \equiv (\bar{\partial}_\mu + \mathrm{i}eA_\mu)\phi \qquad (17.140)$$

$$\bar{\mathrm{D}}_\mu\phi^* \equiv (\bar{\partial}_\mu - \mathrm{i}eA_\mu)\phi^*. \qquad (17.141)$$

$\bar{F}_{\mu\nu}$ is as in (17.47), $\bar{\partial}_\mu$ as in (17.21) and the fields functions of $\bar{x}$ of (17.26) rather than x. We have included the Faddeev–Popov ghosts η because they are needed to cancel contributions to the free energy from unphysical degrees of freedom of the gauge field A_μ, as discussed in §17.3. To obtain the finite temperature effective potential, we first shift the scalar fields by the expectation value (thermal average)

$$\frac{\phi_c}{\sqrt{2}} \equiv \langle\hat{\phi}(\bar{x})\rangle \qquad (17.142)$$

which has been taken real without loss of generality, because of gauge invariance, and assumed constant. (The factor $1/\sqrt{2}$ has no significance, but has simply been introduced for convenience.) Thus, we write

$$\phi = \frac{1}{\sqrt{2}}(\phi_c + \phi_1 + \mathrm{i}\phi_2) \qquad (17.143)$$

where ϕ_1 and ϕ_2 are real (shifted) fields. The quadratic terms in the shifted

Lagrangian are

$$\mathscr{L}_{\text{quad}}=\frac{1}{2}(\bar{\partial}_\mu\phi_1)^2+\frac{1}{2}(\bar{\partial}_\mu\phi_2)^2-\frac{1}{2}\left(m^2+\frac{3\lambda}{4}\phi_c^2\right)\phi_1^2$$

$$-\frac{1}{2}\left(m^2+\frac{\lambda}{4}\phi_c^2\right)\phi_2^2-\frac{1}{4}\bar{F}_{\mu\nu}\bar{F}^{\mu\nu}+\frac{e^2}{2}\phi_c^2A_\mu A^\mu$$

$$-\frac{1}{2\xi}(\bar{\partial}_\mu A^\mu)^2+\bar{\partial}_\mu\eta^*\bar{\partial}^\mu\eta \tag{17.144}$$

where we have adopted Landau gauge, $\xi\to 0$, so as to remove an $A^\mu\bar{\partial}_\mu\phi_2$ cross-term. In the notation of (17.119), we have

$$\hat{M}_S^2(\phi_c)=\text{diag}\left\{m^2+\frac{3\lambda}{4}\phi_c^2,\, \text{m}^2+\frac{\lambda}{4}\phi_c^2\right\} \tag{17.145}$$

$$\hat{M}_V^2(\phi_c)=e^2\phi_c^2. \tag{17.146}$$

Provided that $e^4\ll\lambda$, we may drop the zero temperature one-loop contribution to the effective potential (17.132) compared with the tree terms,

$$\bar{V}_0(\phi_c)=\frac{m^2}{2}\phi_c{}^2+\frac{\lambda}{16}\phi_c{}^2. \tag{17.147}$$

The temperature dependent one-loop contribution is obtained from (17.138), in the high temperature limit, and is

$$\bar{V}_1^T(\phi_c)=-\frac{4\pi^2T^4}{90}+\frac{(\lambda+3e^2)T^2\phi_c^2}{24} \tag{17.148}$$

where we have also assumed that $T^2\gg -m^2$. Thus, the one-loop approximation effective potential is

$$\bar{V}(\phi_c)=\frac{m^2(T)}{2}\phi_c^2+\frac{\lambda}{16}\phi_c^4-\frac{4\pi^2T^4}{90} \tag{17.149}$$

where we have defined a temperature dependent effective mass by

$$m^2(T)=m^2+\frac{(\lambda+3e^2)}{12}T^2. \tag{17.150}$$

The expression (17.149) is valid for temperatures large compared with the masses of all shifted fields,

$$T^2\gg\lambda\phi_c^2,\, e^2\phi_c^2,\, -m^2 \tag{17.151}$$

and

$$e^4\ll\lambda \tag{17.152}$$

so that the zero temperature radiative correction may be dropped. There is no difficulty in relaxing the approximation (17.151), by using (17.133) instead of (17.138), though the high temperature approximation is adequate for our purposes in this section. When (17.152) is not satisfied, the zero temperature radiative correction may be included, with interesting consequences, as we shall see in §17.13.

The expectation value (thermal average) of the field is obtained by minimising (17.149). For $m^2(T)$ negative,

$$\frac{\partial \bar{V}}{\partial \phi_c}=0 \tag{17.153}$$

has two possible solutions:

$$\phi_c=0 \tag{17.154}$$

and

$$\phi_c^2=-4m^2(T)/\lambda. \tag{17.155}$$

For $m^2(T)$ positive, only the solution (17.154) is possible. In (17.150), m^2 is negative, and so there is a temperature (the critical temperature) for which

$$m^2(T_c)=0 \tag{17.156}$$

namely that given by

$$T_c^2=-12m^2/(\lambda+3e^2). \tag{17.157}$$

For T greater than T_c, $m^2(T)$ is positive and, at the minimum of the effective potential, $\phi_c=0$. We then say that the system is in the symmetric phase (no spontaneous symmetry breaking). For T less than T_c, $m^2(T)$ is negative, the minimum at $\phi_c=0$ turns into a maximum, and the system is in the asymmetric phase given by (17.155). (See figure 17.1.) The system passes in a continuous fashion from one phase to the other at $T=T_c$ and there is a second-order phase transition.

17.11 Electroweak theory at finite temperature

The finite temperature (effective) Lagrangian is as in §14.1 and §14.2 with the usual replacements of ∂_μ by $\bar{\partial}_\mu$, x by $\bar{x}$ etc. An expectation value (thermal average) for the Higgs doublet is introduced by

$$\langle\hat{\phi}(\bar{x})\rangle=\begin{pmatrix}0\\ \phi_0/\sqrt{s}\end{pmatrix}. \tag{17.158}$$

Then a calculation exactly analogous to that of §17.10 leads to the one-loop approximation effective potential, when the temperature is large compared

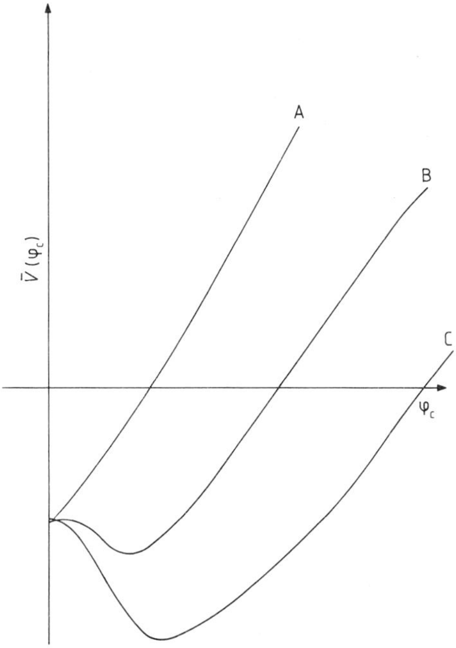

Figure 17.1 The finite temperature effective potential for the Higgs model when $e^4 \ll \lambda$. Curves A, B, C are for $T > T_c$, $T \approx T_c$ and $T < T_c$, respectively.

with all masses (see problem 17.8)

$$\bar{V}(\phi_c) = \tfrac{1}{2}m^2(T)\phi_c^2 + (\lambda/4)\phi_c^4 - (N_B + \tfrac{7}{8}N_F)\pi^2 T^4/90$$
$$+ B\phi_c^4(\ln(\phi_c^2/M^2) - 25/6) \qquad (17.159)$$

with N_B and N_F the number of bosonic and fermionic degrees of freedom, respectively, and

$$m^2(T) = m^2 + \left(\frac{\lambda}{2} + \frac{e^2(1 + 2\cos^2\theta_W)}{4\sin^2 2\theta_W} + \sum_f \frac{G_f^2}{12}\right)T^2 \qquad (17.160)$$

with the weak mixing angle θ_W and the Yukawa couplings of the fermions G_f as in §14.2. The last term in (17.159) is the zero temperature radiative correction to the effective potential, renormalised at mass M as in §13.9, and B has the value

$$B = \frac{3}{64}\left(\frac{e^2}{4\pi}\right)^2 \frac{(2 + \sec^4\theta_W)}{\sin^4\theta_W} - \frac{1}{64\pi^2}\sum_f G_f^4 \qquad (17.161)$$

where we have dropped order λ^2 contributions, which are always perturbatively negligible compared with the tree terms.

For the case $e^4 \ll \lambda$, the $T = 0$ radiative correction is negligible, and the discussion is exactly analogous to §17.10, with a critical temperature T_c for the second-order phase transition given by

$$T_c^2 = -m^2 \Bigg/ \left(\frac{\lambda}{2} + \frac{e^2(1+2\cos^2\theta_W)}{4\sin^2 2\theta_W} + \sum_f \frac{G_f^2}{12} \right). \tag{17.162}$$

(The case $e^4 \gtrsim \lambda$ is discussed in §17.13.)

For T greater than T_c, the system is in the symmetric phase, in which $\phi_c = 0$, and the gauge bosons are all massless. For T less than T_c, the system is in the asymmetric phase for which $\phi_c \neq 0$, the $W^\pm$ and Z^0 bosons acquire masses, and the symmetry is reduced from $SU_L(2) \times U(1)$ to $U_Q(1)$. The critical temperature T_c is of the same order of magnitude as the value at $T=0$ of ϕ_c at the (asymmetric) minimum of the effective potential, and the $W^\pm$ and Z^0 masses are of order $e\phi_c$. Thus, T_c should be of the order of 100 GeV.

17.12 Grand unified theory at finite temperature

In this section we construct the finite temperature effective potential for the SU(5) grand unified theory disucssed in Chapter 16. If we want to study the grand unified phase transition from the SU(5) symmetric phase to the $SU_C(3) \times SU_L(2) \times U(1)$ symmetric phase, we need only retain the Higgs scalars responsible for this particular symmetry breaking, and may drop the Higgs scalars responsible for the breaking of electroweak symmetry. (The argument given at the end of §17.11 leads us to expect that the critical temperature for the grand unified phase transition will be of the order of 10^{15} GeV and at such temperatures the expectation values of the electroweak Higgs scalars, which are of the order of 100 GeV, will be negligible.) Thus, we need only keep the Higgs scalars Φ belonging to the 24-dimensional adjoint representation of SU(5)

$$\Phi = \sum_{a=1}^{24} \phi_a T_a \tag{17.163}$$

where $T_a = \lambda_a/2$ are the generators of the five-dimensional representation of SU(5) as in §16.2, and appendix D.

The finite temperature (effective) Lagrangian is

$$\begin{aligned} \mathscr{L} = &-m_1^2 \,\mathrm{Tr}\,\Phi^2 - \lambda_1(\mathrm{Tr}\,\Phi^2)^2 - \lambda_2\,\mathrm{Tr}\,\Phi^4 \\ &+ \mathrm{Tr}(\bar{\mathrm{D}}_\mu \Phi)^2 - \tfrac{1}{2}\,\mathrm{Tr}(\bar{F}_{\mu\nu}\bar{F}^{\mu\nu}) \\ &- \xi^{-1}\,\mathrm{Tr}(\bar{\partial}_\mu A^\mu)^2 + 2\,\mathrm{Tr}(\bar{\partial}_\mu \eta^* \bar{\mathrm{D}}^\mu \eta) \end{aligned} \tag{17.164}$$

where have dropped fermions (apart from their contribution to the T^4 term in $\bar{V}(\phi_c)$) since the masses of known fermions are negligible on the scale of 10^{15} GeV, and they therefore make a negligible contribution to the ϕ_c dependent part of the effective potential. In (17.164), all quantities are defined as in §16.5, but with the fields now functions of $\bar{x}$ of (17.26) rather than x, and ∂_μ replaced by $\bar{\partial}_\mu$ of (17.21).

For the breaking of SU(5) to $SU_C(3) \times SU_L(2) \times U(1)$ we take the expectation value (thermal average)

$$\langle\hat{\Phi}\rangle = \frac{\phi_c}{\sqrt{15}} \operatorname{diag}\{1, 1, 1, -\tfrac{3}{2}, -\tfrac{3}{2}\}. \tag{17.165}$$

A calculation analogous to §17.10 gives the one-loop approximation to the effective potential for temperatures large compared with all masses (see problem 17.9)

$$\bar{V}(\phi_c) = \frac{1}{2} m_1^2(T)\phi_c^2 + \frac{(\lambda_1 + \frac{7}{30}\lambda_2)}{4}\phi_c^4 - \frac{(N_B + \frac{7}{8}N_F)\pi^2 T^4}{90} + B\phi_c^4\left(\ln(\phi_c^2/M^2) - \frac{25}{6}\right) \tag{17.166}$$

where N_B and N_F are the number of bosonic and fermionic degrees of freedom respectively, and

$$m_1^2(T) = m_1^2 + \frac{(130\lambda_1 + 47\lambda_2 + 75g_G^2)T^2}{60}. \tag{17.167}$$

The last term in (17.166) is the zero temperature radiative correction to the effective potential, renormalised at mass M as in §13.9, and B has the value (neglecting order λ^2):

$$B = 25g_G^4/256\pi^2. \tag{17.168}$$

In deriving (17.166), the identities for SU(N) Gell-Mann matrices given by MacFarlane *et al.*[5] can be useful.

For $g_G^4 \ll \lambda_1, \lambda_2$ we may drop the zero temperature radiative correction, and the discussion is exactly analogous to §17.10, with a critical temperature for the second-order phase transition given by

$$T_c^2 = -60m_1^2/(130\lambda_1 + 47\lambda_2 + 75g_G^2). \tag{17.169}$$

For T greater than T_c, the system is in the SU(5) symmetric phase, for which $\phi_c = 0$, and all gauge bosons are massless. For T less than T_c, ϕ_c is non-zero, the system is in the $SU_C(3) \times SU_L(2) \times U(1)$ symmetric phase, and only the electroweak gauge bosons are massless. By the argument at the end of §17.11, T_c should be of the order of 10^{15} GeV.

17.13 First-order phase transitions

In §§17.10–17.12, we have always assumed that the fourth power of the gauge coupling constant is very much less than the ϕ^4 coupling constant, and have then found a second-order phase transition. In this section we discuss what happens for larger values of the gauge coupling constant[6,7,8]. We shall find two differences. First, that we cannot use the high temperature approximation, that T is very much bigger than the masses of all fields, for the asymmetric minimum. Second, that we cannot always neglect the zero temperature radiative correction to the effective potential. The result of these differences is to produce a first-order phase transition.

For definiteness we shall discuss the Higgs model, though everything we say applies with very minor changes to electroweak and grand unified theory. Including the zero temperature radiative correction, as in (13.253), the finite temperature effective potential is

$$\bar{V}(\phi_c)=\frac{m^2}{2}\phi_c^2+\frac{\lambda}{16}\phi_c^4+B\phi_c^4\left[\ln\left(\frac{\phi_c^2}{M^2}\right)-\frac{25}{6}\right]+\bar{V}_1^T(\phi_c) \qquad (17.169)$$

where, from (13.266) with λ^2 negligible compared with e^4,

$$B=3e^4/64\pi^2. \qquad (17.170)$$

The temperature dependent one-loop contribution $\bar{V}_1^T(\phi_c)$ is given by (17.133) with the mass matrix of (17.145) and (17.146). (We are not now necessarily going to make the high T approximation of (17.138).) Thus

$$\begin{aligned}\bar{V}_1^T(\phi_c)=\frac{T^4}{2\pi^2}\int_0^\infty & \mathrm{d}y\, y^2\{\ln[1-\exp(-\sqrt{y^2+T^{-2}(m^2+3\lambda\phi_c^2/4)})]\\ &+\ln[1-\exp(-\sqrt{y^2+T^{-2}(m^2+\lambda\phi_c^2/4)})]\\ &+3\ln[1-\exp(-\sqrt{y^2+T^{-2}e^2\phi_c^2})]-\ln(1-\mathrm{e}^{-y})\}. \qquad (17.171)\end{aligned}$$

The first question we ask is whether it is correct to use the high temperature approximation for $\bar{V}_1^T(\phi_c)$ at the critical temperature, when $e^4 \gg \lambda$.

From §17.10, the critical temperature for a second-order transition, T_c, is given by

$$T_c^2=-4m^2/e^2 \qquad (17.172)$$

for $e^2 \gg \lambda$. However, at the zero temperature asymmetric minimum (neglecting the radiative correction for the moment)

$$\phi_c^2=v^2=-4m^2/\lambda. \qquad (17.173)$$

Thus

$$T_c^2 \gg m^2+\frac{\lambda}{4}v^2=0 \qquad (17.174)$$

and

$$T_c^2 \gg m^2+\frac{3\lambda}{4}v^2 \qquad (17.175)$$

but

$$T_c^2 \ll e^2 v^2 \qquad \text{for } e^4 \gg \lambda. \tag{17.176}$$

It is therefore not correct to use the high temperature approximation to study the critical temperature when $e^4 \gg \lambda$. At $\phi_c = 0$, the high temperature approximation is valid, and from (17.138) we have

$$\bar{V}_1(\phi_c = 0) = -4\pi^2 T^4/90. \tag{17.177}$$

However, at the asymmetric minimum $\phi_c = v$, the contribution to $\bar{V}_1^T(\phi_c)$, involving the gauge field mass is exponentially suppressed (because of (17.176)), as discussed after (17.135), but the high T approximation may be used for the other terms. Thus, dropping the zero temperature radiative correction for the moment,

$$\bar{V}(\phi_c = v) = -\frac{m^4}{\lambda} - \frac{\pi^2 T^4}{90}. \tag{17.178}$$

(We are working in Landau gauge. In other gauges, all terms in (17.171) except the first one are exponentially suppressed, because of the appearance of an additional mass term $\xi e^2 \phi_c^2$ as can be seen from (12.262). The result is the same, as it must be.)

The symmetric minimum is at a lower value of the effective potential than the asymmetric minimum for

$$T > T_{c1} = (30/\pi^2 \lambda)^{1/4} m. \tag{17.179}$$

The temperature T_{c1} is the temperature for a first-order phase transition, and the development of the asymmetric minimum with temperature is as in figure 17.2. The transition is first-order becuase of the discontinuous change in the expectation value of the field when the phase transition occurs. (The observant reader will have noticed that we have assumed that the value of the effective potential at the asymmetric minimum is the same as at zero temperature, apart from the T^4 term. He will be reassured to know that it can be shown that this is correct apart from corrections of higher order in e^2. See problem 17.10.)

When zero temperature radiative corrections are taken into account, it is convenient to cast the effective potential in terms of the mass of the physical Higgs particle. Thus, by analogy with (12.151)–(12.253),

$$\bar{V}(\phi_c) = B\left(\frac{\alpha}{2} v^2 \phi_c^2 - \frac{(\alpha+2)}{4} \phi_c^4 + \phi_c^4 \ln(\phi_c^2/v^2)\right) + \bar{V}_1^T(\phi_c) \tag{17.180}$$

where the renormalisation has been carried out at the zero temperature asymmetric minimum according to

$$\left.\frac{d^2 V}{d\phi_c^2}\right|_{\phi_c = 0} = m^2 \tag{17.181}$$

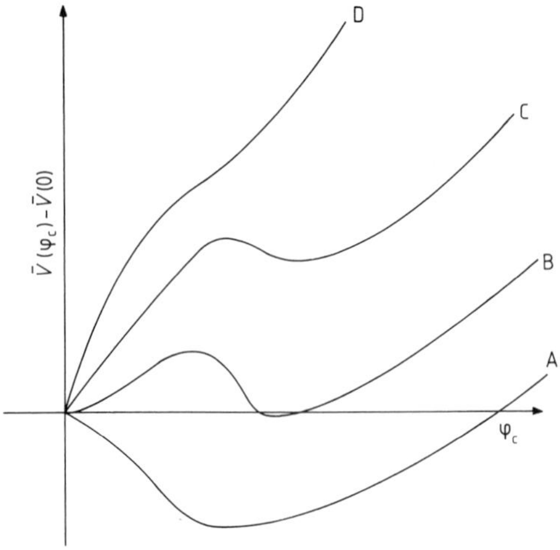

Figure 17.2 Development of asymmetric minimum with temperature in Higgs model for $4\pi^2\lambda/11 \gg e^4 \gg \lambda$. Curve A is at zero temperature, curve B is at T_{c1}, the critical temperature for the first-order phase transition, and C and D correspond to higher temperatures.

$$\left.\frac{d^4V}{d\phi_c^4}\right|_{\phi_c=v}=\tfrac{3}{2}\lambda. \tag{17.182}$$

The physical Higgs scalar mass m_H is given by

$$m_H^2=\left.\frac{d^2V}{d\phi_c^2}\right|_{\phi_c=v}=2Bv^2(4-\alpha). \tag{17.183}$$

The Coleman–Weinberg case of (13.286) ($\alpha=0$) corresponds to

$$m_H^2=m_{CW}^2=8Bv^2. \tag{17.184}$$

In terms of the original parameters of the Lagrangian

$$\alpha=2B^{-1}\left(\frac{22}{3}B-\frac{1}{8}\lambda\right). \tag{17.185}$$

For $m_H^2<m_{CW}^2$ ($\alpha>0$), the situation is qualitatively different from figure 17.2 because there is a (local) minimum of the effective potential at $\phi_c=0$ due to radiative corrections already present at $T=0$. Then we must take account of zero temperature radiative corrections and the development of the effective potential with temperature is as in figure 17.3. The present case, $\alpha>0$, corresponds to

$$e^4>4\pi^2\lambda/11 \tag{17.186}$$

with λ defined as in (17.182).

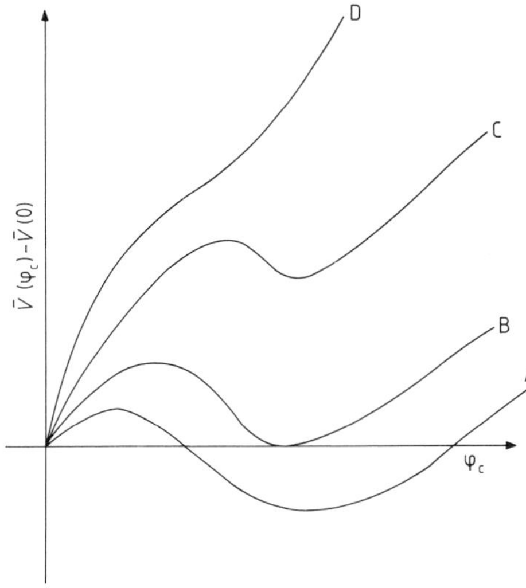

Figure 17.3 Development of asymmetric minimum with temperature in Higgs model for $e^4 > 4\pi^2\lambda/11$. Curve A is at zero temperature, curve B is at T_{c1}, the critical temperature for the first-order phase transition, and C and D correspond to higher temperatures.

In practice, the phase transition to the asymmetric phase may occur at a much lower temperature than T_{c1}, because of a very slow rate of tunnelling through the potential barrier between the symmetric minimum and the asymmetric minimum[8–11]. However, detailed discussion of nucleation at the first-order phase transition is outside our scope here.

Problems

17.1 Show that $\delta(\tau' - \tau)$, defined through (17.33) and (17.25), acts like a Dirac delta function for functions periodic with period β.

17.2 Show that

$$\mathrm{Tr}\ln(-\bar{\not p}+m)=2\ln(m^2-\bar{p}^2)$$

as required for (17.69).

17.3 Calculate the free energy density for free massless spin 1/2 fermions.

17.4 Derive the Feynman rules for temperature Green functions of §17.7.

17.5 Derive the expansion (17.116) of the finite temperature effective potential in terms of temperature Green functions.

17.6 Derive the finite temperature operator traces (17.126), (17.127), (17.128) and (17.129).

17.7 Derive the fermion contribution to (17.138) for massless spin 1/2 fermions.

17.8 Derive the finite temperature one-loop effective potential (17.159) for electroweak theory.

17.9 Derive the finite temperature one-loop effective potential (17.166) for SU(5) grand unified theory.

17.10 Show that the finite temperature effective potential at the asymmetric minimum for the Higgs model is the same, apart from the additive T^4 term, as at $T=0$, correct to leading order in e^2.

References

1 Kirzhnits D A and Linde A D 1972 *Phys. Lett.* **42B** 471
2 Weinberg S 1974 *Phys. Rev.* D **9** 3320
3 Dolan L and Jackiw R 1974 *Phys. Rev.* D **9** 3357
4 Bernard C W 1974 *Phys. Rev.* D **9** 3312
5 MacFarlane A J, Sudbery A and Weisz P H 1968 *Commun. Math. Phys.* **11** 77
6 Kirzhnits D A and Linde A D 1976 *Ann. Phys., NY* **101** 195
7 Iliopoulos J and Papanicolaou N 1976 *Nucl. Phys.* B **111** 209
8 Linde A D 1979 *Rep. Prog. Phys.* **42** 389 and references therein
9 Guth A H and Weinberg E J 1980 *Phys. Rev. Lett.* **45** 1131
10 Sher M A 1980 *Phys. Rev.* D **22** 2989
11 Witten E 1981 *Nucl. Phys.* B **177** 477

APPENDIX A

FEYNMAN INTEGRALS IN 2ω-DIMENSIONAL SPACE

$$\int \frac{d^{2\omega}k}{(2\pi)^{2\omega}} (k^2)^{-n}[(k+p)^2]^{-m}$$

$$= \frac{i(-1)^{m+n}}{(4\pi)^{\omega}} \frac{\Gamma(m+n-\omega)B(\omega-n,\omega-m)}{\Gamma(m)\Gamma(n)} (-p^2)^{\omega-m-n} \quad (A.1)$$

$$\int \frac{d^{2\omega}k}{(2\pi)^{2\omega}} k_\mu (k^2)^{-n}[(k+p)^2]^{-m}$$

$$= \frac{i(-1)^{m+n}}{(4\pi)^{\omega}} \frac{\Gamma(m+n-\omega)B(\omega-n+1,\omega-m)}{\Gamma(m)\Gamma(n)} (-p^2)^{\omega-m-n}(-p_\mu) \quad (A.2)$$

$$\int \frac{d^{2\omega}k}{(2\pi)^{2\omega}} k_\mu k_\nu (k^2)^{-n}[(k+p)^2]^{-m}$$

$$= \frac{i(-1)^{m+n}}{(4\pi)^{\omega}} \frac{(-p^2)^{\omega-m-n}}{\Gamma(m)\Gamma(n)} [p_\mu p_\nu \Gamma(m+n-\omega)B(\omega-m,\omega-n+2)$$

$$+\tfrac{1}{2}g_{\mu\nu}p^2\Gamma(m+n-\omega-1)B(\omega-m+1,\omega-n+1)] \quad (A.3)$$

$$\int \frac{d^{2\omega}k}{(2\pi)^{2\omega}} k_\lambda k_\mu k_\nu (k^2)^{-n}[(k+p)^2]^{-m}$$

$$= \frac{i(-1)^{m+n}}{(4\pi)^{\omega}} \frac{(-p^2)^{\omega-m-n}}{\Gamma(m)\Gamma(n)} [-p_\lambda p_\mu p_\nu \Gamma(m+n-\omega)B(\omega-n+3,\omega-m)$$

$$-\tfrac{1}{2}p^2(g_{\lambda\mu}p_\nu + g_{\mu\nu}p_\lambda + g_{\nu\lambda}p_\mu)\Gamma(m+n-\omega-1)B(\omega-n+2,\omega-m+1)]$$

$$(A.4)$$

$$\int \frac{d^{2\omega}k}{(2\pi)^{2\omega}} k_\lambda k_\mu k_\nu k_\rho (k^2)^{-n}[(k+p)^2]^{-m}$$

$$\equiv \frac{i(-1)^{m+n}}{(4\pi)^{\omega}} \frac{(-p^2)^{\omega-m-n}}{\Gamma(m)\Gamma(n)} [p_\lambda p_\mu p_\nu p_\rho \Gamma(m+n-\omega)B(\omega-n+4,\omega-m)$$

$$-\tfrac{1}{2}p^2(g_{\lambda\mu}p_\nu p_\rho + g_{\nu\rho}p_\lambda p_\mu + g_{\mu\nu}p_\lambda p_\rho + g_{\lambda\rho}p_\mu p_\nu + g_{\lambda\nu}p_\mu p_\rho + g_{\mu\rho}p_\lambda p_\nu)$$

$$\times \Gamma(m+n-\omega-1)B(\omega-n+3,\omega-m+1)$$

$$+\tfrac{1}{4}(p^2)^2(g_{\lambda\mu}g_{\nu\rho} + g_{\lambda\nu}g_{\mu\rho} + g_{\lambda\rho}g_{\mu\nu})\Gamma(m+n-\omega-2)B(\omega-n+2,\omega-m+2)].$$

$$(A.5)$$

APPENDIX B

S-MATRIX ELEMENTS ARE INDEPENDENT OF ξ

We have seen that in order to quantise field theories possessing a local gauge invariance it is necessary to introduce into the Lagrangian a gauge fixing term and the associated contribution from Faddeev–Popov ghosts. Thus the effective quantum Lagrangian is

$$\mathscr{L}_{\mathrm{Q}}=\mathscr{L}_{\mathrm{YM}}+\mathscr{L}_{\mathrm{F}}-\frac{1}{2\xi}(\partial_\mu A_a^\mu)^2+\mathscr{L}_{\mathrm{FP}} \tag{B.1}$$

where $\mathscr{L}_{\mathrm{YM}}$ and $\mathscr{L}_{\mathrm{F}}$ are given in (10.59) and (10.62) respectively, and

$$\mathscr{L}_{\mathrm{FP}}=\partial_\mu\eta_a^*(\partial^\mu\eta_a+gf_{abc}\eta_bA_c^\mu) \tag{10.58a}$$

$$\equiv(\partial_\mu\eta_a^*)(D_{ab}^\mu\eta_b). \tag{10.58b}$$

Since $\mathscr{L}_{\mathrm{Q}}$ has been used to derive the Feynman rules with which we calculate the Green functions, it is clear that in general a Feynman diagram will depend upon the particular value of the parameter ξ which is chosen. However, we have attached no *physical* significance to ξ, and since it was introduced to deal with the technical problems associated with gauge invariance, one feels that no physical observable should depend upon ξ. In fact this is true. Physically observable quantities all derive from S-matrix elements, and we can show that S-matrix elements are independent of ξ. Other Green functions, which are not on-the-mass-shell, in general depend upon ξ.

The proof of this statement which we shall present uses the fact that the effective quantum action

$$S_{\mathrm{Q}}=\int \mathrm{d}^4x\,\mathscr{L}_{\mathrm{Q}} \tag{B.2}$$

is invariant under certain transformations. These transformations are called the 'BRS transformations', after their authors Becci, Rouet and Stora[1]. Under the transformations

$$A_a^\mu\rightarrow A_a^\mu+\delta A_a^\mu \tag{B.3a}$$

where

$$\delta A_a^\mu=\theta D_{ab}^\mu\eta_b. \tag{B.3b}$$

The changes in the other fields are

$$\delta\psi=-\mathrm{i}g\theta t_a\eta_a\psi \tag{B.4a}$$

$$\delta\eta_a^* = -\xi^{-1}\theta(\partial_\mu A_a^\mu) \tag{B.4b}$$

$$\delta\eta_a = -\tfrac{1}{2}g\theta f_{abc}\eta_b\eta_c \tag{B.4c}$$

where θ is a constant real Grassmann number; it follows that $\theta^2=0$, so θ is effectively infinitesimal. The reason for introducing θ is so that the transformations do not alter the character of the fields being transformed. η_a and η_a^* are Grassmann variables and their transformed fields maintain this property. We note that (B.3) is just the usual transformation (9.18) of the gauge field, but with

$$\Lambda_a = \theta\eta_a. \tag{B.5}$$

Since $\mathscr{L}_{\text{YM}}$ is independent of the other fields (ψ, η_a, η_a^*), it follows that the gauge invariance of $\mathscr{L}_{\text{YM}}$ ensures the BRS invariance of $\mathscr{L}_{\text{YM}}$. Similarly $\mathscr{L}_{\text{F}}$ is BRS invariant, since (B.4a) is the usual gauge transformation (9.16) of ψ, again with Λ_a given by (B.5). By design, the gauge fixing term is *not* gauge invariant, and therefore not BRS invariant. In fact

$$\delta\left(-\frac{1}{2\xi}(\partial_\mu A_a^\mu)^2\right) = -\xi^{-1}\theta(\partial_\nu A_a^\nu)\partial_\mu D_{ab}^\mu\eta_b. \tag{B.6}$$

The change in $\mathscr{L}_{\text{FP}}$ is

$$\begin{aligned}\delta\mathscr{L}_{\text{FP}} &= \delta(\partial_\mu\eta_a^*)D_{ab}^\mu\eta_b + \partial_\mu\eta_a^*\delta(D_{ab}^\mu\eta_b)\\ &= -\xi^{-1}\theta[\partial_\mu(\partial_\nu A_a^\nu)]D_{ab}^\mu\eta_b + \partial_\mu\eta_a^*\delta(D_{ab}^\mu\eta_b)\end{aligned} \tag{B.7}$$

and we notice that (B.6) and the first term on the right-hand side of (B.7) add to give a total divergence $\partial_\mu[(\partial_\nu A_a^\nu)D_{ab}^\mu\eta_b]$, so that these two terms also leave S_{Q} invariant. We leave it as an exercise (problem B.1) to verify that $\delta(D_{ab}^\mu\eta_b)=0$ using (B.4a,c) and the Jacobi identity. It is essential to remember at all stages that the order of the Grassmann variables can only be changed with the *anti*commutation relation.

Now, it follows from the defining properties (8.4), (8.5) of integration over Grassmann variables, that

$$\mathsf{I} = \int \mathscr{D}A^\mu\mathscr{D}\psi\mathscr{D}\bar{\psi}\mathscr{D}\eta^*\mathscr{D}\eta\,\eta_a^*(x)\exp\mathrm{i}\left(S_{\text{Q}} + \int \mathrm{d}^4y[J_b^\mu A_{b\mu} + \bar{\psi}\sigma + \bar{\sigma}\psi]\right) = 0 \tag{B.8}$$

since S_{Q} contains η and η^* only bilinearly. We next perform the BRS transformation on all field variables. We leave it as (another) exercise (problem B.2) to verify that the measure $\mathscr{D}A^\mu\mathscr{D}\psi\mathscr{D}\bar{\psi}\mathscr{D}\eta^*\mathscr{D}\eta$ is BRS invariant, and we have just shown that S_{Q} is BRS invariant. Thus the only terms which are affected are η_a^* and the various fields attached to the source terms. To simplify the equations let us drop the fermions from the field theory; the essential features of the ensuing argument are unaffected. Then the only source term which we keep is that attached to the gauge fields. It follows that

$$0=\delta\mathbf{I}=\int\mathscr{D}A_\mu\mathscr{D}\eta^*\mathscr{D}\eta[\xi^{-1}\partial_\mu A_a^\mu(x)+\mathrm{i}J_{\mu b}\eta_a^*D_{bc}^\mu\eta_c(x)]$$

$$\times\exp\mathrm{i}\left(S_\mathrm{Q}+\int\mathrm{d}^4y\,J_{b\nu}A_b^\nu\right). \tag{B.9}$$

This is the generalised Ward–Takahashi identity first proved by Slavnov and Taylor[2]. It is a direct consequence of the BRS invariance of the quantum action S_Q, which itself follows from the (non-Abelian) gauge invariance of the original classical Lagrangian, $\mathscr{L}_\mathrm{YM}+\mathscr{L}_\mathrm{F}$. (In order to define the quantum field theory, this gauge invariance had to be broken, by the addition of the gauge fixing and ghost contributions.) Thus the Slavnov–Taylor identity just expresses the gauge invariance of the original theory. We shall use it to prove that physical S-matrix elements are independent of the gauge-fixing parameter ξ.

So consider the generating functional associated with $\mathscr{L}_\mathrm{Q}$ given in (B.1):

$$W_\xi[J]=\int\mathscr{D}A^\mu\mathscr{D}\eta^*\mathscr{D}\eta\,\exp\mathrm{i}\left[\int\mathrm{d}^4x\left(\mathscr{L}_\mathrm{YM}-\frac{1}{2\xi}(\partial_\mu A_a^\mu)^2\right.\right.$$

$$\left.\left.+\mathscr{L}_\mathrm{FP}+J_a^\mu A_{a\mu}\right)\right]. \tag{B.10}$$

If we change the parameter ξ by the infinitesimal amount $\mathrm{d}\xi$, then the change in $W_\xi[J]$ is

$$\Delta W_\xi[J]=\int\mathscr{D}A^\mu\mathscr{D}\eta^*\mathscr{D}\eta\int\mathrm{d}^4x\left(\frac{\mathrm{id}\xi}{2\xi^2}(\partial_\mu A_a^\mu)^2\right)\exp\mathrm{i}\left(S_\mathrm{Q}+\int\mathrm{d}^4y\,J_{b\nu}A_b^\nu\right). \tag{B.11}$$

Now we use the Slavnov–Taylor identity (B.9). First we operate on it with

$$\delta(x-z)\partial_{\lambda z}\frac{\delta}{\delta J_\lambda^a(z)}$$

and then integrate with respect to x and z. The derivative acts on both the exponential and the pre-factor. On the pre-factor we are left with a total divergence, which integrates to zero. The remaining differentiation then gives

$$0=\int\mathscr{D}A^\mu\mathscr{D}\eta^*\mathscr{D}\eta\int\mathrm{d}^4x(\partial_\lambda A_a^\lambda)[\xi^{-1}\partial_\mu A_a^\mu(x)+\mathrm{i}J_{\mu b}\eta_a^*D_{bc}^\mu\eta_c(x)]$$

$$\times\exp\mathrm{i}\left(S_\mathrm{Q}+\int\mathrm{d}^4y\,J_{b\nu}A_b^\nu\right). \tag{B.12}$$

Using this in (B.11) and then adding (B.10) gives

$$W_\xi[J]+\Delta W_\xi[J]$$

$$=\int \mathscr{D}A^\mu \mathscr{D}\eta^* \mathscr{D}\eta \left[\left(1+\frac{\mathrm{d}\xi}{2\xi}\int \mathrm{d}^4x(\partial_\lambda A_a^\lambda)J_{\mu b}\eta_a^* D_{bc}^\mu \eta_c(x)\right)\right.$$

$$\left.\times \exp \mathrm{i}\left(S_\mathrm{Q}+\int \mathrm{d}^4y\, J_{b\nu}A_b^\nu\right)\right]$$

$$=\int \mathscr{D}A^\mu \mathscr{D}\eta^* \mathscr{D}\eta \exp \mathrm{i}\left[S_\mathrm{Q}+\int \mathrm{d}^4x J_{b\mu}\left(A_b^\mu-\frac{\mathrm{i}\,\mathrm{d}\xi}{2\xi}(\partial_\nu A_a^\nu)\eta_a^* D_{bc}^\mu \eta_c\right)\right]. \tag{B.13}$$

Thus the generating functional $W_{\xi+\mathrm{d}\xi}[J]$ associated with the gauge fixing parameter $\xi+\mathrm{d}\xi$ is just the original generating functional $W_\xi[J]$ with a different field attached to the external source. In general, therefore, the Green functions *are* modified by the change of gauge parameter. However, the S-matrix elements are obtained by rescaling the Green functions so that the propagator poles associated with the external legs have resdue *unity*. As we saw in Chapter 5, all non-pole contributions are projected to zero by the factor $q^2-\mu^2$ as $q^2 \to \mu^2$. Changing the field associated with the external source will change the residues of these pole terms, but this will merely alter the rescaling necessary to arrange that the poles have unit residue. Thus the S-matrix elements are unaffected by changing ξ to $\xi+\mathrm{d}\xi$.

The reader may recall that the particular choice of gauge fixing term (10.31) was relatively arbitrary, and was also accorded no physical significance. So we should be able to change the gauge fixing *function* $F_a[A_b^\mu]$, not merely the associated parameter ξ, and also show that S-matrix elements are unaffected by the variation. This too can be shown using the Slavnov–Taylor identity (at least for linear gauge functions[3]); in this case it is necessary to include also the change in $W[J]$ induced by the corresponding change in $\mathscr{L}_\mathrm{FP}$, which, as we showed in (10.55), depends upon the gauge-fixing function F_a. (The interested reader is invited to construct her own proof of the more general result, and, should she fail, look in the article by B W Lee[4] for assistance.)

The Slavnov–Taylor identity which we have used was for the *un*renormalised theory, and we have shown that a small variation of the unrenormalised gauge-fixing function leaves the S-matrix invariant. The theory we have been studying is, in fact, renormalisable. Thus the renormalised theory has the same structure as the unrenormalised one, and consequently there is a renormalised Slavnov–Taylor identity. It follows by the same argument as before that the S-matrix is invariant with respect to variation of the *renormalised* parameters in the gauge fixing function. Since this variation changes the masses of the 'unphysical' particles (scalars, ghosts), it follows from the invariance of S that these particles really are unphysical. They decouple completely from the physical ones in S-matrix elements, and the physical states are therefore complete. Thus the S-matrix is unitary.

Problems

B.1 Show that $D^{\mu}_{ab}\eta_b$ is BRS invariant.

B.2 Show that the functional integration measures $\mathscr{D}A^{\mu}\mathscr{D}\psi\mathscr{D}\bar{\psi}\mathscr{D}\eta^*\mathscr{D}\eta$ is BRS invariant.

References

1 Becchi C, Rouet A and Stora R 1976 *Ann. Phys., NY* **98** 287
2 Taylor J C 1971 *Nucl. Phys.* B **33** 436
Slavnov A A 1972 *Theor. Math. Phys.* **10** 99
3 Baulieu L and Mieg J T 1982 *Nucl. Phys.* B **197** 477
4 Lee B W in *Methods in Field Theory; Les Houches 1975* ed R Balian and J Zinn-Justin (Amsterdam: North Holland) p. 80
Lee B W and Zinn-Justin J 1973 *Phys. Rev.* D **7** 1049

APPENDIX C

C.1 Vector–vector–scalar–scalar vertices

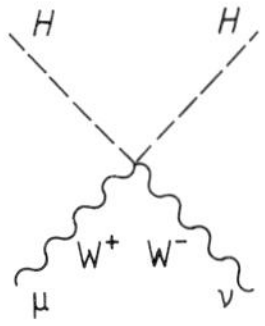

: $\mathrm{i}\tfrac{1}{2}g^2 g_{\mu\nu}$ (C.1)

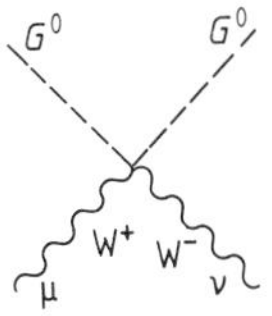

: $\mathrm{i}\tfrac{1}{2}g^2 g_{\mu\nu}$ (C.2)

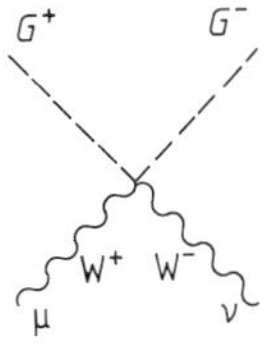

: $\mathrm{i}\tfrac{1}{2}g^2 g_{\mu\nu}$ (C.3)

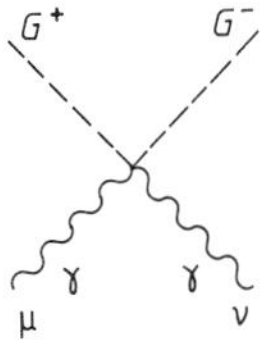

: $\mathrm{i}2e^2 g_{\mu\nu}$ (C.4)

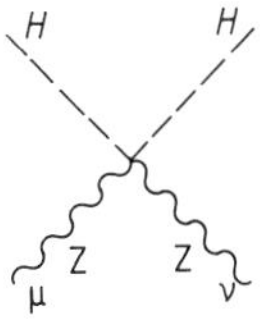

: $\mathrm{i}\tfrac{1}{2}g^2 \sec^2\theta_{\mathrm{w}} g_{\mu\nu}$ (C.5)

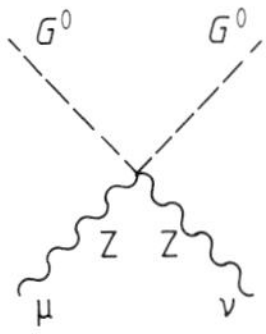

: $\mathrm{i}\tfrac{1}{2}g^2 \sec^2\theta_{\mathrm{W}} g_{\mu\nu}$ (C.6)

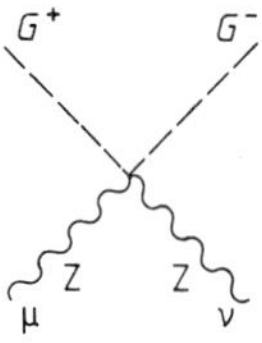

: $\mathrm{i}\frac{1}{2}g^2 \sec^2\theta_{\mathrm{W}} \cos^2 2\theta_{\mathrm{W}}\, g_{\mu\nu}$ (C.7)

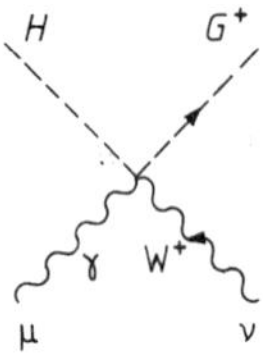

: $\mathrm{i}\frac{1}{2}eg\, g_{\mu\nu}$ (C.8)

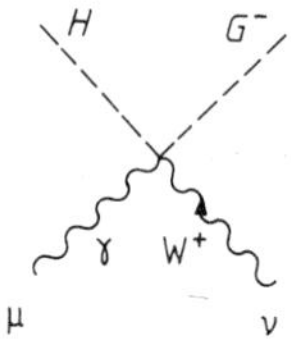

: $\mathrm{i}\frac{1}{2}eg\, g_{\mu\nu}$ (C.9)

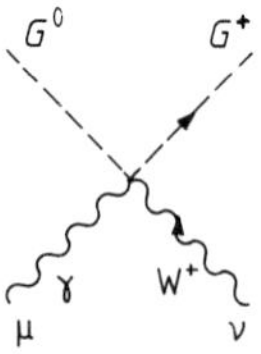

: $-\frac{1}{2}eg\, g_{\mu\nu}$ (C.10)

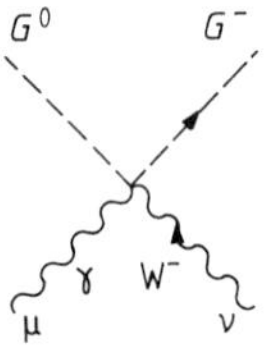

: $\frac{1}{2}eg\, g_{\mu\nu}$ (C.11)

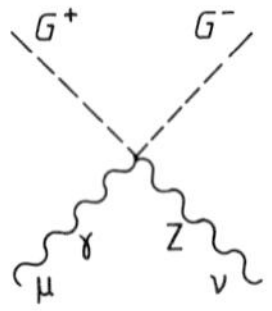

: $\mathrm{i}eg \sec\theta_{\mathrm{W}} \cos 2\theta_{\mathrm{W}}\, g_{\mu\nu}$ (C.12)

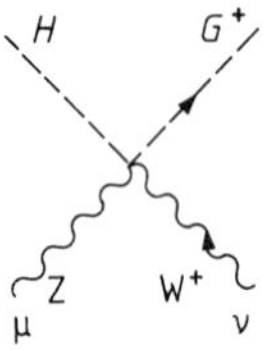

: $\mathrm{i}\frac{1}{2}g^2 \sec\theta_{\mathrm{W}}(\frac{1}{2}\cos 2\theta_{\mathrm{W}} - 1)g_{\mu\nu}$ (C.13)

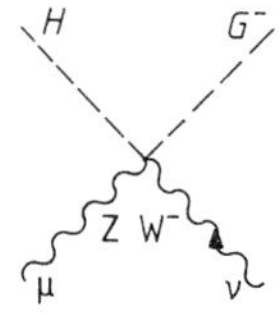

$$: \quad \mathrm{i}\tfrac{1}{2}g^2 \sec\theta_{\mathrm{W}}(\tfrac{1}{2}\cos 2\theta_{\mathrm{W}} - 1)g_{\mu\nu} \qquad \text{(C.14)}$$

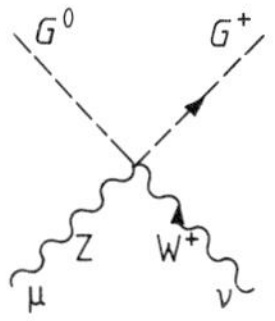

$$: \quad -\tfrac{1}{2}g^2 \sec\theta_{\mathrm{W}}(\tfrac{1}{2}\cos 2\theta_{\mathrm{W}} - 1)g_{\mu\nu} \qquad \text{(C.15)}$$

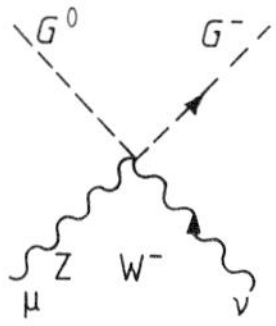

$$: \quad \tfrac{1}{2}g^2 \sec\theta_{\mathrm{W}}(\tfrac{1}{2}\cos 2\theta_{\mathrm{W}} - 1)g_{\mu\nu} \qquad \text{(C.16)}$$

C.2 Vertices involving four scalars

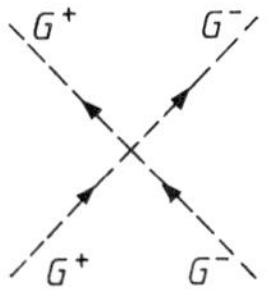

$$: \quad -\mathrm{i}\tfrac{1}{2}g^2 m_{\mathrm{H}}^2/m_{\mathrm{W}}^2 \qquad \text{(C.17)}$$

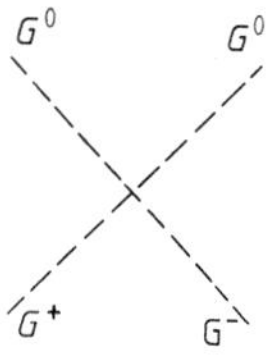

$$: \quad -\mathrm{i}\tfrac{1}{4}g^2 m_{\mathrm{H}}^2/m_{\mathrm{W}}^2 \qquad \text{(C.18)}$$

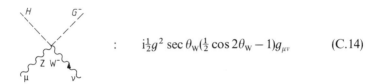

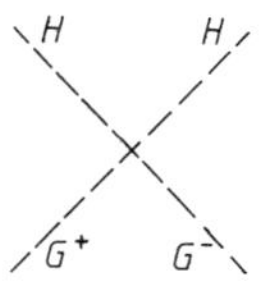

$$: \quad -\mathrm{i}\tfrac{1}{4}g^2 m_{\mathrm{H}}^2/m_{\mathrm{W}}^2 \qquad \text{(C.19)}$$

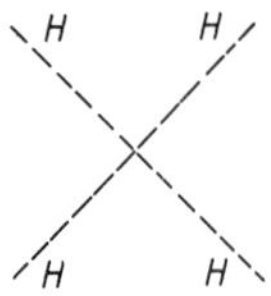

$$: \quad -i\tfrac{3}{4}g^2 m_H^2/m_W^2 \qquad \text{(C.20)}$$

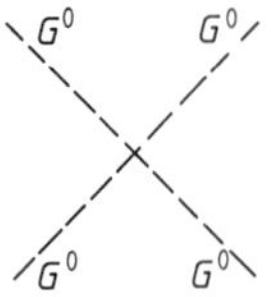

$$: \quad -i\tfrac{3}{4}g^2 m_H^2/m_W^2 \qquad \text{(C.21)}$$

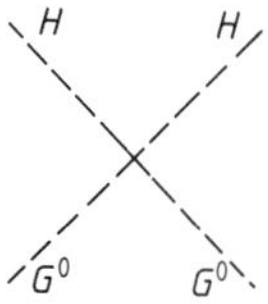

$$: \quad -i\tfrac{1}{4}g^2 m_H^2/m_W^2 \qquad \text{(C.22)}$$

APPENDIX D

SU(5) λ MATRICES

$$\lambda^1 = \begin{pmatrix} 0 & 1 & 0 & 0 & 0 \\ 1 & 0 & 0 & 0 & 0 \\ 0 & 0 & 0 & 0 & 0 \\ 0 & 0 & 0 & 0 & 0 \\ 0 & 0 & 0 & 0 & 0 \end{pmatrix} \qquad \lambda^2 = \begin{pmatrix} 0 & -i & 0 & 0 & 0 \\ i & 0 & 0 & 0 & 0 \\ 0 & 0 & 0 & 0 & 0 \\ 0 & 0 & 0 & 0 & 0 \\ 0 & 0 & 0 & 0 & 0 \end{pmatrix}$$

$$\lambda^3 = \begin{pmatrix} 1 & 0 & 0 & 0 & 0 \\ 0 & -1 & 0 & 0 & 0 \\ 0 & 0 & 0 & 0 & 0 \\ 0 & 0 & 0 & 0 & 0 \\ 0 & 0 & 0 & 0 & 0 \end{pmatrix} \qquad \lambda^4 = \begin{pmatrix} 0 & 0 & 1 & 0 & 0 \\ 0 & 0 & 0 & 0 & 0 \\ 1 & 0 & 0 & 0 & 0 \\ 0 & 0 & 0 & 0 & 0 \\ 0 & 0 & 0 & 0 & 0 \end{pmatrix}$$

$$\lambda^5 = \begin{pmatrix} 0 & 0 & -i & 0 & 0 \\ 0 & 0 & 0 & 0 & 0 \\ i & 0 & 0 & 0 & 0 \\ 0 & 0 & 0 & 0 & 0 \\ 0 & 0 & 0 & 0 & 0 \end{pmatrix} \qquad \lambda^6 = \begin{pmatrix} 0 & 0 & 0 & 0 & 0 \\ 0 & 0 & 1 & 0 & 0 \\ 0 & 1 & 0 & 0 & 0 \\ 0 & 0 & 0 & 0 & 0 \\ 0 & 0 & 0 & 0 & 0 \end{pmatrix}$$

$$\lambda^7 = \begin{pmatrix} 0 & 0 & 0 & 0 & 0 \\ 0 & 0 & -i & 0 & 0 \\ 0 & i & 0 & 0 & 0 \\ 0 & 0 & 0 & 0 & 0 \\ 0 & 0 & 0 & 0 & 0 \end{pmatrix} \qquad \lambda^8 = \frac{1}{\sqrt{3}} \begin{pmatrix} 1 & 0 & 0 & 0 & 0 \\ 0 & 1 & 0 & 0 & 0 \\ 0 & 0 & -2 & 0 & 0 \\ 0 & 0 & 0 & 0 & 0 \\ 0 & 0 & 0 & 0 & 0 \end{pmatrix}$$

$$\lambda^9 = \begin{pmatrix} 0 & 0 & 0 & 1 & 0 \\ 0 & 0 & 0 & 0 & 0 \\ 0 & 0 & 0 & 0 & 0 \\ 1 & 0 & 0 & 0 & 0 \\ 0 & 0 & 0 & 0 & 0 \end{pmatrix} \qquad \lambda^{10} = \begin{pmatrix} 0 & 0 & 0 & -i & 0 \\ 0 & 0 & 0 & 0 & 0 \\ 0 & 0 & 0 & 0 & 0 \\ i & 0 & 0 & 0 & 0 \\ 0 & 0 & 0 & 0 & 0 \end{pmatrix}$$

$$\lambda^{11} = \begin{pmatrix} 0 & 0 & 0 & 0 & 0 \\ 0 & 0 & 0 & 1 & 0 \\ 0 & 0 & 0 & 0 & 0 \\ 0 & 1 & 0 & 0 & 0 \\ 0 & 0 & 0 & 0 & 0 \end{pmatrix} \qquad \lambda^{12} = \begin{pmatrix} 0 & 0 & 0 & 0 & 0 \\ 0 & 0 & 0 & -i & 0 \\ 0 & 0 & 0 & 0 & 0 \\ 0 & i & 0 & 0 & 0 \\ 0 & 0 & 0 & 0 & 0 \end{pmatrix}$$

$$\lambda^{13} = \begin{pmatrix} 0 & 0 & 0 & 0 & 0 \\ 0 & 0 & 0 & 0 & 0 \\ 0 & 0 & 0 & 1 & 0 \\ 0 & 0 & 1 & 0 & 0 \\ 0 & 0 & 0 & 0 & 0 \end{pmatrix} \qquad \lambda^{14} = \begin{pmatrix} 0 & 0 & 0 & 0 & 0 \\ 0 & 0 & 0 & 0 & 0 \\ 0 & 0 & 0 & -i & 0 \\ 0 & 0 & i & 0 & 0 \\ 0 & 0 & 0 & 0 & 0 \end{pmatrix}$$

$$\lambda^{15} = \frac{1}{\sqrt{6}} \begin{pmatrix} 1 & 0 & 0 & 0 & 0 \\ 0 & 1 & 0 & 0 & 0 \\ 0 & 0 & 1 & 0 & 0 \\ 0 & 0 & 0 & -3 & 0 \\ 0 & 0 & 0 & 0 & 0 \end{pmatrix} \qquad \lambda^{16} = \begin{pmatrix} 0 & 0 & 0 & 0 & 1 \\ 0 & 0 & 0 & 0 & 0 \\ 0 & 0 & 0 & 0 & 0 \\ 0 & 0 & 0 & 0 & 0 \\ 1 & 0 & 0 & 0 & 0 \end{pmatrix}$$

$$\lambda^{17} = \begin{pmatrix} 0 & 0 & 0 & 0 & -i \\ 0 & 0 & 0 & 0 & 0 \\ 0 & 0 & 0 & 0 & 0 \\ 0 & 0 & 0 & 0 & 0 \\ i & 0 & 0 & 0 & 0 \end{pmatrix} \qquad \lambda^{18} = \begin{pmatrix} 0 & 0 & 0 & 0 & 0 \\ 0 & 0 & 0 & 0 & 1 \\ 0 & 0 & 0 & 0 & 0 \\ 0 & 0 & 0 & 0 & 0 \\ 0 & 1 & 0 & 0 & 0 \end{pmatrix}$$

$$\lambda^{19}=\begin{pmatrix}0&0&0&0&0\\0&0&0&0&-i\\0&0&0&0&0\\0&0&0&0&0\\0&i&0&0&0\end{pmatrix}\qquad \lambda^{20}=\begin{pmatrix}0&0&0&0&0\\0&0&0&0&0\\0&0&0&0&1\\0&0&0&0&0\\0&0&1&0&0\end{pmatrix}$$

$$\lambda^{21}=\begin{pmatrix}0&0&0&0&0\\0&0&0&0&0\\0&0&0&0&-i\\0&0&0&0&0\\0&0&i&0&0\end{pmatrix}\qquad \lambda^{22}=\begin{pmatrix}0&0&0&0&0\\0&0&0&0&0\\0&0&0&0&0\\0&0&0&0&1\\0&0&0&1&0\end{pmatrix}$$

$$\lambda^{23}=\begin{pmatrix}0&0&0&0&0\\0&0&0&0&0\\0&0&0&0&0\\0&0&0&0&-i\\0&0&0&i&0\end{pmatrix}\qquad \lambda^{24}=\frac{1}{\sqrt{10}}\begin{pmatrix}1&0&0&0&0\\0&1&0&0&0\\0&0&1&0&0\\0&0&0&1&0\\0&0&0&0&-4\end{pmatrix}$$

These matrices are normalised so that

$$\mathrm{Tr}(\lambda^a\lambda^b)=2\delta^{ab}.$$

APPENDIX E

MATSUBARA FREQUENCY SUMS

In the case of bosons, it may be shown by contour integration that

$$\sum_n \frac{1}{(i\omega_n - x)} = \frac{-\beta}{(e^{\beta x} - 1)} \tag{E.1}$$

where ω_n is given by (17.29). Thus

$$\sum_n \frac{x}{\omega_n^2 + x^2} = \frac{\beta}{2} \frac{(e^{\beta x} + 1)}{(e^{\beta x} - 1)}. \tag{E.2}$$

But

$$\frac{d}{dx}\left(\sum_n \ln(\omega_n^2 + x^2)\right) = 2 \sum_n \frac{x}{(\omega_n^2 + x^2)}. \tag{E.3}$$

Performing the x integration we find

$$\sum_n \ln(\omega_n^2 + x^2) = \beta x + 2 \ln(1 - e^{-\beta x}) + (x\text{-independent constant}). \tag{E.4}$$

The constant in (E.4) is temperature dependent and infinite. Fortunately, it cancels against the temperature dependent part of $\tilde{N}(\beta)$ when we evaluate Z.

For fermions, the corresponding results are

$$\sum_n \frac{1}{(i\omega_n - x)} = \frac{\beta}{(e^{\beta x} - 1)} \tag{E.5}$$

where ω_n is given by (17.63), and

$$\sum_n \ln(\omega_n^2 + x^2) = \beta x + 2 \ln(1 + e^{-\beta x}) + (x\text{-independent constant}). \tag{E.6}$$

The temperature dependent constant cancels against $N'(\beta)$ when Z is evaluated.

More details may be found in Fetter A L and Walecka J D 1971 *Quantum Theory of Many-Particle Systems* (New York: McGraw-Hill) p. 248, and in Bernard C W 1974 *Phys. Rev.* D **9** 3312.

INDEX